Methods in Enzymology

Volume 361
BIOPHOTONICS
Part B

METHODS IN ENZYMOLOGY

EDITORS-IN-CHIEF

John N. Abelson   Melvin I. Simon

DIVISION OF BIOLOGY
CALIFORNIA INSTITUTE OF TECHNOLOGY
PASADENA, CALIFORNIA

FOUNDING EDITORS

Sidney P. Colowick and Nathan O. Kaplan

Methods in Enzymology

Volume 361

Biophotonics

Part B

EDITED BY

Gerard Marriott

UNIVERSITY OF WISCONSIN
MADISON, WISCONSIN

Ian Parker

UNIVERSITY OF CALIFORNIA
IRVINE, CALIFORNIA

ACADEMIC PRESS

An imprint of Elsevier Science

Amsterdam   Boston   London   New York   Oxford   Paris
San Diego   San Francisco   Singapore   Sydney   Tokyo

Academic Press
An Elsevier Science Imprint.
525 B Street, Suite 1900, San Diego, California 92101-4495, USA
http://www.academicpress.com

Academic Press
84 Theobald's Road, London WC1X 8RR, UK
http://www.academicpress.com

International Standard Book Number: 0-12-182264-8

PRINTED IN THE UNITED STATES OF AMERICA
03   04   05   06   07   08   9   8   7   6   5   4   3   2   1

Table of Contents

Contributors to Volume 361

Article numbers are in parentheses following the names of contributors.
Affiliations listed are current.

KENGO ADACHI (11), *Center for Integrative Bioscience, Okazaki National Research Institutes, Okazaki, Aichi 444-8585, Japan*

DANIEL AXELROD (1), *Department of Physics and Biophysics Research Division, University of Michigan, Ann Arbor, Michigan 48109*

AMIT K. BHATT (16), *Department of Molecular and Cellular Pharmacology, University of Wisconsin, Madison, Wisconsin 53706*

JAN WILLEM BORST (5), *MicroSpectroscopy Center, Laboratory of Biochemistry, Wageningen University, 6703 HA Wageningen, The Netherlands*

CARLOS BUSTAMANTE (7), *Departments of Chemistry, Molecular and Cell Biology, and Physics, Howard Hughes Medical Institute, University of California, Berkeley, California 94720*

PAUL J. CAMPAGNOLA (3), *Department of Physiology, Center for Biomedical Imaging Technology, University of Connecticut Health Center, Farmington, Connecticut 06030*

WALTER CARRINGTON (13), *Biomedical Imaging Group, Department of Physiology, University of Massachusetts Medical School, Worcester, Massachusetts 01605*

YAN CHEN (4), *Department of Physics and Astronomy, University of Minnesota, Minneapolis, Minnesota 55455*

LAWRENCE B. COHEN (20), *Department of Cellular and Molecular Physiology, Yale University School of Medicine, New Haven, Connecticut 06520*

YUJIA CUI (7), *Department of Cellular and Molecular Pharmacology, University of California, San Francisco, California 94143*

MICAH DEMBO (10), *Department of Biomedical Engineering, Boston University, Boston, Massachusetts 02215*

ERIK W. DENT (18), *Department of Biology, Massachusetts Institute of Technology, Cambridge, Massachusetts 02139*

MAJA DJURISIC (20), *Department of Cellular and Molecular Physiology, Yale University School of Medicine, New Haven, Connecticut 06520*

MICHAEL R. DUCHEN (17), *Life Sciences Imaging Cooperative and Mitochondrial Biology Group, Department of Physiology, University College London, London WC1E 6BT, United Kingdom*

CHUN X. FALK (20), *Department of Cellular and Molecular Physiology, Yale University School of Medicine, New Haven, Connecticut 06520*

ANDREA M. FEMINO (13), *Department of Anatomy and Structural Biology, Albert Einstein College of Medicine, Bronx, New York 10461*

KEVIN FOGARTY (13), *Biomedical Imaging Group, Department of Physiology, University of Massachusetts Medical School, Worcester, Massachusetts 01605*

JONATHAN E. FORMAN (24), *Zyomyx Incorporated, Hayward, California 94545*

GÜNTHER GERISCH (15), *Max-Planck-Institute for Biochemistry, D-82152 Martinsried, Germany*

ERIN M. GILL (21), *Department of Biomedical Engineering, University of Wisconsin, Madison, Wisconsin 53706*

TIMOTHY M. GÓMEZ (19), *Department of Anatomy, University of Wisconsin Medical School, Madison, Wisconsin 53706*

ENRICO GRATTON (4), *Laboratory for Fluorescence Dynamics, Urbana, Illinois 61801*

STEVEN P. GROSS (8), *Departments of Developmental and Cell Biology, and Physics, University of California, Irvine, California 92697*

MARK A. HINK (5), *MicroSpectroscopy Center, Laboratory of Biochemistry, Wageningen University, 6703 HA Wageningen, The Netherlands*

HAYDEN HUANG (22), *Cardiovascular Division, Brigham and Women's Hospital, Boston, Massachusetts 02115*

ANNA HUTTENLOCHER (16), *Departments of Pediatrics and Pharmacology, University of Wisconsin Medical School, Madison, Wisconsin 53706*

EHUD ISACOFF (14), *Department of Molecular and Cell Biology, Physical Biosciences Division, LBNL, University of California, Berkeley, California 94720*

YOSHIHARU ISHII (12), *Single Molecule Processes Project, ICORP, JST, Osaka 562-0035, Japan*

JAKE JACOBSON (17), *Life Sciences Imaging Cooperative and Mitochondrial Biology Group, Department of Physiology, University College London, WC1E 6BT London, United Kingdom*

KATHERINE KALIL (18), *Department of Anatomy, University of Wisconsin, Madison, Wisconsin 53706*

KAZUHIKO KINOSITA, JR, (11), *Center for Integrative Bioscience, Okazaki National Research Institutes, Okazaki, Aichi 444-8585, Japan*

KAZUO KITAMURA (12), *Single Molecule Processes Project, ICORP, JST, Osaka 562-0035, Japan*

B. CHRISTOFFER LAGERHOLM (9), *Science and Technology Center, Carnegie Mellon University, Pittsburgh, Pennsylvania 15213*

FREDERICK LANNI (9), *Science and Technology Center and Department of Biological Sciences, Carnegie Mellon University, Pittsburgh, Pennsylvania 15213*

AARON LEWIS (3), *Division of Applied Physics, Hadassah Department of Ophthalmology, Interdisciplinary Center for Neural Computation, Hebrew University of Jerusalem, Jerusalem 91904, Israel*

LAWRENCE M. LIFSHITZ (13), *Biomedical Imaging Group, Department of Physiology, University of Massachusetts Medical School, Worcester, Massachusetts 01605*

LESLIE M. LOEW (3), *Department of Physiology, Center for Biomedical Imaging Technology, University of Connecticut Health Center, Farmington, Connecticut 06030*

WILLIAM A. MARGANSKI (10), *Department of Biomedical Engineering, Boston University, Boston, Massachusetts 02215*

ANDREW C. MILLARD (3), *Department of Physiology, Center for Biomedical Imaging Technology, University of Connecticut Health Center, Farmington, Connecticut 06030*

WILLIAM MOHLER (3), *Department of Genetics and Developmental Biology, Center for Biomedical Imaging Technology, University of Connecticut Health Center, Farmington, Connecticut 06030*

JOACHIM D. MÜLLER (4), *Department of Physics and Astronomy, University of Minnesota, Minneapolis, Minnesota 55455*

ANNETTE MÜLLER-TAUBENBERGER (15), *Max-Planck-Institute for Biochemistry, D-82152 Martinsried, Germany*

HIROYUKI NOJI (11), *Precursory Research for Embryonic Science and Technology, Institute of Industrial Science, University of Tokyo, Meguro-ku, Tokyo 153-8505, Japan*

GREGORY M. PALMER (21), *Department of Biomedical Engineering, University of Wisconsin, Madison, Wisconsin 53706*

THOMAS J. PURCELL (6), *Department of Biochemistry, Stanford University School of Medicine, Stanford, California 94305*

TIMOTHY M. RAGAN (22), *Department of Mechanical Engineering and Division of Biological Engineering, Massachusetts Institute of Technology, Cambridge, Massachusetts 02139*

NIRMALA RAMANUJAM (21), *Department of Biomedical Engineering, University of Wisconsin, Madison, Wisconsin 53706*

SARAH E. RICE (6), *Department of Biochemistry, Stanford University School of Medicine, Stanford, California 94305*

ESTUARDO ROBLES (19), *Department of Anatomy, Neuroscience Training Program, University of Wisconsin Medical School, Madison, Wisconsin 53706*

ROBERT H. SINGER (13), *Department of Anatomy and Structural Biology, Albert Einstein College of Medicine, Bronx, New York 10461*

STEVEN B. SMITH (7), *Department of Physics, Howard Hughes Medical Institute, University of California, Berkeley, California 94720*

PETER T. C. SO (22), *Department of Mechanical Engineering and Division of Biological Engineering, Massachusetts Institute of Technology, Cambridge, Massachusetts 02139*

ALOIS SONNLEITNER (14), *Department for Biomedical Nanotechnology, Upper Austrian Research GmbH, A-4020 Linz, Austria*

J. RICHARD SPORTSMAN (23), *Molecular Devices Corporation, Sunnyvale, California 94089*

JAMES A. SPUDICH (6), *Department of Biochemistry, Stanford University School of Medicine, Stanford, California 94305*

ALEXANDER SURIN (17), *Institute of General Pathology and Pathophysiology, Russian Academy of Medical Sciences, 125315 Moscow, Russia*

AUDREY D. SUSENO (24), *Zyomyx Incorporated, Hayward, California 94545*

HIROTO TANAKA (12), *Single Molecule Processes Project, ICORP, JST, Osaka 562-0035, Japan*

D. LANSING TAYLOR (9), *Department of Biological Sciences, Carnegie Mellon University, Pittsburgh, Pennsylvania, 15213; Cellomics, Inc., Pittsburgh, Pennsylvania 15219*

STEVEN VANNI (9), *ChemIcon Inc., Pittsburgh, Pennsylvania 15208*

ANTONIE J. W. G. VISSER (5), *MicroSpectroscopy Center, Laboratory of Biochemistry, Wageningen University, 6703 HA Wageningen, The Netherlands; Department of Structural Biology, Vrije Universiteit, 1081 HV Amsterdam, The Netherlands*

MATT WACHOWIAK (20), *Department of Biology, Boston University, Boston, Massachusetts 02215*

PETER WAGNER (24), *Zyomyx Incorporated, Hayward, California 94545*

YU-LI WANG (10), *Department of Physiology, University of Massachusetts Medical School, Worcester, Massachusetts 01605*

IGOR WEBER (2), *Department of Molecular Genetics, Rudjer Boskovic Institute, HR-10000 Zabreb, Croatia; Cell Dynamics Group, Max-Planck-Institute for Biochemistry, D-82152 Martinsried, Germany*

Toshio Yanagida (12), *Single Molecule Processes Project, ICORP, JST, Osaka 562-0035, Japan; Department of Physiology and Biosignaling, Osaka University, Osaka 562-0871, Japan*

Dejan Zecevic (20), *Department of Cellular and Molecular Physiology, Yale University School of Medicine, New Haven, Connecticut 06520*

Michal Zochowski (20), *Department of Physics, University of Michigan, Ann Arbor, Michigan 48109; Center for Complex Systems, Warsaw School of Advanced Social Psychology, Warsaw, Poland*

Preface

The use of optical techniques in experimental biology dates back over 300 years, and has advanced through a series of technological breakthroughs exemplified by the development of diffraction-limited microscopy at the end of the nineteenth century. We are now, once again, at a time of explosive progress, and the newly christened field of biophotonics is emerging from innovative research at the interface of the physical, biological, medical, and engineering sciences. Biophotonics encompasses a broad range of techniques and methodologies developed in areas as diverse as photophysics, photochemistry, optical spectroscopy, and microscopy. It is thus highly interdisciplinary in nature. For example, an imaging technique based on concepts and principles borrowed from physics, chemistry, and engineering may be used to investigate a biological system at the level of a single molecule on an engineered surface within a living cell or even within an animal.

Distinct from previous publications devoted to specific topics, such as imaging or spectroscopy, this 2-volume work of *Methods in Enzymology* is the first to attempt a comprehensive coverage of the broad field of biophotonics research. We believe it will prove a valuable resource for researchers interested in developing and applying photonic technologies to solve biological problems at all levels, from single biomolecules to the living organism. The chapters are written by internationally renowned researchers, and provide technically detailed coverage of the basic principles of optical spectroscopy and microscopy with numerous applications of specialized techniques and probes in biology, medicine, and biotechnology.

Given the interdisciplinary nature of biophotonics, we failed in our initial attempt to classify the chapters within defined categories of technology, applications, and biological systems. Nevertheless, chapters in the two volumes are roughly equally divided among three main areas. Biophotonics, Part A (Volume 360) covers the basic principles and practice of biological spectroscopy and imaging microscopy, with supporting chapters on photophysics, optical design, and biological and synthetic probes. Biophotonics, Part B (Volume 361) focuses on the development and application of imaging technologies to understand the mechanisms underlying biomolecular structure, function, and dynamics in diverse molecular and cellular systems. It also illustrates how biophotonic technologies are being used to power new breakthroughs in medical imaging and diagnosis and in biotechnology.

We would like to thank Mary Ellen Perry for her professional administrative skills and Shirley Light for her excellent editorial assistance.

GERARD MARRIOTT
IAN PARKER

METHODS IN ENZYMOLOGY

VOLUME 91. Enzyme Structure (Part I)
Edited by C. H. W. HIRS AND SERGE N. TIMASHEFF

VOLUME 92. Immunochemical Techniques (Part E: Monoclonal Antibodies and General Immunoassay Methods)
Edited by JOHN J. LANGONE AND HELEN VAN VUNAKIS

VOLUME 93. Immunochemical Techniques (Part F: Conventional Antibodies, Fc Receptors, and Cytotoxicity)
Edited by JOHN J. LANGONE AND HELEN VAN VUNAKIS

VOLUME 94. Polyamines
Edited by HERBERT TABOR AND CELIA WHITE TABOR

VOLUME 95. Cumulative Subject Index Volumes 61–74, 76–80
Edited by EDWARD A. DENNIS AND MARTHA G. DENNIS

VOLUME 96. Biomembranes [Part J: Membrane Biogenesis: Assembly and Targeting (General Methods; Eukaryotes)]
Edited by SIDNEY FLEISCHER AND BECCA FLEISCHER

VOLUME 97. Biomembranes [Part K: Membrane Biogenesis: Assembly and Targeting (Prokaryotes, Mitochondria, and Chloroplasts)]
Edited by SIDNEY FLEISCHER AND BECCA FLEISCHER

VOLUME 98. Biomembranes (Part L: Membrane Biogenesis: Processing and Recycling)
Edited by SIDNEY FLEISCHER AND BECCA FLEISCHER

VOLUME 99. Hormone Action (Part F: Protein Kinases)
Edited by JACKIE D. CORBIN AND JOEL G. HARDMAN

VOLUME 100. Recombinant DNA (Part B)
Edited by RAY WU, LAWRENCE GROSSMAN, AND KIVIE MOLDAVE

VOLUME 101. Recombinant DNA (Part C)
Edited by RAY WU, LAWRENCE GROSSMAN, AND KIVIE MOLDAVE

VOLUME 102. Hormone Action (Part G: Calmodulin and Calcium-Binding Proteins)
Edited by ANTHONY R. MEANS AND BERT W. O'MALLEY

VOLUME 103. Hormone Action (Part H: Neuroendocrine Peptides)
Edited by P. MICHAEL CONN

VOLUME 104. Enzyme Purification and Related Techniques (Part C)
Edited by WILLIAM B. JAKOBY

VOLUME 105. Oxygen Radicals in Biological Systems
Edited by LESTER PACKER

VOLUME 106. Posttranslational Modifications (Part A)
Edited by FINN WOLD AND KIVIE MOLDAVE

VOLUME 107. Posttranslational Modifications (Part B)
Edited by FINN WOLD AND KIVIE MOLDAVE

VOLUME 266. Computer Methods for Macromolecular Sequence Analysis
Edited by RUSSELL F. DOOLITTLE

VOLUME 267. Combinatorial Chemistry
Edited by JOHN N. ABELSON

VOLUME 268. Nitric Oxide (Part A: Sources and Detection of NO; NO Synthase)
Edited by LESTER PACKER

VOLUME 269. Nitric Oxide (Part B: Physiological and Pathological Processes)
Edited by LESTER PACKER

VOLUME 270. High Resolution Separation and Analysis of Biological Macromolecules (Part A: Fundamentals)
Edited by BARRY L. KARGER AND WILLIAM S. HANCOCK

VOLUME 271. High Resolution Separation and Analysis of Biological Macromolecules (Part B: Applications)
Edited by BARRY L. KARGER AND WILLIAM S. HANCOCK

VOLUME 272. Cytochrome P450 (Part B)
Edited by ERIC F. JOHNSON AND MICHAEL R. WATERMAN

VOLUME 273. RNA Polymerase and Associated Factors (Part A)
Edited by SANKAR ADHYA

VOLUME 274. RNA Polymerase and Associated Factors (Part B)
Edited by SANKAR ADHYA

VOLUME 275. Viral Polymerases and Related Proteins
Edited by LAWRENCE C. KUO, DAVID B. OLSEN, AND STEVEN S. CARROLL

VOLUME 276. Macromolecular Crystallography (Part A)
Edited by CHARLES W. CARTER, JR., AND ROBERT M. SWEET

VOLUME 277. Macromolecular Crystallography (Part B)
Edited by CHARLES W. CARTER, JR., AND ROBERT M. SWEET

VOLUME 278. Fluorescence Spectroscopy
Edited by LUDWIG BRAND AND MICHAEL L. JOHNSON

VOLUME 279. Vitamins and Coenzymes (Part I)
Edited by DONALD B. MCCORMICK, JOHN W. SUTTIE, AND CONRAD WAGNER

VOLUME 280. Vitamins and Coenzymes (Part J)
Edited by DONALD B. MCCORMICK, JOHN W. SUTTIE, AND CONRAD WAGNER

VOLUME 281. Vitamins and Coenzymes (Part K)
Edited by DONALD B. MCCORMICK, JOHN W. SUTTIE, AND CONRAD WAGNER

VOLUME 282. Vitamins and Coenzymes (Part L)
Edited by DONALD B. MCCORMICK, JOHN W. SUTTIE, AND CONRAD WAGNER

VOLUME 283. Cell Cycle Control
Edited by WILLIAM G. DUNPHY

VOLUME 284. Lipases (Part A: Biotechnology)
Edited by BYRON RUBIN AND EDWARD A. DENNIS

[1] Total Internal Reflection Fluorescence Microscopy in Cell Biology

By DANIEL AXELROD

Features and Applications

Total internal reflection fluorescence (TIRF) microscopy (also called "evanescent wave microscopy") provides a means to selectively excite fluorophores in an aqueous or cellular environment very near a solid surface (within ≤ 100 nm) without exciting fluorescence in regions farther from the surface.[1] Fluorescence excitation by this thin zone of electromagnetic energy (called an "evanescent field") results in images with very low background fluorescence, virtually no out-of-focus fluorescence, and minimal exposure of cells to light at any other planes in the sample. Figure 1 shows an example of TIRF on intact living cells in culture, compared with standard epifluorescence (EPI). The unique features of TIRF have enabled numerous applications in biochemistry and cell biology as follows.

1. *Selective visualization of cell–substrate contact regions.* TIRF can be used qualitatively to observe the position, extent, composition, and motion of contact regions even in samples in which fluorescence elsewhere would otherwise obscure the fluorescent pattern.[2] A variation of TIRF to identify cell–substrate contacts involves doping the solution surrounding the cells with a nonadsorbing and nonpermeable fluorescent volume marker; focal contacts then appear relatively dark.[3,4] Although TIRF cannot view deeply into thick cells, it can display with high contrast the fluorescence-marked submembrane filament structure at the substrate contact regions.[5]

2. *Visualization and spectroscopy of single-molecule fluorescence near surface.*[6–12] The purpose of single-molecule detection is to avoid the ensemble

[1] D. Axelrod, *J. Cell Biol.* **89,** 141 (1981).

[2] R. M. Weis, K. Balakrishnan, B. Smith, and H. M. McConnell, *J. Biol. Chem.* **257,** 6440 (1982).

[3] D. Gingell, O. S. Heaven, and J. S. Mellor, *J. Cell Sci.* **87,** 677 (1987).

[4] I. Todd, J. S. Mellor, and D. Gingell, *J. Cell Sci.* **89,** 107 (1988).

[5] T. Lang, I. Wacker, I. Wunderlich, A. Rohrbach, G. Giese, T. Soldati, and W. Almers, *Biophys. J.* **78,** 2863 (2000).

[6] R. D. Vale, T. Funatsu, D. W. Pierce, L. Romberg, Y. Harada, and T. Yanagida, *Nature* **380,** 451 (1996).

[7] S. Khan, D. Pierce, and R. D. Vale, *Curr. Biol.* **10,** 927 (2000).

[8] R. M. Dickson, D. J. Norris, Y.-L. Tzeng, and W. E. Moerner, *Science* **274,** 966 (1996).

[9] R. M. Dickson, D. J. Norris, and W. E. Moerner, *Phys. Rev. Lett.* **81,** 5322 (1998).

[10] Y. Sako, S. Miniguchi, and T. Yanagida, *Nature Cell Biol.* **2,** 168 (2000).

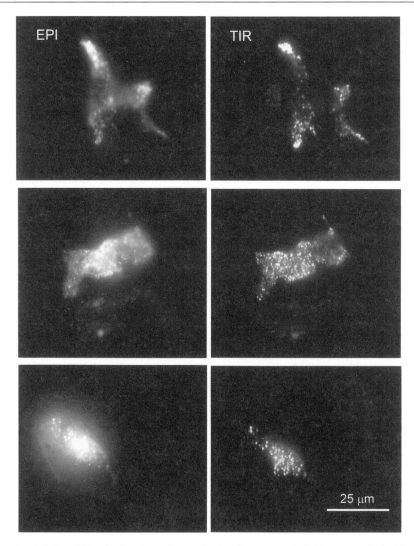

Fig. 1. EPI vs. TIR prismless (through-the-objective) fluorescence digital images, excited with an argon laser beam of wavelength 488 nm entering the side illumination port of an Olympus IX-70 microscope as depicted in Fig. 6A, and viewed through an Olympus 1.45 NA 60× objective. Three different bovine chromaffin cells are depicted, all containing secretory granules marked with GFP-atrial naturetic protein. The images were recorded by a cooled monochrome CCD camera (Photometrics Star-1). [Reproduced with permission from D. Axelrod, *Traffic* **2**, 764 (2001).]

averaging inherent in standard spectroscopies on bulk materials and thereby enable the detection of kinetic features and states that otherwise are obscured. TIRF provides the very dark background needed to observe single fluorophores. Related to single-molecule detection is the capability of seeing fluorescence fluctuations as fluorescent molecules enter and leave the thin evanescent field region in the bulk. These fluctuations (which are visually obvious in TIRF) can be quantitatively autocorrelated in a technique called fluorescence correlation spectroscopy (FCS) to obtain kinetic information about the molecular motion.[13]

3. *Tracking of secretory granules in intact cells before and during secretory process.* The thin evanescent field allows small intensity changes to be interpreted as small motions of granules in the direction normal to the substrate with a precision as small as 2 nm, much smaller than the light microscope resolution limit. In some cases, dispersal of granule contents can be observed and interpreted as exocytosis.[14–29]

4. *Measurements of kinetic rates of binding of extracellular and intracellular proteins to cell surface receptors and artificial membranes.* Some of these studies combine TIR with fluorescence recovery after photobleaching (FRAP) or

[11] A. E. Knight and J. E. Molloy, *in* "Molecular Motors, Essays in Biochemistry" (S. J. Higgins and G. Banting, eds.), Vol. 35, p. 200. Portland Press, London, 2000.

[12] T. J. Ha, A. Y. Ting, J. Liang, W. B. Caldwell, A. A. Deniz, D. S. Chemla, P. G. Schultz, and S. Weiss, *Proc. Natl. Acad. Sci. U.S.A.* **96,** 893 (1999).

[13] T. E. Starr and N. L. Thompson, *Biophys. J.* **80,** 1575 (2001).

[14] T. Lang, I. Wacker, J. Steyer, C. Kaether, I. Wunderlich, T. Soldati, H.-H. Gerdes, and W. Almers, *Neuron* **18,** 857 (1997).

[15] J. A. Steyer and W. Almers, *Biophys. J.* **76,** 2262 (1999).

[16] D. Toomre, J. A. Steyer, P. Keller, W. Almers, and K. Simons, *J. Cell Biol.* **149,** 33 (2000).

[17] D. P. Zenisek, J. A. Steyer, and W. Almers, *Biophys. J.* **78,** 1538 (2000).

[18] J. A. Steyer and W. Almers, *Nature Rev. Mol. Cell Biol.* **2,** 268 (2001).

[19] T. Lang, D. Bruns, D. Wenzel, D. Riedel, P. Holroyd, C. Thiel, and R. Jahn, *EMBO J.* **20,** 2202 (2001).

[20] M. Oheim, D. Loerke, W. Stuhmer, and R. H. Chow, *Eur. Biophys. J.* **27,** 83 (1998).

[21] M. Oheim, D. Loerke, W. Stuhmer, and R. H. Chow, *Eur. Biophys. J.* **28,** 91 (1999).

[22] M. Oheim and W. Stuhmer, *J. Memb. Biol.* **178,** 163 (2000).

[23] W. Han, Y.-K. Ng, D. Axelrod, and E. S. Levitan, *Proc. Natl. Acad. Sci. U.S.A.* **96,** 14577 (1999).

[24] J. Schmoranzer, M. Goulian, D. Axelrod, and S. M. Simon, *J. Cell Biol.* **149,** 23 (2000).

[25] L. M. Johns, E. S. Levitan, E. A. Shelden, R. W. Holz, and D. Axelrod, *J. Cell Biol.* **153,** 177 (2001).

[26] T. Tsuboi, C. Zhao, S. Terakawa, and G. A. Rutter, *Curr. Biol.* **10,** 1307 (2000).

[27] T. Tsuboi, T. Kikuta, A. Warashina, and S. Terakawa, *Biochem. Biophys. Res. Commun.* **282,** 621 (2001).

[28] A. Rohrbach, *Biophys. J.* **78,** 2641 (2000).

[29] D. Toomre and D. J. Manstein, *Trends Cell Biol.* **11,** 298 (2001).

FCS.[30–42] TIR/FRAP additionally can be used to measure lateral surface diffusion coefficients along with on/off kinetics of reversibly adsorbed fluorescent molecules.[30,31,34,39,42]

5. *Micromorphological structures and dynamics on living cells.* By utilizing the unique polarization properties of the evanescent field of TIR, endocytotic or exocytotic sites, ruffles and other submicroscopic irregularities can be highlighted.[43] By combining TIRF with atomic force microscopy, stress–strain relationships can be directly measured on living cells.[44]

6. *Long-term fluorescence movies of cells during development in culture.* Because the cells are exposed to TIR excitation light only at their cell–substrate contact regions but not through their bulk, they tend to survive longer under observation, thereby enabling time-lapse recording of a week in duration. During this time, newly appearing cell surface receptors can be immediately marked by fluorescent ligand that is continually present in the full cell culture medium while maintaining a low fluorescence background.[45]

7. *Comparison of membrane-proximal ionic transients with simultaneous transients deeper in cytoplasm.* Because TIRF is completely compatible with standard epifluorescence, bright-field, dark-field, or phase-contrast illumination, these methods of illumination can be switched back and forth rapidly by electrooptic devices.[46]

Theoretical Principles

The thin layer of illumination is an "evanescent field" produced by an excitation light beam in a solid (e.g., a glass coverslip or tissue culture plastic) that is incident at a high angle upon the solid–liquid surface at which the sample (e.g., single molecules or cells) adhere. The incidence angle θ, measured from the normal,

[30] N. L. Thompson, T. P. Burghardt, and D. Axelrod, *Biophys. J.* **33**, 435 (1981).

[31] T. P. Burghard and D. Axelrod, *Biophys. J.* **33**, 455 (1981).

[32] N. L. Thompson and D. Axelrod, *Biophys. J.* **43**, 103 (1983).

[33] E. Hellen and D. Axelrod, *J. Fluor.* **1**, 113 (1991).

[34] R. M. Fulbright and D. Axelrod, *J. Fluor.* **3**, 1 (1993).

[35] A. L. Stout and D. Axelrod, *Biophys. J.* **67**, 1324 (1994).

[36] A. M. McKiernan, R. C. MacDonald, R. I. MacDonald, and D. Axelrod, *Biophys. J.* **73**, 1987 (1997).

[37] S. E. Sund and D. Axelrod, *Biophys. J.* **79**, 1655 (2000).

[38] E. Kalb, J. Engel, and L. K. Tamm, *Biochemistry* **29**, 1607 (1990).

[39] R. Gilmanshin, C. E. Creutz, and L. K. Tamm, *Biochemistry* **33**, 8225 (1994).

[40] P. Hinterdorfer, G. Baber, and L. K. Tamm, *J. Biol. Chem.* **269**, 20360 (1994).

[41] B. C. Lagerholm, T. E. Starr, Z. N. Volovyk, and N. L. Thompson, *Biochemistry* **39**, 2042 (2000).

[42] R. D. Tilton, A. P. Gast, and C. R. Robertson, *Biophys. J.* **58**, 1321 (1990).

[43] S. E. Sund, J. A. Swanson, and D. Axelrod, *Biophys. J.* **77**, 2266 (1999).

[44] A. B. Mathur, G. A. Truskey, and W. M. Reichert, *Biophys. J.* 1725 (2000).

[45] M. D. Wang and D. Axelrod, *Dev. Dynam.* **201**, 29 (1994).

[46] G. M. Omann and D. Axelrod, *Biophys. J.* **71**, 2885 (1996).

must be greater than some "critical angle" for the beam to totally internally reflect rather than refract through the interface. TIR generates a very thin electromagnetic field in the liquid with the same frequency as the incident light, exponentially decaying in intensity with distance from the surface. This field is capable of exciting fluorophores near the surface while avoiding excitation of a possibly much larger number of fluorophores farther out in the liquid.

Infinite Plane Waves

The simplest case of TIR is that of an "infinitely" extended plane wave incident upon a single interface (i.e., a beam width many times the wavelength of the light, which is a good approximation for unfocused or weakly focused light); see Fig. 2. When a light beam propagating through a transparent medium 3 of high index of refraction (e.g., glass) encounters a planar interface with medium 1 of lower index of refraction (e.g., water), it undergoes total internal reflection for

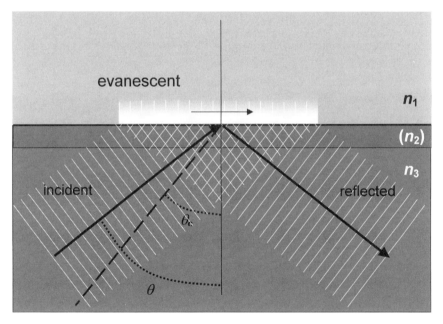

FIG. 2. Geometric scheme of TIR. Refractive index n_3 must be greater than n_1. The intermediate layer (consisting of metal or a dielectric material of refractive index n_2) is not necessary for TIR to occur, but can introduce some useful optical properties as explained in the text. Most applications of TIRF do not use an intermediate layer. The incidence angle θ must be larger than the critical angle θ_c for TIR to occur. The exponentially decaying evanescent field in the n_1 material is used to excite fluorophores in TIRF.

incidence angles θ (measured from the normal to the interface) greater than the "critical angle" θ_c given by

$$\theta_c = \sin^{-1}(n_1/n_3) \tag{1}$$

where n_1 and n_3 are the refractive indices of the liquid and the solid, respectively. Ratio n_1/n_3 must be less than unity for TIR to occur. (A refractive index n_2 will refer to an optional intermediate layer to be discussed below.) For incidence angle $\theta < \theta_c$, most of the light propagates through the interface into the lower index material with a refraction angle (also measured from the normal) given by Snell's law. (Some of the incident light also internally reflects back into the solid.) But for $\theta > \theta_c$, all of the light reflects back into the solid. Even in this case, some of the incident light energy penetrates through the interface and propagates parallel to the surface in the plane of incidence. The field in the liquid (sometimes called the evanescent "wave") is capable of exciting fluorescent molecules that might be present near the surface.

The intensity I of the evanescent field at any position is the squared amplitude of the complex electric field vector $\mathbf{E}$ at that position:

$$I(z) = \mathbf{E}(z) \cdot \mathbf{E}^*(z) \tag{2}$$

For an infinitely wide beam, the intensity of the evanescent wave (measured in units of energy/area/second) exponentially decays with perpendicular distance z from the interface:

$$I(z) = I(0)e^{-z/d} \tag{3}$$

where

$$\begin{aligned}
d &= \frac{\lambda_o}{4\pi}\left(n_3^2 \sin^2\theta - n_1^2\right)^{-1/2} \\
&= \frac{\lambda_o}{4\pi n_3}(\sin^2\theta - \beta^2)^{-1/2}
\end{aligned} \tag{4}$$

Parameter β here is defined as the ratio of the lower to higher refractive index:

$$\beta \equiv n_1/n_3 < 1 \tag{5}$$

Parameter λ_o is the wavelength of the incident light in vacuum. Depth d is independent of the polarization of the incident light and decreases with increasing θ. Except for $\theta \to \theta_c$ (where $d \to \infty$), d is in the order of λ_o or smaller. A physical picture of refraction at an interface shows TIR to be part of a continuum, rather than a sudden new phenomenon appearing at $\theta = \theta_c$. For small θ, the refracted light waves in the liquid are sinusoidal, with a certain characteristic period noted as one moves normally away from the surface. As θ approaches θ_c, that period becomes longer as the refracted rays propagate increasingly parallel to the surface. At exactly $\theta = \theta_c$, that period is infinite, as the wavefronts of the refracted light are themselves normal to the surface. This situation corresponds to $d = \infty$. As

θ increases beyond θ_c, the period becomes mathematically imaginary. Physically, this corresponds to the exponential decay of Eq. (3).

The local intensity I of the evanescent field at any point is proportional to the probability rate of energy absorption by a fluorophore that can be situated at that point. Although the emission from a fluorophore excited by an evanescent field as viewed by a microscopic objective might be expected to follow an exponential decay with z according to Eq. (3), this is not precisely true. Fluorescence emission near a dielectric interface is rather anisotropic and the degree of anisotropicity is itself z dependent.[47,48] One cause of the anisotropicity of emission is simply partial reflection from the interface and consequent interference between the direct and reflected emission beams. Another more subtle cause is the interaction of the "near field" of the fluorophore with the interface and its consequent conversion into light propagating at high angles into the higher index n_3 material (the solid substrate). In general, the closer a fluorophore is to the surface, the larger the proportion of its emitted light will enter the substrate. If a sufficiently high aperture objective is used to gather the near-field emission, then the effective thickness of the surface detection zone is reduced beyond surface selectivity generated by the evanescent field excitation of TIR. Low-aperture objectives will not produce this enhanced effect because they miss capturing the near-field light. On its own, the near-field emission capture effect can be the basis of surface-selective fluorescence detection and imaging, as discussed in a separate section below.

The polarization (i.e., the vector direction of the electric field $\mathbf{E}$) of the evanescent wave depends on the incident light polarization, which can be either "p-pol" (polarized in the plane of incidence formed by the incident and reflected rays, denoted here as the x–z plane) or "s-pol" (polarized normal to the plane of incidence).

For p-pol incident light, the evanescent electric field vector direction remains in the plane of incidence, but it "cartwheels" along the surface with a nonzero longitudinal component (see Fig. 3):

$$\mathbf{E}_p(z) = 2\cos\theta(\beta^4\cos^2\theta + \sin^2\theta - \beta^2)^{-1/2}e^{-i\delta_p}e^{-z/2d}$$
$$\times [-i(\sin^2\theta - \beta^2)^{1/2}\hat{\mathbf{x}} + \sin\theta\,\hat{\mathbf{z}}]$$

(6)

The evanescent wavefronts travel parallel to the surface in the x-direction. Therefore, the p-pol evanescent field is a mix of transverse (z) and longitudinal (x) components; this distinguishes the p-pol evanescent field from freely propagating subcritical refracted light, which has no component longitudinal to the direction of travel. The longitudinal component of the p-pol evanescent field diminishes to zero amplitude as the incidence angle is reduced from the supercritical range back toward the critical angle.

[47] E. H. Hellen and D. Axelrod, *J. Opt. Soc. Am. B* **4**, 337 (1987).
[48] J. Mertz, *J. Opt. Soc. Am. B* **17**, 1906 (2000).

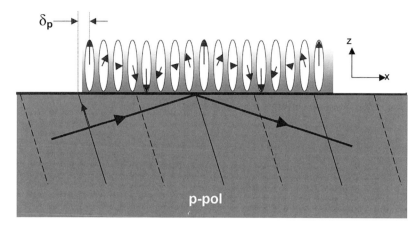

FIG. 3. Schematic drawing of the evanescent polarization resulting from p-pol incident light. The incident light wavefronts (with the intervals from solid to dashed wavefront lines representing one-half of a wavelength in the glass) determine the spacing of the wavefronts in the evanescent field. Reflected light is not shown. The p-pol evanescent field is elliptically polarized in the x–z plane as shown (primarily polarized in the z-direction with a weaker x-component at a relative phase of $\pi/2$). For pictorial clarity, only two periods of evanescent electric field oscillation are shown; in reality, the evanescent region is much more extended and contains many more periods of oscillation in the x-direction. The exact phase relationship between the incident field and the evanescent field is a function of incidence angle and is represented by δ_p here. [Reproduced with permission from D. Axelrod, *Traffic* **2**, 764 (2001).]

For s-pol incident light, the evanescent electric field vector direction remains purely normal to the plane of incidence:

$$\mathbf{E}_s(z) = 2\cos\theta(1-\beta^2)^{-1/2}e^{-i\delta_s}e^{z/2d}\hat{\mathbf{y}} \tag{7}$$

In Eqs. (6) and (7), the incident electric field amplitude in the substrate is normalized to unity for each polarization, and the phase lags relative to the incident light are

$$\delta_p = \tan^{-1}\left[\frac{(\sin^2\theta - \beta^2)^{1/2}}{\beta^2\cos\theta}\right] \tag{8}$$

$$\delta_s = \tan^{-1}\left[\frac{(\sin^2\theta - \beta^2)^{1/2}}{\cos\theta}\right] \tag{9}$$

The corresponding evanescent intensities in the two polarizations (assuming incident intensities normalized to unity) are

$$I_p(z) = \frac{(4\cos^2\theta)(2\sin^2\theta - \beta^2)}{\beta^4\cos^2\theta + \sin^2\theta - \beta^2}e^{-z/d} \tag{10}$$

$$I_s(z) = \frac{(4\cos^2\theta)}{1-\beta^2}e^{-z/d} \tag{11}$$

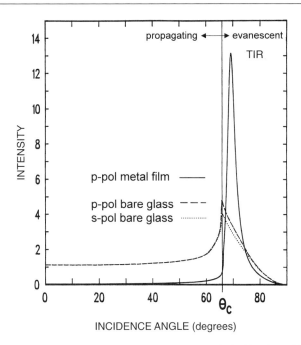

FIG. 4. Transmitted intensities $I_{p,s}$ at $z = 0$ vs. θ, assuming the incident intensities in the glass are set equal to unity. At angles $\theta > \theta_c$, the transmitted light is evanescent; at angles $\theta < \theta_c$, it is propagating. Both s- and p-polarizations are shown. Refractive indices $n_3 = 1.46$ (fused silica) and $n_1 = 1.33$ are assumed here, corresponding to $\theta_c = 65.7°$. Also shown is the evanescent intensity that would be obtained with a thin (20-nm) aluminum film coating, as discussed in the section on Intermediate Films. [Reproduced with permission from D. Axelrod, *Traffic* **2**, 764 (2001).]

Intensities $I_{p,s}(0)$ are plotted vs. θ in Fig. 4. The evanescent intensity approaches zero as $\theta \rightarrow 90°$. On the other hand, for supercritical angles within $10°$ of θ_c, the evanescent intensity is as great or greater than the incident light intensity. The plots can be extended without breaks to the subcritical angle range, where the intensity is that of the freely propagating refracted light in medium 1. The subcritical intensity might at first be expected to be slightly *less* than the incident intensity (accounting for some reflection at the interface) but certainly not *more* as shown. The discrepancy arises because the intensity in Fig. 3 refers to EE^* alone rather than to the actual energy flux of the light, which involves a product of EE^* with the refractive index of the medium in which the light propagates.

Regardless of polarization, the spatial period of the evanescent electric field is $\lambda_o/(n_3 \sin \theta)$ as it propagates along the surface. Unlike the case of freely propagating light, the evanescent spatial period is not affected by the medium 1 in which it resides. It is determined only by the spacing of the incident light wavefronts in medium 3 as they intersect the interface. This spacing can be important experimentally because it determines the spacing of interference fringes produced when two coherent TIR beams illuminate the same region of the surface from different directions.

Finite-Width Incident Beams

For a finite-width beam, the evanescent wave can be pictured as the partial emergence of the beam from the solid into the liquid, its travel for some finite distance along the surface, and then its reentrance into the solid. The distance of propagation along the surface is measurable for a finite-width beam and is called the Goos–Hanchen shift. The Goos–Hanchen shift ranges from a fraction of a wavelength at $\theta = 90°$ to infinite at $\theta = \theta_c$, which of course corresponds to the refracted beam skimming along the interface. A finite incident beam can be expressed mathematically as an integral of infinite plane waves approaching at a range of incidence angles. In general, the intensity profile of the resulting evanescent field can be calculated from the mathematical form for the evanescent wave that arises from each infinite plane wave incidence angle, integrated over all the constituent incident plane wave angles. For a TIR Gaussian laser beam focused with a typically narrow angle of convergence, the experimentally observed evanescent illumination is approximately an elliptical Gaussian profile, and the polarization and penetration depth are approximately equal to those of a single infinite-plane wave.

Intermediate Layers

In actual experiments in biophysics or cell biology, the interface may not be a simple interface between two media, but rather a stratified multilayer system. One example is the case of a biological membrane or lipid bilayer interposed between glass and an aqueous medium. Another example is a thin metal film coating, which can be used to quench fluorescence within the first $\sim$10 nm of the surface. We discuss here the TIR evanescent wave in a three-layer system in which incident light travels from medium 3 (refractive index n_3) through the intermediate layer (n_2) toward medium 1 (n_1). Qualitatively, the following features can be noted.

1. Insertion of an intermediate layer never thwarts TIR, regardless of the intermediate layer's refractive index n_2. The only question is whether TIR takes place at the n_3–n_2 interface or the n_2–n_1 interface. Because the intermediate layer is likely to be very thin (no deeper than several tens of nanometers) in many applications, precisely which interface supports TIR is not important for qualitative studies.

2. Regardless of n_2 and the thickness of the intermediate layer, the evanescent wave's profile in medium 1 will be exponentially decaying with a characteristic decay distance given by Eq. (3). However, the intermediate layer affects the overall distance of penetration of the field measured from the surface of medium 3.

3. Irregularities in the intermediate layer can cause scattering of incident light that then propagates in all directions in medium 1. Experimentally, scattering appears not be a problem on samples even as inhomogeneous as biological cells. Direct viewing of incident light scattered by a cell surface lying between the glass substrate and an aqueous medium confirms that scattering is many orders of magnitude dimmer than the incident or evanescent intensity, and will thereby excite a correspondingly dim contribution to the fluorescence.

A particularly interesting kind of intermediate layer is a metal film. Classic electromagnetic theory[47] shows that such a film will reduce the s-polarized evanescent intensity to nearly zero at all incidence angles. But the p-polarized behavior is quite different. At a certain sharply defined angle of incidence θ_p ("the surface plasmon angle"), the p-polarized evanescent intensity becomes an order of magnitude brighter than the incident light at the peak (see Fig. 4). This strongly peaked effect is due to a resonant excitation of electron oscillations at the metal/water interface. For an aluminum, gold, or silver film at a glass/water interface, θ_p is greater than the critical angle θ_c for TIR. The intensity enhancement is rather remarkable since a 20-nm-thick metal film is almost opaque to the eye.

There are some potentially useful experimental consequences of TIR excitation through a thin metal film coated on glass:

1. The metal film will almost totally quench fluorescence within the first 10 nm of the surface, and the quenching effect is virtually gone at a distance of 100 nm. Therefore, TIR with a metal-film-coated glass can be used to selectively excite fluorophores in the 10-nm to 200-nm distance range.

2. A light beam incident on a 20-nm aluminum film from the glass side at a glass/aluminum film/water interface evidently does not have to be collimated to produce TIR. Those rays that are incident at the surface plasmon angle will create a strong evanescent wave; those rays that are too low or high in incidence angle will create a negligible field in the water. This phenomenon may ease the practical requirement for a collimated incident beam in TIR and make it easier to set up TIR with a conventional arc light source.

3. The surface plasmon angle is strongly sensitive to the index of refraction on the low-density side of the interface. With a sample of spatially varying refractive index (such as an adhered cell), some areas will support the surface plasmon resonance whereas neighboring areas will not. The locations supporting a resonance will show a much stronger light scattering. This effect can be used as the basis of "surface plasmon microscopy."[49]

4. The metal film leads to a highly polarized evanescent wave, regardless of the purity of the incident polarization.

[49] M. G. Somekh, S. G. Liu, T. S. Velinov, and C. W. See, *Appl. Opt.* **39**, 6270 (2001).

Another type of intermediate film is now coming into commercial use in bioassay surfaces based on TIRF: the thin film planar waveguide. An ordinary glass (or plastic) surface is etched or embossed with closely spaced grooves that behave like a diffraction grating so that some of the light coming through the glass toward the surface will diffract away from the normal at the surface. Coated on the grooved glass is a thin intermediate film of a material (such as titanium dioxide) with refractive index higher than that of either the glass substrate or the water at its boundaries. The diffracted incident light passing from the glass into the intermediate film has a sufficient angle not only to totally reflect first at the thin film/aqueous interface but also to totally reflect subsequently at the thin film/glass interface, thereby trapping the light in the film. At a very particular angle (which depends on the wavelength and the film thickness), the film acts like an optical waveguide where the reflections at the two surfaces create wavefronts that constructively add in-phase with each other. The result is a very strong field in the high index film and a corresponding very strong evanescent field in the aqueous medium that can be used to excite fluorescence in biosensor devices.[50] The thin film waveguide surface preparation should also be applicable to observation of living cells in a microscope.

Combination of TIR with Other Fluorescence Techniques

TIR illumination can be combined with other standard fluorescence spectroscopy techniques. Brief comments here point out certain unique properties of the combinations.

1. *TIR-FRET* (fluorescence resonance energy transfer) provides an opportunity to observe real-time changes in the conformation of single molecules attached to the substrate due to the much reduced background provided by TIR.

2. *TIR-fluorescence lifetime measurements* should give results somewhat different from results of the same fluorophores in solution. Proximity to the surface may directly perturb the molecular state structure, thereby affecting lifetimes. In addition, the near-field capture effect generally increases the rate of transfer of energy out of the fluorophore and thereby decreases lifetimes.

3. In *polarized TIR,* a p-polarized evanescent field can be uniquely utilized to highlight submicroscopic irregularities in the plasma membrane of living cells,[43] as shown schematically in Fig. 5. The effect depends on the incorporation of the dye into the membrane with a high degree of orientational order. In the case of the phospholipid-like amphipathic carbocyanine dye 3,3'-dioctadecylindocarbocyanine (diI), the direction of absorption dipole of the fluorophores is known to be parallel to the local plane of the membrane and free to rotate in it.[51]

---

[50] G. L. Duveneck and A. P. Abel, *Proc. SPIE* **3858,** 59 (1999).
[51] D. Axelrod, *Biophys. J.* **26,** 557 (1979).

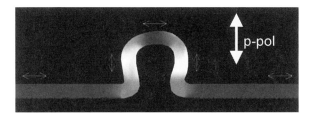

FIG. 5. Schematic drawing of the excitation probability of oriented carbocyanine fluorophores embedded in a membrane in a z-polarized evanescent field (the dominant direction of a p-polarized evanescent field). The membrane is depicted in cross section with a curved region corresponding to a bleb or an invagination. The direction of absorption dipole of the fluorophores is shown with bidirectional arrows. Higher excitation probability is depicted by lighter shades. The z-component of the electric field selectively excites the regions of oblique membrane orientation. [Reproduced with permission from D. Axelrod, *Traffic* **2**, 764 (2001).]

4. In *TIR-FRAP* (fluorescence recovery after photobleaching), a bright flash of the evanescent field bleaches the surface-proximal fluorophores, and the subsequent rate of recovery of fluorescence excited by dimmer evanescent illumination provides a quantitative measure of the desorption kinetic rate. [30,31]

5. In *TIR-FCS* (fluorescence correlation spectroscopy), the thin evanescent field combined with an image plane diaphragm that passes light only from a small (typically submicron) region, an extremely small observation volume ($\sim$0.02 μm^3) can be defined in the sample. Random fluctuations in the number of fluorophores in this volume provide information about the diffusive rate in the bulk solution near the surface, the kinetic desorption rate, and the absolute concentration of surface-proximal fluorophores. [13,30]

6. In *multiphoton TIRF*, excitation by an intense flash of evanescent infrared light excites a fluorescent response that is proportional to the square (or higher order) of the incident light intensity. This might seem at first to have some advantages for further decreasing the effective evanescent wave depth d. Although d for two-photon TIR excitation should be half of the single photon case described by Eq. (3), note that the infrared wavelength is double what would have been used in "single-photon" TIRF, so the effect on the depth d would be minimal (apart from chromatic dispersion effects in the substrate and water, which would affect the index of refraction ratio β). On the other hand, scattering of the evanescent field by inhomogeneities in the sample would likely not be bright enough to excite much multiphoton fluorescence, so multiphoton TIRF may produce a cleaner image, less affected by scattering artifacts.

7. *Optical trapping with TIRF* should be possible as the gradient of the evanescent intensity can be made comparable to the gradient of intensity in standard focused spot optical traps. The evanescent optical gradient force will always be directed toward the interface. The analog of a "photon pressure" force may be directed parallel to the substrate in the plane of incidence; it might be canceled out by the use of oppositely directed TIR beams.

Surface Near-Field Emission Imaging

Because TIR selectively excites only surface-proximal fluorophores, the special emission behavior of such fluorophores should be considered. This emission behavior also suggests designs for potentially useful microscope modifications.

The classic model for an excited fluorophore in electromagnetic theory is an oscillating dipole. The radiation emission pattern of an oscillating dipole can be expressed (by spatial Fourier transforming) as a continuous superposition of plane waves traveling in all directions from the fluorophore. Some (but not all) of these plane waves have wavelengths given by $\lambda = c/(n\nu)$ as expected for propagating light, where ν is the frequency (color) of the light, n is the refractive index of the liquid in which the fluorophore resides, and c is the speed of light in vacuum. However, this restricted set of plane waves is not sufficient to describe the actual radiation emission pattern; other plane waves with shorter wavelengths must be included. Because the frequency of the light is invariant as determined by the color, the only way to obtain shorter wavelengths is for the plane waves to be exponentially decaying in one of the three spatial directions.

In a slightly more mathematical view,[47] a plane wave has a wave vector $\mathbf{k}$ given by

$$\mathbf{k} = k_x\hat{\mathbf{x}} + k_y\hat{\mathbf{y}} + k_z\hat{\mathbf{z}} \qquad (12)$$

where the z-direction is chosen as the normal to the interface, and the square scalar amplitude of $\mathbf{k}$ is fixed at a constant value by the frequency ν according to

$$k^2 = (2\pi n\nu/c)^2 = k_x^2 + k_y^2 + k_z^2 \qquad (13)$$

Short wavelengths are obtained by using plane waves with $(k_x^2 + k_y^2) > k^2$, which forces k_z^2 to be negative. This forces k_z to be imaginary, which corresponds to plane waves that exponentially decay in the z-direction. The exponential decay "starts" at the z position of the fluorophore. The fluorophore's "near field" is defined as this set of exponentially decaying plane waves with $(k_x^2 + k_y^2) > k^2$. Clearly, this set is not a single exponential but superposition (actually a continuous integral) of exponentially decaying waves of a range of characteristic decay lengths each given by $2\pi/|k_z|$. These near-field waves have wavefronts more closely spaced than would be expected for propagating light in the liquid medium surrounding the fluorophore. Where the "tails" of the exponentially decaying wavefronts touch the surface of the glass, refraction converts some of the near-field energy into propagating light in the glass.

Because the periodicity of the wavefronts must match at both sides of the boundary, the near-field-derived propagating light in the glass is directed in a hollow cone *only* into angles greater than some critical angle ("supercritical"). (That critical angle is the same as the TIR critical angle for the same frequency of light passing from the solid toward the liquid.) *None* of the far-field propagating energy, for which $(k_x^2 + k_y^2) \leq k^2$, is cast into that high-angle hollow cone; it is all cast into "subcritical" angles. Therefore, collection and imaging of just the super-critical high angle hollow cone light should select only those fluorophores that are sufficiently close to the surface for the surface to capture their near fields. An implementation of this principle is discussed in the section on Optical Configurations below.

Note that the light-collecting advantage of very high-aperture objectives (>1.4) resides purely in their ability to capture supercritical near-field light. The "extra" numerical aperture does not help in gathering far-field emission light because none of it propagates in the glass at the supercritical angles (and then into the corresponding high apertures).

The capture of the fluorophore near field and its conversion into light propagating into the substrate at supercritical angles cause deviations from the expected exponential decay of the TIR evanescent field implied by Eq. (3). It is true that the *excitation* rate follows the exponential decay in z, but the *emission* pattern depends on the distance the fluorophore resides from the surface in a complicated manner. Therefore, because every objective gathers only part of this emission pattern, the dependence of gathered emission on z will be nonexponential and will depend on the numerical aperture of the objective (as well as on the orientational distribution of the fluorophore dipoles, the refractive indices bordering the interface, and the polarization of the excitation). For objectives of moderate numerical aperture (NA < 1.33) so that no near-field light is gathered, the deviation from the simple exponentiality is generally only 10–20%. For very high-aperture objectives that do gather some near-field emission, the deviation is larger and generally leads to a steeper z dependence. This corrected z dependence can be approximated as an exponential with a shorter decay distance, but the exact shape is not truly exponential.

A metal-coated surface affects the emission, both the intensity and the angular pattern. Far-field light for which $(k_x^2 + k_y^2) \leq k^2$ reflects off the metal surface and very little penetrates through. But for a near-field component with a very particular $(k_x^2 + k_y^2) > k^2$, a surface plasmon will form in the metal and then produce a thin-walled hollow cone of emission light propagating into the glass substrate.[52] This hollow cone of light potentially could be captured for imaging by a very high-aperture objective. The rapid conversion of near-field light into either a surface plasmon or heat (via the quenching effect mentioned earlier) leads to a much reduced excited state lifetime for fluorophores near a surface with a consequent great reduction in photobleaching rate. These effects could form the basis of a new sensitive means of single fluorophore detection with minimal photobleaching.

Measurement of Distances from Surface

In principle, the distance z of a fluorophore from the surface can be calculated from the fluorescence intensity F, which as an approximation might be considered

[52] W. H. Weber and C. F. Eagen, *Opt. Lett.* **4**, 236 (1979).

to be proportional to the evanescent intensity $I(z)$ as given by Eq. (2). In practice, the situation is more complicated for several reasons.

1. The proportionality factor between F and I depends on efficiencies of absorption, emission, and detection, and is generally not well known.

2. $F(0)$ may not be known if a fluorophore positioned exactly at $z = 0$ (the substrate surface) cannot be identified. In this circumstance, Eq. (3) can be used to calculate only relative distances. That is, if a fluorophore moves from some (unknown) distance z_1 with observed intensity I_1 to another distance z_2 with intensity I_2, then assuming (as an approximation) an exponentially decaying $F(z)$,

$$\Delta z = z_1 - z_2 = d \ln(I_2/I_1) \tag{14}$$

This relationship is (approximately) valid even if the fluorescent structure consists of multiple fluorophores and is large or irregularly shaped (e.g., a secretory granule[25]). Again assuming an exponentially decaying $F(z)$, a motion of Δz for the whole structure causes each fluorophore in the structure to change its fluorescence by the same multiplicative factor, and the whole structure therefore changes its fluorescence by that same factor.

3. The z dependence of the emission pattern (because of near-field capture by the substrate) causes $F(z)$ to deviate from exponential decay behavior, as discussed above. Because the deviation depends on the numerical aperture, it may be possible (with an exact theory) to calculate absolute z distances by comparing the emission intensity gathered by moderate and very high-aperture objectives on the same sample.

4. For TIRF studies on biological cells, the distance of a fluorophore or organelle from the plasma membrane is likely to be of more interest than its distance from the substrate surface. The absolute distance between the plasma membrane and the substrate can be deduced by adding a membrane impermeant fluorescent dye to the medium of the cell culture under TIRF illumination. (For viewing labeled organelles at the same time as this impermeant dye, fluorophores of distinctly different spectra should be selected.) In off-cell regions, the fluorescence $F_{offcell}$ will appear uniformly bright, arising from the full depth of the evanescent field. In cell–substrate contacts, the dye will be confined to a layer thinner than the evanescent field. The fluorescence F in contact regions will be darker by a factor that can be converted to separation distance h between the substrate and the plasma membrane gap according to

$$h = -d \ln [1 - F/F_{offcell}] \tag{15}$$

This formula is an approximation because it assumes an exponential decay of gathered fluorescence intensity vs. z and thereby neglects the near-field effects discussed earlier.

Optical Configurations

A wide range of optical arrangements for TIRF can been employed. Some configurations use a high numerical aperture (NA > 1.4) microscope objective for both TIR illumination and emission observation, and others use a prism to direct the light toward the TIR interface with a separate objective for emission observation. This section gives examples of these arrangements. For concreteness in the descriptions, we assume that the sample consists of fluorescence-labeled cells in culture adhered to a glass coverslip.

"Prismless" TIR through High-Aperture Objective

By using an objective with a sufficiently high NA, supercritical angle incident light can be cast on the sample by illumination through the objective.[53,54] Although an arc lamp can be used as the light source, the general features are best explained with reference to a laser source. The optical system has the following general features:

1. The laser beam used for excitation is focused (by an external focusing lens) to a point at the back focal plane of the objective so that the light emerges from the objective in a collimated form (i.e., the "rays" are parallel to each other). This ensures that all the rays are incident on the sample at the same angle θ with respect to the optical axis.

2. The point of focus in the back focal plane is adjusted to be off-axis. There is a one-to-one correspondence between the off-axis radial distance ρ and the angle θ. At a sufficiently high ρ, the critical angle for TIR can be exceeded. Further increases in ρ serve to reduce the characteristic evanescent field depth d in a smooth and reproducible manner.

The beam can emerge into the immersion oil (refractive index n_{oil}) at a maximum possible angle θ_m measured from the optical axis) given by

$$NA = n_{oil} \sin \theta_m \qquad (16)$$

Because $n \sin \theta$ is conserved (by Snell's law) as the beam traverses through planar interfaces from one material to the next, the right side of Eq. (16) is equal to $n_3 \sin \theta_3$ (where subscript 3 refers to coverslip substrate on which the cells grow). For total internal reflection to occur at the interface with an aqueous medium of refractive index n_1, θ_3 must be greater than the critical angle θ_c as calculated from

$$n_1 = n_3 \sin \theta_c \qquad (17)$$

From Eqs. (16) and (17), it is evident that the NA must be greater than n_1, preferably

[53] D. Axelrod, *J. Biomed. Opt.* **6**, 6 (2001).
[54] A. L. Stout and D. Axelrod, *Appl. Opt.* **28**, 5237 (1989).

by a substantial margin. This is no problem for an interface with water with $n_1 = 1.33$ and an NA = 1.4 objective. But for viewing the inside of a cell at $n_1 = 1.38$, an NA= 1.4 objective will produce TIR at just barely above the critical angle. The evanescent field in this case will be quite deep, and dense heterogeneities in the sample (such as cellular organelles) will convert some of the evanescent field into scattered propagating light.

Fortunately, objectives are now available with NA > 1.4. The highest aperture available is an Olympus 100× NA = 1.65; this works very well for through-the-lens TIR on living cells. However, that objective requires the use of expensive 1.78 refractive index coverslips made of either LAFN21 glass (available from Olympus) or SF11 glass (custom cut by VA Optical Co., San Anselmo, CA). SF11 glass is the less expensive of the two but it may have a chromatic dispersion not perfectly suited to the objective, thereby requiring slight refocusing for different fluorophores. The 1.65 objective also requires special $n = 1.78$ oil (Cargille), which is volatile and leaves a crystalline residue. Four other objectives that are now available circumvent these problems: an Olympus 60× NA = 1.45 a Nikon 60× NA = 1.45, a Nikon 100× NA = 1.45, and a Zeiss 100× NA = 1.45. The 1.45 objectives use standard glass (1.52 refractive index) coverslips and standard immersion oil and yet have an aperture adequate for TIR on cells. The 1.45 objective is probably the method of choice for TIR except for cells that have particularly dense organelles. Dense organelles tend to scatter the evanescent field, and this effect is less prominent with the higher angles of incidence accessible through higher aperture objectives.

3. The angle of convergence/divergence of the laser beam cone at the back focal plane is proportional to the diameter of the illuminated region at the sample plane. Large angles (and consequent large illumination regions) can be produced by use of a beam expander placed just upbeam from the focusing lens.

4. The orientation of the central axis of the laser beam cone at the back focal plane determines whether the TIR-illuminated portion of the field of view is centered on the microscope objective's optical axis.

A microscope can be configured in several variations for through-the-lens TIRF excited by a laser beam by use of either commercial accessories or fairly simple custom-built add-on modifications (Fig. 6A–C). An arc-lamp illumination system, rather than a laser, can also be configured for TIRF illumination by use of an opaque disk of the correct diameter inserted in a plane equivalent (but upbeam) from the objective back focal plane (Fig. 6D). This allows only rays at significantly off-axis radii in the back focal plane to propagate through to the TIR sample plane, on which they are incident at supercritical angles. Switching back and forth between EPI and TIR can be done simply by placing or removing the opaque disk as shown. Arc illumination has the advantages of easy selection of excitation colors with filters and freedom from coherent light interference fringes, but it is somewhat dimmer because much of the arc lamp power directed toward the sample at subcritical angles is necessarily blocked. Somewhat more of the incident light can be utilized rather than wasted by employing a conical prism in the arc lamp light path.[54] This kind of prism converts the illumination profile, normally brightest in the center, into a dark-centered ring.

The evanescent illumination is not "pure" with through-the-lens prismless TIRF: a small fraction of the illumination of the sample results from excitation light scattered within the objective, and a small fraction of the observed fluorescence arises from luminescence of the objective's internal elements.

Here is a practical protocol for setting up prismless TIRF with laser illumination through an inverted microscope, and verifying that the setup works correctly. The description is most relevant to the setup in Fig. 6A, but modifications to Fig. 6B–D (in which the microscope already has lenses installed in the illumination beam path) are straightforward.

1. Prepare a test sample consisting of a film of fluorescent material adsorbed to a coverslip of the appropriate type of glass ($n = 1.52$ for NA = 1.45 objectives; $n = 1.78$ for NA = 1.65 objectives). A convenient and durable uniform film is made from 3,3′-dioctadecylindocarbocyanine (also known as "diI," available from Molecular Probes, Eugene, OR). Dissolve the diI at about 0.5 mg/ml in ethanol, and place a single droplet of the solution on a glass coverslip. Then, before it dries, rinse off the coverslip with distilled water. A monolayer of diI fluorophore will remain adhered to the glass. When excited by laser light at 488 nm and observed through a long-pass filter set appropriate for fluorescein, the diI adsorbed to the surface will appear orange. Above the diI-coated surface, place a bulk layer of aqueous fluorescein solution, which will appear green in the same spectral system. As an alternative to the diI/fluorescein test sample, a suspension of fluorescent microbeads in water can be used. Some of the beads will adhere to the surface. Although the same color as the suspended beads, the surface-adhered beads will be immobile whereas the bulk suspended beads will be jittering with Brownian motion. Place the sample on the microscope stage and raise the objective with the focusing knob to make optical contact through the appropriate immersion oil.

2. Be sure to remove the excitation filter in the filter cube if it was present for arc illumination, leaving only the dichroic mirror and the emission barrier filter appropriate for the excitation wavelength (typically 488 nm for fluorescein).

3. Remove all obstructions between the test sample and the ceiling. Allow a collimated laser beam (the "raw" beam) to enter the side or rear epiillumination port along the optical axis. A large area of laser illumination will be seen on the ceiling, roughly straight up. (Take care to keep eyes out of the beam path.)

4. Place a focusing lens (plano- or double-convex, about 100 mm focal length and mounted on a microtranslator so its lateral position can be adjusted precisely) at a position ∼20 cm "upbeam" from where the beam enters the microscope illumination port. The illuminated region on the ceiling will now be a different size, probably smaller.

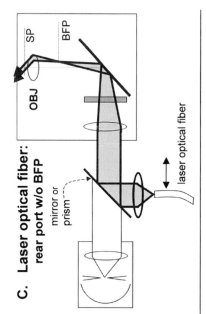

C. Laser optical fiber: rear port w/o BFP

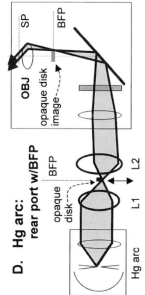

D. Hg arc: rear port w/BFP

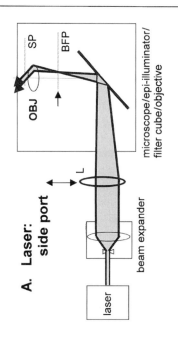

A. Laser: side port

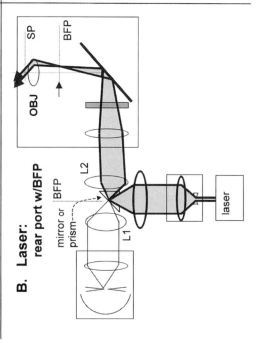

B. Laser: rear port w/BFP

5. Move the focusing lens longitudinally (along the axis of the laser beam) to minimize the illuminated region on the ceiling. This will occur where the converging lens focal point falls exactly at the objective's back focal plane. At this position, the beam is thereby also focused at the objective's actual back focal plane and emerges from the objective in a roughly collimated form.

6. Fine tune the lateral position of the focusing lens so that the beam spot on the ceiling moves down a wall to the right or left. The inclined path of the beam through the fluorescent aqueous medium in the sample will be obvious to the eye. Continue to adjust the focusing lens lateral position until the beam traverses almost horizontally through the fluorescent medium and then farther, past where it just disappears. The beam is now totally reflecting at the substrate/aqueous interface.

7. View the sample through the microscope eyepieces. The diI/fluorescein sample should appear orange, not green, as only the surface is being excited; the microbead sample should show only immobilized dots. Back off the focusing lens to where the beam reappears in the sample (i.e., subcritical incidence). When viewed through the eyepieces, the diI/fluorescein sample should now appear very bright green and the microbead sample very bright with most beads jittering laterally and longitudinally in and out of focus.

8. If the illuminated region in the field of view is not centered, adjust the lateral position of the "raw" laser beam before it enters the focusing lens and repeat the

---

FIG. 6. Four arrangements for prismless (through-the-objective) TIRF in an inverted microscope. In all these configuration, OBJ refers to the objective, SP refers to sample plane, and BFP refers to the objective's back focal plane or its equivalent planes (also called "aperture planes"). Components drawn with heavier lines need to be installed; components in lighter lines are possibly preexisting in the standard microscope. (A) Laser illumination through a side port (requires a special dichroic mirror cube facing the side, available for the Olympus IX-70 microscope). The beam is focused at the back focal plane at a radial position sufficient to lead to supercritical angle propagation into the coverslip. Moving the "focusing" lens L transversely changes the angle of incidence at the sample and allows switching between subcritical (EPI) and supercritical (TIR) illumination. This is how Figure 1 was produced. (B) Laser illumination in microscope systems containing an equivalent BFP in the rear path normally used by an arc lamp. The laser beam is focused at the BFP where the arc lamp would normally be imaged. The Zeiss Axiovert 200 and the Nikon TE 2000-U provide this BFP, marked as an "aperture plane." If (as in the Olympus IX-70) an aperture plane does not exist in the indicated position, it can be created with the pair of lens L1 and L2. (C) Laser illumination introduced by an optical fiber through the rear port normally used by the arc lamp. This scheme is employed by the commercial Olympus TIRF device. (D) Arc lamp TIR illumination with no laser at all. The goal is to produce a sharp-edged shadow image of an opaque circular disk at the objective back focal plane such that only supercritical light passes through the objective. The actual physical opaque disk (ideally made of aluminized coating on glass) must be positioned at an equivalent upbeam BFP, which, in Kohler illumination, also contains a real image of the arc. The Zeiss Axiovert 200 and the Nikon TE 2000-U provide this BFP, marked as an "aperture plane." If (as in the Olympus IX-70) an aperture plane does not exist in the indicated position, it can be created with the pair of lens L1 and L2. The illumination at the back focal plane is a circular annulus; it is shown as a point on one side of the optical axis for pictorial clarity only. The through-the-lens arc-lamp TIRF configuration D can be switched easily to laser TIRF configuration C by insertion of the reflecting prism in the arc lamp light path. [Reproduced with permission from D. Axelrod, *Traffic* **2**, 764 (2001).]

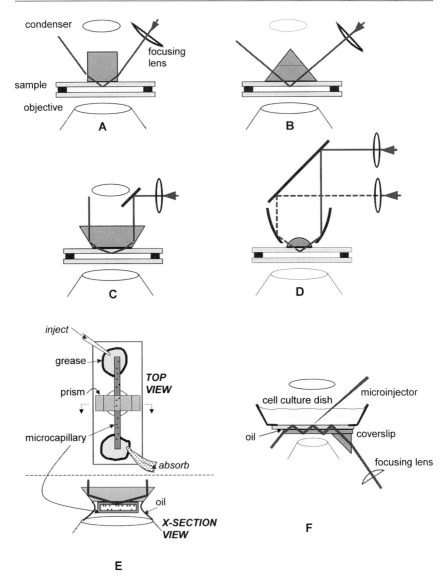

FIG. 7. Schematic drawings for prism-based TIR in an inverted microscope, all using a laser as a light source. The vertical distances are exaggerated for clarity. The first five configurations (A–E) use a TIR prism above the sample. (A–D) The buffer-filled sample chamber sandwich consists of a lower bare glass coverslip, a spacer ring (made of 60-μm-thick Teflon or double-stick cellophane tape), and the cell coverslip inverted so the cells face down. The upper surface of the cell coverslip is put in optical contact with the prism lowered from above by a layer of immersion oil or glycerol. The lateral position of the prism is fixed but the sample can be translated while still maintaining optical contact. The lower coverslip can be oversized and the spacer can be cut with gaps so that solutions can be changed by

two steps above. If the illuminated region thereby approaches the center, continue to readjust the raw beam in the same direction and repeat the steps until centering is achieved. If the illuminated region moves even farther to the side, then adjust the raw beam in the opposite direction and repeat the steps. The final position will be a laser beam entering the focusing lens parallel to its optical axis but somewhat off-center.

9. If the illuminated region is too small, increase the width of the beam before it enters the focusing lens with a beam expander or long focal length diverging lens.

10. Replace the test sample with the sample of interest and focus the microscope. As the lateral position of the external focusing lens is adjusted through the critical angle position, the onset of TIRF should be obvious as a sudden darkening of the background and a flat two-dimensional look to the features near the surface such that the entire field of view has only one plane of focus. With increased experience, the test sample can be skipped and all the adjustments made directly on the actual sample of interest.

11. If interference fringe TIR is desirable, use a beam splitter and arrange the second beam to enter the focusing lens off-center, parallel to the optic axis, but at a different azimuthal angle around it. A relative azimuthal angle of 180° will give the closest spaced fringes. Be sure that any difference in path length of the two beams from the beam splitter to the focusing lens is less than the coherence length of the laser (a few millimeters or centimeters); otherwise, no interference will occur.

TIRF with Prism

Although a prism may restrict sample accessibility or choice of objectives in some cases, prism-based TIR is very inexpensive to set up and produces a "cleaner" evanescent-excited fluorescence (i.e., less excitation light scattering in the optics) than prismless TIR. Figure 7 shows several schematic drawings for setting up laser/prism-based TIR in an inverted microscope.

---

capillary action with entrance and exit ports. (D) Two incident beams split from the same laser intersect at the TIR surface, thereby setting up a striped interference pattern on the sample, which is useful in studying surface diffusion. (E) Rectangular cross section microcapillary tube (Wilmad Glass Co., Buena, NJ) instead of a coverslip sandwich. With the ends of the microcapillary tube immersed in droplet-sized buffer baths delimited by silicon grease rings drawn on a support (one for supply and one for draining by absorption into a filter paper tab), very rapid and low-volume solution changes during TIRF observation can be accomplished. If an oil immersion objective is used here, the entire region outside the microcapillary tube between the objective and the prism can be filled with immersion oil. (F) Prism is placed below the sample and depends on multiple internal reflections in the substrate. This configuration thereby allows complete access to the sample from above for solutions changing and/or electrophysiology studies. However, only air or water immersion objectives may be used because oil at the substrate's lower surface will thwart the internal reflections. [Reproduced with permission from D. Axelrod, *Traffic* **2,** 764 (2001).]

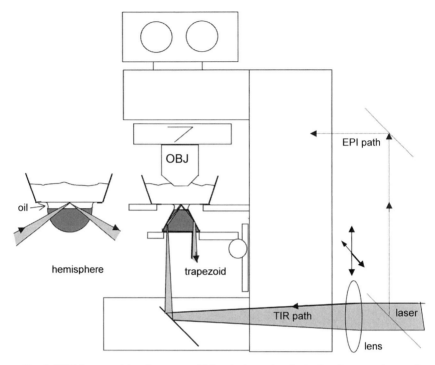

FIG. 8. TIRF for an upright microscope utilizing the integral optics in the microscope base and a trapezoidal prism on the vertically movable condenser mount. The position of the beam is adjustable by moving the external lens. An alternative hemispherical prism configuration for variable incidence angle is also indicated to the left. Vertical distances are exaggerated for clarity. An extra set of mirrors can be installed to deflect the beam into an epiillumination light path (shown with dashed lines). [Reproduced with permission from D. Axelrod, *Traffic* **2,** 764 (2001).]

Figure 8 shows an exceptionally convenient (and low cost) prism-based TIRF setup for an upright microscope. The laser beam is introduced in the same port in the microscope base as intended for the transmitted light illuminator (which should be removed), thereby utilizing the microscope's own in-base optics to direct the beam vertically upward. The prism, in the shape of a trapezoid, is mounted on the microscope's condenser mount, in optical contact (through a layer of immersion oil) with the underside of the glass coverslip sample plane. An extra lens just upbeam from the microscope base allows adjustment of the incident beam to totally reflect one of the sloping sides of the prism, from which the beam continues up at an angle toward the sample plane where it is weakly focused. This system gives particularly high-quality images if a water immersion objective is employed and submerged directly into the buffer solution in an uncovered cell chamber. Samples with cells adhering directly on the bottom of tissue culture plastic dishes rather than on coverslips can also be used; the plastic/cell interface is then the site of

TIR. If the objective has a long enough working distance, reasonable accessibility to micropipettes is possible.

In this configuration with the trapezoidal prism, flexibility in incidence angle (to obtain a range of evanescent field depths) is sacrificed in exchange for convenience. However, a set of various-angled trapezoids will allow various discrete incidence angles to be employed. The most inexpensive approach is to start with a commercially available equilateral triangle prism (say 1 inch × 1 inch × 1 inch sides, and 1 inch length), cleave off and polish one of the vertices to form a trapezoid, and slice the length of the prism to make several copies. Note, however, this prism will provide an incidence angle of only $60°$ at the top surface of the prism. If the prism is made from ordinary $n = 1.52$ glass, that angle is insufficient to achieve TIR at an interface with water. [Recall that we need $(n \sin 60°)$ to exceed 1.33, the refractive index of water.] However, equilateral prisms made from flint glass ($n = 1.648$) are commercially available (Rolyn Optics) and these will provide a sufficiently high $(n \sin 60°)$ for TIR to occur at the aqueous interface. In an alternative approach for varying incidence angles over a continuous range, a hemispherical prism can be substituted for the trapezoidal prism.[55] The incident laser beam is directed along a radius line at an angle set by external optical elements.

Choice of optical materials for the prism-based methods is somewhat flexible, as follows.

1. The prism used to couple the light into the system and the (usually disposable) slide or coverslip in which TIR takes place need not be matched exactly in refractive index.

2. The prism and slide may be optically coupled with glycerol, cyclohexanol, or microscope immersion oil, among other liquids. Immersion oil has a higher refractive index (thereby avoiding possible TIR at the prism/coupling liquid interface at low incidence angles) but it tends to be more autofluorescent (even the "extremely low" fluorescence types).

3. The prism and slide can both be made of ordinary optical glass for many applications (except as noted above for the configuration in Fig. 8) unless shorter penetration depths arising from higher refractive indices are desired. Optical glass does not transmit light below about 310 nm and also has a dim autoluminescence with a long (several hundred microsecond) decay time, which can be a problem in some FRAP experiments. The autoluminescence of high-quality fused silica (often called "quartz") is much lower. Tissue culture dish plastic (particularly convenient as a substrate in the upright microscope setup) is also suitable, but tends to have a significant autofluorescence compared to ordinary glass. More exotic high n_3 materials such as sapphire, titanium dioxide, and strontium titanate (with n as high

[55] D. Loerke, B. Preitz, W. Stuhmer, and M. Oheim, *J. Biomed. Opt.* **5,** 23 (2000).

as 2.4) can yield exponential decay depths d as low as $\lambda_0/20$, as can be calculated from Eq. (4).

In all the prism-based methods, the TIRF spot should be focused to a width no larger than the field of view; the larger the spot, the more that spurious scattering and out-of-focus fluorescence from the immersion oil layer between the prism and coverslip will increase the generally very low fluorescence background attainable by TIRF. Also, the incidence angle should exceed the critical angle by at least a couple of degrees. At incidence angles very near the critical angle, the cells cast a noticeable "shadow" along the surface.

Here is a practical protocol for setting up prism-based TIRF.

1. Mount the prism on the condenser mount carrier if possible. This need not be done in a precise fashion, but only accurate enough so a usable area at the sample-contacting surface of the prism lies directly in the optical axis of the microscope objective. The mounting may require some custom machining of plexiglass or brass plates and use of a glue (e.g., Duco cement) that can be easily cracked off and reglued if necessary. If a standard condenser will be used for simultaneous phase contrast or dark field and the condenser mount cannot hold two separate carriers, then the prism must be mounted on a separate holder with the capability of vertical motion. If the microscope focuses by moving the stage up and down, then this separate holder must be fixed to the microscope stage itself. Otherwise, the prism holder can be fixed directly to the optical table.

2. Depending on the configuration, a system of mirrors with adjustable angle mounts fixed to the table must be used to direct the beam toward the prism. One of these mirrors (or a system of shutters) should be movable and placed near the microscope so switching between standard epiillumination and TIR is possible without interrupting viewing.

3. Place a test sample (e.g., a diI-coated coverslip, see the description in the prismless TIR section) in the same kind of sample holder as will be used for cell experiments. An aqueous medium, possibly containing some fluorescein, can be used to fill the sample chamber. Alternatively, a test sample of fluorescent microbeads can be used as previously described.

4. With the test sample on the stage, focus on the fluorescent surface with transmitted (usually tungsten) illumination. Usually, dust and defects can be seen well enough to assay the focus. Fluorescent epiillumination can also be used to find the focus because only at the focal position are laser interference fringes seen sharply.

5. Place a small droplet of immersion oil on the non-diI surface of the sample coverslip or directly on the prism (depending on which one faces upward in the chosen configuration) and carefully translate the prism vertically so it touches and spreads the oil but does not squeeze it so tightly that lateral sliding motion is

inhibited. Too much oil will bead up around the edges of the prism and possibly interfere with the illumination path.

6. By naked eye (perhaps with safety goggles to attenuate errant reflections) and without any focusing lens in place, adjust the unfocused ("raw") collimated laser beam position with the mirrors so that TIR occurs directly in line with the objective's optical axis. This can usually be seen by observing the scattering of the laser light as it traverses through the prism, oil, and TIR surface.

7. Insert the focusing lens so that the focus is roughly at the TIR area under observation. Again by naked eye, adjust its lateral position with translators on the focusing lens so that the TIR region occurs directly in line with the objective. To guide this adjustment, look for three closely aligned spots of scattered light, corresponding to where the focused beam first crosses the immersion oil layer, where it totally reflects off the sample surface, and where it exits by recrossing the oil.

8. The TIR region should now be positioned well enough to appear in view in the microscope when viewed as test sample fluorescence with the standard filters in place. In general, the TIR region will appear as a yellow-orange ellipse or streak with a diI test sample and a region of discrete tiny bright dots with a microbead test sample . Make final adjustments with the focusing lens to center this area. The TIR area can be distinguished from two out-of-focus blurs past either end of the ellipse or streak (arising from autofluorescence of the immersion oil) because the TIR spot contains sharply focused images of defects in the diI coating or sharp dots of adsorbed microbeads. The focusing lens can be moved forward or backward along the laser optical path to achieve the desired size of the TIR area. If fluorescein solution was used to fill the sample chamber, the characteristic green color of fluorescein should *not* be seen with successful TIR. If the alignment is not correct and propagating light reaches the solution, then a bright featureless blur of green will be seen.

9. With the optics now correctly aligned for TIR, translate the prism vertically to remove the diI sample, and replace it with the actual cell sample. Relower the prism to make optical contact. Although the TIR region will not be exactly in the same spot (because of irreproducibility in the prism height), it will be close enough to make final adjustments with the focusing lens while observing fluorescence from the cell sample.

TIR from Multiple Directions

Configurations involving a single laser beam that produces TIR from one incidence direction (i.e., from one azimuthal angle around the optical axis) are the simplest ones to set up, but configurations involving multiple or a continuous range of azimuthal angles incident on the sample plane have some unique advantages. A single illumination direction tends to produce shadows on the "downstream" side of cells because of evanescent field scattering by the cells; these shadows are less apparent at higher incidence angles even with a single beam. However, a hollow cone of illumination over all the azimuthal angles around the optical axis virtually eliminates this shadow artifact. The prismless TIR configuration that uses an arc lamp for its illumination system (Fig. 6D) automatically provides such a hollow cone. A hollow cone can also be produced from a laser beam by positioning a conical lens in the incident beam path exactly concentric with the laser beam, in a fashion similar to that already described for arc illumination.[54] Alternatively, a single laser beam can be scanned with electrooptical devices to trace a circle where it focuses at the objective back focal plane.

Illumination by two mutually coherent TIR laser beams produces a striped interference fringe pattern in the evanescent field intensity in the sample plane. For a relative azimuthal angle of ϕ between the two beams, both with an incidence angle of θ, the node-to-node spacing s of the fringes is given by

$$s = \lambda_o / [2n_3 \ \sin\theta \ \sin(\phi/2)] \qquad (18)$$

The spacing s is not dependent on the refractive index n_1 of the medium (or cell). For beams coming from opposite azimuthal directions ($\phi = \pi$), $s = \lambda_o/2$ at the critical angle of incidence and $s = \lambda_o/2n_3$ at glancing incidence ($\theta = \pi/2$). For a typical glass substrate with $n_3 = 1.5$, this latter spacing is smaller than the Raleigh resolution limit of the microscope and can barely be discerned by the imaging system.

These interference stripes can be used in combination with TIR/FRAP (see above) to bleach antinode regions but not node regions. The subsequent fluorescence recovery will then provide information about surface diffusion of fluorophores.[30,34]

The spatial phase of the node/antinode intensity pattern can be easily controlled by retarding the phase of one of the interfering beams with an extra glass plate inserted into its light path. By illuminating with four discrete coherent TIR beams at relative azimuthal angles of $\phi = 0, \pi/2, \pi$, and $3\pi/2$ a checkerboard evanescent pattern in both the x and y lateral directions of the sample plane can be produced. Then by imaging a sample at lateral node/antinode spatial phase steps of $0°$, $120°$, and $240°$, images with superresolution well below the Raleigh limit in the sample plane can be computed.[56,57]

Rapid Chopping between TIR and EPI

Regardless of the method chosen to produce TIR in a microscope, it is often desirable to switch rapidly between illumination of the surface (by TIR) and deeper

[56] F. Lanni and T. Wilson, *in* "Imaging Neurons—A Laboratory Manual" (R. Yuste, F. Lanni, and A. Konnerth, eds.), Chap. 8. Cold Spring Harbor Laboratory Press, Cold Spring Harbor, NY, 1999.
[57] G. E. Cragg and P. T. C. So, *Opt. Lett.* **25,** 46 (2000).

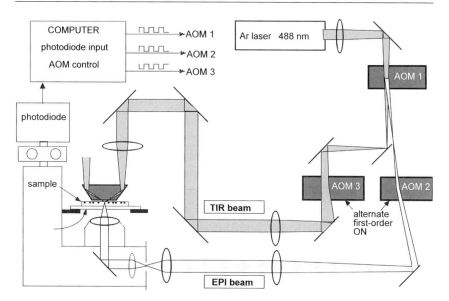

FIG. 9. Rapid alternation between TIR and EPI illumination. Three acoustooptic modulators (AOM) are used, with driving voltage phases as shown. The resulting chopped output of the single-channel photodetector can be separated into two traces vs. time by software.

illumination of the bulk (by standard epifluorescence). For example, a transient process may involve simultaneous but somewhat different fluorescence changes in both the submembrane and the cytoplasm, and both must be followed on the same cell in response to some stimulus.[46] For spatial resolved images, the switching rate is often limited by the readout speed of the digital camera or by photon shot noise. Single-channel integrated intensity readings over a region defined by an image plane diaphragm can be performed much more rapidly. Figure 9 shows a method using computer-driven acoustooptic modulators by which very rapid chopping (with a switching time as small as 10 μs) can be done.

Surface Near-Field Imaging

As discussed in the Theoretical Principles section, the technique for surface-selective emission imaging does not depend on TIR excitation, and in fact it works with standard epiillumination and, in principle, should even work with nonoptical excitation such as chemiluminescence. Near-field light from surface-proximal fluorophores propagates into the glass exclusively at supercritical hollow cone angles. Figure 10 shows a method for blocking all subcritical light emanating from the fluorophore far field while passing a range of angles of supercritical light from the near field by use of an opaque disk at a back focal plane in the emission path. Any

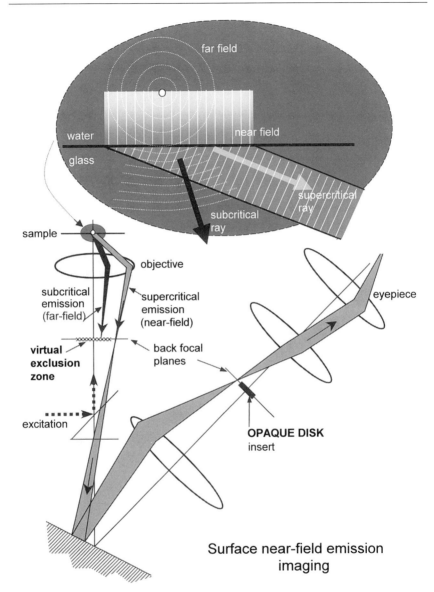

FIG. 10. Schematic diagram of the surface-selective near-field emission in an Olympus IX-70 micro-scope. *Top:* Magnified view of the sample region shows the electric field of the near field corresponding to one particular k_z as an exponential decay (shown with graded shading) extending in the z-direction away from a single fluorophore (shown as a black dot). The lateral spacing (left to right here) of the wavefronts for that one contribution to the near field is shown in a series of vertical stripes. If the fluorophore is close enough to the glass, the tails of the decaying near-field wavefronts are captured

one particular supercritical hollow cone angle of emission corresponds to a single exponential near-field decay, but typically a wider range of supercritical angles is collected by the objective and passed outside the periphery of the opaque disk. The resulting surface selectivity corresponds to a weighted integral of exponentials with a range of decay lengths. An actual implementation of this modification has been demonstrated to work on both artificial systems of surface-bound free fluorophores and on living biological cells, albeit with a loss of some lateral resolution.[52]

General Experimental Considerations

Laser Source

A laser with a total visible output in the 100 mW or greater range should be adequate for most TIRF applications. Air-cooled argon or diode lasers of less than 100 mW are adequate for some purposes, but probably marginal for dim samples or for samples where a weaker laser line (e.g., the 457 nm line of argon) may be desired to excite a shorter wavelength fluorescent marker (such as cyan fluorescent protein, CFP). Laser illumination produces interference fringes, which are manifested as intensity variations over the sample area. For critical applications, it may be advisable to rapidly jiggle the beam (e.g., in a commercially available optical fiber phase scrambler) or to compute a normalization of sample digital images against a control digital image of a uniform concentration of fluorophores.

---

by the glass and converted into light propagating away at a particular supercritical angle. (A complete picture of the near field would show a continuous spectrum of decay lengths, each corresponding to different wavefront spacings, with shorter decay lengths associated with closer wavefront spacings. The total near field is a superposition of all such different k_z contributions, with their relative wavefronts all aligned exactly in phase at the lateral position of the fluorophore so that the region laterally nearest the fluorophore is the brightest with a rapid decay to either side. Each different k_z contribution gives rise to light in the glass propagating at a different supercritical angle.) Far-field light is shown as a series of circular wavefronts emanating from the fluorophore and refracting into the glass at subcritical angles only. *Bottom:* Fate of the super- and subcritical light rays in the microscope. An opaque disk is inserted in an accessible region at a plane complimentary to the objective back focal plane, concentric with the optical axis and with a radius just sufficient to block out all the subcritical light. The correct radius of the opaque disk can be determined by observing this plane directly (with the eyepiece removed) with a diI/fluorescein preparation (as described in the prismless TIR section) on the sample stage; the subcritical light will appear green, surrounded by an orange annulus. When placed at the correct longitudinal position along the optical axis, the opaque disk will show no parallax relative to the orange annulus. The emission light forms a hollow cylindrically symmetric cone around the optical axis after the opaque disk; emission light is shown traversing one off-axis location for pictorial clarity only. Subcritical light is actually blocked only at the real opaque disk; the diagram shows the blockage at the image of the disk at the objective's back focal plane as a "virtual exclusion zone" to illustrate the principle.

Functionalized Substrates

TIRF experiments often involve specially coated substrates. A glass surface can be chemically derivatized to yield special physi- or chemiabsorptive properties. Covalent attachment of certain specific chemicals are particularly useful in cell biology and biophysics, including: poly(L-lysine) for enhanced adherence of cells; hydrocarbon chains for hydrophobicizing the surface in preparation for lipid monolayer adsorption; and antibodies, antigens, or lectins for producing specific reactivities. Derivatization generally involves pretreatment of the glass by an organosilane.[34]

A planar phospholipid coating (possibly with incorporated proteins) on glass can be used as a model of a biological membrane. Methods for preparing such model membranes on planar surfaces suitable for TIR have been reviewed.[58]

Aluminum coating (for surface fluorescence quenching) can be accomplished in a standard vacuum evaporator; the amount of deposition can be made reproducible by completely evaporating a premeasured constant amount of aluminum. After deposition, the upper surface of the aluminum film spontaneously oxidizes in air very rapidly. This aluminum oxide layer appears to have some chemical properties similar to the silicon dioxide of a glass surface; it can be derivatized by organosilanes in much the same manner.[34]

Photochemistry at the Surface

Illumination of surface-adsorbed proteins can lead to apparent photochemically induced crosslinking. This effect is observed as a slow, continual, illumination-dependent increase in the observed fluorescence. It can be inhibited by deoxygenation (aided by the use of an O_2-consuming enzyme/substrate system such as protocatachuic deoxygenase/protocatachuic acid or a glucose/glucose oxidase system), or by 0.05 M cysteamine.[34]

TIRF versus Other Optical Section Microscopies

Confocal microscopy achieves exclusion of out-of-focus emitted light with a set of image plane pinholes. It has the clear advantage in versatility; its method of optical sectioning works at any plane of the sample, not just at an interface between dissimilar refractive indices. However, other differences exist that, in some special applications, can favor the use of TIRF.

1. The depth of the optical section in TIRF is typically ≤ 0.1 μm whereas in confocal microscopy it is a relatively thick ~ 0.6 μm.

---

[58] N. L. Thompson, K. H. Pearce, and H. V. Hsieh, *Eur. Biophys. J.* **22**, 367 (1993).

2. In some applications (e.g., FRAP, FCS, or on cells whose viability is damaged by light), illumination and not just detected emission is best restricted to a thin section; this is possible only with TIRF.

3. Because TIRF can be adapted to and made interchangeable with existing standard microscope optics, even with "home-made" components, it is much less expensive than confocal microscopy. Laser-based TIRF microscopy kits are also now available commercially from Olympus, Nikon, and Till Photonics.

4. Unlike confocal microscopy, TIRF is suitable not only for microscopic samples but also for macroscopic applications; in fact those were the first TIRF studies. A previous review covers much of that work.[59]

Two-photon (or more generally, multiphoton) microcopy has many desirable features, including true optical sectioning, whereby the plane of interest is the only one that is actually excited, as in TIRF. Multiphoton microscopy is not restricted to the proximity of an interface, but its optical section depth is still several times thicker than that of TIRF. The setup expense of multiphoton microscopy for an infrared laser with sufficient pulse peak intensity can also be a consideration. Both multiphoton and confocal microscopy are necessarily scanning techniques; TIRF microscopy is a "wide-field" technique and is thereby not limited in speed by the scanning system hardware or image reconstruction software.

Cell–substrate contacts can be located by a nonfluorescence technique completely distinct from TIRF, known as "internal reflection microscopy." Using conventional illumination sources, IRM visualizes cell–substrate contacts as dark regions. Internal reflection microscopy has the advantage that it does not require the cells to be labeled, but the disadvantages that it contains no information of biochemical specificities in the contact regions and that it is less sensitive to changes in contact distance (relative to TIRF) within the critical first 100 nm of the surface.

Acknowledgments

This work was supported by NIH Grant 5 R01 NS38129. The author wishes to thank Drs. Ron Holz and Edwin Levitan for the chromaffin cells and the GFP construct, respectively, depicted in Fig. 1, and Dr. Geneva Omann for reviewing a draft of this chapter. The author also thanks all the past and present graduate students and postdoctoral researchers who have contributed to aspects of the TIRF work described here: Thomas Burghardt, Nancy Thompson, Edward Hellen, Andrea Stout, Ariane McKiernan, Michelle Dong Wang, Robert Fulbright, Laura Johns, Susan Sund, and Miriam Allersma.

[59] D. Axelrod, E. H. Hellen, and R. M. Fulbright, in "Fluorescence Spectroscopy: Principles and Applications, Vol. 3: Biochemical Applications" (J. Lakowicz, ed.), p. 289. Plenum Press, New York, 1992.

[2] Reflection Interference Contrast Microscopy

By IGOR WEBER

Introduction

The method by which the light reflected from a semitransparent, epiillumi-nated specimen contributes to the image formation in a microscope has been given different names, depending on details of the optical configuration of the micro-scope and on the particular application. It was introduced into cell biology by Curtis[1] using the name interference reflection microscopy (IRM), and this name, or its variant RIM, is often used in studies of adhesion of living cells. Ploem[2] used the term reflection contrast microscopy (RCM) when he introduced the antiflex method to increase image contrast. The term RCM is still mostly used in studies in which the technique is applied to visualize fixed specimens stained with dyes or labeled with colloidal gold particles. The term reflection interference contrast mi-croscopy (RICM) will be used here to cover all optical configurations and practical applications of the method.

The image in reflection interference contrast microscopy is generated by inter-ference of light beams reflected from interfaces between media with different in-dices of refraction. Boundaries between lipid bilayers and aqueous solutions rep-resent such interfaces. In RICM of living cells, reflections from the cell plasma membrane and intracellular membranes contribute to the image formation by inter-fering with each other and with the light reflected from the interface between the un-derlying glass surface and the liquid medium. The vertical distance between the closely apposed surfaces is encoded in distribution of light intensity in the image plane. For example, an image of the lower surface of a lipid vesicle residing on a hor-izontal glass substratum will appear as an interference pattern consisting of alter-nating dark and bright rings, resembling Newton's fringes. In principle, it should be possible to reconstruct the topology of the cell membranes using RICM, but in prac-tice independent knowledge of the basic parameters of the cell shape is required.

Practical RICM of living cells is mostly restricted to monitoring the shape of the ventral cell membrane and its interaction with the underlying surface during cell movement. Pioneering studies of fibroblastic cells by RICM led to discov-ery of focal contacts, i.e., adhesion plaques, which represent specialized sites of interaction of these cells with a substratum.[3,4] Certain stains, enzyme precipitates,

[1] A. S. G. Curtis, *J. Cell Biol.* **20**, 199 (1964).

[2] J. S. Ploem, *in* "Mononuclear Phagocytes in Immunity, Infection and Pathology" (R. Van Furth, ed.), p. 405. Blackwell, Oxford, 1975.

[3] A. Abercrombie and G. A. Dunn, *Exp. Cell Res.* **92**, 57 (1975).

[4] C. S. Izzard and L. R. Lochner, *J. Cell Sci.* **21**, 129 (1976).

and gold particles show an improved contrast in reflected light microscopy, and this is the basis for applications of RICM in immunocytochemistry and *in situ* hybridization studies. RICM is also being increasingly used as a probe detection method in scanning probe microscopy and other biophysical applications.

Image Formation in RICM

The basic principle of RICM, as applied to unstained biological specimens, is illustrated in Fig. 1. A semitransparent sample, for instance a cell, is observed under monochromatic epiillumination. The incoming light beam with intensity I_0 is partly reflected at the interface between the glass coverslip and the liquid medium. The transmitted part is again partly reflected at the interface between the medium and the cell plasma membrane. The reflection coefficients for the light beam components polarized in the directions perpendicular and parallel to an interface between two layers with different indices of refraction are given by the Fresnel equations:

$$r_{ij}^{\perp} = \frac{n_i \cos \theta_i - n_j \cos \theta_j}{n_i \cos \theta_i + n_j \cos \theta_j}, \qquad r_{ij}^{\parallel} = \frac{n_j \cos \theta_i - n_i \cos \theta_j}{n_j \cos \theta_i + n_i \cos \theta_j} \tag{1}$$

where n_i and n_j are the refractive indices of layers i and j, whereas θ_i and θ_j are angles of incidence and refraction at the (i, j) interface, respectively. The intensities of the light beams reflected from $(0, 1)$ and $(1, 2)$ interfaces, I_1 and I_2, respectively, are related to the reflection coefficients at the interfaces by

$$I_1 = r_{01}^2 I_0, \qquad I_2 = \left(1 - r_{01}^2\right) r_{12}^2 I_0 \tag{2}$$

It can be seen from Eqs. (1) and (2) that the intensity of reflected light at an interface between two media depends on the square of the difference between their

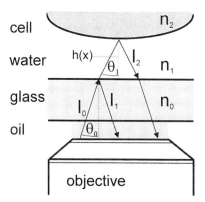

FIG. 1. Schematic presentation of the basic principle of RICM. See text for details.

respective indices of refraction. It follows from the values for refraction indices of glass ($n_g = 1.52$), water ($n_w = 1.34$), and the cell membrane ($1.4 < n_m < 1.5$), that both I_1 and I_2 are on the order of $10^{-3} I_0$. The two reflected beams interfere with each other, and the angular dependence of the observed total intensity imaged by the microscope objective onto the image plane is described by

$$I(\vartheta, x, y) = I_1 + I_2 + 2\sqrt{I_1 I_2} \cos[2kh(x, y)\cos(\vartheta) + \delta], \qquad k = \frac{2\pi n_1}{\lambda} \quad (3)$$

where $h(x, y)$ is the vertical object–substratum distance as a function of the planar projection of a reflection locus at the water–cell interface. The argument of the cosine function represents the difference of the optical path lengths between the two reflected beams. For reflection of a beam at an optically denser object (I_2), a phase reversal occurs ($\delta = \pi$). From Eq. (3) it follows that the RICM image represents a cosine transform of the ventral cell surface profile.

In conventional reflection microscopy light sources with a small coherence length such as a high-pressure mercury lamp are commonly used. To calculate the intensity distribution in the image plane of an RIC microscope, rays of light described by Eq. (3) have to be incoherently summed over the range of angles ϑ defined by the illumination numerical aperture (INA). This problem has been approached by analytical and numerical methods,[5–8] and the results have shed light on the well-known phenomenon that visibility of the interference pattern in an RICM image depends on INA. Generally, the amplitude of the periodic interference pattern decays with increasing object–substratum distance, and this decay is more rapid at larger values of INA. This means that in RICM it is possible to regulate the depth of the focus by changing the opening of the aperture diaphragm.

It has been shown that this so-called finite aperture effect can be interpreted as a case of partial coherence between the light rays reflected from the two vertically displaced surfaces.[6] Two light beams that originate from two point sources will interfere with each other depending of their mutual coherence. In the case of RICM imaging, the extent of coherence between the two reflected beams I_1 and I_2 is described by their mutual coherence function Γ_{12}. Γ_{12} depends on the distance between the two points $h(x, y)$ and on the intensity distribution of the illuminating light in the aperture plane, which is determined by INA. According to the Van Citter–Zernike theorem, Γ_{12} is given by the amplitude spread function near the focus in the direction of the microscope optical axis.[9] Direct application of this theorem leads to the same result as an incoherent integration of Eq. (3), and the final expression for the intensity distribution in the image plane of an RIC microscope,

[5] D. Gingell and I. Todd, *Biophys. J.* **26**, 507 (1979).
[6] J. Rädler and E. Sackmann, *J. Phys. France II* **3**, 727 (1993).
[7] M. Kühner and E. Sackmann, *Langmuir* **12**, 4866 (1996).
[8] G. Wiegand, K. R. Neumaier, and E. Sackmann, *Appl. Opt.* **37**, 6892 (1998).
[9] M. Born and E. Wolf, "Principles of Optics," 6th Ed. Pergamon, Oxford, 1993.

as a result of reflection from a surface profile $h(x, y)$, reads:

$$I(x, y) = 4\pi \sin^2 \alpha/2 \left\{ I_1 + I_2 + 2\sqrt{I_1 I_2} \frac{\sin[2kh(x, y)\sin^2 \alpha/2]}{2kh(x, y)\sin^2 \alpha/2} \right.$$

$$\left. \times \cos[2kh(x, y)(1 - \sin^2 \alpha/2) + \delta] \right\} \tag{4}$$

where α is related to INA by $\alpha = \sin^{-1}(INA/n_1)$. Intensity profiles corresponding to reflections from a sphere and a wedge made of material with $n = 1.45$, corresponding approximately to the refraction index of the cell plasma membrane, as derived from Eq. (4), are shown in Fig. 2. Damping of higher interference

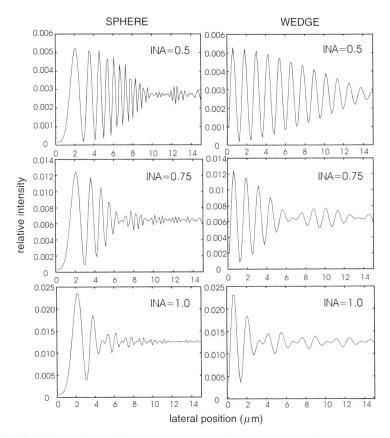

FIG. 2. Solutions of Eq. (4) for reflections from a sphere ($r = 20~\mu$m) and a wedge ($\varphi = 10°$), calculated for three different values of INA. Relative intensity in the image plane is plotted along a horizontal coordinate x. The vertical coordinate h is related to x by $h(x) = r - \sqrt{r^2 - x^2}$ in the case of a sphere, and by $h(x) = x \cdot tg(\varphi)$ in the case of a wedge. It was assumed that green light ($\lambda = 546$ nm) is reflected from a glass–water and a lipid–water interface ($n_g = 1.52, n_w = 1.34, n_1 = 1.45$).

orders is due to the partial coherence effect. The damping is faster for higher INA values, but contrast of the still visible orders increases. Additional damping in the case of a sphere is due to the surface curvature, and the simple theory fails to reproduce measured intensity profiles for angles larger than $10°$. Corrections for the difference between the approximate and the exact theory for reflection from inclined surfaces have been published for wedge-like and spherical geometries.[8]

Equation (4) has been derived under the following simplifying assumptions: (1) Angular dependence of the reflection coefficients r_{ij} [Eq. (1)] was neglected, which means that the normal incidence was assumed. This approximation is justified by the weak dependence of the reflection coefficients on the angle of incidence for circularly polarized light usually used in RICM, and is particularly appropriate for small values of INA and small inclination angles of the reflecting surfaces. A theory of RICM image formation has been published that explicitly takes into account nonnormal incidence and reflections from nonplanar interfaces.[8] The nonlocal effects arise primarily from the superposition of interfering rays with different optical path lengths resulting from different angles of incidence, and from lateral displacement of the loci of interference of the rays reflected at nonplanar interfaces. (2) Reflection from only two interfaces has been taken into account. Multiple reflections on planar surfaces can be handled by a transfer matrix formalism, but for arbitrary surface orientation exact calculations become quite elaborate.[10] Additionally, the number and shape of the contributing surfaces have to be known in advance, a condition fulfilled only in simple experimental situations. In living cells, the position and configuration of intracellular membranes are not known, but appreciation of the multiple reflection effects helps in qualitative interpretation of observed intensity distributions (Fig. 3).

Lateral resolution in RICM is determined by classic criteria, for instance the Rayleigh criterion, and depends on the sum of observational and illumination numerical apertures. Vertical resolution in RICM is extremely high, and absolute resolution in the range of several nanometers has been reached in measurements of the lipid film thickness.[6] Relative positions along the optical axis can be determined with an even higher precision, and values with an estimated precision of ± 0.2 nm have been reported.[7] In the case of steeply inclined surfaces the two resolutions cannot be strictly separated, as the interference fringes that encode the information about vertical distances merge with each other laterally. To account for such effects, the three-dimensional optical transfer function of the imaging system needs to be known. Calculated and experimentally determined maximal slopes that can still be resolved in RICM are about $40°$ for a glass–water–lipid system.[8] The maximal resolvable slope depends on the difference between observational and illumination numerical apertures.[6]

[10] G. Wiegand, T. Jaworek, G. Wegner, and E. Sackmann, *J. Colloid Interface Sci.* **196,** 299 (1997).

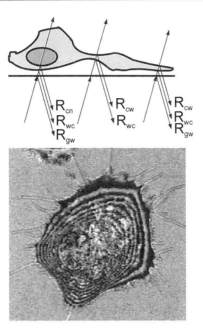

FIG. 3. Beams partially reflected from closely apposed interfaces interfere with each other and contribute to an RICM image. *Top*: Reflections from cytoplasm–nucleus (cn), water–cytoplasm (wc), cytoplasm–water (cw), and glass–water (gw) interfaces are shown. *Bottom*: RICM image of a fibroblast beginning to spread on a fibronectin-coated glass surface is shown. Interference fringes originate from the light reflected at the upper cell plasma membrane. Reflections from intracellular structures are visible in the central region of the cell.

Methods of Contrast Enhancement

Because the intensity of the light reflected from the water–cell interface is on the order of one-thousandth of the incoming light intensity, stray light reflected from the optical surfaces within the microscope, mostly from the rear objective lens, and nonspecific reflections from the specimen can seriously degrade the image quality. There are several ways of suppressing these reflections and the most important is the antiflex technique,[2] which filters out the stray light (Fig. 4). A polarizer linearly polarizes the incoming light and its greater part, which exits the microscope objective, is converted into circularly polarized light by a quarter-lambda plate attached to the front objective lens. At the water–cell interface, the sense of circular polarization is reversed by reflection from an optically denser medium. The reflected light is then reconverted into a linear polarization by the quarter-lambda plate, in an orientation perpendicular to the incoming light. Finally, a crossed analyzer in the observation optical path lets only the reflected light through and suppresses the stray light, which has retained its original sense of polarization.

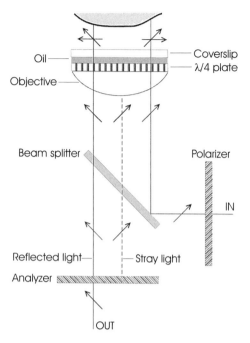

FIG. 4. Schematic presentation of the antiflex configuration, which suppresses the stray light reflected from optical surfaces within the objective. See text for details.

Insertion of a circular aperture diaphragm with a central (or Stach) stop into the epiillumination path at a plane conjugate to the entrance pupil of the objective serves two purposes. First, it reduces the disturbing reflections within the objective, which originate largely from the central portions of the lens. Further, in the case of thin specimens that are often used in immunocytochemistry, reflections from the glass–air interface at the distal side of a chamber or a microscope slide are often strong enough to obscure the RICM signal. An antiflex device does not help in this case, as the reflected light comes from outside of the objective. An aperture diaphragm with a central stop blocks these reflections, but it also reduces the effective range of illuminating angles and thereby reduces resolution. In addition, the possibility of changing the depth of focus by changing the illumination numerical aperture and of quantitative interpretation of intensities in an image is lost. An alternative way to suppress reflections from the distal glass surface of a sample is to use a light sink in the form of a glass cap or a prism, which is optically linked to the glass surface by immersion oil. Appropriate devices should deflect the incoming light that passes through a partly transparent specimen away from the objective.[11] In the case of experiments with living cells in a liquid medium,

[11] I. Cornelese-ten Velde, J. Bonnet, H. J. Tanke, and J. S. Ploem, *J. Microsc.* **159,** 1 (1990).

simply increasing the height of the liquid column to at least 3 mm will suppress the very strong reflection from the upper medium–air interface. Care should be taken that this interface is not curved, as it will act as a concave mirror and focus the reflected light back to the objective.

As discussed in the section on image formation, relatively low contrast in an image of a cell closely apposed to a substratum is due to the low reflectivity at the glass–water and the water–cell interfaces. Although it is not possible to lower the refraction index of the aqueous environment of living cells, adding large amounts of the osmotically inert protein albumin to the solution can increase it. Addition of bovine serum albumin (BSA) at millimolar concentrations can even lead to contrast reversal, i.e., cell–substratum contact areas will appear brighter then the background. Also, coating glass surfaces with dielectric films by vapor-deposition techniques can enhance the sensitivity of RICM. By adjusting the composition and thickness of surface coatings, the intensity and phase relationships between the reflected beams can be fine-tuned to optimize the visibility of the interference pattern. Numerical simulations are used in the design of such multilayer systems.[10] A material that proved to be valuable in measurements of the thickness of lipid bilayers and monolayer films is magnesium fluoride (MgF_2).[6,8] It has an index of refraction comparable to water, and can be coated with a thin layer of SiO_2 to mimic the surface properties of the original glass surface.

Practical RICM

Any fluorescence microscope with epiillumination optics can be easily adapted for RICM, but inverted microscopes are better suited to work with living cells. Because of the low reflectivity of the cell surfaces, a strong source of illumination is needed. High-pressure mercury or xenon lamps are usually used. Heat absorption and ultraviolet (UV) filters are needed to protect the polarizer, and for quantitative work with quasimonochromatic illumination a narrow-band bandpass filter should be used. Instead of a fluorescence filter block, a block containing a polarizer, a neutral beam splitter, and an analyzer in an extinction position relative to the polarizer must be used. An oil-immersion objective equipped with a quarter-lambda plate on the front lens should be used. Objectives with a revolving quarter-lambda plate can be obtained commercially, and the position of the plate should be adjusted to achieve the maximal contrast.

Epiillumination optics of fluorescent microscopes are usually not equipped with both aperture and field diaphragms that enable the full Köhler illumination. If only a field stop is provided, it is possible to use an auxiliary lens that projects the image of the field diaphragm onto a plane corresponding to the entrance pupil of the objective optics. In this way, the field stop effectively assumes the role of the aperture diaphragm, enabling control of the depth of focus and contrast in the RICM image. A field diaphragm in addition to the aperture diaphragm is still useful, as it makes it possible to narrow the illumination field to the object of interest. In this way the image quality increases because the stray light reflected obliquely from elsewhere in the specimen is blocked off. For studies of cell adhesion and the ventral cell membrane, an adjustable aperture diaphragm is not needed, as the maximal INA must be used to suppress reflections from more dorsal structures in the cell.

To obtain the most intense monochromatic illumination, the 546.1 nm line of the high-pressure mercury lamp is usually used. Living cells are, however, sensitive to such an intense green illumination, and for studies that require long-term exposure of cells to light, other types of illumination are more suitable. Cells are least sensitive to far-red or near-infrared light, and can be observed over hours or even days without harmful effects.[12] Also, absorption losses in thick specimens are minimized when red light is used for illumination. White-light illumination is sometimes used for qualitative work, mainly for better visualization of reflections from fixed and labeled specimens. Colored interference patterns result from reflection of the white light from cell membranes, and can be interpreted by using interference color charts that are available from manufacturers of interference microscopes. This method, or its simpler variant double-color RICM, which uses two simultaneous quasimonochromatic illuminations, offers the possibility of internal calibration of optical path differences between light rays reflected within a sample. It is straightforward to implement double-color RICM when using a confocal scanning microscope.

Because precision of focusing is of critical importance for RICM, a motorized focusing control or a piezoelectric focusing device is strongly recommended. Viscous resistance and temperature fluctuations within the immersion oil both contribute to the unsteady position of the focus. Therefore, control of the temperature at the focus, for example, by thermostating the objective, and use of a small amount of oil are recommended for high-precision and long-term experiments. Self-focusing algorithms can be used to keep the focus at a constant plane. For the same reason, small and stably affixed coverslips should be used to minimize mechanical vibrations.

In most cases it is sufficient to use a simple CCD camera with a dynamic range of 8 bits and adjustable level and gain controls for detection. If the intensity of the illuminating light has to be kept low, cameras with the integrated capability of dynamic averaging or more sensitive cooled CCD cameras are recommended. For work using near-infrared illumination, cameras equipped with chips that have an enhanced sensitivity in the infrared region should be used. Methods of digital image processing are well suited to enhancing the contrast in RICM images and have made possible precise microinterferometric measurements in model systems and living cells. RICM can also be readily combined with transmitted-light and fluorescence microscopy techniques using a conventional or a confocal microscope.

[12] M. S. Zand and G. Albrecht-Bühler, *Cell Motil. Cytoskel.* **13**, 94 (1989).

Confocal RICM

An image of a specimen in reflected light can be obtained simultaneously with the fluorescence image by a confocal microscope equipped with two detection channels without any modification. The reflected light is separated from the fluorescent light by a dichroic mirror and, if required, additionally filtered by a short-pass filter. Because of a long coherence length of the laser light, speckle and scatter from out-of-focus particles and surfaces can interfere with the image. For reflecting or scattering material in the specimen, the intrinsic optical sectioning property of the confocal microscope serves as a remedy. The only way for the stray light outside of an optical section of the specimen to contribute to the image is by multiple scattering. That is, after initial scattering outside the optical section, the light must be scattered again into an appropriate direction within the bundle of imaging rays, which is a rare event. Reflections inside the microscope optics present a more serious problem. Coherence of illumination can be reduced by using several light scrambling methods.[13] Also, a beam splitter, a pair of crossed polarizers, and a quarter-lambda plate in an antiflex configuration can be utilized to filter out the speckle from optical surfaces, analogous to their role in conventional RICM.[14] Although RICM can be carried out with a laser confocal microscope without any modification, use of the antiflex device is recommended for optimal performance.

The main difference between conventional and confocal RICM lies in the illumination pathway. In confocal RICM, the angular range of illuminating rays depends on details of the scanning mechanism and on the zoom factor, and the INA value is not clearly defined. On the other hand, strong coherence of the laser light significantly reduces the contrast-damping effect, resulting in unusually sharp images. It is difficult to separate contributions from reflecting layers within depth of the focus, which makes the cell–substrate separation measurements and microtopographic reconstructions not feasible. Both laser-scanning and spinning-disk configurations can be used for imaging, but a theory of image formation in confocal RICM is still lacking.

Applications of RICM

The older literature on applications of RICM in cell biology has been reviewed by Verschueren.[15] In the present article, a short overview of more recent applications in cell biology and related disciplines will be given.

[13] James B. Pawley, ed., "Handbook of Biological Confocal Microscopy," 2nd Ed. Plenum, New York and London, 1995.
[14] D. H. Szarowski, K. L. Smith, A. Herchenroder, G. Matuszek, J. W. Swann, and J. N. Turner, *Scanning* **14,** 104 (1992).
[15] H. Verschueren, *J. Cell Sci.* **75,** 279 (1985).

RICM of Living Cells

RICM was originally introduced to investigate adhesion of primary chicken heart fibroblasts to glass.[1] Subsequent applications have also concentrated on fibroblastic cells in culture, particularly on the study of focal and close contacts.[3,4] In a series of publications, Gingell and co-workers investigated mechanisms of cell–substratum interaction in red blood cells and *Dictyostelium discoideum*.[16,17] A surface reconstruction method has been devised to determine the three-dimensional shape of erythrocytes based on RICM microinterferometry.[18] Traces that are left behind locomoting fibroblasts on a glass substratum have also been analyzed by RICM.[19]

With the advent of video microscopy and digital image processing, quantification of several aspects of cell dynamics by RICM became possible. Cell flickering, the thermally excited surface undulations of erythrocyte plasma membranes, has been investigated and the bending modules of these membranes determined.[20] The dynamics of the cell–substratum contacts in wild-type and mutant *Dictyostelium* cells that lack actin-binding proteins has been correlated with their shape changes during motility.[21,22] Changes of the contact angle of these cells caused by the flow-induced deformation enabled simultaneous evaluation of their adhesion energy, bending modulus, and membrane lateral tension.[23] Because of a good contrast of the cell–substratum contact areas in some cell types, RICM can be used for automatic cell tracking, and image processing algorithms for this purpose have been developed.[24]

Because of its ease of use, confocal RICM is gradually replacing the more conventional technique in research on cell adhesion in different cell types. Adhesion patterns of endothelial cells growing in capillary glass tubes were characterized by confocal RICM.[25] A video-rate scanning confocal microscope was used in the RIC mode to demonstrate rapid motion of intracellular organelles and the cell periphery in neoplastic cell lines.[26] Using the confocal RICM, it was shown that neurites arise from filopodia that are attached only at their tips, and not from filopodia that

[16] H. Wolf and D. Gingell, *J. Cell Sci.* **63**, 101 (1983).

[17] S. Vince and D. Gingell, *Exp. Cell Res.* **126**, 462 (1980).

[18] F. Pera, *Micron Micr. Acta* **15**, 7 (1984).

[19] E. Richter, H. Hitzler, H. Zimmermann, R. Hagedorn, and G. Fuhr, *Cell Motil. Cytoskel.* **47**, 38 (2000).

[20] A. Zilker, M. Ziegler, and E. Sackmann, *Phys. Rev. A* **46**, 7998 (1992).

[21] I. Weber, E. Wallraff, R. Albrecht, and G. Gerisch, *J. Cell Sci.* **108**, 1519 (1995).

[22] J. Niewöhner, I. Weber, M. Maniak, A. Müller-Taubenberger, and G. Gerisch, *J. Cell Biol.* **138**, 349 (1997).

[23] R. Simson, E. Wallraff, J. Faix, J. Niewöhner, G. Gerisch, and E. Sackmann, *Biophys. J.* **74**, 514 (1998).

[24] I. Weber and R. Albrecht, *Comput. Meth. Prog. Bio.* **53**, 113 (1997).

[25] P. F. Davies, A. Robotewskyj, and M. L. Griem, *J. Clin. Invest.* **91**, 2640 (1993).

[26] P. Vesely, S. J. Jones, and A. Boyde, *Scanning* **15**, 43 (1993).

are attached along their length.[27] Simultaneous confocal fluorescence and RIC microscopy have been used to image focal adhesions in osteoblasts,[28] and to study the dynamics of cell–substratum contacts during keratocyte motility.[29]

RICM of Labeled Specimens in Immunocytochemistry

RICM is increasingly being used as a tool in immunocytochemical research.[30,31] Many stained biological specimens show an increased contrast when viewed in reflected light. Color of the reflected light is largely complementary to the stain color as viewed in transmitted light, so that blue-stained structures appear red in the reflected-light image. Interference effects also contribute to an increased visibility of reflecting material. Early examples of RICM applied on stained cellular structures were the improved images of cytoskeleton components stained with Coomassie blue and several other stains.[32,33] Strong reflectivity of the end product of an immunoperoxidase staining procedure, diaminobenzidine (DAB), was used for in situ hybridization studies.[30] Several other enzyme precipitates also display an increased contrast in RICM, and are of potential use for in situ hybridization studies.[34] RICM offers particularly high detection sensitivity for immunostained material and for small gold particles in ultrathin sections, where small clusters of 15-nm gold particles can be visualized directly, without silver enhancement.[35] All these labels are stable and do not bleach out like immunofluorescent labels. Moreover, the unstained parts of the specimen are often also visible in an RICM image. RICM can also be used for three-dimensional reconstruction of thick tissue sections by sequential focusing through the specimen.[36] Because of its high sensitivity, RICM provides a bridge between light microscopy and electron microscopy in immunocytochemical studies.

RICM in Surface Science and Biophysics

RICM is ideally suited to monitor the dynamics of surface phenomena because of its ability to measure the local optical thickness of thin films, and profiles of nonplanar interfaces with high precision. Coupled to the video acquisition and

[27] C. L. Smith, *J. Neurosci.* **14**, 384 (1994).

[28] Y. Usson, A. Guignandon, N. Laroche, M. H. Lafage-Proust, and L. Vico, *Cytometry* **28**, 298 (1997).

[29] K. I. Anderson and R. Cross, *Curr. Biol.* **10**, 253 (2000).

[30] J. S. Ploem, I. Cornelese-ten Velde, F. A. Prins, and J. Bonnet, *Proc. R. Microsc. Soc.* **30**, 185 (1995).

[31] T. J. Filler and E. T. Peuker, *J. Pathol.* **190**, 635 (2000).

[32] M. Opas and V. I. Kalnins, *J. Microsc.* **133**, 291 (1984).

[33] S. W. Paddock, *J. Cell Sci.* **93**, 143 (1989).

[34] E. J. Speel, M. Kamps, J. Bonnet, F. C. Ramaekers, and A. H. Hopman, *Histochemistry* **100**, 357 (1993).

[35] F. A. Prins, R. van Diemen-Steenvoorde, J. Bonnet, and I. Cornelese-ten Velde, *Histochemistry* **99**, 417 (1993).

[36] T. J. Filler, C. H. Rickert, U. K. Fassnacht, and F. Pera, *Histochemistry* **101**, 375 (1994).

image processing techniques, it offers the possibility of studying dynamic phenomena with a temporal resolution of 40 ms and better. RICM has been used to measure local adhesion energies and elastic properties of phospholipid vesicles.[37] Phase separation in supported lipid monolayers and dynamics of their spreading have been studied by RICM.[6,38] Wetting and dewetting phenomena have been investigated, and contact angles on chemically heterogeneous surfaces have been measured by RICM.[10] Structural, viscous, and elastic properties of soft polymer films have been studied by measuring fluctuations of the distance between the surface and colloidal latex beads.[7] Latex beads have also been used as microprobes to measure vertical distances by RICM in a force apparatus, which was designed to investigate molecular mechanisms of adhesion at biological interfaces.[39] Analogously, separation between a scanning probe and a sample in force measurements by scanning force microscopy has been independently measured by RICM.[40]

Caveats

Deduction of absolute cell–substratum distances from RICM images has to be interpreted with caution. Even when an exact theory of image formation is applied, a parameter that decisively influences the fitting procedure, the index of refraction of the plasma membrane, is either not known or it must be measured independently. Even then, it can vary spatially and temporally within the membrane, and also from cell to cell. A study using fluorescence interference contrast (FLIC) microscopy has indicated that even the focal adhesions, which have long been considered to be regions of intimate cell–substratum contact, owe their high reflectivity to a locally increased index of refraction, which is due to the presence of a large protein complex closely apposed to the intracellular side of the plasma cell membrane.[41]

Similar problems arise from multiple reflections at intracellular organelles, or at the upper cell membrane. It was long believed that leading lamellipods of moving cells are in especially close contact to the supporting surface, as they appeared darker than the rest of the ventral cell membrane in RICM images. A study by total internal reflection fluorescence microscopy (TIRFM), which measures cell–substratum separation in an independent way, has shown that its dark appearance is due to its thinness and a strong negative contribution to the total intensity of the light reflected from the dorsal membrane.[42]

[37] J. O. Rädler, T. J. Feder, H. H. Strey, and E. Sackmann, *Phys. Rev. E.* **51,** 4526 (1995).
[38] J. Rädler, H. Strey, and E. Sackmann, *Langmuir* **11,** 4539 (1995).
[39] E. Evans, K. Ritchie, and R. Merkel, *Biophys. J.* **68,** 2580 (1995).
[40] J. K. Stuart and V. Hlady, *Biophys. J.* **76,** 500 (1999).
[41] Y. Iwanaga, D. Braun, and P. Fromherz, *Eur. Biophys. J.* **30,** 17 (2001).
[42] I. Todd, J. S. Mellor, and D. Gingell, *J. Cell Sci.* **89,** 107 (1988).

Conclusions

RICM, either used as a quantitative microinterferometric method or to visualize stained and labeled microscopic structures, offers wide possibilities for application in investigating living cells and model systems in the life sciences. High spatial and temporal resolution makes it particularly suited for measurement of dynamic events with high precision. The main advantages of RICM are simplicity of setup in comparison to other interferometric techniques, high lateral and out-of-plane resolutions, and suitability for dynamic measurements when coupled with digital detection and signal processing methods. RICM can be implemented using a conventional or a confocal microscope, and readily combined with transmission and fluorescence microscopy techniques.

[3] Second Harmonic Imaging Microscopy

By Andrew C. Millard, Paul J. Campagnola, William Mohler, Aaron Lewis, and Leslie M. Loew

Introduction

Over the past three decades, the physical phenomenon of optical second harmonic generation (SHG) has been used to study interfaces between materials and has been adapted for the purposes of microscopy. SHG is a nonlinear optical process that can take place in a microscope that uses illumination from ultrafast (near-infrared) laser light. As in the case of two-photon excitation[1] the probability of SHG is proportional to the square of the incident light intensity. The idea that two-photon excited fluorescence (2PF) and SHG might each be used for nonlinear microscopy was first proposed by Sheppard and co-workers[2,3] and the application of mode-locked lasers to make 2PF microscopy practical was demonstrated in 1990 by Denk *et al.*[4] Whereas 2PF involves the near-simultaneous absorption of two photons to excite a fluorophore, followed by relaxation and noncoherent emission, SHG is a nearly instantaneous process in which two photons are converted into a single photon of twice the energy, emitted coherently. Furthermore, SHG is confined to loci lacking a center of symmetry; this constraint

[1] M. Göppert-Mayer, *Ann. Phys.* **9,** 273 (1931).

[2] C. J. R. Sheppard, R. Kompfner, J. Gannaway, and D. Walsh, *IEEE J. Quan. Electron.* **13E,** 100D (1977).

[3] T. Wilson and C. J. R. Sheppard, "Theory and Practice of Scanning Optical Microscopy." Academic Press, New York, 1984.

[4] W. Denk, J. H. Strickler, and W. W. Webb, *Science* **248,** 73 (1990).

is readily satisfied at cellular membranes in which SHG-active constituents are asymmetrically distributed. One of the first demonstrations of SHG from a biological specimen was of bacteriorhodopsin in a membrane preparation.[5,6] More recently, SHG microscope images have been obtained when one leaflet of the lipid bilayer of the cell membrane has been stained with a dye that enhances SHG.[7-10] Furthermore, there are numerous supramolecular structures within cells and tissues that can produce SHG; signals from collagen, for example, were also the basis of early demonstrations of SHG from biological specimens.[11] Because it is a nonlinear process, SHG can be confined to the region of greatest power density at the focus of the microscope, resulting in intrinsic three-dimensional sectioning without the use of a confocal aperture[4] and greatly reducing out-of-plane photobleaching and phototoxicity. SHG is a less efficient process than 2PF, but can be significantly resonance enhanced. However, because 2PF still results in some photobleaching at the focus, the best wavelengths for SHG are on the edges of two-photon excitation bands, enhancing SHG while reducing absorption.

Several features make this form of microscopy very powerful. Excitation uses near-infrared wavelengths, allowing excellent depth penetration, and hence this method is well-suited for studying intact tissue samples.[12] For example, we have acquired optical sections throughout 550 μm of mouse muscle tissue.[13] The increased penetration depth and other advantages of working at longer wavelengths, including minimized one-photon absorption, also make it easier to increase laser power and hence signal while avoiding photodamage. Information about the organization of chromophores, including dyes and structural proteins, at the molecular level can be extracted from SHG imaging data in several ways. SHG signals have well-defined polarizations, and hence SHG polarization anisotropy can be used to determine the absolute orientation and degree of organization of proteins in tissues. In addition, 2PF images can be collected in a separate data channel simultaneously with SHG. Correlation between the SHG and 2PF images provides the basis not only for molecular identification of the SHG source but also for probing the radial and lateral symmetry within structures of interest. Ratiometric SHG : 2PF techniques allow for the quantitation of chiral enhancement and voltage sensitivity,

[5] J. Huang, Z. Chen, and A. Lewis, *J. Phys. Chem.* **93**, 3314 (1989).

[6] T. Rasing, J. Huang, A. Lewis, T. Stehlin, and Y. R. Shen, *Phys. Rev. A* **40**, 1684 (1989).

[7] O. Bouevitch, A. Lewis, I. Pinevsky, J. P. Wuskell, and L. M. Loew, *Biophys. J.* **65**, 672 (1993).

[8] I. Ben-Oren, G. Peleg, A. Lewis, B. Minke, and L. M. Loew, *Biophys. J.* **71**, 1616 (1996).

[9] G. Peleg, A. Lewis, M. Linial, and L. M. Loew, *Proc. Natl. Acad. Sci. U.S.A.* **96**, 6700 (1999).

[10] P. J. Campagnola, M. D. Wei, A. Lewis, and L. M. Loew, *Biophys. J.* **77**, 3341 (1999).

[11] I. Freund, M. Deutsch, and A. Sprecher, *Biophys. J.* **50**, 693 (1986).

[12] A. K. Dunn, V. P. Wallace, M. Coleno, M. W. Berns, and B. J. Tromberg, *Appl. Opt.* **39**, 1194 (2000).

[13] P. J. Campagnola, A. C. Millard, M. Terasaki, P. E. Hoppe, C. J. Malone, and W. Mohler, *Biophys. J.* **81**, 493 (2002).

while normalizing out irrelevant parameters arising from laser fluctuations, sample movement, and nonuniform staining.

Theory

In general, the polarization for a material can be expressed as a Taylor series in the electric field:

$$\mathbf{P} = \chi^{(1)} * \mathbf{E} + \chi^{(2)} * \mathbf{E} * \mathbf{E} + \chi^{(3)} * \mathbf{E} * \mathbf{E} * \mathbf{E} + \cdots$$

$\mathbf{P}$ is the induced polarization vector, $\chi^{(i)}$ is the ith order nonlinear susceptibility tensor, and $\mathbf{E}$ is the electric field vector, all of which are frequency dependent; the operator $*$ therefore represents both Einstein summation over tensor indices and integration over frequency. At the molecular level, the polarization is replaced by the dipole moment and the molecular (hyper)polarizabilities replace the susceptibilities,[14] with the first molecular hyperpolarizability β giving rise to the bulk second-order nonlinear susceptibility $\chi^{(2)}$. The first term in the series describes normal absorption and reflection of light. The second term describes sum and difference frequency generation; SHG is the degenerate case of sum frequency generation. The third term describes two-photon absorption, the probability of which is linearly proportional to the imaginary part of the third-order nonlinear susceptibility tensor,[15] as well as four wave mixing, degenerate cases of which are third harmonic generation[16-18] and voltage-dependent SHG, and stimulated Raman processes. The portion of the polarization giving rise to SHG is therefore

$$\mathbf{P}^{(2)} = \chi^{(2)} * \mathbf{E} * \mathbf{E}$$

If the medium of interest has inversion symmetry, the incoming electric fields $\mathbf{E}$ and $-\mathbf{E}$ should induce polarizations of $\mathbf{P}^{(2)}$ and $-\mathbf{P}^{(2)}$; however, this is not consistent with the equation above unless $\chi^{(2)} = 0$, indicating that SHG is prohibited. (The general properties of the permutation and spatial symmetries of nonlinear susceptibilities have been treated by Shen[15] and Boyd.[19]) Inversion symmetry must therefore be broken to allow SHG, a criterion that can be satisfied at mismatched interfaces[20] and within birefringent crystals such as potassium dihydrogen phosphate (KDP) and beta barium borate (BBO). SHG imaging is of particular use

[14] D. J. Williams, *Angew. Chem. Int. Ed. Engl.* **23,** 690 (1984).

[15] Y. R. Shen, "The Principles of Non-Linear Optics." John Wiley & Sons, New York, 1984.

[16] M. Müller, J. A. Squier, K. R. Wilson, and G. J. Brakenhoff, *J. Microsc.* **191,** 266 (1998).

[17] D. Yelin, Y. Silberberg, Y. Barad, and J. S. Patel, *Phys. Rev. Lett.* **82,** 3046 (1999).

[18] A. C. Millard, P. W. Wiseman, D. N. Fittinghoff, K. R. Wilson, J. A. Squier, and M. Müller, *Appl. Opt.* **38,** 7393 (1999).

[19] R. W. Boyd, "Non-Linear Optics." Academic Press, San Diego, 1992.

[20] Y. R. Shen, *Nature* **337,** 519 (1989).

in biology given the prevalence of lipid bilayers and chiral proteins: asymmetric staining of both intracellular organelle membranes and cellular plasma membranes produces an environment lacking inversion symmetry, while chiral enhancement of SHG has been observed for monolayers of molecules,[21] for supramolecular arrays,[22] and for dyes in plasma membranes.[10]

The time-averaged SHG signal is expected to scale quadratically with the average power of the fundamental and with $\chi^{(2)}$ (manifested as a quadratic dependence of SHG on molecular concentration) and inversely with the focal area and with the pulse width. The theoretical basis for these dependencies has been treated in greater depth by Moreaux et al.[23,24] The inverse linear dependence on the pulse width (which is shared with 2PF) arises because although the instantaneous signal scales as the square of the peak power, SHG occurs only for the duration of the laser pulse. Similarly, although the signal at a point is proportional to the square of the intensity, SHG is localized to the focal volume. Note that SHG and two-photon excitation probabilities have the same dependence on average power, inverse pulse width, and focal area.

The voltage sensitivity of SHG may be expressed including a third-order term from the Taylor series as follows:

$$\mathbf{P}^{(2)} = (\chi^{(2)} + \chi^{(3)} * \mathbf{E}_{DC}) * \mathbf{E} * \mathbf{E} = \chi_{eff}^{(2)} * \mathbf{E} * \mathbf{E}$$

where the effective second-order nonlinear susceptibility[25,26]

$$\chi_{eff}^{(2)} = \chi^{(2)} + \chi^{(3)} * \mathbf{E}_{DC}$$

has the same symmetry constraints as $\chi^{(2)}$ and is hence restricted to surfaces. Third-order susceptibilities are typically four to five orders of magnitude smaller than second-order susceptibilities, so that the DC-field term $\chi^{(3)} * \mathbf{E}_{DC}$ is usually negligible; for typical transmembrane electric fields of 10^7 V/m, however, that term can be significant for cells and hence detectable. Because the time-averaged SHG signal depends quadratically on the effective second-order susceptibility, SHG will in general show a combination of linear and quadratic dependence on $\mathbf{E}_{DC}$, depending on the relative magnitudes of $\chi^{(2)}$ and $\chi^{(3)}$. It should be noted that this mechanism is different from that of electric field-induced second harmonic (EFISH) generation, in which an applied field aligns a random distribution of molecules and is responsible for a net change in susceptibility: in the case of membrane-bound dyes,

[21] J. D. Byers, H. I. Lee, T. Petralli-Mallow, and J. M. Hicks, *Phys. Rev. B* **49,** 14643 (1994).
[22] T. Verbiest, S. V. Elshocht, M. Kauranen, L. Hellemans, J. Snauwaert, C. Nuckolls, T. J. Katz, and A. Persoons, *Science* **282,** 913 (1998).
[23] L. Moreaux, O. Sandre, and J. Mertz, *J. Opt. Soc. Am. B* **17,** 1685 (2000).
[24] L. Moreaux, O. Sandre, S. Charpak, M. Blanchard-Desce, and J. Mertz, *Biophys. J.* **80,** 1568 (2001).
[25] J. C. Conboy and G. L. Richmond, *J. Phys. Chem. B* **101,** 983 (1997).
[26] E. C. Y. Yan, Y. Liu, and K. B. Eisenthal, *J. Phys. Chem. B.* **102,** 6331 (1998).

the molecules are already aligned because of their placement in the membrane, as indicated by a nonzero, steady-state SHG signal.

History

Nonlinear optical techniques have revolutionized microscopy in the past decade, particularly with the widespread application of two-photon fluorescence (2PF) microscopy since its first implementation in 1990 by Denk et al.[4] With greatly reduced out-of-plane photobleaching and phototoxicity, 2PF microscopy has gained considerable popularity as a means for imaging living cells. The application of this technology has been facilitated by advances in mode-locked ultrafast laser technology, particularly the advent of titanium-doped sapphire as a femtosecond lasing medium. 2PF techniques have proved useful in diverse biomedical fields, with specific implementations employing green fluorescent protein (GFP),[27–29] imaging in tissues and turbid media[30,31] and active, whole brains,[32] or analyzing data with correlation spectroscopies.[33–35] Similarly, three-photon fluorescence techniques have been applied for biomedical imaging.[36,37] The resolution demonstrated in multiphoton applications has been comparable to or slightly less than that achievable by ordinary confocal microscopy[38]: although the Rayleigh range is larger, due to the longer excitation wavelength, the resolution is improved due to the cooperative nature of nonlinear excitation. With image deconvolution and 4π microscopy, Hell et al. have achieved a resolution of $\sim$100 nm for 2PF microscopy.[39] Despite the advantages, new problems can arise from nonlinear optical schemes. The high peak powers necessary for multiphoton absorption can generate undesired and toxic side effects, including three-photon absorption by nucleic acids and proteins and ionizing breakdown. Furthermore, in-plane absorption and eventual photobleaching still generate toxic free radicals in the same manner as one-photon excitation. Acceptable exposure limits given these forms of photodamage have been determined

[27] K. D. Niswender, S. M. Blackman, L. Rohde, M. A. Magnuson, and D. W. Piston, J. Microsc. **180,** 109 (1995).

[28] S. M. Potter, C. M. Wang, P. A. Garrity, and S. E. Fraser, Gene **173,** 25 (1996).

[29] G. H. Patterson, S. M. Knobel, W. D. Sharif, S. R. Kain, and D. W. Piston, Biophys. J. **73,** 2782 (1997).

[30] D. Kleinfeld, P. P. Mitra, F. Helmchen, and W. Denk, Proc. Natl. Acad. Sci. U.S.A. **95,** 15741 (1998).

[31] M. Maletic-Savatic, R. Malinow, and K. Svoboda, Science **283,** 1923 (1999).

[32] F. Helmchen, M. S. Fee, D. W. Tank, and W. Denk, Neuron **31,** 903 (2001).

[33] K. G. Heinze, A. Koltermann, and P. Schwille, Proc. Natl. Acad. Sci. U.S.A. **97,** 10377 (2000).

[34] P. W. Wiseman, J. A. Squier, M. H. Ellisman, and K. R. Wilson, J. Microsc. **200,** 14 (2000).

[35] E. Gratton, N. P. Barry, S. Beretta, and A. Celli, Methods **25,** 103 (2001).

[36] D. L. Wokosin, V. E. Centonze, S. Crittenden, and J. White, Bioimaging **4,** 208 (1996).

[37] S. Maiti, J. B. Shear, R. M. Williams, W. R. Zipfel, and W. W. Webb, Science **275,** 530 (1997).

[38] M. Gu and C. J. R. Sheppard, J. Microsc. **177,** 128 (1995).

[39] S. Hell, M. Schrader, and H. T. M. van der Voort, J. Microsc. **187,** 1 (1997).

for Chinese ovarian hamster (CHO) cells by König et al. using the monitoring of cell division.[40,41] By contrast, techniques based on optical harmonic generation, rather than fluorescence, do not arise from absorption, so in-plane photodamage of labeling chromophores can be greatly reduced if the laser frequency is off resonance.

Optical SHG has been developed over the past three decades as a technique for studying surfaces and interfaces[20,42] and many of the approaches used to probe bulk material properties can be extended to microscopy. In 1974 Hellwarth and Christensen used SHG in an optical microscope to visualize the microscopic crystal structure in polycrystalline ZnSe.[43] The concept of SHG microscopy was also demonstrated in 1977 by Sheppard et al.[2] and more recently with modern imaging equipment and laser sources by Gauderon et al.[44] In a series of rigorous experiments in 1986, Freund et al. used SHG microscopy to study endogenous collagen structure in rat tail tendon at a resolution of approximately 50 μm.[11] In the late 1990s, Guo et al. used stage-scanning laser excitation in a reflection-mode system to image SHG within muscle and connective tissue, where frame rates of several hours were required.[45,46] Most recently Campagnola et al. have used SHG to obtain high-resolution, three-dimensional images of endogenous arrays of collagen, actomyosin, and tubulin in a wide variety of cell and tissue types from various species.[13]

As well as more recent interest by other research groups,[23,47] SHG microscopy has been central to a long-running program of research in the Lewis and Loew laboratories that aims to understand and exploit the membrane biophysics of endogenous chromophores and of voltage-sensitive styryl dyes. In 1988 Huang et al. characterized SHG signals from di-n-ASPSS and di-4-ANEPPS in Langmuir–Blodgett monolayers[48] and Bouevitch et al. extended this work in 1993 to determine the dependence on potential of other styryl dyes in hemispherical bilayers of oxidized cholesterol.[7] SHG was used to investigate the purple membrane protein bacteriorhodopsin in 1989 by Huang et al.[5,6,49] and for the purposes of functional imaging of protein–chromophore interactions in 1999 by Lewis et al.[50] In 1996

[40] K. König, P. T. C. So, W. W. Mantulin, and E. Gratton, Opt. Lett. 22, 135 (1997).

[41] K. König, T. W. Becker, P. Fischer, I. Riemann, and K. J. Halbhuber, Opt. Lett. 22, 113 (1999).

[42] K. B. Eisenthal, Chem. Rev. 96, 1343 (1996).

[43] R. Hellwarth and P. Christensen, Opt. Commun. 12, 318 (1974).

[44] R. Gauderon, P. B. Lukins, and C. J. R. Sheppard, Opt. Lett. 23, 1209 (1998).

[45] Y. Guo, P. P. Ho, H. Savage, D. Harris, P. Sacks, S. Schantz, F. Liu, N. Zhadin, and R. R. Alfano, Opt. Lett. 22, 1323 (1997).

[46] Y. Guo, H. Savage, F. Liu, N. Zhadin, S. Schantz, P. P. Ho, and R. R. Alfano, Proc. Natl. Acad. Sci. U.S.A. 96, 10854 (1999).

[47] L. Moreaux, O. Sandre, M. Blanchard-Desce, and J. Mertz, Opt. Lett. 25, 320 (2000).

[48] Y. A. Huang, A. Lewis, and L. M. Loew, Biophys. J. 53, 665 (1988).

[49] J. Huang and A. Lewis, Biophys. J. 55, 835 (1989).

[50] A. Lewis, A. Khatchatouriants, M. Treinin, Z. Chen, G. Peleg, N. Friedman, O. Bouevitch, Z. Rothman, L. M. Loew, and M. Sheres, Chem. Phys. 245, 133 (1999).

the membrane physiology of living *Musca* photoreceptor cells was probed with SHG by Ben-Oren *at al.,*[8] demonstrating changes in SHG signal on illuminating the cells. Peleg *et al.* in 1996 observed considerable enhancement of SHG from styryl dyes by complexing or placing in proximity silver or gold nanoparticles[51] and in 1999 used SHG from styryl dyes, enhanced by gold nanoparticles conjugated by antibodies to membrane proteins, to probe membrane potential in P19 neuronal cells.[9] Most recently, Clark *et al.* have extended the idea of metal-enhanced SHG by demonstrating strong enhancements from polymer-encapuslated gold nanoparticles that had been covalently linked to a small number of styryl dye molecules.[52] The sensitivity to membrane potential of SHG from GFP-labeled neuronal and mechanosensory cells in *Caenorhabditis elegans* has recently been investigated by Lewis and co-workers,[50,53] while Campagnola *et al.* have obtained SHG images from a variety of cell types stained with styryl dyes, characterizing the high sensitivity of the technique as a means of measuring membrane potential.[10]

Apparatus

In our laboratory we use a laser-scanning transmission-mode microscope, extending SHG imaging to higher resolution ($\sim$400 nm) and much higher rates of image acquisition ($\sim$1 frame/sec) than previously possible. With this system we obtain bright, high-resolution three-dimensional SHG images from stained cells and tissues and from unstained endogenous structural proteins (in connective tissues, muscle, and mitotic spindles), obtained at confocal-like frame rates. A schematic of the nonlinear microscope is shown in Fig. 1. The microscope is designed to be sufficiently flexible so that a range of specimens can be imaged with the same system, requiring only appropriate choices in excitation and collection optics, in $xyzt$-scanning and in modes of detection.

Our 2PF/SHG imaging/quantitation experiments are performed on a modified Bio-Rad (Hercules, CA) MRC600 confocal scan-head mounted on a Zeiss Axioscop upright microscope. The laser system currently consists of an 8-W argon ion laser (Coherent, Santa Clara, CA; Innova 310) pumping a femtosecond titanium-doped sapphire oscillator (Coherent, Mira 900-F), characterized by a pulse width of $\sim$100 fs at a repetition rate of 76 MHz, tunable between 700 and 950 nm. Average powers at the sample range between 1 and 50 mW. A number of objectives are used for excitation, including a 40 $\times$ 1.3 numerical aperture (NA) oil immersion objective (Zeiss, Oberkochen, Germany; Fluar) with a working distance of 0.14 mm, a 40 $\times$ 0.8 NA water immersion objective (Zeiss, IR-Achroplan)

[51] G. Peleg, A. Lewis, O. Bouevitch, L. M. Loew, D. Parnas, and M. Linial, *Bioimaging* **4,** 215 (1996).

[52] H. A. Clark, P. J. Campagnola, J. P. Wuskell, A. Lewis, and L. M. Loew, *J. Am. Chem. Soc.* **122,** 10234 (2000).

[53] A. Khatchatouriants, A. Lewis, Z. Rothman, L. M. Loew, and M. Treinin, *Biophys. J.* **79,** 2345 (2000).

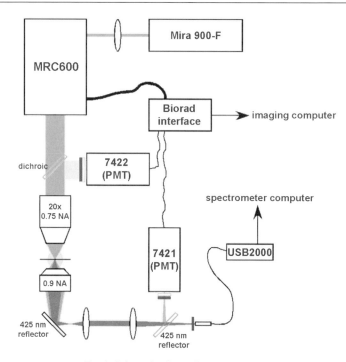

FIG. 1. Schematic of a nonlinear microscope.

with a working distance of 3.6 mm, a 20 × 0.75 NA air objective (Zeiss, Fluar) with a working distance of 0.7 mm, and a 5 × 0.25 NA air objective (Zeiss, Fluar) with a working distance of 12.5 mm. Figure 2 shows a schematic of the focus of the microscope. A 0.9 NA condenser (Zeiss) is used for collection of transmitted light with the high NA objectives, and it can be configured to 0.32 NA, with a much longer working distance, for use with the low NA objective. The divergence of the Ti:sapphire oscillator is sufficiently compensated by passage of the beam through a singlet lens on its way to the scan-head. Given the pulse width, there is essentially no additional dispersion of ~850-nm pulses and hence no external precompensation is used to compensate for the minimal group delay arising within the microscope.

A number of modes of detection are possible, with 2PF and SHG signals being detected simultaneously. For imaging, the 2PF signal is descanned within the scan-head and collected with the pinhole aperture fully opened, using the scan-head's built-in photomultiplier tube (PMT) and electronics. For tissues and other turbid media, the nondescanned signal may be greater than the descanned signal due to losses from absorption by the mirrors in the scan-head and from exclusion by the confocal aperture in the scan-head: although opened as fully as possible, it still excludes some nonballistic, scattered photons. Therefore, the 2PF signal can also be detected directly by an external PMT [Hamamatsu (Hamamatsu City, Japan) Photo

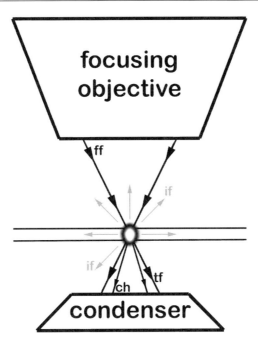

FIG. 2. Details of focus of a nonlinear microscope. ff, Focused fundamental; tf, transmitted fundamental; ch, coherent harmonic; if, incoherent fluorescence.

Sensor Module 7422-40] for quantitation, or the signal from the external PMT can be inverted and delivered to the Bio-Rad acquisition electronics for imaging. Because SHG is a coherent process, the signal copropagates with the fundamental and is collected in transmission, although in tissues and other turbid media, scattering may also allow significant SHG to be detected in the backward direction. The SHG signal is first reflected with either a 425- or 475-nm hard reflector (50 nm bandwidth) or a broad-band aluminum mirror, then collimated and refocused with a lens doublet. Spectra (of SHG and, optionally, 2PF) can be measured using a fiber-based spectrometer (Ocean Optics, Dunedin, FL; USB2000), setting the scanner to a high zoom (100× or larger) setting, using a BG 39 glass filter to discriminate against the fundamental, and choosing an appropriate integration time that can range from milliseconds to seconds. For imaging, the SHG signal is redirected by a (second) hard reflector into a photon-counting PMT [Hamamatsu (Hamamatsu City, Japan) Photon Counting Head 7421-40], using a 450- or 500-nm short-pass filter and, optionally, BG 39 glass filters to discriminate against the fundamental. The TTL pulses from this module are either integrated by the Bio-Rad acquisition electronics for imaging or counted by a counter/timer (Stanford Research Systems, Sunnyvale, CA; SR400) for quantitation. Data acquisition times for images range between 1 and 4 sec per 768 by 512 pixel frame, and images can be acquired either

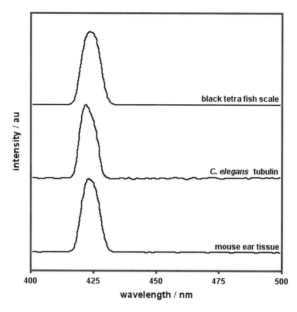

FIG. 3. SHG spectra from various tissues.

individually or as a result of a number of Kalman averages. Image sets can be acquired either as time series or as z-section stacks using a focus motor controlled by
the scanning software and at various ways—low or high zoom or using line scans
or point excitations—depending on the desired temporal and signal resolution.

Controls

It is important to perform control measurements as part of an experiment to verify that signal arises only from SHG and not from the residual fundamental or from
autofluorescence. The simplest control measurements consist of taking the laser out
of mode-lock; if the signal vanishes, it arises from a nonlinear process, although it
is important to note that the converse is not necessarily true as nonlinear processes
can take place under continuous wave laser illumination of sufficient intensity.[54]
The power of the fundamental can also be varied (using either neutral density filters
or half-waveplate and polarizer) to verify the quadratic dependence on power.

The SHG channel should generally use low-pass or band-pass filters to remove any signal from autofluorescence. That this is taking place can be verified by analyzing the light with a spectrometer. Figure 3 shows such spectra for

[54] S. W. Hell, M. Booth, S. Wilms, C. M. Schnetter, A. K. Kirsch, D. J. Arndt-Jovin, and T. M. Jovin,
    Opt. Lett. 23, 1238 (1998).

three tissues excited at 850 nm, clearly indicating that the signal arose exclusively at the expected 425-nm wavelength, without any autofluorescence. Furthermore, all the spectra have the same ~10-nm bandwidth, consistent with the use of 100-fs pulses. Still using the spectrometer, the laser can be scanned across a range of wavelengths to ensure that the spectra track appropriately. Figure 4 shows such tracking for SHG from the pharyngeal muscle of *C. elegans* when the fundamental is scanned between 820 and 880 nm, indicating that the spectra do indeed follow the excitation with the proper bandwidth, and are again free of any autofluorescent contamination. Finally, a monochromator can be used to select particular emission wavelengths before imaging. Without descanning the beam, this can be very inefficient, but again the signal from SHG uncontaminated by autofluorescence should be present at only half of the wavelength of the fundamental and track as the excitation wavelength is scanned. An upper bound (typically one in 100 at most) can be placed on the amount of fluorescence in the sample by comparing two images taken at the same monochromator wavelength—the first image (with SHG) at an excitation wavelength of twice the monochromator wavelength and the second (without SHG) at an excitation wavelength displaced from the monochromator wavelength by at least the sum of the spectral bandwidth of the fundamental and the resolution of the monochromator.

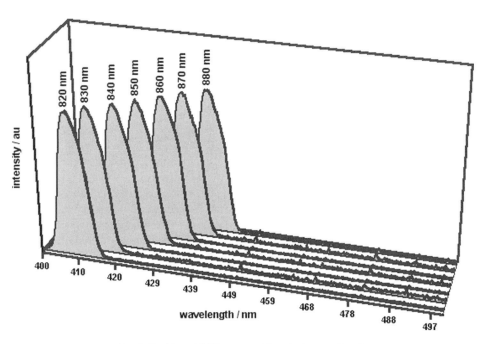

Fig. 4. Tracking of SHG spectra with excitation wavelength.

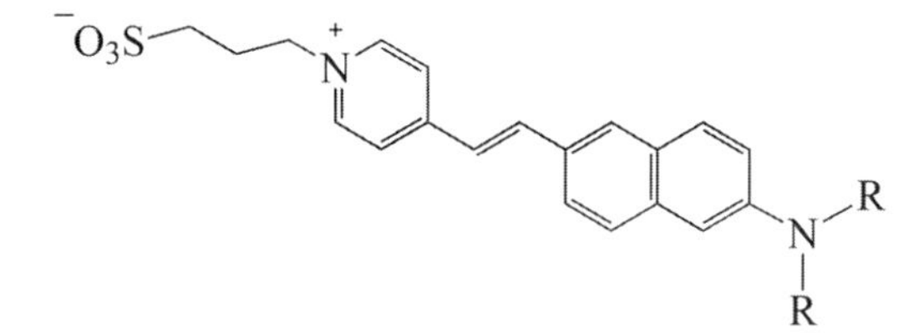

a) R = (CH$_2$)$_3$CH$_3$

b) R = (CH$_2$)$_7$CH$_3$

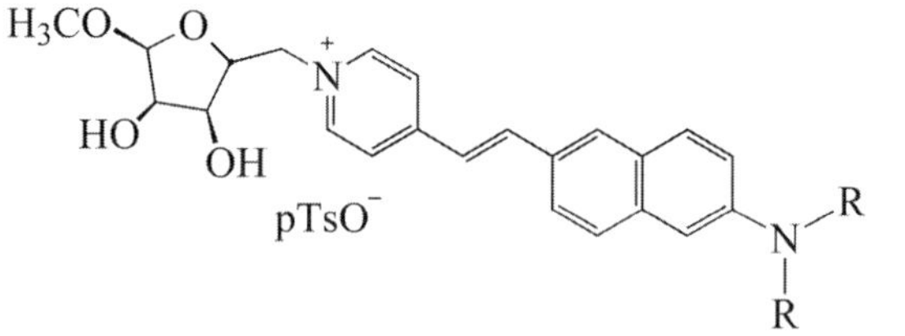

c) R = (CH$_2$)$_3$CH$_3$

d) R = (CH$_2$)$_7$CH$_3$

FIG. 5. Structures of some styryl dyes that are used for labeling membranes: (a) di-4-ANEPPS, (b) di-8-ANEPPS, (c) di-4-ANEPMRF, and (d) di-8-ANEPMRF.

Practical Considerations

Lipophilic Dyes

A convenient feature of SHG microscopy is that lipid-soluble dyes can readily be used to label membranes, greatly increasing contrast. The fluorescent dyes shown in Fig. 5 have similar styryl (ANEP) chromophores, with large second-order optical nonlinearities, but have different alkyl chains and chiralities. They also exhibit solvato- and electrochromism, and are hence sensitive to the potential across the membrane in which they reside, as used in ratiometric fluorescence methods.[55,56] Such dyes can be used to investigate the constrast mechanisms of single- and double-leaflet staining, to determine the enhancement of SHG due to a chiral center and to monitor membrane potential—in some cases with greater sensitivity than is possible via fluorescence imaging. For these molecules, the second-order response[48] is an order of magnitude larger than that of rhodamine

[55] V. Montana, D. L. Farkas, and L. M. Loew, *Biochemistry* **28**, 4536 (1989).

[56] J. Zhang, R. M. Davidson, M. D. Wei, and L. M. Loew, *Biophys. J.* **74**, 48 (1998).

$6G^{20}$ while the 2PF cross section is $\sim$20 times larger, making these dyes excellent probes for both SHG and 2PF studies. Although the specific chemical features leading to large nonlinear susceptibilities are being investigated as part of an active area of research,[57–62] as a rule large $\chi^{(2)}$ and $\chi^{(3)}$ dyes feature networks of conjugated π bonds, aromatic heteroatom (nitrogen or sulfur) substitution, and electron donor/acceptor pairs with large changes in the dipole moment between the ground and excited states. The ANEP chromophores, with the strongly donating dialkylamino group and pyridinium acceptor moiety, satisfy all of these criteria, and are sufficiently bright to permit rapid acquisition of SHG images with reasonably low laser power. Other styryl dyes have also been reported to have large second-order nonlinear susceptibilities.[58,63]

Resonance Enhancement

SHG can be resonance-enhanced when the second harmonic transition overlaps with an electronic absorption band,[64] in which case the total second-order response is a sum of resonant and nonresonant contributions,

$$\chi^{(2)} = \chi^{(2)}_{\text{res}} + \chi^{(2)}_{\text{non-res}}$$

where $\chi^{(2)}_{\text{res}}$ arises from a sum-over-states expression.[65] Depending on the chromophore, the resonant contribution can dominate, resulting in enhancement of SHG by an order of magnitude or more.

Specifically, the resonance-enhanced molecular hyperpolarizability,[14] which determines the resonance contribution, is given by

$$\beta = \frac{4\mu_{\text{eg}}^2(\mu_{\text{e}} - \mu_{\text{g}})}{3h^2(\nu_{\text{eg}}^2 - \nu^2)(\nu_{\text{eg}}^2 - 4\nu^2)}$$

where h is Planck's constant, μ_{g} and μ_{e} are the dipole moments of the ground and excited states, respectively, μ_{eg} is the transition moment, ν is the frequency of the incident light, and $h\nu_{\text{eg}}$ is the energy for the electronic transition from the ground to the excited state. For the very broad absorption bands of organic dyes in

[57] S. R. Marder, J. W. Perry, and W. P. Schaeffer, *Science* **245,** 626 (1989).

[58] S. R. Marder, D. N. Beratan, and L.-T. Cheng, *Science* **252,** 103 (1991).

[59] C. M. Whitaker, E. V. Patterson, K. L. Kott, and R. J. McMahon, *J. Am. Chem. Soc.* **118,** 9966 (1996).

[60] I. D. L. Albert, T. J. Marks, and M. A. Ratner, *J. Am. Chem. Soc.* **120,** 11174 (1998).

[61] R. R. Tykwinski, U. Gubler, R. E. Martin, F. Diederich, C. Bosshard, and P. Gunter, *J. Phys. Chem. B.* **102,** 4451 (1998).

[62] L. Ventelon, L. Moreaux, J. Mertz, and M. Blanchard-Desce, *Chem. Commun.* **1999,** 2055 (1999).

[63] L. M. Loew and L. Simpson, *Biophys. J.* **34,** 353 (1981).

[64] T. F. Heinz, C. K. Chen, D. Ricard, and Y. R. Shen, *Phys. Rev. Lett.* **48,** 478 (1982).

[65] M. D. Levenson and S. S. Kano, "Introduction to Non-Linear Laser Spectroscopy." Academic Press, San Diego, 1982.

solution, the resonance condition where $2v$ overlaps v_{eg} will pertain over a large range of wavelengths. Other consequences of this simple expression are that β is enhanced when there is a large difference in the ground and excited state electron distribution, i.e., a large μ_{eg}, and when the dye has a large extinction coefficient. The electrochromic styryl dyes that have been developed in our laboratory as potentiometric probes display these features, since electrochromicity also requires them.

When SHG is resonance enhanced, collateral two-photon absorption is increased and in-plane photobleaching and phototoxicity can be considerable. With our system, an average power of 50 mW at 880 nm is observed to bleach styryl dyes with a lifetime of about a minute, without any obvious physical damage; however, images of live cells can be obtained at a lower average power of 10 mW, with no reduction in SHG signal being observed. Conversely, attempts to image at 780 nm resulted in rapid, readily visible damage to the cells. Because it is not necessary to match the fundamental to a specific chromophore, wavelengths further into the infrared (i.e., $\lambda > 900$ nm) may be used, avoiding most endogenous two- and three-photon (ultraviolet) absorption and minimizing photodamage. To confirm the lack of photodamage, live cell/dead cell assays and monitoring of cell division can be used; for instance, we have observed uninterrupted spindle separation during the mitosis of a *C. elegans* embryo while imaging.

Although not related to the absorption bands of dyes, enhancement of SHG from dyes due to colocalization with metal nanoparticles has also been observed, both for SHG from styryl dyes in proximity to silver and gold nanoparticles,[9,51] and for SHG from dyes linked to polymer-coated gold nanoparticles.[52] Such enhancement is believed to arise through a combination of surface plasmon resonance and the corona effect.[66]

Signal Levels

We give the following figures for the purpose of estimating signal levels in SHG imaging measurements. The molecular hyperpolarizability[14] of typical styryl dyes has been determined[23,48] to be $\beta = 6 \times 10^{-28}$ esu or 2×10^{-48} C m^3 V^{-2}, giving a molecular second harmonic scattering cross section[24] at an excitation wavelength of 850 nm of $\sigma_{SHG} = 2 \times 10^{-64}$ m^4 s photon^{-1}. (For comparison, the two-photon absorption cross section of rhodamine B is[67] $\sim 2 \times 10^{-56}$ m^4 s photon^{-1}.) Focused with an objective of 0.75 NA, a diffraction-limited focus has a width of $w_0 = 420$ nm, and for an average power at the sample of 50 mW, a repetition rate of 76 MHz and a pulse duration of 100 fs, the peak intensity (as a photon flux density) at the focus is $I_0 = 2 \times 10^{35}$ photons m^{-2} s^{-1}. We follow

[66] G. T. Boyd, T. Rasing, R. R. Leite, and Y. R. Shen, *Phys. Rev. B* **30**, 519 (1984).
[67] C. Xu and W. W. Webb, *J. Opt. Soc. Am. B* **13**, 481 (1996).

Moreaux et al.[24] in calculating the second harmonic photon generation rate as

$$\mathbf{P}_{SHG} = \frac{1}{2} \Theta_2 N^2 \sigma_{SHG} I_0^2$$

where N is the effective number of radiating molecules and Θ_2 is a factor to correct for the imperfect phasing of the molecules due to their spread over the focus; the product $\Theta_2 N^2$ may be reasonably approximated as $0.8 \times w_0^4 N_S^2$, with $N_S = 2 \times 10^{18}$ m^{-2} being the surface density of molecules. Hence, $\mathbf{P}_{SHG} = 5 \times 10^{17}$ photons s^{-1} or 5×10^4 photons per pulse. A pixel dwell time of 1 μs will integrate photons from 76 pulses and assuming that the collection and detection efficiencies are 50% and 10%, respectively, a membrane that is stained at a molecular density of 1% will yield 20 counts per pixel. In our experiments, a typically bright pixel corresponds to 10 counts, in rough agreement with the estimate. It should be noted that the photon-counting PMT module saturates at 30 counts per pixel, although for most tissues the practical limit is imposed by the speed of scanning.

The most important factors affecting signal strength are resonance enhancement and dye density, on which SHG has a quadratic dependence. The signal is relatively insensitive to numerical aperture for excitation[23,24]: a tighter focus increases the intensity at the focus $\sim NA^2$ and improves the phasing across the focus $\sim NA^2$, but reduces the number of molecules within the focus $\sim NA^{-3}$ and hence $\mathbf{P}_{SHG} \sim NA^2 (NA^{-3})^2 (NA^2)^2 \sim NA^0$. However, the numerical aperture for collection is a more important factor because, unlike fluorescence where light is emitted incoherently at all angles, second harmonic light is emitted coherently in two lobes at $\sim NA/\sqrt{2}$ and requires correct placement of the condenser or other collection optics.

Voltage Sensitivity

One of the areas of interest in SHG microscopy concerns its use as a new optical method for probing membrane potential. For fast membrane-staining potentiometric dyes like the ANEPPS series, fluorescence measurements are relatively insensitive, showing changes of only $\sim$10% for a 100 mV change. It is therefore a demanding task to obtain accurate quantitative measurements with such a small dynamic range. In contrast, we have demonstrated that SHG in a model membrane is strongly modulated by an applied electric field[7] and that SHG is highly sensitive to changes in membrane potential in living cells.[10] In the latter work, we measured SHG and 2PF from samples of L1210 lymphocytes in both normal ($\sim$5 mM [K$^+$]) and high potassium ($\sim$135 mM [K$^+$]) buffers. Confocal fluorescence imaging of TMRE, a Nernstian indicator,[68,69] shows this corresponds to a

[68] L. M. Loew, *Methods Cell Biol.* **38,** 194 (1993).

[69] L. M. Loew, *in* "Cell Biology: A Laboratory Handbook" (J.E. Cellis, ed.), Vol. 3, p. 375. Academic Press, Orlando, 1998.

membrane depolarization of $\sim$25 mV. Using combined spectroscopic and imaging data from 30 samples, we found that the change of buffer changed the ratio of SHG signal to 2PF signal by a factor of 0.46 ± 0.07; in other words, a $\sim$25-mV depolarization results in a halving of the effective second-order susceptibility. These results, from measurements on our laser-scanning microscope at physiologically relevant time scales, corroborate those from our earlier low-resolution work indicating that SHG from these dyes is a more sensitive probe of membrane potential than fluorescence-based methods.[8,9]

Endogenous "Harmonophores"

The protein bacteriorhodopsin forms the purple membrane of *Halobacterium salinarum* and is similar to the visual pigment rhodopsin. It has a very large nonlinear susceptibility,[5,49] with a molecular hyperpolarizability of $\beta = \sim10^{-26}$ esu or $\sim10^{-46}$ C m^3 V^{-2}, and has been used to probe the origins of SHG in organic systems such as photochemical pathways and to develop retinal derivatives with large optical nonlinearities.[50] Extensive conjugation is also a characteristic of such porphyrin-based molecules as hemoglobin and chlorophyll, which have been observed to enhance third harmonic generation.[18] However, a large molecular hyperpolarizability is not always necessary: water at a charged interface produces an SHG signal,[70,71] and has been used to monitor cytochrome *c* adsorption to silica surfaces and negatively charged supported phospholipid bilayers.[72]

Endogenous structural proteins can also enhance SHG and allow imaging that is not restricted to interfaces. Guo *et al.* recently demonstrated SHG imaging of chicken muscle tissue in which an acquisition time of several hours was required to collect a single frame,[45] whereas Kim *et al.* have studied the effects of thermal denaturation, glycooxidative damage, and enzymatic cleavage on SHG signals from collagen.[73] The contrast mechanisms for SHG from endogenous proteins are currently being researched. From polarization microscopy, muscle and connective tissues are known to be highly birefringent, but a correspondence cannot be directly drawn with frequency doubling in crystals such as KDP and BBO, where the only nonvanishing components of $\chi^{(2)}$ require that the fundamental and second harmonic waves are polarized orthogonally.[74] In the cases we have investigated,[13] the second harmonic and the fundamental are polarized parallel and, further, no Type I phase-matching condition exists for the proteins of interest,[75] although

[70] S. Ong, X. Zhao, and K. B. Eisenthal, *Chem. Phys. Lett.* **191**, 327 (1992).
[71] X. Zhao, S. Ong, H. Wang, and K. B. Eisenthal, *Chem. Phys. Lett.* **214**, 203 (1993).
[72] J. S. Salafsky and K. B. Eisenthal, *J. Phys. Chem. B* **104**, 7752 (2000).
[73] B.-M. Kim, J. Eichler, K. M. Reiser, A. M. Rubinchik, and L. B. Da Silva, *Lasers. Surg. Med.* **27**, 329 (2000).
[74] A. Yariv, "Quantum Electronics." John Wiley & Sons, New York, 1989.
[75] F. P. Bolin, L. E. Preuss, R. C. Taylor, and R. J. Ference, *Appl. Opt.* **28**, 2297 (1989).

phase-matching requirements are relaxed at high NA.[76] Furthermore, proteins such as collagen and tubulin are organized into chiral structures, leading to enhancement of SHG as described in a later subsection.

In our examination of SHG from structural proteins[13] such as myosin, tubulin, and collagen, we observed signal strengths that were much greater, by one or up to two orders of magnitude, than were obtained while imaging membrane-bound dyes. This is a striking result in that the second-order non-linear susceptibilities for such dyes should be much greater than those of proteins, which lack the networks of conjugated π bonds generally required for large susceptibilities.[61] For example, collagen typically has a low density of aromatic residues and the polarisability of single bonds is significantly less than that of double bonds; to further verify that we were imaging SHG and not 2PF, we have compared images taken at the same wavelength for detection but at different wavelengths for excitation, placing an upper bound on autofluorescence from fish scale collagen of one in 500. However, these structural proteins exist at millimolar concentrations, whereas cellular membranes can only be stained at micromolar concentrations if aggregation and toxicity are to be avoided. Thus it is primarily the quadratic dependence of SHG on concentration and the highly organised nature of the proteins that is leading to intense images.

Symmetry Effects

As bilayers, cell membranes are a form of interface fundamentally different from that typically studied with SHG methods. A feature of central importance is that either one or both of the leaflets can be stained, determining the degree of inversion symmetry at the membrane. The choice of cell is a determining factor for the observed contrast, as is the choice of dye. For example, NIH 3T3 fibroblasts are flat cells that internalize lipophilic dyes relatively rapidly, resulting in equal staining of both leaflets. Additionally, shorter alkyl chain dyes such as di-4-ANEPPS and di-4-ANEPMRF [structures (a) and (c) in Fig. 5] are known to internalize faster than their octyl analogs [structures (b) and (d) in Fig. 5, respectively].[77] Hence, 3T3 fibroblasts stained with di-4-ANEPPS give rise to strong 2PF images but with essentially no detectable SHG signal.[10]

As a general principle, SHG is inhibited if two antiparallel regions are separated by a distance less than the optical coherence length, L_c. This limit is on the order of the excitation wavelength,

$$L_c = 2/\Delta k$$

where Δk is the wavevector mismatch $k_{2\omega} - 2k_\omega$, and $k\omega$ and $k_{2\omega}$ are the wavevectors for the fundamental and for the second harmonic, respectively, both

[76] D. N. Fittinghoff, A. C. Millard, J. A. Squier, and M. Müller, *IEEE J. Quant. Electron.* **35,** 479 (1999).
[77] L. M. Loew, *Neuroprotocols* **5,** 72 (1994).

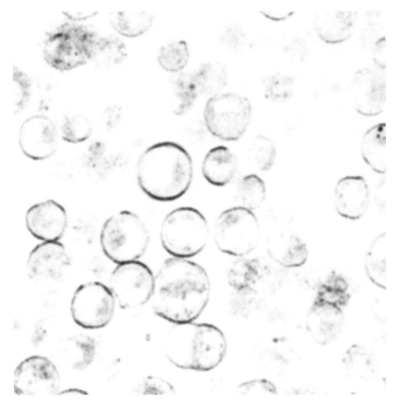

FIG. 6. Inverse SHG images of L1210 cells stained with di-4-ANEPPS; arcs indicate dependence of SHG on polarization of fundamental.

inversely proportional to their respective wavelengths in the medium. For the usual dependence of refractive index on frequency, known as normal dispersion,[78] Δk as defined is positive; the ideal condition for SHG in bulk noncentrosymmetric media would be that of perfect phase matching,[19] where Δk vanishes and L_c is effectively infinite. In the case of equally stained membrane leaflets, second harmonic waves from each leaflet will be equal in magnitude but out of phase, interfering destructively so that there is no net SHG, as observed in 3T3 fibroblasts. In contrast, L1210 lymphocytes show little internalization of di-4-ANEPPS, suggesting that only the outer leaflet is stained, and strong SHG signals are seen from the cell membranes as expected. A typical image for L1210 cells is shown in Fig. 6. As a rule of thumb, more rounded cells seem to internalize dyes more slowly.[77] For intermediate cases, the decay of SHG signal has been used to determine the rate of

[78] A. E. Siegman, "Lasers." University Science Books, Sausalito, 1986.

internalization: of the order of a minute to an hour for malachite green in $\sim$230-nm dioleoylphosphatidylglycerol liposomes[79] and about 2 hr for di-6-ASPBS in giant unilamellar vesicles,[24] both at room temperature.

As a further example of this principle, small phosphatidylcholine vesicles made using sonication have not given rise to any observable SHG. Such vesicles are in the size range of 20–100 nm, smaller than the coherence length, and thus second harmonic waves from opposite sides of the vesicles interfere destructively so that there is no net SHG. In contrast, L1210 lymphocytes have a diameter of $\sim$10 μm, larger than the coherence length, and hence second harmonic waves from opposite sides of the cells do not cancel out. Similar size-dependent behavior has been observed for 1.05-μm polystyrene sulfate beads and for 220-nm oil droplets[26] and has been employed to study the transfer of dye molecules between microparticles of polystyrene sulfate and montmorillonite clay.[80]

A more complicated case of the symmetry dependence of SHG is exhibited by the mitotic spindles of the eggs of the nematode worm *C. elegans*,[13] which also demonstrates how useful biophysical information may be deduced. A typical image is shown in Fig. 7. It is known that microtubules extend from the poles of the spindle and interdigitate in the mid-zone, and the distribution of SHG is consistent with this. Comparing 2PF arising from β-tubulin::GFP with SHG from the tubulin itself, bright 2PF in the center of the spindle indicates that microtubules are abundant whereas the lack of SHG indicates that the microtubules are arranged antiparallel. In contrast, more distal portions of the spindle, where microtubules are aligned, are bright in both 2PF and SHG. Similarly, the centrosomal regions of the spindle poles show little SHG, due to cancellation of signal arising from the spherically symmetric array of microtubules. No such cancellation is seen in the 2PF channel, nor is there any apparent dependence on angle: the GFP domain is flexibly tethered and hence randomly oriented, generating fluorescence at all polarizations.

Chirality Effects

The dyes di-4-ANEPPS and di-8-ANEPPS [structures (a) and (b) in Fig. 5] were developed as voltage-sensitive probes. The chiral sugar group in di-4-ANEPMRF and di-8-ANEPMRF [structures (c) and (d) in Fig. 5] was added specifically to enhance the second-order nonlinear susceptibility by increasing the molecular asymmetry without affecting the chromophore. Typical methods for determining $\chi^{(2)}$ values involve spin-casting the dye onto a substrate to form a monolayer, but from the perspective of biophysical applications, it is more relevant to make such measurements in a biological membrane. Given their lack of

---

[79] A. Srivastava and K. B. Eisenthal, *Chem. Phys. Lett.* **292,** 345 (1998).

[80] E. C. Y. Yan, Y. Liu, and K. B. Eisenthal, *J. Phys. Chem. B* **105,** 8531 (2001).

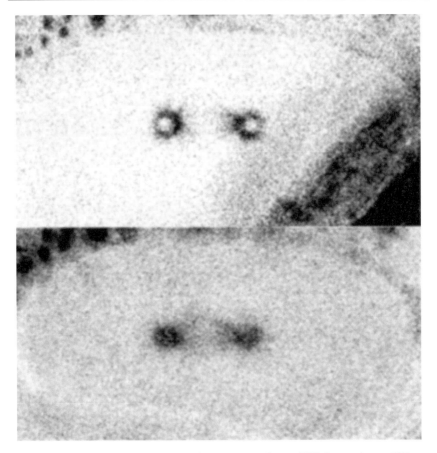

FIG. 7. Images of mitotic spindles in *C. elegans* egg. *Top:* inverse SHG; *bottom:* inverse 2PF.

differentiation (and hence their uniform spherical shape) and their very slow internalization of dye, L1210 lymphocytes are convenient substrates for this purpose.

For the specific case of the dyes mentioned here, the similarity of the chromophores between the chiral and achiral analogs results in indistinguishable 2PF spectra that at concentrations low enough so that self-quenching is not of concern are linear with concentration, whereas SHG spectra should have a quadratic dependence. Hence the ratio of SHG signal to 2PF signal provides a measure of the "magnitude" of $\chi^{(2)}$ (bearing in mind that this is really a third-rank tensor), conveniently normalizing out the other dependencies on average power, pulse width, and focal area. We have previously found[10] that at 800 nm di-4-ANEPMRF and di-4-ANEPPS have a relative susceptibility of 1.5 ± 0.1; in other words, the addition of the chiral sugar group to the dye results in a more than twofold increase in

FIG. 8. Inverse SHG image of collagen matrix in a black tetra fish scale.

SHG signal due to squaring of the relative susceptibility. Furthermore, we found that 3T3 fibroblasts stained with di-4-ANEPMRF show internalization of the dye under 2PF imaging, as with di-4-ANEPPS, but unlike the achiral dye, the chiral version produces significant SHG from the membrane, even though both leaflets are stained. In other words, the chiral sugar group on each dye molecule introduces enough asymmetry so that the equally stained leaflets are no longer good mirror images of each other, and hence no longer inhibiting SHG. This is a simple case of the enhancement of nonlinear properties by supramolecular chirality.[22]

In recent work[13] we found that the collagen matrix of scales of black tetra fish, *Gymnocorymbus ternetzi,* gives rise to very bright SHG images. A typical image is shown in Fig. 8. The coiled-coils in collagen exhibit supramolecular chirality: although each helix in the triple-helix structure is expected to produce second harmonics, the three helices will produce much larger signals. By contrast, myosin has two coiled-coils and is expected to be less SHG efficient. This is consistent our qualitative observations and with work by Kim *et al.* where in bulk (nonimaging) measurements they compared relative efficiencies of several forms of connective and muscle tissues and found ratios in the range of 50 : 1.[73,81]

Conclusion

We have introduced SHG microscopy as a powerful technique that is being developed to exploit biophysical features of membranes and endogenous chromophores, providing new and improved information compared with existing optical imaging methods. We have described our laser and imaging system, constructed from commercially available components, used in some of our recent work. In particular we have discussed the roles of lipophilic dyes and native proteins in SHG microscopy, in particular the effects of resonance, symmetry, and chirality.

Given the large intrinsic SHG signal levels in most cells and tissues, the limit of data acquisition can be imposed by the current scanning system. Such conventional xy scanners cannot scan faster than about 1 MHz—of the order of one frame per second for a typical screen size—without introducing large mechanical instabilities. This imaging modality could greatly benefit from a faster scanning system. One possibility is to use a resonant galvanometer system, which can operate faster than 30 frames per second (video rate).[82] Another is to use an acoustooptical (AO) system, which can operate at even higher rates of the order of a hundred frames per second. Although AO modulators and deflectors can lead to some pulse broadening, for $\sim$100-fs pulses at $\sim$850 nm, such temporal effects will be no worse than those of other optical components in the microscope. Spatial effects are generally greater but can be compensated.[83]

For SHG signals from exogenously applied dyes, improvements in signal strength are possible by engineering the molecular structures of the probes to tune the resonance condition. Probes with larger charge redistributions between the ground and excited states or larger transition moments will have larger resonance-enhanced second-order susceptibilities. Both of these improvements can be achieved with styryl dyes that have larger π-systems with greater separations between the donor and acceptor moieties. Such dyes would also be likely to exhibit linear absorption at longer wavelengths, possibly necessitating the application of pulsed lasers with outputs greater than the 1 μm limit imposed by the Ti:sapphire oscillator. Another improvement might be realised by incorporating chirality within the probe chromophore itself. This could be achieved through the design of long spiral chromophores with a specified helicity. Finally, SHG signals

[81] B.-M. Kim, J. Eichler, and L. B. Da Silva, *Appl. Opt.* **38,** 7145 (1999).

[82] M. J. Sanderson and I. Parker, *Methods Enzymol.* **360,** 447 (2003).

[83] J. D. Lechleiter, D. Lin, and I. Sienaert, *in* "Multiphoton Microscopy in the Biomedical Sciences" (A. Periasamy and P. T. So, eds.), Proc. SPIE Vol. 4262, p. 111. International Society for Optical Engineering, 2001.

could be genetically targeted to specific sites in cells through transfection of cells with constructs that contain appropriate signaling sequences and also code for SHG-enhancing proteins.

Acknowledgments

We thank Prof. Ann Cowan, Dr. Kurt Hoffacker, Prof. Gary Leach, Dr. Gadi Peleg, and Prof. Mark Terasaki for helpful technical discussions, and we thank Ivo Kalajzic, Prof. Vladimir Rodionov, and Prof. David Rowe for providing tissue samples. The term "harmonophore" was coined by Prof. Kent Wilson (1937–2000) to describe molecules or moieties that enhance harmonic generation. We gratefully acknowledge financial support under Office of Naval Research Grant N0014-98-1-0703, National Institutes of Health, National Institute of General Medical Sciences Grant R01-GM35063, the National Science Foundation Academic Research Infrastructure Grant DBI-9601609, the State of Connecticut Critical Technology program, and the Muscular Dystrophy Association.

[4] Fluorescence Correlation Spectroscopy

By Joachim D. Müller, Yan Chen, and Enrico Gratton

Introduction

Spontaneous, microscopic fluctuations are an integral part of every fluorescence measurement and add a noise component to the observed fluorescence signal. Fluorescence correlation spectroscopy (FCS) extracts information from this noise and characterizes the kinetic processes that are responsible for the signal fluctuations. For instance, the dynamic equilibrium between a fluorescent and a nonfluorescent state of a fluorophore introduces fluctuations. Another example is Brownian motion, which leads to the stochastic appearance and disappearance of fluorescent molecules in a small observation volume. FCS characterizes any kinetic process that leads to changes in the fluorescence, because the spontaneous fluctuations at thermodynamic equilibrium are governed by the same laws that describe the kinetic relaxation of a system to equilibrium. Thus, FCS offers a very convenient method for determining kinetic properties at equilibrium without requiring a physical perturbation of the sample. This is especially important for systems in which the use of perturbation techniques in extremely difficult and challenging, such as measurements in living cells.

The concept of measuring signal fluctuations has direct consequences for the experimental realization of FCS. Statistics tells us that the relative fluctuation amplitude of a signal is inversely proportional to the number of molecules simultaneously measured. Thus, the presence of a large number of molecules, as typically encountered in the macroscopic world, suppresses the effects of fluctuations and only the ensemble average is observed. This simple statistical argument illustrates that FCS measurements require signals from a single or a very small number of molecules in order not to mask signal fluctuations. Although the concept of FCS was presented in 1972,[1] the technical challenge of single molecule detection has been overcome only in the past decade. The most important contributions are the development of stable laser light sources, the availability of photodetectors with high sensitivity, and most importantly the introduction of new microscope techniques.

The development of single molecule sensitivity has been linked to the use of optical microscopes for two reasons. First, the microscope is a very efficient optical instrument. High numerical aperture (NA) objectives capture a significant fraction of the fluorescent light that is emitted randomly in all spatial directions. Second, the microscope optics allows the generation of very small observation volumes. High numerical aperture objective focus an incoming laser beam down to a diffraction-limited spot. The radial size of the diffraction-limited spot is determined by the laws of optics, but is given in good approximation by the wavelength of the light. Although the radial dimension of the incoming light is confined by the optics of the objective, the axial direction of the light is not confined. The light leaving the focusing optics of the objective describes the shape of a double cone and fluorescence is excited everywhere within the double cone of light. Optical tricks are used to confine the light in the axial direction. In a confocal microscope a pinhole is placed in front of the detector to block fluorescence light coming from sections other than the focal plane (Fig. 1).[2] In two-photon microscopy a nonlinear optical phenomenon is used to reduce the excitation of fluorescence to the focal region.[3] A molecule that is excited with visible light at 400 nm is also excited by the simultaneous absorption of two photons of near-infrared light at 800 nm. In both cases the excited molecule returns to the ground state via emission of fluorescence (Fig. 2). Because two individual photons need to be absorbed at the location of the fluorophore, the probability of excitation is proportional to the square of the photon flux. The objective focuses the laser light and the photon flux (or intensity) is highest in the focal region. Because of the quadratic intensity dependence the two-photon excitation is restricted to the focal region and drops off drastically outside of the focal region. Consequently, no pinhole before the detector is needed for two-photon FCS, because the excitation of fluorophores is limited to the focal region by the inherent optical sectioning effect of two-photon absorption.

The optical observation volume generated by confocal and two-photon microscopy is less than 1 fl. Conventional, cuvette-based fluorescence instrumentation employs observation volumes on the order of 10 μl. The reduction of the illuminated volume by about 10 orders of magnitude drastically reduces the number

[1] D. Magde, E. Elson, and W. W. Webb, *Phys. Rev. Lett.* **29**, 705 (1972).

[2] J. B. Pawley, ed., "Handbook of Biological Confocal Microscopy." Plenum Press, New York, 1995.

[3] W. Denk, J. H. Strickler, and W. W. Webb, *Science* **248**, 73 (1990).

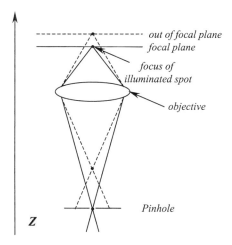

FIG. 1. Illustration of the confocal principle. Fluorescence that emerges from the focus of the illuminated spot in the sample is collected by the objective and passes through the pinhole. Fluorescence light that is excited outside of the focal region is also collected by the objective, but cannot pass through the pinhole.

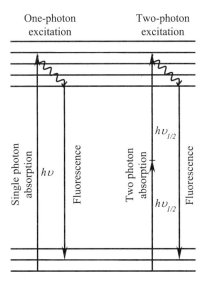

FIG. 2. Jablonski diagram of one-photon versus two-photon excitation. In one-photon excitation the absorption of a single photon of energy $h\nu$ promotes the molecule from the electronic ground state into an excited state. In two-photon excitation two photons of half the energy, $h\nu/2$, are absorbed simultaneously to reach the excited state. The molecule returns in both cases via emission of fluorescence to its ground state.

of molecules present in the observation volume. A concentration of 1 nM corresponds to an average of less than one molecule in the volume generated by the microscope. In addition, the subfemtoliter observation volumes efficiently suppress background signal from the bulk of the sample. Due to the introduction of these technological innovations, FCS measurements are now routinely performed on biological samples with single-molecule sensitivity.

Instrumentation

FCS instruments are commercially available from a number of companies, but it is relatively straightforward to build your own instrument. A typical setup of an FCS instrument is shown in Fig. 3. The light of a commercial laser source passes through a beam expander, is reflected by a dichroic mirror, passes through the objective, and is focused onto the sample. The fluorescence excited by the laser light is collected by the same objective. The dichroic mirror transmits the fluorescence signal and thereby separates the excitation light from the fluorescence of the sample. A barrier filter is added to suppress additional scattered light from the laser. The tube lens of

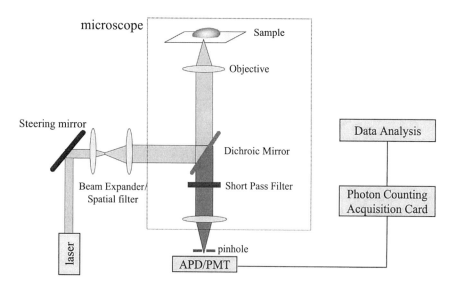

FIG. 3. Schematic diagram of an FCS instrument. The laser light passes through a beam expander, reflects on the dichroic mirror, and is focused by the objective into the sample. The fluorescence light excited in the sample is collected by the objective and passes through the dichroic mirror. The microscope tube lens focuses the fluorescence light onto a photodetector. A pinhole is placed in front of the detector for confocal FCS. Two-photon FCS does not require a pinhole. An electronic data acquisition board processes the signal from the photodetector and a computer is used to analyze the FCS data.

the microscope focuses the fluorescence light. A pinhole is placed at the location of the focus of the fluorescence light in the confocal arrangement to suppress out of focal signal contributions. The photodetector is placed directly behind the pinhole and converts the impinging fluorescent light into electronic pulses that are fed to a data acquisition board. The board works together with computer software to calculate and display the autocorrelation function. In a two-photon microscope the pinhole is not needed, because the fluorescence light originates only from the focal spot of the microscope.

The individual components of the FCS instrument will be considered next. The main requirement for the laser source is the stability of its intensity. Otherwise the laser light introduces intensity fluctuations, which cannot be separated from the fluorescence fluctuations caused by the sample. A Gaussian shape of the laser beam is recommended, because it leads to the smallest focus, which is limited only by diffraction. One-photon excitation with confocal FCS requires very little laser power; less than 1 mW of power at the sample is sufficient. Argon ion and HeNe (helium) lasers are widely used for confocal FCS. Two-photon FCS requires lasers in the near-infrared with ultrashort pulses and a high repetition frequency. The short pulses bunch photons together temporally and create at the time of the pulse an enormously high flux of photons that is required for efficient two-photon absorption. Although a number of different laser sources are available for two- and multiphoton excitation spectroscopy, almost every two-photon system uses the titanium–sapphire laser. It provides wide wavelength tunability (700–1000 nm), high average power (about 1 W), high repetition frequency (80 MHz), and short pulse width ($\approx$100 fs). Titanium–sapphire lasers have good intensity stability and are essentially turnkey systems.

Any research-grade biological microscope can serve as the body of an FCS instrument. The back aperture of the microscope objective needs to be overfilled to get the best optical performance. However, the beam diameter of most lasers is only about 1 mm wide. To overfill the back aperture, the laser beam diameter is magnified by passing through a beam expander. A spatial filter can be used as a beam expander as well and has the additional advantage that the beam is cleaned up. Dichroic mirrors and barrier filters need to be considered for each experiment separately and are available from many commercial sources.

The microscope manufacturers offer a wide range of objectives. Most FCS experiments require a high NA objective for good signal statistics. Chromatic corrections are important for FCS studies. This is particularly true for two-photon FCS, in which the wave lengths of the excitation light and the fluorescence are separated by more than 100 nm. Because FCS measurements are performed on the optical axis, field flatness is less of a concern. However, every objective behaves differently and the experimenter should characterize its performance. Pinholes are available in many different sizes and the optimal pinhole size for confocal measurements has been discussed in the literature.[2] Avalanche photodiodes (APD) and photomultipliers (PMT) are the detectors of choice for FCS experiments. APDs have a higher quantum yield in the visible than photomultipliers and their sensitivity extends to the near infrared. However, the sensitivity of the APD drops off drastically in the blue part of the spectrum. Here, photomultipliers perform better than APDs. Two types of data acquisition cards are available for processing the detected photon counts from FCS experiments. Traditional boards take the signal from the detector and calculate the autocorrelation function electronically on board. Today these boards are all based on the multiple-tau(τ) correlator design, which calculates the autocorrelation function for evenly logarithmic spaced sampling frequencies.[4] A disadvantage of these data acquisition schemes is that they provide access only to the autocorrelation function and not to the complete time sequence of photon counts. Electronic cards that provide a time-stamp for every photon event represent a different data acquisition strategy and are commercially available.[5] The autocorrelation function is subsequently determined from the recorded photon events by software. Computers today are fast enough to calculate the autocorrelation online. The advantage of this approach is that the complete sequence of photon events is available, which allows new and different analysis techniques on the raw data, such as moment analysis, higher order correlation functions, and histogram analysis.[6–8]

Autocorrelation Function

Fluctuation measurements require statistical analysis methods to extract the information hidden in the data. FCS uses the autocorrelation function $g(\tau)$ of the fluorescence signal to analyze the intensity fluctuations,

$$g(\tau) = \frac{\langle F(t)F(t+\tau)\rangle - \langle F\rangle^2}{\langle F\rangle^2} \tag{1}$$

$F(t)$ is the fluorescence intensity at time t, the brackets $\langle\ \rangle$ indicate a time average over the fluorescence signal, $\langle F\rangle$ is the average fluorescence intensity, and τ represents the lag time. The autocorrelation function measures how long fluctuations persist and the autocorrelation curve describes the temporal decay of memory as a function of the lag time τ. The value $g(\tau)$ characterizes the residual persistence of a fluctuation after a time τ has passed. The autocorrelation function offers a very convenient way to separate processes with short and long memory.

Models are needed to extract the dynamic information encoded in the shape of the autocorrelation function. The theory of FCS and the autocorrelation function

[4] K. Schäzel, *Inst. Phys. Conf. Ser.* **77**, 175 (1985).
[5] J. S. Eid, J. D. Müller, and E. Gratton, *Rev. Sci. Instrum.* **71**, 361 (2000).
[6] H. Qian and E. L. Elson, *Biophys. J.* **57**, 375 (1990).
[7] A. G. Palmer and N. L. Thompson, *Proc. Natl. Acad. Sci. U.S.A.* **86**, 6148 (1989).
[8] Y. Chen, J. D. Müller, P. T. So, and E. Gratton, *Biophys. J.* **77**, 553 (1999).

have been reviewed in detail.[9] Here, we consider the simplest model, a single, freely diffusing species (such as a solution of fluorescein), to illustrate the use of the autocorrelation function. Before we look more closely at the autocorrelation function we want to stress that the functional form of the autocorrelation depends on the point spread function (PSF). The PSF describes the spatial intensity profile seen by the detector of the FCS instrument. The PSF depends on the microscope optics and the properties of the excitation light source. Three different PSFs have been widely used in the FCS literature. (1) The two-dimensional Gaussian PSF describes a radially symmetric Gaussian intensity profile. This PSF is a good approximation of the intensity distribution in the focal plane of the microscope. It is primarily used to describe processes on surfaces, such as on lipid membranes. (2) The three-dimensional Gaussian PSF has like the two-dimensional Gaussian PSF a Gaussian intensity profile in the radial direction. The intensity distribution in the third dimension along the optical axis is also given by a Gaussian. The three-dimensional Gaussian PSF approximately describes the spatial intensity distribution of confocal FCS experiments. The Gaussian intensity distribution along the optical axis approximates the contribution of light from out-of-focal planes in a confocal detection arrangement. (3) The Gaussian–Lorentzian PSF has been used to describe two-photon FCS. Again, a two-dimensional Gaussian describes the light intensity distribution in a plane perpendicular to the optical axis. A Lorentzian function characterizes the intensity profile along the optical axis in the absence of a confocal pinhole before the detector. Note that the intensity distribution needs to be squared, because of the quadratic intensity dependence of two-photon absorption.

The autocorrelation function for a single, freely diffusing fluorescent species is given by Eq. (2) for a two-dimensional Gaussian PSF and by Eq. (3) in the case of a three-dimensional Gaussian PSF. The autocorrelation function $g_{GL}(\tau)$ for the Gaussian–Lorentzian PSF is not shown here. It cannot be written in closed form and is given by an integral expression.[10]

$$g_{2DG}(\tau) = \frac{\gamma_{2DG}}{N} \left[\frac{1}{1 + (\tau/\tau_D)} \right] \tag{2}$$

$$g_{3DG}(\tau) = \frac{\gamma_{3DG}}{N} \left[\frac{1}{1 + (\tau/\tau_D)} \right] \left[\frac{1}{\sqrt{1 + (\omega_0/z_0)^2(\tau/\tau_D)}} \right] \tag{3}$$

The shape of the autocorrelation function $g_{2DG}(\tau)$ for the two-dimensional Gaussian PSF is given by a hyperbolic function (Fig. 4). Here, the diffusion time

[9] N. L. Thompson, *in* "Topics in Fluorescence Spectroscopy" (J. R. Lakowicz, ed.), Vol. 1, p. 337. Plenum Press, New York, 1991.
[10] K. M. Berland, P. T. C. So, and E. Gratton, *Biophys. J.* **68,** 694 (1995).

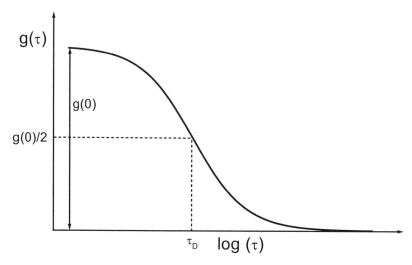

Fɪɢ. 4. Autocorrelation function for a two-dimensional Gaussian PSF. The fluctuation amplitude $g(0)$ is inversely proportional to the number of molecules N in the observation volume of the microscope. At the diffusion time τ_D the autocorrelation function has decayed to one-half of its value.

τ_D is the time, where the autocorrelation function has decayed to one-half of its value. The fluctuation amplitude $g(0)$ for a single species is inversely proportional to the number of molecules N in the observation volume,

$$g(0) = \frac{\gamma}{N} \qquad (4)$$

The shape of the PSF determines the value of the factor γ. Its numerical value is $\gamma_{2DG} = 0.5$ for the two-dimensional Gaussian PSF, $\gamma_{3DG} = 0.35$ for the three-dimensional Gaussian PSF, and $\gamma_{GL} = 0.076$ for the Gaussian–Lorentzian PSF. The functional shape of the autocorrelation functions characterized by Eqs. (2) and (3) is very similar. Mathematically, the only difference between the functions is an additional multiplicative factor for the three-dimensional Gaussian autocorrelation function. This factor takes the influence of the Gaussian beam profile along the axial direction into account. The beam waist ω_0 characterizes the width of the Gaussian in the radial direction and the beam waist z_0 describes the width in the axial direction.

The diffusion time τ_D characterizes the average time it takes for a molecule to diffuse through the radial part of the observation volume of the microscope. However, the diffusion time is not a constant and changes with the size of the observation volume, which depends on the laser wavelength and the optics of the instrument. The diffusion coefficient D, on the other hand, is a property of a molecule in a given solvent and thus much better suited to characterize the experimental data than the diffusion time. The relationship between the diffusion time and the diffusion coefficient is given by

$$\tau_D = \frac{\omega_0^2}{4D} \tag{5}$$

which simply is a consequence of the Gaussian beam profile and the laws of diffusion. In the case of two-photon FCS the diffusion time is reduced by a factor of two, $\omega_0^2/8D$. It is important to remember that the diffusion coefficient depends linearly on the viscosity of the solvent. However, most experiments are performed in aqueous solution at room temperature and under these conditions the viscosity is with good approximation a constant.

We briefly compare the autocorrelation function for each PSF. The autocorrelation function $g_{GL}(\tau)$ for the Gaussian-Lorentzian PSF and its best approximation by the function $g_{3DG}(\tau)$ for the three-dimensional Gaussian PSF are shown in Fig. 5A. The differences are so minute that experimentally the two PSFs cannot be distinguished by autocorrelation analysis. For that reason two-photon FCS experiments are mostly analyzed assuming a three-dimensional Gaussian PSF instead of the mathematically more complex equations required by the Gaussian–Lorentzian PSF. Figure 5B displays the differences between the autocorrelation functions for the two-dimensional and three-dimensional Gaussian PSF. The autocorrelation functions overlap almost perfectly at early times. However, the tail of the two autocorrelation functions is clearly different. This difference is of course due to the extension of the PSF into the third dimension. Molecules have an additional path to escape from the observation volume, which leads to a faster decay of the autocorrelation function for the three-dimensional Gaussian PSF. The spatial resolution of confocal and two-photon microscopy is worse along the axial direction than in radial direction. In other words, the ratio of the beam waists ω_0/z_0 is less than one and the multiplicative factor in Eq.(3) decays much more slowly than the first term. Consequently, the three-dimensional Gaussian autocorrelation function [Eq.(3)] approaches the two-dimensional Gaussian autocorrelation function [Eq.(2)] as the axial beam waist z_0 increases. In the limiting case, when the axial beam waist is infinity, the three-dimensional autocorrelation function $g_{3DG}(\tau)$ reduces to the two-dimensional Gaussian autocorrelation function $g_{2DG}(\tau)$. Thus, if the axial beam waist z_0 is extended, then the two-dimensional Gaussian model is a fair approximation for FCS measurements in solution. However, if the radial and axial beam waists are of similar size, then the two-dimensional Gaussian model introduces a misfit of the experimental autocorrelation function and the three-dimensional PSF must be used.

Data evaluation of the autocorrelation function requires choosing models with which to fit the data. A statistical criterion is needed to judge the quality of a particular model. The reduced chi-squared (χ^2) is the most widely used tool to judge model-dependent fits of data. A χ^2 value of 1 indicates a perfect fit of the data, whereas χ^2 values larger than three typically are interpreted as a rejection of the

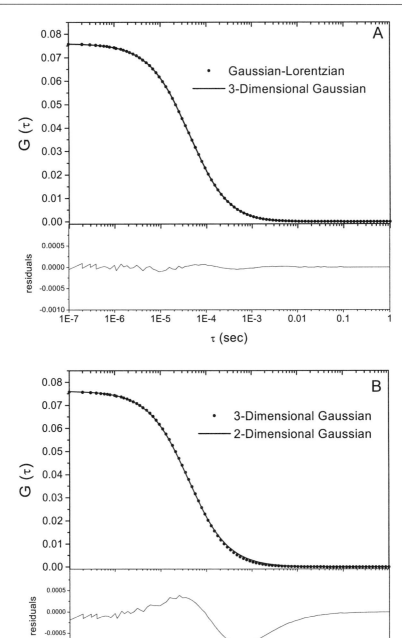

model by the data. However, χ^2-based analysis of the data requires the assignment of the correct experimental errors to each data point. The standard deviation (SD), which is a measure of the experimental error, is not trivial to determine in FCS. The correlation function alone does not provide enough information to calculate the SD and error analysis has been mostly neglected in FCS. However, especially in FCS knowledge of the experimental error is almost indispensable, because different models often lead only to minor changes in the autocorrelation function. Data evaluation based on error analysis establishes an objective procedure by which models are accepted or rejected. A detailed discussion and several methods to determine the SD from FCS data have been recently described in the literature.[11]

We have used the following strategy with good success to determine the SD of FCS data. Because it is not straightforward to calculate the SD of the autocorrelation function, we measured it experimentally. The data acquisition card used allows access to the complete sequence of photon counts and the following procedure was chosen to analyze the autocorrelation function from the raw data. The complete data set is evenly divided into records of equal length. The number of records n depends on the total length of the photon sequence, but is always larger than 10 to have enough records to determine its statistics. Each record is treated as an independent and individual experiment and the autocorrelation function of each record is determined by software. The autocorrelation functions from all individual records allow a straightforward calculation of the standard deviation of a single record $\sigma_{\text{Record}}(\tau)$. The SD $\sigma(\tau)$ of the autocorrelation function for the full data set is related to the SD $\sigma_{\text{Record}}(\tau)$ of a single record by $\sigma(\tau) = \sigma_{\text{Record}}(\tau)/\sqrt{n}$. The last equation is valid, if the n data records used to calculate the SD are statistically independent. This condition is easily verified by inspecting the autocorrelation functions calculated from each record. If these autocorrelation functions decay to zero, then all memory is lost within a single record and individual records are statistically independent from one another. Figure 6 shows the autocorrelation function of the dye rhodamine 110 and its standard deviation measured by two-photon FCS. The autocorrelation function is fit within statistical error to a model of a single species model with a three-dimensional Gaussian PSF ($\chi^2 = 1.1$).

Autocorrelation Function of Multiple Species

A major interest of FCS experiments is the detection of molecular interaction between biomolecules, such as binding between two macromolecules. These

[11] T. Wohland, R. Rigler, and H. Vogel, *Biophys. J.* **80,** 2987 (2001).

FIG. 5. Comparison between autocorrelation functions for different PSFs. (A) The autocorrelation function for a Gaussian–Lorentzian beam profile (●) and its approximation by an autocorrelation function for a three-dimensional Gaussian PSF (solid lines). (B) Comparison between the autocorrelation function for a three-dimensional Gaussian PSF (●) and its best approximation using a two-dimensional Gaussian model (solid line).

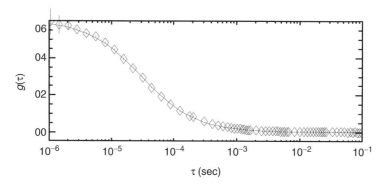

FIG. 6. The autocorrelation function of a rhodamine 110 solution was fit to a single species model for a three-dimensional Gaussian PSF.

interactions are chemical reactions between different species. In a reaction a molecule goes generally from one state into another state, for example from a free ligand to a bound complex. Chemical reaction or other kinetic processes have to occur within the diffusion time of the molecule for FCS to see them. Once a molecule has left the observation volume we have no means of measuring its state, and any other molecule entering the observation volume is statistically independent from the molecule that left the volume. Thus, for soluble proteins the direct measurement of reactions is limited to processes that are faster than approximately a few milliseconds. FCS experiments require nanomolar sample concentrations and the observation of binding at these concentrations requires dissociation coefficients K_D with values, which are nanomolar or lower. The sum of the on and off rates is under these conditions much smaller than the reaction rate limit imposed by the FCS technique. In other words, the probability of observing a biomolecular complex dissociate during its passage time through the observation volume is essentially zero. Although the binding kinetics are typically not visible in FCS measurements, the presence of two species, the free ligand and the bound complex, is reflected in the autocorrelation data. An understanding of the autocorrelation function for multiple, noninteracting species is in most cases sufficient to address binding equilibria between biomolecules with FCS.

We will briefly discuss how the presence of more than one species affects the autocorrelation function. For simplicity, we will consider only the case of two, noninteracting species. The generalization to more than two species is straightforward. Let $g_1(\tau)$ and $g_2(\tau)$ be the autocorrelation functions of species 1 and of species 2, respectively. The autocorrelation function of the mixture of two species is given by

$$g(\tau) = f_1^2 g_1(\tau) + f_2^2 g_2(\tau) \tag{6}$$

The fractional intensity f_i is defined as

$$f_i = \langle k_i \rangle / \sum_j \langle k_j \rangle \qquad (7)$$

where $\langle k_i \rangle$ is the fluorescence intensity of the ith species. The fractional intensity is simply the fractional contribution of a species to the total fluorescence intensity. The autocorrelation function of a mixture is a superposition of the individual autocorrelation functions of each species. Each autocorrelation function is scaled by the square of the corresponding fractional intensity. This nonlinear scaling of the autocorrelation function has practical consequences for FCS experiments. Consider two species with the same concentration, but a difference of 2 in their molecular brightness. The fractional contribution, $f_1^2/(f_1^2 + f_2^2)$, of the dim species to the autocorrelation function is only 20%. This simple analysis illustrates that the detection of a dim species in a sample with a bright species can be difficult experimentally.

A difference in the molecular weight of the two species gives rise to different diffusion coefficients. The autocorrelation function of such a mixture contains a fast and a slow decaying fraction (Fig. 7). The fast decay originates from the species with the larger diffusion coefficient, whereas the slow decay characterizes the species with the smaller diffusion coefficient. Fitting the autocorrelation function of the mixture to Eq. (6) separates the two components. However, a direct separation of the autocorrelation function into its two components requires

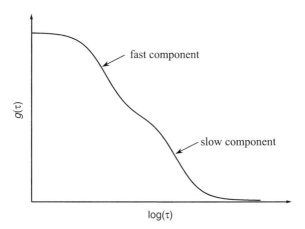

FIG. 7. Theoretical autocorrelation function for a binary mixture. The autocorrelation function for two species with vastly different diffusion coefficients is shown. The fast decay component represents the species with the higher diffusion coefficient. The slow decay characterizes the slowly diffusing species. The amplitude of both components depends on the fluctuation amplitude of each species and their fractional intensities.

a minimum difference of their molecular weight ratio of 5 to 8.[12] If the molecular weight difference between the species is insufficient for a direct separation of the two components from the autocorrelation function, other approaches that are not based on the diffusion coefficient are possible. The analysis of the fluctuation amplitude $g(0)$ of a titration experiment allows the characterization of binding equilibria.[13,14] The photon count distribution offers another statistical tool to separate a mixture of species.[8,15,16] Finally, two-color FCS is an elegant technique for separating species that have been labeled with differently colored fluorophores.[17]

Calibration of Instrument

Calibration of Observation Volume

Data evaluation of the autocorrelation function requires knowledge about the dimensions of the observation volume. The radial and axial beam waists ω_0 and z_0 parameterize the PSF in the case of the three-dimensional Gaussian intensity distribution. For the Gaussian–Lorentzian PSF only the knowledge of the radial beam waist ω_0 is required. One approach to attain these parameters is the direct measurement of the PSF. A number of techniques are available for determining the PSF of the instrument. However, these methods are time consuming and often require equipment not available on a standard FCS instrument.

A relatively simple and fast, but indirect method for finding the beam waists is the calibration of the instrument with a sample of known concentration and diffusion coefficient. Experimentally, we do not directly measure diffusion coefficients, but rather measure the residence time τ_D of a molecule inside the observation volume, which is given by $\omega_0^2/4D$ (or by $\omega_0^2/8D$ in the case of two-photon FCS). Fitting the autocorrelation function of the calibration sample, while keeping the diffusion coefficient D fixed to the known value, determines the radial beam waist ω_0. If the data are fit with a three-dimensional Gaussian PSF the axial beam waist z_0 is determined as well. After the instrument has been calibrated all further experiments are analyzed with the beam waist parameters recovered from the standard sample. A sample with known diffusion coefficient is required for the calibration of the instrument. In the past, we used fluorescent spheres of known diameter for this purpose. The radius of the sphere and the viscosity of the water determine the diffusion coefficient according to the Stokes–Einstein equation. However, we found that spheres tend to aggregate as a function of time. Depending on the size

[12] U. Meseth, T. Wohland, R. Rigler, and H. Vogel, *Biophys. J.* **76,** 1619 (1999).

[13] K. M. Berland, P. T. C. So, Y. Chen, W. W. Mantulin, and E. Gratton, *Biophys. J.* **71,** 410 (1996).

[14] Y. Chen, J. D. Müller, S. Y. Tetin, J. D. Tyner, and E. Gratton, *Biophys. J.* **79,** 1074 (2000).

[15] P. Kask, K. Pälo, D. Ullmann, and K. Gall, *Proc. Natl. Acad. Sci. U.S.A.* **96,** 13756 (1999).

[16] J. D. Müller, Y. Chen, and E. Gratton, *Biophys. J.* **78,** 474 (2000).

[17] P. Schwille, F. J. Meyer-Almes, and R. Rigler, *Biophys. J.* **72,** 1878 (1997).

and the concentration of the spheres, they aggregate on a time scale of minutes to hours. Spheres also adsorb to many other materials, such as test tubes, glass slides, and biological cells. Frequently, the fluorescent intensity from the sphere samples decreases during measurements due to the adsorption of spheres to the walls of the sample holder. We found that fluorescent dyes are much better suited for the calibration procedure. Most fluorescent dyes are stable, do not tend to aggregate, and their sample preparation is straightforward.

Rhodamine 6G, which is a very bright and photostable dye, is a good choice for calibrating the observation volume. Its diffusion coefficient in aqueous solution is 280 μm^2/sec in aqueous solution. We often use fluorescein in high pH buffer (50 mM tris(hydroxymethyl)aminomethane buffer, pH 10) for calibration. Fluorescein is a pH-sensitive dye, and its spectroscopic properties vary drastically from pH 7.5 to 2. At pH > 7.5, fluorescein has a constant quantum yield and very good water solubility. The diffusion coefficient of fluorescein in aqueous solution at room temperature is 300 μm^2/sec.[18] An advantage of using fluorescein is its lack of interactions with surfaces. We have found no evidence of adsorption to container walls for a wide variety of sample holders. Another dye we frequently used is rhodamine 110; rhodamine 110 has lower water solubility than fluorescein, but on the two-photon instrument is almost a factor of two brighter than fluorescein when excited at 780 nm.

A fit of the autocorrelation function to a three-dimensional Gaussian model [Eq. (3)] also yields the axial beam waist z_0. Yet, the accurate determination of the axial beam waist z_0 is much more difficult than finding the radial beam waist ω_0. This simply reflects the fact that the axial beam profile has only a minor influence on the shape of the autocorrelation function. Autocorrelation data of excellent quality are needed to acquire the statistics necessary for an accurate determination of the axial beam waist. The larger the axial beam waist is in comparison to the radial beam waist, the harder it is to determine the parameter z_0 from FCS data, because the experimental autocorrelation function rapidly approaches the shape of the two-dimensional Gaussian correlation function $g_{2DG}(\tau)$ [Eq.(2)], as previously discussed. Analysis of the shape of the PSF for confocal microscopy predicts a beam waist ratio z_0/ω_0 of 2 to 3. However, experimentally often a ratio of 4 to 5 is found for confocal FCS, which is most likely due to aberrations. In two-photon FCS the autocorrelation function based on a Gaussian–Lorentzian beam profile is approximated by an autocorrelation function for a three-dimensional Gaussian model with a ratio of the beam waists of approximately 5.

FCS not only characterizes dynamic processes, but also measures the concentration of the sample. The fluctuation amplitude $g(0)$ of the autocorrelation function is given by the ratio of the gamma factor γ and the number of molecules N in the observation volume V [see Eq. (4)]. FCS has the remarkable property

[18] Y. Chen, University of Illinois at Urbana-Champaign (1999).

that it determines particle concentrations from statistical fluctuations, which are governed by fundamental physical principles. Traditional techniques that measure concentration require the knowledge of a molecular property, such as the extinction coefficient. In contrast, the number of fluctuations within a small observation volume that is in contact with a large surrounding bath is Poisson distributed. This law holds universally for noninteracting particles, independent of specific molecular properties. Thus, in principle, FCS offers a very attractive way for determining concentrations. However, the determination of the average number of molecules N requires knowledge of the γ factor. The value of the γ factor depends on the PSF of the instrument. As already discussed, the autocorrelation function is not a sensitive measure for distinguishing PSFs. For example, the correlation functions based on the three-dimensional Gaussian and the Gaussian–Lorentzian model are essentially identical, but their γ factor differs by a factor of 4.6. Image formation of high numerical aperture objectives is complex and the PSFs discussed so far are mathematical idealizations of the real PSF. A direct measurement of the PSF by imaging or other techniques is difficult and the proper parameterization of the measured PSF is not obvious either. Calibration of the instrument with a sample of known concentration is a practical approach to address the problem. The volume of the PSF is defined as $V = \int \text{PSF}(\mathbf{r})d\mathbf{r}/\text{PSF}(\mathbf{0})$, where $\text{PSF}(\mathbf{r})$ is the value of the PSF at the spatial location $\mathbf{r}$. This volume definition is not a measure of the geometric extent of the PSF, but represents an effective volume of the PSF. For a Gaussian–Lorentzian PSF the volume is given by $V_{\text{GL}} = \pi \omega_0^4/\lambda$ and for the three-dimensional Gaussian the volume is given by $V_{\text{3DG}} = (\pi^{3/2}/8)\omega_0^2 z_0$, where λ is the wavelength of the excitation light source.

The definition of the molar concentration is the ratio of the number of molecules per volume $c = N/V N_{\text{A}}$ and Avogardo's number N_{A} converts the particle number concentration into a molar concentration. The experimental γ factor is determined by $\gamma = g(0)c V/N_{\text{A}}$ and allows a comparison of the measured γ factor with the theoretical γ factor of the assumed PSF. The concentration of the dye used for calibration is measured by absorption spectroscopy and its concentration determined from the known extinction coefficient according to Beer's law. Micromolar dye concentrations, which are required for absorption spectroscopy, are too high for FCS experiments and the sample solution is diluted to nanomolar concentrations. The dye used for calibration should not adsorb to surfaces, so that the sample can be diluted reliably.

Another approach taken by some researchers is a procedure where the γ factor is simply set to a fixed value of one. Such a procedure requires a different definition of the observation volume. A γ factor of one describes any PSF, which is constant within a volume of arbitrary shape and vanishes everywhere outside of the volume, such as a small cylindrical-shaped volume element. A simple geometric interpretation of the observation volume is then given by $V = 1/[g(0)c N_{\text{A}}]$. Whenever relative concentrations instead of absolute concentrations are of concern, no calibration procedure is necessary. The ratio of the number of molecules measured by FCS reflects their concentration ratio, because the γ factor stays constant, if the instrument optics and excitation light are unchanged.

Molecular Brightness

Another important quantity of FCS measurements is the molecular brightness of a fluorophore. FCS uses photon counting for detecting the fluorescence signal and the average intensity $\langle k \rangle$ is measured in counts per second (cps). The average photon count rate $\langle k \rangle$ is the product of the number of molecules N in the observation volume V and the fluorescence brightness ε of a single molecule,

$$\langle k \rangle = \varepsilon N \tag{8}$$

The molecular brightness ε is a measure of the detected fluorescence intensity of a single molecule. Molecular brightness is expressed in counts per second and per molecule (cpsm). The value of the molecular brightness is not a constant, but depends on the excitation intensity, the optical filters, the microscope optics, the quantum yield of the detector, and the molecular properties of the dye. However, the molecular brightness allows a meaningful evaluation of the performance of the instrument as long as the same experimental conditions are used.

It is good practice to measure a calibration sample under the same instrumental conditions at the beginning of each experiment. We typically start by measuring a sample of fluorescein at a particular wavelength and laser power. The fluorescein sample serves for calibrating the two-photon observation volume. In addition, we calculate the molecular brightness of the sample according to Eq. (8) and its value is fairly reproducible. A reduction in the molecular brightness from its usual value indicates a problem with the instrument, such as a misalignment of the optics, and the problem can be addressed right away. This is an important issue, because the molecular brightness is a crucial parameter in FCS measurements. The signal-to-noise ratio of FCS measurements depends on the square of the molecular brightness.[19] A reduction of the molecular brightness by a factor of 2 decreases the signal-to-noise ratio by a factor of 4. A fourfold increase of the data acquisition time is required to offset the reduced signal-to-noise ratio.

Dilution Study with FCS

The fluctuation amplitude $g(0)$ is inversely proportional to the concentration of fluorophores. The fluctuation amplitude decreases until an upper concentration limit is reached, where instrumental noise overtakes the fluorescence fluctuations of the sample. Similarly a lower concentration limit exists, where background counts overwhelm the signal counts. The exact value of the concentration limits

[19] D. E. Koppel, *Phys. Rev. A* **10,** 1938 (1974).

depends on the molecular brightness of the fluorophore and many instrumental parameters, but to get a better feel for the FCS instrument, the performance of the instrument at different sample concentrations should be experimentally determined. We probed the range of concentrations that can be measured on the two-photon FCS instrument by performing a dilution experiment. Figure 8 shows the result of measurements on fluorescein diluted from 275 to 0.27 nM. The measured fluorescence intensity $\langle k \rangle$ and the number of molecule N in the observation volume are plotted as a function of the fluorescein concentration. Both fluorescence intensity $\langle k \rangle$ and the number of molecules N exhibit a linear behavior as a function of concentration. A closer inspection of Fig. 8 indicates that both curves deviate slightly from the ideal linear curve. Plotting of the fluorescence intensity versus the average number of molecules yields a straight line and therefore suggests a slight systematic error in the successive dilution of the sample. The fluorophore concentration range measured in this experiment covers three orders of magnitude. The upper concentration limit of this experiment was about 275 nM. Signal-to-noise considerations set the lower concentration limit. Two sources of noise, the dark counts of the detector and the background counts of the buffer, determine the lower concentration limit of FCS experiments. Ideally, both the dark and background counts are uncorrelated noise sources, which add to the average photon counts measured. The APD used in this study has about 50 dark counts per second and the buffer contributed an additional 70 cps. Compared with a photon count rate of 1600 cps for fluorescein at 0.275 nM, the dark and background counts do not yet contribute significantly to the overall fluorescence signal. Yet a calculation of the fractional intensity [Eq. (7)] already indicates a suppression of the fluctuation amplitude by about 14%. We estimate based on these values that by properly taking the background and dark counts into account, the number of molecules in the sample can be measured down to about 70 pM, where the uncorrected fluctuation amplitude is half of its nominal value and uncertainties in the correction procedure are becoming a concern.

Sample Preparation

Contaminations

The extreme sensitivity of FCS allows measurements at nano- and subnanomolar concentrations. For such experiments the presence of fluorescent contaminants poses a severe problem. The water, any buffer, or cosolvent used in experiments should be checked for contaminants by FCS measurements. We once experienced contamination from our filtered, deionized water source. The molecular brightness of some contaminants rivals that of bright fluorophores and gives rise to autocorrelation curves of excellent quality. The concentration of contaminants is typically subnanomolar. Solvents, such as ethanol or dimethyl sulfoxide (DMSO), often contain fluorescent contaminants even if the solvents purchased are of spectroscopic

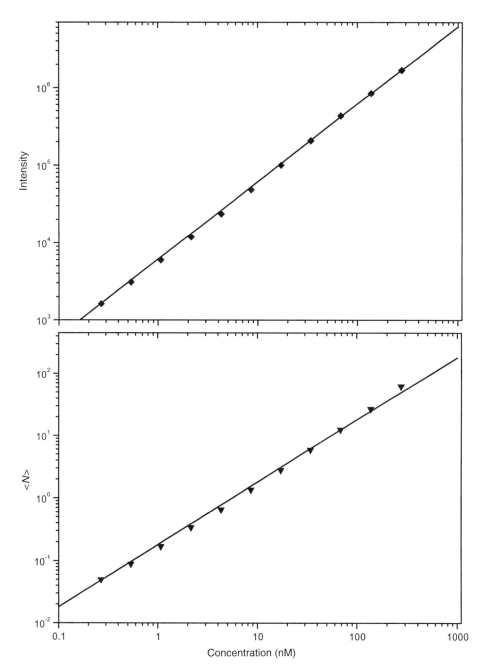

FIG. 8. Dilution study of fluorescein by FCS. Fluorescein in Tris buffer (pH 9.5) was successively diluted from 275 n*M* to 0.27 n*M*. The intensity (diamonds) and the average number of molecules *N* (triangles) are plotted as a function of fluorescein concentration. The solid line is a fit of the data to a straight line.

or higher grade. It is helpful to compare solvents from different sources and select the one best suited for FCS experiments. Contaminants can also be introduced by contact with glassware, pipette tips, and sample holders that are "dirty." If experiments are performed on surfaces a much greater effort is required to eliminate contaminants than for experiments in the bulk phase, because a number of contaminants are surfactants and specifically stick to surfaces. Surface substrates, such as fused silica, should be cleaned meticulously, just as done for the preparation of single molecule experiments. After solution has been added to the cleaned surface, it should be examined for the presence of fluorescent surfactants.

Sample Loss in Fluorescence Fluctuation Measurements

Some biological samples are very precious and the amount available for experiments is often very limited. In these situations it is prudent to reduce the volume and the concentration of a sample used in measurements as much as possible. In FCS experiments the observation volume is of the order of 10^{-1} μm^3 and sample concentrations range from micromoler to picomolar. It therefore is quite easy to perform fluorescence fluctuation measurements on a few microliters of a highly diluted sample. Unfortunately, there are some experimental complications that might occur and disturb the measurement.

The adsorption of a small amount of sample to the sample holder results in a loss of concentration. This is especially important for samples with large surface-to-volume ratios at low concentrations, which can lead to a significant fraction of adsorbed sample at the container walls. Hanging drop glass microscope slides with 22×22-mm square coverslips are widely used as sample holders for microscope experiments. They accommodate 90 μl of sample, and have a surface area close to 10 cm^2. We have observed sample loss due to adsorption with this assembly. We also use a commercial sample holder for cell cultures (Nalge Nunc International, Naperville, IL). This sample holder has a surface area of 3 cm^2 and holds a solution of 500 μl, which reduces the surface-to-volume ratio by at least one order of magnitude as compared to the hanging drop slide. However, the surface-to-volume ratio is only one of the variables affecting sample loss. Some biomolecules preferentially stick to particular materials. It might be necessary to try different sample cells and coat surfaces to minimize adsorption.

Determining Sample Loss

Whenever we start to work with a new fluorescent dye, we check the surface adsorption properties of the dye by performing a simple dilution experiment. Depending on the way the dilution is carried out, different results are observed if adsorption is present. For example, consider a dilution experiment of rhodamine B in water. If fresh sample holders are used for each dilution step, then the fluorescence intensity is linear with the rhodamine B concentration at submicromolar concentrations, but displays sublinear behavior at lower concentrations. However, if the same dilution is done from high to low concentrations in the same sample holder, a superlinear dependence of the intensity as a function of concentration is observed. These results are consistent with the Langmuir isotherm of binding to surfaces. The molecules on the surface are in dynamic equilibrium with the molecules in solution. When a fresh sample holder is used for each concentration, the adsorption of fluorescent molecules to the unoccupied binding sites of the surface removes molecules from solution. When the dilution is done in the same sample holder, the molecules adsorbed to the container walls at high concentration reappear according to Le Chatelier's principle in the solution at low concentrations.

Dilution experiments are very useful for detecting sample adsorption of fluorescent molecules. However, it is very difficult to recognize adsorption when the sample is nonfluorescent. Consider the titration of a nonfluorescent receptor with a fluorescent ligand, where the receptor absorbs to surfaces. The failure to identify the adsorption of the protein leads to a misinterpretation of the experimental results. Thus, great care should be exercised when performing titration studies.

Aggregation of Macromolecules

Fluorescence fluctuation spectroscopy is an extremely sensitive technique for the detection of aggregates. Aggregates are easily identified at the molecular level through their different intensity and diffusion coefficient. An aggregated protein appears brighter, because it contains many fluorescent labels. In addition, the aggregate stays inside the excitation volume longer due to its increased mass. Consequently, the measured fluorescence intensity increases above its average value when an aggregate passes through the illumination volume. The sensitivity of FCS to detect aggregates lies in the small observation volume of the technique. If an aggregate happens to move through the observation volume, then its properties dominate the fluorescence signal detected. Even if aggregates are very rare (for example, a single event in 200 sec), a single passage through the excitation volume is sufficient to produce a clear signature. In conventional fluorometry the observation volume is about a factor of 10^{10} larger than in FCS and aggregates are very hard to detect, because their molecular characteristics are obscured once averaged over an ensemble of millions of molecules. In fact, people were surprised to learn that their sample contains aggregates, because the conventional, cuvette-based instruments they had used has shown no evidence of their presence. It is important to stress that the presence of even one single large aggregate during an FCS measurement is enough to affect the shape of the autocorrelation function. No meaningful analysis of the autocorrelation function is possible, because a single or a few events do not provide the statistics required for FCS analysis. These experiments have to be repeated under conditions in which no aggregates are present. If the data are taken with a data acquisition board that records the sequence of photon counts; the occurrence of an aggregate can then be cut out from the original data and an autocorrelation analysis of the processed data is now possible. We have found this procedure to be very convenient and useful. Although FCS is quite sensitive in detecting large and very bright aggregates, it provides very little sensitivity in detecting the presence of small or less bright aggregates. For example, the detection of dimers or tetramers in a solution of monomers with FCS is challenging. Two-color FCS is useful for detecting small oligomers; another alternative is photon counting histogram analysis, which determines the molecular brightness heterogeneity of a sample.

FCS Measurements *in Vivo*

FCS is an attractive technique for intracellular applications. First, FCS determines kinetic processes from equilibrium fluctuations. Thus, no external perturbation is required to obtain kinetic information. Second, FCS provides excellent spatial resolution. The subfemtoliter observation volumes allow the investigator to probe specific organelles and other local regions within a living cell.

Experiments that extend FCS into the cellar environment involve a few challenges not encountered by measurements *in vitro*. (1) Cells are essentially cuvettes with a volume of a few picoliter. Photobleaching of fluorescent probes in the out-of-focal region of the laser beam is a serious problem for confocal FCS, because of the limited amount of fluorophores available in the tiny volumes of cellular compartments. We will focus here on two-photon FCS, because of its advantages over confocal FCS for *in vivo* applications. The nonlinear nature of the excitation process limits two-photon absorption to the focal volume of the microscope. This inherent three-dimensional sectioning effect of two-photon excitation eliminates photodamage outside of the focal volume of the microscope. This reduction in photobleaching is the principal advantage of two-photon over conventional single-photon excitation for *in vivo* applications.[20] (2) Cells contain molecules with intrinsic fluorescence. This autofluorescence adds a background contribution to any fluorescence measurement in the intracellular environment. In contrast to *in vitro* measurements, where background fluorescence can be avoided by careful sample preparation, the autofluorescence is always present and has to be considered by the investigator. (3) A living cell is a nonequilibrium system. The influence of the cellular environment, such as cellular motion, on FCS measurements is currently not sufficiently understood. For example, do we expect to see simple or anomalous diffusion inside of cells? How do other processes, such as active transport, contribute to the autocorrelation function? However, a number of FCS studies applied to cells have been reported and demonstrate that *in vivo* FCS experiments are feasible.[21,22]

---

[20] P. Schwille, U. Haupts, S. Maiti, and W. W. Webb, *Biophys. J.* **77,** 2251 (1999).

[21] R. Brock, M. A. Hink, and T. M. Jovin, *Biophys. J.* **75,** 2547 (1998).

[22] J. C. Politz, E. S. Browne, D. E. Wolf, and T. Pederson, *Proc. Natl. Acad. Sci. U.S.A.* **95,** 6043 (1998).

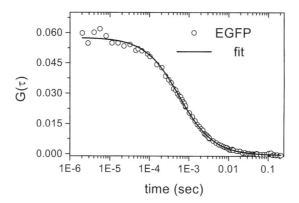

FIG. 9. Autocorrelation function of EGFP inside the nucleus. The autocorrelation function (○) was fitted to a model of a single, freely diffusing species (solid line).

When working with FCS it is necessary to carefully choose the right fluorescent dyes, particularly for *in vivo* measurements. A fluorescent reporter group is needed, which should merely label the protein of interest, but not take part in any interactions with the cellular environment. Thus, the properties of a fluorophore should first be studied under *in vivo* conditions, before using it as a reporter group in an intracellular study. Green fluorescent protein (GFP) has become a vital tool for cell biology and is widely used as a reporter group for imaging and functional studies of proteins in living cells. We therefore characterized EGFP protein, which is a fairly bright fluorophore, in HeLa cells. The diffusion coefficient of EGFP *in vivo* is slowed down by a factor of 3 as compared to its value in aqueous solution. The slowing of diffusion is simply due to the increased viscosity in the cellular environment. We found that the autocorrelation function of EGFP *in vivo* is described within experimental error by a simple diffusion process (Fig. 9). We conclude from these results that EGFP does not stick to cellular components in the nucleus and cytoplasm, where the measurements were performed. Thus, EGFP is a good fluorescent reporter group for FCS measurements *in vivo*.[23]

Most of the autofluorescence in mammalian cell lines comes from NAD, flavins, and lipofuscin.[24–26] The two-photon cross section of FMN and NADH has been determined *in vitro*.[27] NADH is excited at wavelengths between 700 and

[23] Y. Chen, J. D. Müller, Q. Q. Ruan, and E. Gratton, *Biophys. J.* **82,** 133 (2002).

[24] J. E. Aubin, *J. Histochem. Cytochem.* **27,** 36 (1979).

[25] R. C. Benson, R. A. Meyer, M. E. Zaruba, and G. M. McKhann, *J. Histochem. Cytochem.* **27,** 44 (1979).

[26] H. Andersson, T. Baechi, M. Hoechl, and C. Richter, *J. Microsc.* **191,** 1 (1998).

[27] C. Xu, W. Zipfel, J. B. Shear, R. M. Williams, and W. W. Webb, *Proc. Natl. Acad. Sci. U.S.A.* **93,** 10763 (1996).

800 nm, whereas the excitation spectrum of FMN is very broad and is excited over the whole tuning range of the titanium–sapphire laser (700–1000 nm). The two-photon excitation spectrum of lipofuscin is not known. We have measured *in vivo* at two different excitation wavelengths (780 and 895 nm). We noticed that while measuring with the same laser power (1.75 mW at the sample) in the nucleus the response of the cells was wavelength dependent. The autofluorescence intensity increases as a function of time when exciting at 780 nm, but stays constant when excited at 895 nm. Exposure to intense light sources causes oxidative stress that damages cells.[28] A telltale sign of cellular stress is the increase of its autofluorescence. We have observed dramatic and rapid increases in the autofluorescence intensity of cells after exposure to laser light. The power threshold depends on the type of cell, the wavelength of the laser light, and the laser repetition frequency. At equal power, light of longer wavelength apparently is less damaging to cells than light of shorter wavelengths. These observations illustrate the importance of ensuring that the power and wavelength of the excitation light are benign for the cells studied.

The autofluorescence intensity varies strongly between the different cellular compartments of a cell. Inside the nucleus the autofluorescence intensity is typically weak and homogeneous, whereas its intensity in the cytoplasm is stronger and spatially more heterogeneous. When performing FCS measurements inside the nucleus, we found that the fluorescence intensity was very stable. In the cytoplasm, however, the fluorescence intensity depends on the spatial location, and sometimes strong fluctuations, which persist for a few seconds, are observed, whereas at other times the intensity is as stable as inside the nucleus. These differences in the autofluorescence properties make measurement of fluorescently tagged biomolecules inside the cytoplasm much more challenging that inside the nucleus. The concentration range accessible by FCS *in vivo* is less than under *in vitro* conditions, because of the presence of autofluorescence and intensity fluctuations caused by the cell. At very low fluorophore concentrations, the autofluorescence dominates the fluorescence signal. The lowest concentration we measured was around 5 nM. The fluctuation amplitude of EGFP is small at high fluorophore concentrations and any other source of noise, such as intensity fluctuations caused by the cell or instrumental noise, starts to strongly influence the autocorrelation function. The highest concentration we were still able to measure was 300 nM.

---

[28] K. Konig, P. T. So, W. W. Mantulin, B. J. Tromberg, and E. Gratton, *J. Microsc.* **183**, 197 (1996).

[5] Fluorescence Correlation Spectroscopy of GFP Fusion Proteins in Living Plant Cells

By MARK A. HINK, JAN WILLEM BORST, and ANTONIE J. W. G. VISSER

Introduction

Fluorescence Correlation Spectroscopy

Until recently fluorescence correlation spectroscopy (FCS) was mainly applied to well-defined *in vitro* systems. However, the possibility of monitoring molecular dynamics under equilibrium conditions at a single-molecule level makes FCS an attractive technique for intracellular studies. This article presents an overview of FCS measurements inside living plant cells. FCS was introduced in the early 1970s[1] to measure transport properties and concentrations via autocorrelation of fluorescence fluctuations arising from fluctuations in the occupation number in the system (for an extensive overview of theory and applications see Rigler and Elson[2] and Thompson[3]). In FCS a focused laser beam illuminates a subfemtoliter volume element. Fluorescently labeled molecules present in the volume element will emit photons. The fluorescence photons pass through a pinhole and are detected by a highly sensitive detector. The signal-to-noise ratio achieved by this method is very high, since signal interference from scattered laser light, background fluorescence, and Raman emission can be largely eliminated. This allows measurements at the single-molecule level. Typical fluorophore concentrations used in FCS measurements are in the nanomolar range, which makes FCS ideally suited to study biomolecules at physiologically relevant concentrations. The intensity signals (I) are autocorrelated over time resulting in the normalized autocorrelation curve $G_{ij}(\tau)$, being dependent on the fluctuating intensity δI:

$$G_{ij}(\tau) = \frac{\langle \delta I_i(t)\, \delta I_j(t+\tau)\rangle}{\langle I_i\rangle \langle I_j\rangle} \qquad \text{with } i = j \text{ for autocorrelation} \qquad (1)$$

Any process that will result in temporal changes of the fluorescence intensity may be represented in $G(\tau)$. Diffusion of molecules entering and leaving the volume element is one of the most studied sources of fluctuation as well as other processes such as triplet state dynamics,[4] isomerization,[5]

[1] D. Magde, E. L. Elson, and W. W. Webb, *Phys. Rev. Let.* **29**, 705 (1972).
[2] R. Rigler and E. S. Elson, eds., "Fluorescence Correlation Spectroscopy. Theory and Applications." Springer Verlag, Berlin, 2001.
[3] N. L. Thompson, *in* "Topics in Fluorescence Spectroscopy" Vol. 1 (J. R. Lakowicz, ed.), p. 337. Plenum Press, New York, 1991.
[4] J. Widengren, Ü. Mets, and R. Rigler, *J. Phys. Chem.* **99**, 13368 (1995).
[5] J. Widengren and P. Schwille, *J. Phys. Chem.* **104**, 6416 (2000).

quenching,[6] or protonation of the chromophore.[7,8] The possibility of monitoring molecular interactions with FCS has resulted in numerous publications on protein–protein, protein–lipid, and ligand–receptor interactions.[9–15] In these studies a small fluorescently tagged molecule interacts with a nonlabeled molecule having a much larger mass. Autocorrelation of the fluorescence signal then distinguishes smaller (faster diffusing) from larger (slower diffusing) molecules and the fractions (F) of free and bound molecules of a binding equilibrium can be determined according to a three-dimensional (3D) diffusion fitting model:

$$G_D(\tau) = 1 + \frac{1}{N}\left(\sum_{i=1}\frac{F_i\eta_i^2}{\left(\sum_{i=1}F_i\eta_i\right)^2} Diff_i\right)$$

$$\text{with } Diff_{i,3D} = \left\{\frac{1}{1+(\tau/\tau_i)}\frac{1}{\sqrt{1+[\tau/((\frac{z_0}{\omega_0})^2\tau_i)]}}\right\} \tag{2}$$

Here, N is the total number of fluorescent particles in the detection volume element, which corresponds to $1/[G_D(0)-1]$. Parameters z_0 and ω_0 represent the e^{-2} radii of the axial and equatorial axis of the detection volume element, respectively, and η_i is the molecular brightness of species i. Note that each species contributes to the curve by the squared value of its brightness, which means that a species with a brightness twice that of another species will contribute four times more to the correlation curve. The diffusion time for species $i(\tau_i)$, is related to the translational diffusion constant according to

$$D_i = \frac{\omega_0^2}{4\tau_i} \tag{3}$$

Fluorescence Cross-Correlation Spectroscopy

The application of FCS to study molecular interactions is limited by the fact that the diffusion time scales only to the power of one-third of the molecular

[6] G. Bonnet, O. Krichevsky, and A. Libchaber, *Proc. Natl. Acad. Sci. U.S.A.* **95**, 8602 (1998).
[7] U. Haupts, S. Maiti, P. Schwille, and W. W. Webb, *Proc. Natl. Acad. Sci. U.S.A.* **95**, 13573 (1998).
[8] J. Widengren, Ü. Mets, and R. Rigler, *Chem. Phys.* **250**, 171 (1999).
[9] K. M. Berland, P. T. C. So, Y. Chen, W. W. Mantulin, and E. Gratton, *Biophys J.* **71**, 410 (1996).
[10] B. Rauer, E. Neumann, J. Widengren, and R. Rigler, *Biophys. Chem.* **58**, 3 (1996).
[11] U. Trier, Z. Olah, B. Kleuser, and M. Schäfer-Korting, *Pharmazie* **54**, 263 (1999).
[12] T. Wohland, K. Friedrich, R. Hovius, and H. Vogel, *Biochemistry* **38**, 8671 (1999).
[13] A. Pramanik, P. Thyberg, and R. Rigler, *Chem. Phys. Lipids* **104**, 35 (2000).
[14] J. Goedhart, H. Röhrig, M. A. Hink, A. van Hoek, A. J. W. G. Visser, T. Bisseling, and T. W. J. Gadella, *Biochemistry* **38**, 10898 (1999).
[15] E. van Craenenbroeck and Y. Engelborghs, *Biochemistry* **38**, 5082 (1999).

mass. Therefore discrimination between single fluorescently labeled molecules and complexed molecules will be difficult in case of almost equally sized components such as those occurring in studies of homodimer receptors. Meseth *et al.*[16] examined the resolving power of FCS to distinguish between different molecular sizes. In the case of an unchanged fluorescence yield on binding, the diffusion times of the bound and unbound forms have to differ at least 1.6 times, which corresponds to a fourfold mass increase, which is required to distinguish both species without prior knowledge of the system. Although small mass differences can be visualized in the autocorrelation traces, this requires detailed knowledge about the photophysical properties of the molecules involved. For most experimental systems and especially for measurements in living cells, these properties are hard to obtain. To overcome these limitations Schwille *et al.*[17] have developed dual-color fluorescence cross-correlation spectroscopy (FCCS). In FCCS studies interacting molecules can be tagged by spectrally different fluorescent groups, e.g., green and red emitting dyes. Interaction can be studied by following the fluctuations in fluorescence intensity of both labeled molecules. Therefore the emission light is split into two different detectors by which the two dyes can be monitored simultaneously. Cross-correlation curves are analyzed according to Eq. (2) with τ_{dif} representing the weighted diffusion time of the doubly labeled molecules, N_{gr}:

$$\tau_{\mathrm{dif,gr}} = \frac{\omega_{0,\mathrm{g}}^2 + \omega_{0,\mathrm{r}}^2}{8D_{\mathrm{gr}}} \tag{4}$$

The time-independent part, $G_{\mathrm{D}}(0)$ is not equal to $1 + 1/N_{\mathrm{gr}}$ here, but is also related to the number of molecules emitting in only one of the two channels. Equation (5) corrects for this[18]:

$$G_{\mathrm{D}}(0) = 1 + \frac{N_{\mathrm{g}}(\eta_{\mathrm{rgg}}/\eta_{\mathrm{rrr}}) + N_{\mathrm{gr}}[1 + (\eta_{\mathrm{rgg}}/\eta_{\mathrm{rrr}})]}{(N_{\mathrm{g}} + N_{\mathrm{gr}})[N_{\mathrm{r}} + N_{\mathrm{g}}(\eta_{\mathrm{rgg}}/\eta_{\mathrm{rrr}}) + N_{\mathrm{gr}}[1 + (\eta_{\mathrm{rgg}}/\eta_{\mathrm{rrr}})]]} \tag{5}$$

N_{g} and N_{r} are the numbers of free, singly labeled molecules and $\eta_{\mathrm{em,dye,ex}}$ are the molecular brightness values for the dyes at various excitation and emission wavelengths. In the ideal case the time-dependent part of G_{D} represents only the fluctuation characteristics of the doubly labeled molecule. In practice, however, the green dye will also emit into the red detector (this is called cross-talk) and therefore contribute to the cross-correlation curve as an additional species.

[16] U. Meseth, T. Wohland, R. Rigler, and H. Vogel, *Biophys. J.* **76**, 1619 (1999).
[17] P. Schwille, F. J. Meyer-Almes, and R. Rigler, *Biophys. J.* **72**, 1878 (1997).
[18] R. Rigler, Z. Földes-Papp, F. J. Meyer-Almes, C. Sammet, M. Volcker, and A. Schnetz, *J. Biotech.* **63**, 97 (1999).

FCCS applications that have been reported include enzyme kinetics,[19,20] nucleotide hybridization,[17,18,21] conformational dynamics in DNA,[22] and protein–DNA interactions.[23] From these studies it is clear that FCCS is an attractive technique to observe very specific molecular interactions. Now it is a challenge to apply this technique to cellular systems in order to monitor molecular interactions.

Practical Considerations for FCS Measurements in Living Cells

Only a limited number of intracellular FCS applications have been reported[14,24–34] (for a review see Schwille[35]). Cellular autofluorescence, photobleaching of the dye, cellular damage, and reduced signal-to-noise ratios due to scattering are major points of concern when taking FCS into the living cell. We will briefly discuss these factors.

Autofluorescence

One of the most important problems is the presence of autofluorescent molecules. Endogeneous molecules like NADH, flavins, flavoproteins, and chlorophyll (and some others) will fluoresce strongly in the blue, green, and red spectral regions. The optimal spectral range for fluorescence of probes used in plant cells is roughly between 500 and 600 nm, which is covered by variants of fluorescent proteins (see below). Autofluorescence (more generally the background) can contribute to the correlation curves in two ways. It may be present as a constant,

[19] U. Kettling, A. Koltermann, P. Schwille, and M. Eigen, *Proc. Natl. Acad. Sci. U.S.A.* **95,** 1416 (1998).
[20] A. Koltermann, U. Kettling, J. Bieschke, T. Winkler, and M. Eigen, *Proc. Natl. Acad. Sci. U.S.A.* **95,** 1421 (1998).
[21] Z. Földes-Papp, B. Angerer, P. Thyberg, M. Hinz, S. Wennmalm, W. Ankenbauer, H. Seliger, A. Holmgren, and R. Rigler, *J. Biotech.* **86,** 237 (2001).
[22] M. I. Wallace, L. M. Ying, S. Balasubramanian, and D. Klenerman, *J. Phys. Chem.* **48,** 11551 (2001).
[23] K. Rippe, *Biochemistry* **39,** 2131 (2000).
[24] K. M. Berland, P. T. C. So, and E. Gratton, *Biophys. J.* **68,** 694 (1995).
[25] J. C. Politz, E. S. Brown, D. E. Wolf, and T. Pederson, *Proc. Natl. Acad. Sci. U.S.A.* **95,** 6043 (1998).
[26] R. Brock, M. Hink, and T. M. Jovin, *Biophys. J.* **75,** 2547 (1998).
[27] R. Brock, G. Vamosi, G. Vereb, and T. M. Jovin, *Proc. Natl. Acad. Sci. U.S.A.* **96,** 10123 (1999).
[28] W. J. H. Koopman, M. A. Hink, A. J. W. G. Visser, E. W. Roubos, and B. G. Jenks, *Cell Calcium* **26,** 59 (1999).
[29] P. Schwille, J. Korlach, and W. W. Webb, *Cytometry* **36,** 176 (1999).
[30] P. Schwille, U. Haupts, S. Maiti, and W. W. Webb, *Biophys. J.* **77,** 2251 (1999).
[31] R. Rigler, A. Pramanik, P. Jonasson, G. Kratz, O. T. Jansson, P.-Å. Nygren, S. Ståhl, K. Ekberg, B.-L. Johansson, S. Uhlén, M. Uhlén, H. Jörnvall, and J. Wahren, *Proc. Natl. Acad. Sci. U.S.A.* **96,** 13318 (1999).
[32] M. Wachsmuth, W. Waldeck, and J. Langowski, *J. Mol. Biol.* **298,** 677 (2000).
[33] J. Goedhart, M. A. Hink, A. J. W. G. Visser, T. Bisseling, and T. W. J. Gadella, *Plant J.* **21,** 109 (2000).
[34] R. H. Köhler, P. Schwille, W. W. Webb, and M. R. Hanson, *J. Cell Sci.* **113,** 3921 (2000).
[35] P. Schwille, *Cell Biochem. Biophys.* **34,** 383 (2001).

noncorrelating species, lowering the amplitude of the correlation curve by a factor of $(1 - I_{background}/I_{total})^2$. Autofluorescent molecules may also contribute to the curve as an additional correlating component. In this case the difference in brightness between the autofluorescent molecules and the fluorescent probe used will determine how large the contribution of the autofluorescence will be [see Eq. (2)] and therefore a careful selection of cell type, fluorescent probe, and excitation wavelength is required. Also differences between subcellular locations and the metabolic state of a cell[26] can influence the brightness of autofluorescence. To investigate processes in living plant cells two cell types were selected, which have been widely used as a model system for plant cells: Both Bright Yellow 2 (BY2) tobacco suspension cells and cowpea protoplasts are easy to grow and to manipulate and show a moderate level of autofluorescence. Figure 1A displays the emission spectra for both cell types acquired at an excitation wavelength of 436 nm. The autofluorescence spectra show broad-banded emission peaks without any fine structure. The autofluorescence intensity decreases going toward 600 nm, as is also shown by the FCS intensity traces at various excitation wavelengths (Fig. 1B). The molecular brightness, obtained by analysis of the autocorrelation curves, decreases as well, when excited further in the red (Fig. 1C). However, above 600 nm a steep increase of intensity and brightness is found for the cowpea protoplasts that is due to the presence of chlorophyll in the chloroplasts. The latter emission peak is not present inside the BY2 cells that lack chloroplasts, due to their growth in a dark environment. Table I summarizes the molecular brightness values of the autofluorescent molecules at various subcellular locations using the experimental setup for detection of the fluorescent proteins (see below) CFP (cyan fluorescent protein) and YFP (yellow fluorescent protein). The brightness values are low compared to the ones obtained for CFP and YFP (Table II) and therefore the contribution of autofluorescence to the correlation curve can be neglected except for measurements performed in the cytoplasm of cowpea protoplasts, where many chloroplasts are located. To remove autofluorescent molecules from the volume element, the sample can be prebleached before the real measurement. However, this approach is not very successful in the plant cells studied due to the high mobility of the autofluorescent molecules, which quickly replace the destroyed molecules in the prebleached area.

Cell Culture and Handling

Suspension cells of tobacco BY-2 (*Nicotiana tabacum* L. cv. Bright Yellow 2) were cultured at 22° under gently shaking (13 rpm) in growth medium (4.3 g/liter Murashige and Skoog plant salt base, 255 mg/liter KH_2PO_4, 1 mg/liter thiamin hydrochloride, 0.2 mg/liter 2,4-dichlorophenoxyacetic acid, 30 g/liter sucrose, and 100 g/liter *myo*-inositol, pH 5.8) and weekly subcultured at 50 times dilution with fresh medium. Protoplasts were prepared from the cells cultured after 3 days of subculture by adding 10 ml enzyme solution consisting of 1% cellulase, 0.1%

A

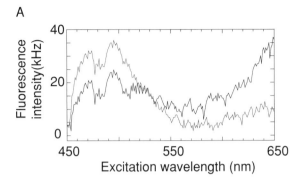

B

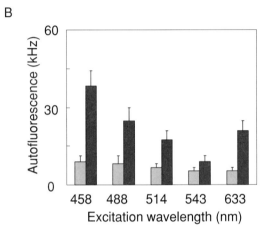

C

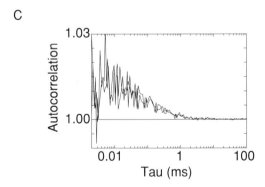

FIG. 1. Autofluorescence characteristics of BY2 suspension cells (gray) and cowpea protoplasts (black). (A) Whole cell emission spectra were obtained by spectral imaging using excitation at 436 nm. (B) Average fluorescence intensities and (C) autocorrelation curves of autofluorescent cytoplasm of cowpea protoplasts measured with various excitation wavelengths during 90 sec.

TABLE I
MOLECULAR BRIGHTNESS AND INTENSITY OF AUTOFLUORESCENT MOLECULES[a]

| | Molecular brightness η (kHz/molecule) | | | | | | | |
| | Nucleus | | Membrane | | Cytoplasm | | Vacuole | |
Cell	CFP	YFP	CFP	YFP	CFP	YFP	CFP	YFP
BY2	0.1	0.1	0.1	0.0	0.2	0.1	0.2	0.1
Cowpea protoplast	0.0	0.1	0.2	0.3	0.5	0.4	0.3	0.3
	Fluorescence intensity I (kHz)							
BY2	6	4	2	2	6	5	3	2
Cowpea protoplast	8	7	8	8	43	29	16	12

[a] At several subcellular locations determined by FCS measurements using the experimental setup to detect CFP or YFP at excitation intensities of 2.5 and 3.1 kW cm^{-2}, respectively.

pectolyase, and 0.4 M mannitol, pH 5.5. After 3 hr of incubation at 28° in a rotating vessel at 60 rpm for 2 hr the cells were washed twice by centrifugation room temperature for 2 min at 850 rpm followed by addition of 10 ml solution containing 125 mM CaCl$_2$, 154 mM NaCl, 5 mM KCl, 5 mM sucrose, and 0.1% morpholino-ethanesulfonic acid (MES), pH 5.5.

Cowpea mesophyll protoplasts are prepared by peeling off the lower epidermis of the primary leaves of 10-day-old *Vigna unguiculata,* using forceps. Three leaves are floated on a 15 ml enzyme solution (0.1% cellulase, 0.05% pectinase, 10 mM CaCl$_2$, and 0.5 M mannitol, pH 5.5) for 3.5 hr at room temperature with gentle shaking. Cells are washed twice by adding 2 ml solution containing 10 mM CaCl$_2$ and 0.5 M mannitol followed by centrifugation for 5 min at 600 rpm.

TABLE II
MOLECULAR BRIGHTNESS OF CFP AND YFP[a]

| | Molecular brightness η (kHz/molecule) | | | |
| | CFP Detector | | YFP Detector | |
Dye	λ_{exc} 458 nm	λ_{exc} 514 nm	λ_{exc} 458 nm	λ_{exc} 514 nm
CFP	5.8	0.0	1.9	0.1
YFP	0.0	0.0	0.3	4.9

[a] Measured in BY2 cytoplasm (intensity) after background correction.

Fluorescent Proteins

Studying molecules with FCS requires that the molecule of interest is fluorescent, but since most natural molecules show only weak autofluorescence, labeling with an external fluorophore is essential. In the case of living cells it is in principle possible to introduce a labeled molecule into the cell, e.g., via uptake of esterase-cleavable dyes, pH-shock methods, electroporation, or microinjection. However, plant cell walls form a large physical barrier for most techniques resulting in a low efficiency of uptake. Therefore, in the case of cellular protein studies, it is far more attractive to add fluorescent tags by genetic approaches. Intrinsic fluorescent proteins (FPs) such as GFP (green fluorescent protein),[36,37] identified in jellyfish *Aequorea victoria,* are especially suitable for this purpose since they can be relatively easily fused to the gene of interest. At present GFP or color variants, like the cyan (CFP) or yellow (YFP) fluorescent proteins, are most often used to fluorescently tag proteins. The sequence of the GFP gene of *A. victoria* has been optimized for plant codon usage and in this way a cryptic splice site has been eliminated.[38] However, in the experiments described here this sequence has not been used. Advantages of the FPs are the high brightness values, a well-protected chromophore that is relatively insensitive to environmental changes, and the fact that FPs do not tend to stick to intracellular structures as some chemical dyes do. However, in recent years it has been shown that for enhanced GFP and YFP additional correlating processes occur that are superimposed on the diffusional fluctuation in the correlation curve.[7,8,39] Due to a photophysical effect called "blinking," or a chemical effect such as protonation of the chromophoric group at low pH (pK_a 5.8), a large fraction of fluorophores can exist in a "dark" state.

CFP–YFP is a widely used pair of dyes to study the colocalization of different proteins or to monitor molecular interactions, making use of fluorescence resonance energy transfer (FRET).[40] FRET, however, requires that both dyes are in close proximity (2–8 nm) of each other. For FCCS this proximity is not required since the technique is not based on physical interaction of the dyes but on the temporal coincidence of both dyes being in the same volume. A disadvantage of the CFP–YFP pair for FCCS is the relatively high cross-talk of the CFP emission in the YFP detection channel. Correction for this effect of the correlation curve of CFP fluorescence in the YFP channel is therefore required.

Because FCS correlates the relative fluorescence fluctuations, an upper limit for observing a measurable correlation curve exists where the intensity fluctuations are

[36] R. Y. Tsien, *Annu. Rev. Biochem.* **67,** 509 (1998).

[37] P. M. Conn, ed., *Methods Enzymol.* **302** (1998).

[38] J. Haselhoff, K. R. Siemering, D. C. Prasher, and S. Hodge, *Proc. Natl. Acad. Sci. U.S.A.* **94,** 2122 (1997).

[39] P. Schwille, S. Kummer, A. A. Heikal, W. E. Moerner, and W. W. Webb, *Proc. Natl. Acad. Sci. U.S.A.* **97,** 151 (2000).

[40] T. W. J. Gadella, G. N. M. van der Krogt, and T. Bisseling, *Trends Plant Sci.* **4,** 287 (1999).

too small as compared to the average intensity [Eq. (1)]. In the setup used here the concentration of fluorescent molecules should not exceed 5 μM corresponding to 800 molecules in the confocal volume element. Therefore, cells should be selected that have only a moderate FP expression level. Because all constructs in our studies were transcribed using the strong 35S-promoter, measurements were performed at 8–12 hr after transfection. Longer incubation times resulted in too high levels of fluorescent protein. An alternative possibility to control the level of expressed protein is the use of inducible promoters such as tetracycline or ethanol promoters.

Constructs

The open reading frame of enhanced CFP and YFP cDNA is amplified by the polymerase chain reaction (PCR) from the full-length cDNA and cloned into the pTYB11 vector (New England Biolabs, Beverly, MA) using the following primers: FPfor (5′ GGTGGTTGCTCTTCCAACATGGTGAGCAAGGGCG 3′) and FPrev (5′ GGTGGTGGATTCTTACTTGTACAGCTCG 3′). The pTYB11-FP constructs are transformed via heatshock into BL21 DE3 *Escherichia coli* bacteria strain for high expression levels. The expression is induced after 3 hr of incubation at 37° by adding 0.3 mM isopropylthiogalactoside (IPTG). The bacteria were grown overnight at 20° to obtain soluble protein for microinjection.

The AtSERK1 construct was amplified by PCR from AtSERK1 full-length cDNA (accession no: A67827) and cloned downstream of the 35S promoter into the *Nco*I site of PMON999-YFP[41] using primers *Nco*I215f (5′ CATGCCATG GTGGAGTCGAGTTATGTGG 3′) and *Nco*I2068 (5′ CATGCCATGGACCTTG GACCAGATAACTC 3′).[42]

For targeting purposes, a 87-bp fragment (5′ ATGTTGTCACTACGTCAATC TATAAGATTTTTCAAGCCAGCCACAAGAACTTTGTGTAGCTCTAGATATC TGCTTCAGCAAAAACCC 3′) encoding for the coxIV mitochondrial targeting sequence from yeast[43] is cloned downstream of the 35S promoter of either into the *Nco*I site of either PMON999-CFP,[41] PMON999YFP or PMON999CFP-(Ala)$_{25}$-YFP plasmid. All constructs are checked by sequence analysis.

Transfection

Ten micrograms purified plasmid in 30 μl water is added to 0.5–1 × 10^6 protoplasts in 75–150 μl solution of 0.6 M mannitol, 10 mM CaCl$_2$, pH 5.5. After gentle mixing 3 ml solution containing 40% (w/v) polyethylene glycol (PEG) 6000, 0.6 M mannitol, 0.1 M Ca(NO$_3$)$_2$ is added. The protoplast suspension is incubated for 10 sec under gentle shaking followed by addition of 4.5 ml washing

[41] H. van Bokhoven, J. Verver, J. Wellink, and A. van Kammen, *J. Gen. Virol.* **74**, 2233 (1993).
[42] K. Shah, T. W. J. Gadella, H. Van Erp, V. Hecht, and S. C. De Vries, *J. Mol. Biol.* **309**, 641 (2001).
[43] R. H. Köhler, W. R. Zipfel, W. W. Webb, and M. R. Hanson, *Plant J.* **11**, 625 (1997).

solution consisting of 0.5 M mannitol, 15 mM MgCl$_2$, and 0.1% MES, pH 5.5, to stop the transfection. After incubation at room temperature for 20 min the cells are washed three times and incubated for 24 hr in petri dishes at room temperature under constant illumination.

Purification of FPs

Fluorescent protein was purified using the IMPACT (New England Biolabs) system, which utilizes the inducible self-cleavage activity of an intein splicing element to separate the target protein from the affinity tag. Transformed bacteria are collected by centrifugation and resuspended in 50 mM Tris, pH 8.0, 120 mM KCl, 1 mM EDTA. Cells are lysed by passage through a French pressure cell. Soluble protein is obtained after centrifugation at 20,000g for 30 min at 4°. The fusion protein is purified by using an affinity column matrix of chitin beads. The fusion protein binds to the chitin and after extensive wash the FPs are eluted from the column by incubating the beads overnight in 50 mM dithiothreitol (DTT). Protein purity is checked on sodium dodecyl sulfate-polyacrylamide gel electrophoresis (SDS-PAGE).

Dye Depletion and Cellular Damage

Because the subcellular compartments are rather small, dye depletion due to photobleaching can be a serious problem. For *in vitro* systems the photobleached molecules will be replaced by fresh material diffusing from outside the volume element. This is often not possible in living cells due to compartmentalization of fluorescent molecules and therefore the limited number present. Photobleaching of the dye gives rise to artifacts in the correlation curves. Because the dye will be bleached during its passage through the detection volume, the apparent diffusion time becomes too small. This problem is clearly present when slowly moving molecules, due to their large size or diffusion restrictions, are being monitored. The prolonged residence time in the excitation volume will result in a higher chance of being photobleached. Comparison of the photostability of the FPs showed that YFP is more sensitive to photobleaching than the others.[42] When YFP is fused to membrane proteins the slower diffusion rates increase the chance to photobleach it further, so that very low laser intensities (<1.0 kW cm^{-2}) should be applied. Figure 2 shows the fluorescence intensity and correlation curves for YFP present in the cytoplasm of BY2 cells when excited with various laser intensities. At 20 and 10 kW cm^{-2} a significant decrease of the fluorescence intensity is present, resulting in an autocorrelation curve with a fast decay time due to the shortening of the diffusion time induced by photobleaching and faster flickering.[39] When sequential measurements are performed, an increase of the correlation amplitude and thus a decrease in number of fluorescent molecules is observed. Below 5.0 kW cm^{-2} no photobleaching is observed. However, for YFP fused to a transmembrane protein (AtSERK1) a significant photobleaching is already observed at intensities higher than 2.2 kW cm^{-2} due to the slow diffusion of the protein.

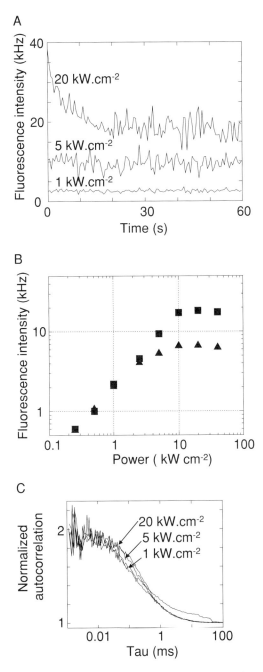

FIG. 2. (A) Photobleaching traces, (B) laser power dependency plot (after a prebleach pulse of 20 sec), and (C) autocorrelation curves (after a prebleach pulse of 20 sec) of YFP localized in the cytoplasm (squares) and AtSERK1-YFP in the cell membrane (triangles) of BY2 cells, excited with various laser powers at 514 nm.

To check the cellular damage induced by laser light, individual cells are exposed to various laser intensities at either 458 or 488 nm and recultured for 8 hr, after which the cells are visually inspected and counted to check the growth rate. Below a laser intensity of 50 kW cm^{-2} no effect is seen, but at 75 and 100 kW cm^{-2} the growth rate is reduced about 15% and some misshapen cells are present. However, at the moderate light intensities used in the studies here no damaging effect is expected.

Signal-to-Noise Ratios

FCS measurements in living cells will in general suffer from significantly lower signal-to-noise (S/N) ratios as compared to *in vitro* measurements. The presence of dense structures within the cell can lead to scattering of both excitation and emission light. Considering all the effects discussed in previous sections it is clear that the signal may be enhanced by increasing the intensity of the excitation source but not at an unlimited high level. The optimal setting will therefore be a compromise between high molecular brightness of the dye and a minimized contribution of effects such as photobleaching, cellular damage, and autofluorescence. Improvement of the S/N ratio can also be achieved by increasing the measurement time (T_m), since S/N improves with the square root of T_m. However, at measurements times above 1 min, special attention should be paid to movement of intracellular structures or even the complete plant cell, which may cause additional fluctuations in the correlation curve.

Fluorescent Proteins in Plant Cells

Microscope

The fluorescence correlation spectroscopic measurements are carried out with a Zeiss microscope system based on an inverted Axiovert microscope equipped with a baseport LSM510 module for collecting confocal laser scanning images and a sideport ConfoCor-2 module to perform fluorescence correlation spectroscopy measurements (Fig. 3). The system contains three laser modules providing excitation light at 458, 488, 514, 543, and 633 nm. In our study CFP is excited at 458 nm and YFP at 514 nm. The excitation light is focused into the sample by a 40× water immersible Apochromat objective lens NA 1.2 (Zeiss). Samples are stored in (borosilicate) glass-bottomed 96-well plates (Whatman, Clifton, NJ) at room temperature. Fluorescence passes through the main dichroic filter, which reflects both 458 and 514 nm excitation light, and a secondary dichroic filter, LP510, to separate the emission in the two different detection channels. Fluorescent light is split by two emission filters (BP470-500 for CFP and BP527-562 for YFP, respectively) and is directed through size-adjustable pinholes (internal diameter 25 μm), which is placed near the image plane, and is fiber coupled to an avalanche photodiode.

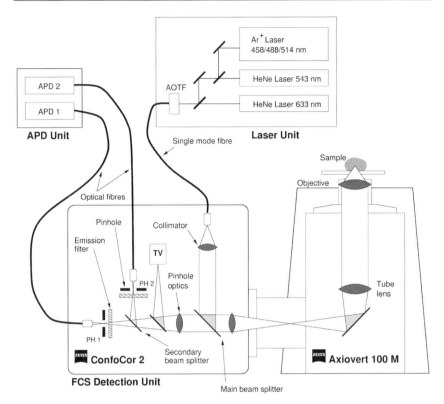

FIG. 3. Schematic overview of the experimental setup for FCCS measurements of CFP and YFP fusion proteins. The fluorescence from the confocal detection volume element is detected via fiber-coupled photodiodes in the ConfoCor2-module. An additional option in this setup is to collect confocal images making use of the LSM510 scanning detection module, coupled to the base of the Axiovert microscope (not shown).

The microscope is equipped with a mercury lamp and a back-illuminated TEK $512 \times 512D$ CCD camera (Princeton, Roper Scientific B.V., The Netherlands) to collect wide-field fluorescence images of cells expressing small amounts of FPs. A microinjection holder is attached to the side of the microscope.

Microinjection

Purified protein is dissolved in 50 mM phosphate-buffered saline (PBS) pH 7.2 to a final concentration of 1 μM and centrifuged at 80,000 rpm for 10 min at 4° to remove possible aggregates. Femtotips (Eppendorf) with an opening diameter of 0.5 ± 0.2 μm are loaded with approximately 10 μl fluorescent protein solution using Microloader tips (Eppendorf). The cells are injected under visual control using an Eppendorf microinjector 5242 and micromanipulator 5170 to control injection parameters ($P_1 = 4000$ hPa, $P_{inj} = 100$ hPa, $P_{backpressure} = 60$ hPa, and $t_{injection} = 0.2$ sec). After injection FCS measurements are repeated twice at intervals of 5 min to check for leakage of the cell. Microinjection in plant cells requires specialized skills and the percentage of successful injections without collapsing the cell, caused by the high internal pressure (turgor), is in our studies ca. 50% for BY2 suspension cells and only 8% for cowpea protoplasts.

Spectral Imaging

Spectral images are acquired using a Leica DMR epifluorescence microscope equipped with a 250IS imaging spectrograph (Chromex) coupled to a back-illuminated 512×512 CH250 CCD camera (Photometrics). The sample is excited by a mercury lamp coupled to an excitation filter wheel containing a 435DF10 excitation filter. Fluorescent light is collected by the $20\times$ Plan Neofluor objective lens, N.A. 0.5 (Leica), and passed through a 430DLCP dichroic filter (Omega) and a LP455 emission filter (Schott). Spectral images are collected during 2 sec using a 150 groove/mm grating set at a central wavelength of 500 nm and a slit entrance width of 200 μm, corresponding to 10 μm in the object plane.

Measurement Protocol

By visual inspection those cells are selected in which the amount of fluorescent protein is just sufficient to be seen by confocal imaging. At lower expression levels wide-field images could be collected by integrating the fluorescence intensity over 30 sec, using the CCD camera. High levels of FPs (above approximately 5 μM) resulted in the presence of too many particles in the volume element and thus a too low correlation amplitude for reliable analysis. Spots of interest were marked in the displayed confocal image by mouse after which the x-, y-, and z-coordinates of the spots are passed to the scanning table so that experiments with the stationary laser beam of the ConfoCor2 could be performed. The laser power was set at 2.5 kW cm^{-2} for the 458 nm laser line and 3.1 kW cm^{-2} for the 514 nm laser line to prevent photobleaching, cellular damage, and photophysical effects but still sufficient to achieve good S/N ratios. In case of FCCS experiments, the cross-talk of CFP fluorescence in the YFP channel also has to be considered. Typical measurement times are 40–90 sec. Fitting parameters were averaged over 15–30 different FCS curves, each obtained in another plant cell.

Data Analysis FCS

The obtained intensity traces are correlated by the software correlator, integrated in the Zeiss AIM software (Zeiss, EMBL Heidelberg). Curves are fitted in a home-developed software package that allows global fitting with several types of fitting models, using Marquardt least-squares fitting algorithms. The quality of the fitting is checked using the minimal value of χ^2 and by visual inspection of fitted and experimental traces and the residuals. Auto- and cross-correlation curves are corrected for uncorrelated background and triplet-state dynamics according to

$$G(\tau) = G_D \left(1 - \frac{I_{\text{background}}}{I_{\text{total}}} \right)^2 \left(1 + \frac{F_T e^{-\tau/\tau_T}}{1 - F_T} \right) \qquad (6)$$

with F_T the fraction of molecules in the triplet state and τ_T the triplet relaxation time. The axial and equatorial axis of the detection volume element are determined by fitting the calibration measurements with rhodamine 6G ($D = 2.8 \times 10^{-10} \text{m}^2 \text{ s}^{-1}$) with Eq. (2). In the analysis of other data sets, these parameters are fixed. In our studies the contribution of a correlating background species could be neglected, but correction for the noncorrelating background, according to Eq. (2), is necessary, especially in the case of cowpea protoplasts (see section on autofluorescence).

Figure 4 gives an example of an FCS experiment of YFP microinjected into a cowpea protoplast. YFP is homogeneously distributed throughout the cytoplasm and nucleus but is not present in the vacuole, chloroplast, and other plastids (Fig. 4A). The detection volume is positioned in the cytoplasm, where an intensity scan along the optical (z axis) confirmed the correct positioning of the volume element along the z axis (Fig. 4B). The obtained correlation curves are analyzed according to Eq. (6) yielding information about the local YFP concentration and the diffusion rate. The translational diffusion constant of $4.1 \pm 2.1 \times 10^{-11}$ $\text{m}^2 \text{ s}^{-1}$($n = 20$) is about half the value obtained in buffer ($D = 9.0 \pm 0.6 \times 10^{-11}$ $\text{m}^2 \text{ s}^{-1}$), which can be explained by the higher viscosity inside the cell. No differences are found between CFP and YFP or between the two cell types used.

Figure 5A displays the localization of mitochondrial-targeted CFP expressed in BY2 protoplasts. Small oval-shaped structures with a diameter of ca. 2 μm can be observed in the confocal image. Colocalization with Mitotracker Red (Molecular Probes, Eugene, OR), a mitochondrial specific dye, confirmed the proper targeting of CFP to the mitochondria, as has been shown before by Köhler et al.[43] The correlation curves are acquired at several locations within the cell (Fig. 5B) and analyzed with several models since the 3D triplet model [Eq. (6)] could not accurately fit the experimental data (Fig. 5C). Due to the limited spatial resolution of confocal imaging it is not clear where in the mitochondria the dye is located. Köhler et al.[34] studied the diffusional characteristics of GFP in BY2 plastid tubules. Their experiments showed that besides the normal 3D diffusion, GFP is transported actively through the tubule. Fitting our results to models including active transport or 2D diffusion (in case of protein diffusion in a planar environment such as a membrane) did not give satisfying results. However, including the condition that the diffusion may not be Brownian, but is restricted due to interactions of the protein with other particles, resulted in acceptable fits (Fig. 5C). The model for fitting correlation

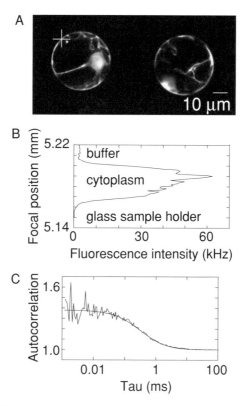

FIG. 4. FCS experiment of YFP microinjected into cowpea protoplasts. After localization of the protein with (A) confocal imaging and (B) intensity profiles along the optical axis, an FCS measurement was performed at the selected spot (cross on A) in the cytoplasm for 20 sec. (C) The experimental curve was fitted according to Eq. (6).

curves for anomalous diffusion[32,44] is similar to Eq. (2) with a slightly altered description of $Diff_i$:

$$Diff_{i,\mathrm{An}} = \left(\frac{1}{1 + (\tau/\tau_i)^\alpha} \frac{1}{\sqrt{1 + \frac{1}{(z_0/\omega_0)^2}(\tau/\tau_i)^\alpha}} \right) \qquad (7)$$

in which α is the coefficient that indicates the degree of restriction. A restriction coefficient of $\alpha = 1$ indicates normal Brownian motion but the smaller α will be, the more restricted the diffusion behavior is. Here a restriction coefficient of 0.51 ± 0.12 ($n = 30$) and a diffusion constant of $3.8 \pm 2.1 \times 10^{-13}$ m^2 s^{-1} is found. Because the diffusion is more than 200 times slower than CFP in the

[44] P. Schwille, J. Korlach, and W. W. Webb, *Cytometry* **36,** 176 (1999).

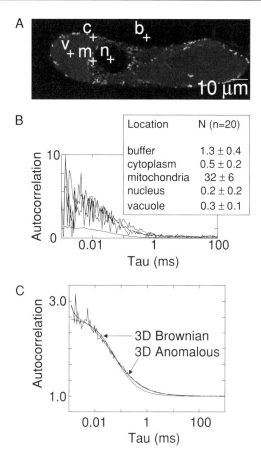

FIG. 5. FCS experiments of mitochondrial-targeted CFP, expressed in cowpea protoplasts. After localization of the protein with (A) confocal imaging, (B) FCS measurements of 60 sec were performed at the selected spots (crosses on A) at several subcellular locations. b, Buffer; c, cytoplasm; m, mitochondrium; n, nucleus; v, vacuole. Analysis of the corrected correlation curves confirmed that the protein was present only within the mitochondria. (C) The experimental curves acquired in the mitochondria were fitted (smooth curves) to several models including a 3D diffusion-triplet model and a 3D anomalous diffusion model.

cytoplasm and α is very small, it is likely that CFP is localized in an environment where diffusion is severely restricted, which can be explained by the high protein density in the mitochondrial matrix.

Data Analysis FCCS

For a correct analysis of FCCS experiments with the chimeric protein CFP-$(Ala)_{25}$-YFP (CAY), control measurements are performed in cells expressing CFP or YFP only. The molecular brightness values obtained (Table II) are used to correct for the cross-talk of the CFP in the YFP detection channel. In this experimental system, having covalently linked dyes, only one green and one red dye molecule is present per protein molecule. However, when a complex is studied in which more than one green or red dye is present, a correction factor should be included as has been described by Földes-Papp et al.[21] This correction can also take into account effects such as homoquenching or donor quenching of the fluorescence caused by FRET. Spectral imaging of CAY does not result in an enhanced YFP/CFP emission ratio compared to control experiments and therefore FRET between CFP and YFP does not take place in CAY, as can be expected with a long linker of 25 alanine residues between the two FPs.

The localization of CAY, transfected into BY2 protoplasts, is similar to the pattern observed in cells transfected with CFP or YFP only. The cross-correlation curves are fitted according to a one-component 3D diffusion-triplet model [Eq. (6)]. The diffusion constant of CAY in the cytoplasm, $3.7 \pm 1.8 \times 10^{-11}$ $m^2 s^{-1}(n = 30)$, corresponds to the value obtained for CFP or YFP alone. As discussed above, this similarity can be explained by the relative insensitivity of FCS to small mass differences. To determine the number of CAY proteins (N_{CAY}) the cross-correlation curve is fitted to

$$G_{D,cross}(0) = 1 + \left\{ N_{CAY} + \left[\frac{1}{G_{D,CFP}(0) - 1} \right] \left(\frac{\eta_{rgg}}{\eta_{rrr}} \right) \right\} / \left\{ \left[\frac{1}{G_{D,CFP}(0) - 1} \right] \left[\frac{1}{G_{D,YFP}(0) - 1} \right] \right\} \quad (8)$$

where $G_{D,CFP}(0)$ and $G_{D,YFP}(0)$ are the amplitudes of the autocorrelation curves for the CFP and YFP channels, respectively, The number of CAY proteins equals the number of molecules found in the CFP channel and (corrected) YFP channel, indicating that all the CFP and YFP molecules are present in the doubly labeled fusion protein and no free CFP or YFP is present.

Cells cotransfected with both CFP and YFP result in identical expression patterns as found for CAY (Fig. 6). When the cross-correlation curve is corrected for CFP cross-talk, no doubly labeled molecules could be found. This experiment shows that although a high amount of colocalized dyes is present, it does not automatically mean that molecular interactions are taking place, at least not on a time scale detectable by our setup (50 ns–10 hr).

Conclusions

The work presented here results from an ongoing effort from our and other laboratories to optimize fluorescence-based techniques such as FCS and FCCS for studying molecular dynamics and interactions in living (plant) cells. To date our

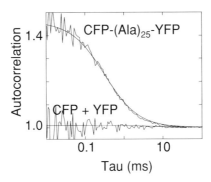

FIG. 6. FCCS experiment of a CFP–YFP fused protein, expressed in BY2 protoplasts. A 90-sec measurement was performed in the cytoplasm. Cross-correlation curves for a cell cotransfected with both CFP and YFP and for a cell transfected with the CFP-$(Ala)_{25}$-YFP construct. The correlation curves were fitted (smooth curves) according to Eq. (6), taking into account the cross-talk of the CFP fluorescence into the YFP detection channel. Fractions of interacting molecules were estimated using Eq. (5).

efforts have been centered on the application of fluorescent proteins, widely used to label the protein of interest by genetic engineering. Although the CFP–YFP pair used in these experiments is not an ideal couple for FCCS experiments, the potential of FCS to monitor interactions between biomolecules at physiological relevant concentrations has been demonstrated. However, to further develop this technique many aspects are subject to improvement. One example is the development of better dyes that have sharper, well-separated absorbance and emission spectra, are more photostable, and preferably are small in size.[45] Also much progress has been made in the field of data analysis, where single molecule approaches are integrated in the analysis to extract more information from the collected data, such as molecular brightness, fluorescence lifetime, and anisotropy.[46–49] Theoreticians are currently working to derive more complex equations required to describe the complicated environmental properties and dynamic processes occurring within the living cell. The ultimate goal in plant science would be to monitor molecular dynamics not only in single living cells but in whole plants. Scattering of the light, however, drastically reduces the signal-to-noise ratio in experiments using a setup as described here. However, Schwille et al.[29] showed that the use of a

[45] B. A. Griffin, S. R. Adams, and R. Y. Tsien, *Science* **281,** 269 (1998).
[46] Y. Chen, J. D. Muller, P. T. C. So, and E. Gratton, *Biophys. J.* **77,** 553 (1999).
[47] K. Palo, Ü. Mets, S. Jäger, P. Kask, and K. Gall, *Biophys. J.* **79,** 2858 (2000).
[48] J. Schaffer, A. Volkmer, C. Eggeling, V. Subramaniam, G. Striker, and C. A. M. Seidel, *J. Phys. Chem. A* **103,** 331 (1999).
[49] J. R. Fries, L. Brand, C. Eggeling, M. Köllner, and C. A. M. Seidel, *J. Phys. Chem. A* **102,** 6601 (1998).

two-photon excitation source could significantly reduce the scattering and therefore improve the signal quality sufficiently to enable FCS measurements in whole plants.

Acknowledgments

This work has been supported by grants from the Netherlands Council of Earth and Life Sciences (ALW-NWO), the Technology Foundation (STW-NWO), and the Mibiton Foundation. Dr. K. Shah, Department of Neurobiology, Harvard Medical School, is gratefully acknowledged for supplying the AtSERK1 construct.

[6] Building and Using Optical Traps to Study Properties of Molecular Motors

By Sarah E. Rice, Thomas J. Purcell, and James A. Spudich

Introduction

In the study of molecular motors, the experiments that are the most direct are also the most telling. The velocity, stall force, step size, and processive run length of molecular motors have all been determined using *in vitro* motility assays, both at the multiple and single molecule level. Optical trapping experiments in particular are used to gain information on stall forces, step sizes, and kinetics of a variety of molecular motors, such as kinesins,[1,2] myosins,[3–6] and processive DNA enzymes[7] (reviewed in Ref. 8). This information places constraints on models for the mechanisms of these molecular motors, and ultimately has led to a vastly increased understanding of how they work.

The barrier for most laboratories to optical trapping has generally been the high cost and difficulty of building, maintaining, and using the devices. However,

[1] K. Svoboda, C. F. Schmidt, B. J. Schnapp, and S. M. Block, *Nature* **365,** 721 (1993).

[2] K. Visscher, M. J. Schnitzer, and S. M. Block, *Nature* **400,** 184 (1999).

[3] J. T. Finer, R. M. Simmons, and J. A. Spudich, *Nature* **368,** 113 (1994).

[4] A. D. Mehta, R. S. Rock, M. Rief, J. A. Spudich, M. S. Mooseker, and R. E. Cheney, *Nature* **400,** 590 (1999).

[5] M. Rief, R. S. Rock, A. D. Mehta, M. S. Mooseker, R. E. Cheney, and J. A. Spudich, *Proc. Natl. Acad. Sci. U.S.A.* **97,** 9482 (2000).

[6] R. S. Rock, S. E. Rice, A. L. Wells, T. J. Purcell, J. A. Spudich, and H. L. Sweeney, *Proc. Natl. Acad. Sci. U.S.A.* **98,** 13655 (2001).

[7] M. D. Wang, M. J. Schnitzer, H. Yin, R. Landick, J. Gelles, and S. M. Block, *Science* **282,** 902 (1998).

[8] A. D. Mehta and J. A. Spudich, *Adv. Struct. Biol.* **5,** 229 (1999).

all of the components for an extremely sensitive optical trap are now commercially available. Optical trapping assays have now been performed on a variety of different systems, using experimental techniques that may apply well to new, uncharacterized molecular motors and other biological macromolecules as well. These developments in both the technology and the assays involved in optical trapping have made the technique more accessible, increasing the likelihood that the optical trap will become a standard microscope used in laboratories that study movement and force in biological systems. Here, we describe methods for using optical traps to study molecular motor proteins, with an emphasis on how to design both the instrument and the assays to characterize the motility of processive and nonprocessive molecular motors. We focus on the particular challenges of determining the maximum force and step size of myosin VI as a model system.

Optical Trapping Instrumentation

Optical Trap Design

At its most basic level, an optical trap consists of a laser that is strongly focused through a lens with a very short focal length, usually a high numerical aperture (NA) microscope objective. As the laser light penetrates an object centered at its focus, it is refracted through the object. Because light has momentum, the laser light exerts forces on the object as it bends through it. For spherical objects, such as the beads referred to in this article, the laser acts to draw the center of the object into its focus, and its force characteristics can be approximated as a linear spring in all directions. The spring constant is referred to as the trap stiffness, which is typically 0.02–0.06 pN/nm for the optical traps used in studying molecular motors although much stronger traps (up to 70 pN/nm) have been designed for some applications.[9] The key to designing an optical trap for measuring single displacements of molecular motors is generally not to design the trap with the highest stiffness possible. Rather, it is to design a detection system that can reliably detect single displacements of objects in the trap on the order of nanometers with millisecond time resolution. A simple and relatively inexpensive optical trap can be created by modifying a high-quality microscope with an epifluorescence port. The method for building such a trap has been completely described by Sterba and Sheetz,[10] and this trap is capable of picking up beads or cells and placing them where desired. The trap discussed in this article has enhanced positioning and detection capabilities over the system described by Sterba and Sheetz,[10] which give it the millisecond and nanometer resolution necessary to detect single motor steps.

[9] S. B. Smith, Y. Cui, and C. Bustamante, *Science* **271**, 795 (1996).
[10] R. E. Sterba and M. Sheetz, *in* "Laser Tweezers in Cell Biology" (M. Sheetz, ed.). Academic Press, San Diego, CA, 1998.

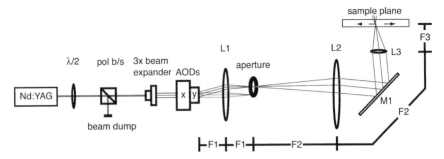

FIG. 1. Optical trap design. Components of the system will be described from left to right here and in the text. For the purposes of drawing ray optics of the laser light in this diagram, it is assumed that the lenses are thin. The Nd:YAG trapping laser (Coherent Compass 1064-2000) is shown on the left side, and the laser beam emitting from it is shown as a black line. Most of the power is split off by the λ/2 waveplate (λ/2) and polarizing beamsplitter (pol b/s) to a beam dump. After a 3× expansion, the beam is deflected by the orthogonal AODs. Only the original beam and one deflected beam are shown, although in reality, four different major beams come out of the pair of AODs. The deflected beams are selected by an aperture at the focal length of lens L1 (described in text). Lenses L1 and L2 serve as a second 3× telescope, which expands the beam sufficiently to fill the back aperture of the objective (L3). The laser light is reflected up to the objective by a dichroic mirror (M1), which is also used to adjust the position and tilt of the optical trap in the sample plane for a coarse alignment.

These modifications come at a significant price but expand the range of uses for the device.

Our present design is shown in Fig. 1. Our first design of a dual-beam optical trap used acousto-optic deflectors (AODs) to perform feedback on the position of the trap, but used mirror translations and separated beam paths for steering the traps.[3] Our present design uses a single set of two commercially available, orthogonal AODs to create multiple traps and to rapidly adjust their position in the sample plane.[3,11,12] This means that many functions of the trap, namely positioning, modulation of trap stiffness, making multiple traps, force clamping, and position clamping can now be quickly and easily controlled by changing the input signal to the AODs. This greatly simplifies the optical layout of the trap from previous designs, making it easier to build in a smaller space, while adding a large number of new capabilities to the instrument. The next section of the text goes through this trap design in detail, from the trapping laser to the sample plane (left to right in Fig. 1). Design considerations at all points are discussed.

Optical trapping systems for the study of molecular motors most commonly use Neodynium:YAG, Neodynium:YVO$_4$ (both 1064 nm), or Neodynium:YLF (1047 nm) trapping lasers. The ~1-μm wavelength of these lasers causes a fairly

[11] K. Visscher, S. P. Gross, and S. M. Block, IEEE J. Sel. Top. Quant. Electr. 2, 1066 (1996).
[12] J. E. Molloy, Methods Cell Biol. 55, 205 (1998).

minimal amount of photodamage to most biological samples.[13–15] These lasers are all diode-pumped solid-state lasers that have good pointing stability for nanometer-precision positioning, and have high power output levels (1–10 W). They can also be air cooled, or if they are water cooled, the diode pumping laser can be fiber coupled into the main cavity so that the cooling system can be placed away from the optical table for reduced noise. The system described here uses a Compass 1064-2000 CW laser (Coherent, Inc., Santa Clara, CA), which is air cooled and has an output of 2 W with a beam diameter of 1 mm.

The trapping laser must always be operated well above its lasing threshold for good beam stability, and trapping lasers have sufficiently high output power at their source to damage some optics. In this design (Fig. 1), the laser power is attenuated controllably by rotating a $\lambda/2$ waveplate to control the polarization of the input light and using a polarizing beam splitter to divert some of the laser power into a beam stop.

A 3× beam expander (T81-3×, Newport, Inc., Irvine, CA) is placed after the polarizing beam splitter to expand the 1-mm-diameter beam to 3 mm in diameter, which is close to the aperture size of the AODs (4 mm diameter). Equivalently, a pair of lenses can be placed at this position to magnify the beam instead of using a beam expander. Expanding the beam as much as possible here, before the AODs, gives the largest possible range of movement for the traps in the sample plane, as explained below. Referring specifically to Fig. 1, after deflection by the AODs the beam must be magnified to fill the back aperture of the objective lens (L3). Deflections of the beam in the plane of L1 by the AODs are demagnified in the sample plane by a factor of F_3/F_2. The range of beam deflection by the AODs in the sample plane is maximized if the expansion of the beam after deflection by the AODs (F_2/F_1) is minimized. A significant disadvantage to maximizing the range of travel of the AODs in the sample plane in this manner is that fluctuations in the position of the AODs due to noise can be magnified to an unacceptable level in the sample plane. In practice, the magnification of the beam before and after the AODs can be manipulated to achieve the maximal range of travel for the trap in the sample plane that still has an acceptable level of noise in the trap position.

The AODs are placed after the 3× beam expander. In an AOD, a crystalline material is bonded to a piezoelectric transducer that converts a high-frequency oscillating voltage (megahertz) into an acoustic wave that is propagated through the crystal. The acoustic wave has the same frequency as the applied voltage, and as it travels through the crystal it creates a standing wave pattern of compressions and expansions parallel to the transducer. This results in periodic changes in the index of refraction of the crystal that periodically shift the phase of the incoming

[13] K. M. Svoboda and S. M. Block, *Annu. Rev. Biophys. Biomol. Struct.* **23,** 247 (1994).
[14] M. W. Berns, J. R. Aist, W. H. Wright, and H. Liang, *Exp. Cell Res.* **198,** 375 (1992).
[15] K. C. Neumann, E. H. Chadd, G. F. Liou, K. Bergman, and S. M. Block, *Biophys. J.* **77,** 2856 (1999).

light. Diffraction off this phase grating causes the output beam to be deflected. If the frequency of the applied voltage is changed, the acoustic wave, the periodicity of the grating, and the angle of deflection of the light are all changed as well. The intensity profile of the diffraction pattern created by this grating depends on the incident angle of the laser light on the crystal, and the AODs are positioned in the optical path so that the first-order deflection through the grating has the maximum light intensity. Higher order deflections occur at large enough angles to be excluded from the light path using an aperture, to be discussed later. Practically speaking, each of the AODs used in this design (model #DTD274, IntraAction Corporation, Bellwood, IL) deflects a maximum of $\sim$75% of the incoming light in the first-order deflection, while some light remains undeflected and a small amount of the light is deflected at higher orders. This means that four zero-and first-order beams are emitted from the set of two orthogonally aligned (X and Y) AODs: one undeflected, one that is deflected only in X, one that is deflected only in Y, and one (maximally about 55% of the total incoming light) that is deflected in both X and Y. The latter is the selected trapping beam. Extremely stable optical mounts are needed for the orthogonal AODs, and 4-axis tilt aligners (model #9071, New Focus, San Jose, CA) work well to enable alignment of the AODs in the optical path and to stabilize them for nanometer-precision measurements. Methods for computer controlling the AODs will be discussed below.

Two lenses placed after the AODs serve as a telescope to expand the beam 3× to a total beam diameter of 9 mm, slightly overfilling the 7-mm-diameter back aperture of the objective. Slightly overfilling the back aperture is preferable to underfilling it. When the objective is underfilled, the incoming light does not focus as tightly, resulting in lower trap stiffness in all directions, but most detrimentally in the direction of the laser beam propagation. The first lens (L1 in Fig. 1) has a focal length of 50 mm, and an aperture at the focal point serves to select the beam that is deflected to first order in both X and Y. The maximum deflection angle by the IntraAction AODs used in this design in one direction is approximately 50 mrad, with a range of 25 mrad. This translates into a 2.5 mm transverse distance in the focal plane of L1 between the centers of each of the four zero- and first-order deflected beams, which will be four focused spots separated in position by a minimum of 1.25 mm in either X or Y. An aperture positioned in the focal plane of L1 at the average position of travel in both X and Y will accommodate the maximum range for first-order AOD deflections in both directions while eliminating the undeflected beam and the two beams deflected only in X and in Y, and higher order deflections. Because the second lens, L2, is placed after the aperture by 150 mm, its focal length, the light is collimated after L2 and is then reflected up to the sample plane through a distance of 150 mm, its focal length. Maintaining this separation between L2 and the focal plane of the objective ensures that the objective is always back-filled, even as the beam is deflected by the AODs in the plane of L2. A dichroic mirror (M1) that reflects wavelengths $>$1000 nm reflects the laser light up to the sample plane.

The microscope objectives used in optical trapping nearly always have a high numerical aperture (≥ 1.2, and most use 1.4) to focus the trapping laser very sharply, because the stiffness of the trap is ultimately determined by the gradient of the trapping laser intensity as it goes through the bead.[16] However, by setting up two counterpropagating traps with low-NA water immersion objectives an extremely stiff optical trapping system (70 pN/nm) can be created.[9] Using this counterpropagating type of setup, a very strong trap is created in the direction of the laser propagation. In single optical traps, such as that described here, the direction of propagation is weakest.

Recording Bead Position on Nanometer Length Scales and Millisecond Time Scales

The imaging system used to record the bead position in the optical trap is critical, as it must be able to make nanometer-precision measurements on millisecond or submillisecond time scales. Quadrant photodiode detectors (QPDs) are used for this purpose, because they are extremely sensitive to motion in two dimensions. The method that is focused on (Fig. 2) involves imaging the trapped bead or beads onto QPDs (S-1557, Hamamatsu, Inc., Bridgewater, NJ). Other nonimaging methods for detecting bead position on QPDs in the back-focal plane of the condenser have also been discussed.[1,11] The system described here (Fig. 2) uses a multimode fiber-coupled diode laser (800 nm, CW, 0.5 W from Laser Diode, Inc., Edison, NJ) operated below its lasing threshold for bright-field illumination of the sample. In general, lasers make poor bright-field light sources because the light coming from them is coherent and therefore interferes with itself after passing through the sample, giving a "speckle" pattern in the image plane. To eliminate this problem, the laser is coupled into a multimode fiber to scramble the phase of the light, and it is operated just below its lasing threshold, where the light emitted is bright enough for imaging but is not spatially coherent. The laser is used simply because it is an extremely bright light source and this greatly reduces the shot noise that results from random fluctuations in the number of photons hitting the QPD in a given amount of time, compared to conventional light sources. Light that refracts through the bead is deflected away from the QPD, leaving the image of a single dark silhouette on the QPD corresponding to the bead position. For maximum resolution of the bead position, the diameter of this silhouette should be on the order of the size of the QPD (1 mm, in the setup discussed here). This magnification is accomplished by using a lens with a very long focal length (750 mm), L4, at a distance from the objective which gives $\sim 600\times$ magnification on the QPD placed at its focal plane. In this setup, the QPD is sensitive to changes in the position of a centered 1-mm-diameter image on the order of 1 μm, which translates to nanometer-scale sensitivity to motions of the bead. A QPD detects motion in two dimensions by

[16] A. Ashkin, *Methods Cell Biol.* **55**, 1 (1998).

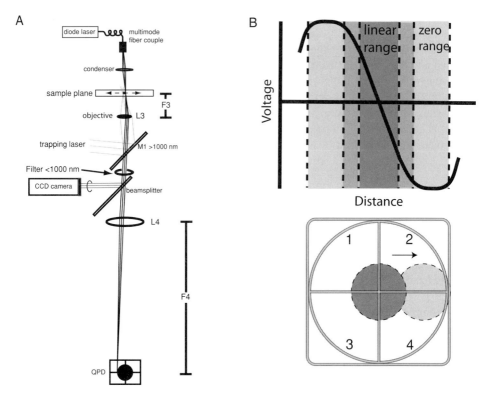

FIG. 2. Tracking of bead movements on a QPD detector. A setup for imaging the 1-μm bead onto the QPD detector is shown in (A). For simplicity the figure is drawn assuming that lenses are thin. A diode laser at 800 nm (Laser Diode, Inc. LCW-200F, 500 mW output) coupled into a multiple mode fiber and pumped just below its lasing threshold is used for a high brightness light source introduced through the condenser. The image of the bead passes through the objective and the dichroic mirror (M1). A <1000-nm filter blocks out any trapping laser light, which is back-reflected off of the water–glass interface in the sample. A beam splitter diverts a small amount of the incoming light to the CCD camera (Sony part #) for bright-field imaging. L4 magnifies the image of the bead onto the QPD, which is set up one focal length away from L4, for nanometer-precision positioning of the trapped bead. (B) The voltage response of the QPD detector as the image of a bead is dragged from left to right through its center. *Top:* Response of the QPD to movement of the bead, described in the text. *Bottom:* Silhouette of the bead on the QPD in the voltage ranges described in the top. The distance scales on the top and bottom are the same. In practice, the linear range of the QPD is about 300 nm.

comparing the voltage response of four different photodiodes to incident light. Referring to Fig. 2B, the outputs of the QPD compare the voltage of the left two versus the right two quadrants, $[(1 + 3) - (2 + 4)]/(1 + 2 + 3 + 4)$, and the top two versus the bottom two quadrants $[(1 + 2) - (3 + 4)]/(1 + 2 + 3 + 4)$. As a bead is moved from the center of the QPD to the right, the QPD output decreases linearly as the left–right signal $[(1 + 3) - (2 + 4)]/(1 + 2 + 3 + 4)$ decreases linearly. When the edge of the bead crosses the center of the detector, the QPD output does not change as the bead is moved, until the bead begins to leave the detector and the left–right signal begins to increase, bringing the QPD voltage–response back up until it reaches zero. In practice, the useful range of the QPD can be extended somewhat by fitting the QPD response to a cubic or higher order odd polynomial to extend the useful linear range of the QPD slightly (Fig. 2B).

The voltage outputs from the QPDs are recorded by computer for later analysis. High-speed data acquisition (DAQ) cards are commonly available and are capable of recording bead position at sufficient time resolutions (megahertz or above). Once recorded, the data can be viewed and analyzed by software packages such as LabView (National Instruments, Austin, TX), Matlab (MathWorks, Natick, MA), or Igor (WaveMetrics, Lake Oswego, OR).

It is advantageous to be able to use bright-field and fluorescence imaging simultaneously while the trap is being used. Many filaments used in optical trapping (i.e., actin, DNA) are not easily detected by light microscopy, and fluorescence imaging is therefore required. The proper choice of dichroic mirrors and placement of components enables the use of a fluorescence light source, a bright-field source, cameras for both types of illumination, and the QPD simultaneously. This has been described previously.[17]

Finding and positioning samples in the optical trap require coarse control of the x, y, and z positions of the microscope stage. This is typically accomplished using joystick-controlled motorized micrometers for x and y and an electrostrictive actuator (#ESA-CSA, Newport, Inc., Irvine, CA) for z. The range of the electrostrictive actuator corresponds to a change in depth of 30 μm for the fine focus adjustment. This is an acceptable working range, because most optical traps using high NA oil-immersion objectives cannot trap particles more than 30 μm in solution due to distortion of the trap by spherical aberrations in water. A three-axis piezoelectric stage can be used for extremely accurate positioning ($\sim$1 nm) if needed. Feedback control mechanisms on these stages have been used to greatly decrease stage drift and measure the positions of system components with less than nanometer accuracy.[18]

Optical Trap Alignment Procedure

A laser can be aligned through any optic by placing an aperture that is closed down to a pinhole in the beam path between the laser and the optic. The back reflection off the optic strikes the aperture and the lens is translated in x and y in the beam path until the back reflection forms a halo around the opening of the aperture. There is a second reflection of the laser light off the front of the lens, which is brought to the center of the aperture by tilting the optic in the plane

[17] A. D. Mehta, J. T. Finer, and J. A. Spudich, *Methods Cell Biol.* **55,** 47 (1998).
[18] W. Steffen, D. Smith, R. M. Simmons, and J. Sleep, *Proc. Natl. Acad. Sci. U.S.A.* **98,** 14949 (2001).

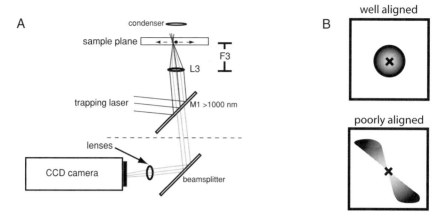

FIG. 3. Alignment of the optical trap. (A) Setup that enables alignment of the trap in the sample plane using the CCD camera to image the back reflection of the trapping light off the glass–water interface. Because the cutoff of the dichroic mirror is 1000 nm, a small amount of the trapping laser light (1064 nm) will be transmitted through it, and this can be imaged onto the CCD camera at high sensitivity. (B) CCD image of a well-aligned (*top*) and a poorly aligned (*bottom*) trap. The trap image appears as an *x* in the center of the field of view, and when the objective is focused in and out of the sample plane, the *x* is blurred out into a symmetric circle about its focal point (*top*). If focusing in and out causes the *x* to blur out to one side and then to the other (*bottom*), the trap is skewed relative to the sample plane and should be realigned by moving any two optical components in the system.

perpendicular to the table. This centers the optic precisely perpendicular to the laser. A modified version of this general technique is useful for checking and adjusting the alignment of the optical trap in the sample plane. In this procedure, the bright-field CCD camera is used to image the back reflection of the laser off the glass–water interface in a sample, enabling a coarse trap alignment without using an infrared viewer (Fig. 3). The setup described here uses a high-resolution Sony Iris CCD camera for bright-field visualization, configured as in Fig. 3A. For bright field illumination, a <1000-nm cutoff filter is placed in the light path to avoid imaging the small amount of light from the trapping laser on the CCD camera and QPD (Fig. 2).[17] For this alignment procedure, the filter is removed and the bright-field light source is blocked so the camera is sensitive to the small amount of trapping laser light back-reflected into it. The back-reflected image of the trap at the glass–water interface of the sample is in focus when the objective is focused near the coverslip surface. By focusing the objective up and down, one can see this image going into and out of focus. When the trap is properly aligned perpendicular to the back-focal plane of the objective, the out of focus image will center around the focal point of the trap. If the trap is significantly out of alignment, this image will drift at an angle to the focal point of the trap as the objective is adjusted in and out of focus (Fig. 2B). If the trap is out of alignment, the beam

can be aligned by systematically adjusting any pair of optics, such as the two 3×
telescope lenses L1 and L2, and testing the alignment as above in a "walk-in"
procedure.[10] An alternative method for performing the same alignment check is
to create a slide with an extremely high density of beads.[10] The traps pull beads
near the center in and push those just outside the trap away, and create a pattern
that should look like a symmetric "ripple" if the trap is centered perpendicular to
the sample plane. If the traps appear to pull beads from one side of the field and
push them toward the other side, the trap is skewed relative to the sample plane
and the same realignment procedure described above should be performed. The
trap stiffness and AODs must be recalibrated after alignment, as the optical path
is affected by this procedure.

Calibration of Trap Components: QPD, AODs, and Trap Stiffness

The deflection of the bead by a given input frequency on the AODs, the magnification of the bead on the QPD, and the trap stiffness must all be computed
before forces and step sizes of molecular motors can be measured in an optical
trap. The calibration procedure described here has four sequential steps. First, the
magnification of the video screen showing the image of the bead on the CCD
camera is found, using a calibration slide (5 or 10 μm scale) to create a scale for
the microscope on the video screen of the CCD camera. Second, the magnitude
of beam deflection by the AODs in the sample plane is quantified by measuring
distance deflected on the video screen versus AOD input signal. The magnification
of the bead on the QPD is found by deflecting the beam a known distance using
the AOD and measuring the QPD response through its entire range. Lastly, the
trap stiffness is calibrated.

The trap stiffness can be calibrated by a number of methods.[11,19,20] Two calibration methods are discussed here, and they are complementary because the
assumptions made by each method and the system parameters on which they
depend are different. The first method determines the trap stiffness by measuring
the position variance of a trapped bead over a long period of time, and the second
method determines stiffness by fitting the power spectrum of the Brownian motion
of the bead over a wide range of frequencies. The variance of the bead position
in one dimension in an optical trap over sufficiently long time scales is given by
$k_b T/\kappa$, where κ is the trap stiffness. This variance should depend only on the
temperature of the sample, assuming that the trap behaves like a linear spring in
any one dimension. This generally holds for the x and y dimensions of the bead's
movement in the trap, regardless of the distance from the bead to the coverslip
surface. However, because the variance of the bead position is the square of the

[19] F. Gittes and C. F. Schmidt, *Methods Cell Biol.* **55,** 129 (1998).
[20] K. Visscher and S. M. Block, *Methods Enzymol.* **298,** 460 (1998).

standard deviation of its movements, a small amount of drift or noise in the system can artificially increase the calculated variance of the bead position in the trap, resulting in an underestimate of the trap stiffness. The power spectrum method of calibrating trap stiffness can serve as a check to the variance method. This method relies on the fact that the motion of a trapped bead over long periods of time is controlled by the restoring force of the trap, which acts as a linear spring, and this motion is described by a single characteristic frequency. Motion over short times is random diffusion, which depends on the drag coefficient of the bead but not the trap stiffness, and is described by an infinite number of frequencies. A fit of the power spectrum of a bead is used to calculate its corner frequency, which is the frequency at which the transition between trap-dominated motion of the bead (at low frequencies) and diffusion-dominated motion (at high frequencies) occurs. The mathematical formulation for calculating the power spectrum is described in detail by Gittes and Schmidt[19] and Svoboda and Block.[13] The corner frequency of the bead (f_c) depends on both the trap stiffness and the drag coefficient of the bead. The power spectrum readily shows sources of noise at specific frequencies that are not related to bead motion (60 Hz is an obvious one that occurs on occasion) and low-frequency noise due to drift. These are assumed to be negligible in calculating the trap stiffness from the variance in the bead position. However, the power spectrum depends on the drag coefficient of the bead and, therefore, the distance from the bead to the coverslip surface, unless the bead is in bulk solution. Because the variance method and the power spectrum method of calibrating trap stiffness make completely different assumptions about the system, the use of both methods should correct for several common types of error in using either method alone. A well-calibrated system will yield the same result using both methods.

Computer Controlling the Optical Trap

The computer-controlled positioning of the trap by the AODs needs to be both very fast (≥ 10 kHz) and very accurate (≤ 1 nm) for effective feedback control of trap position or for modulating the frequency input to the AODs to make multiple traps.[12] Methods for controlling the AODs will be described starting with the frequency synthesizers that generate the input to the AODs, then discussing methods of creating multiple traps or controlling trap position using feedback.

For nanometer-precision positioning of the trap over the large ranges of AOD deflections in the design presented here, digital frequency synthesizers are needed to drive the AODs. A digital frequency synthesizer produces RF signals at 30–80 MHz that are amplified to drive the piezoelectric transducer on the AOD. Digital frequency synthesizers result in extremely accurate positioning of the trap because for a set digital input, the same digital signal is always synthesized, and this always results in the same deflection by the AOD to a level limited by bit noise. Bit noise is caused by fluctuations in the lowest increment by which the signal is digitized, an error that can be minimized by increasing the number of bits used to represent the signal. For a 13-bit digital frequency synthesizer, this noise results in an error of 0.012% ($1/2^{13}$) in the trap position, corresponding to 1.2 nm over the 10-μm range of movement for the AODs in this design. There are 24- and 32-bit digital frequency synthesizers available that would reduce this error to levels well below 1 nm. In contrast, analog frequency synthesizers have an error of roughly $\pm 0.25\%$ in converting the 0–1 V analog input into an RF signal, which is comparable to the sensitivity of most analog voltage measuring devices. A 0.25% error in the deflection of the laser by an AOD with a 10-μm range in the sample plane (as in the present design) corresponds to a position instability of 25 nm, unacceptable for most motor assays. However, if the range of trap movement in the sample plane by deflection of the AODs is decreased, this position instability is decreased as well.

The digital frequency synthesizers available today can update signals with speeds up to 100 kHz or higher, and this is useful for creating multiple traps by quickly rastering the position of the beam or for feedback control of the optical trap. In practice, the scanning rate of the QPD between trap positions has to be about an order of magnitude faster than the corner frequency of the bead (~100–1000 Hz) in order to create multiple traps by rastering the beam back and forth rapidly between two positions.[11,12,20] A scanning frequency of 10 kHz is sufficient for most applications, adjusting the position of the trap every 100 μs. Some degree of caution must be used in updating signals at high speeds because modulating the 17- to 33-MHz RF signal for driving a tellurium dioxide AOD (#DTD-274HA6, IntraAction Corporation, Bellwood, IL) at 10 kHz, for example, induces a carrier frequency of 10 kHz on that signal. An AOD driven in this manner deflects the laser as though it were a convolution of two different diffraction gratings, one from the megahertz input signal and one from the 10-kHz modulation.[12] This results in three different traps, the original first-order deflected beam, and two beams at a distance from the original beam corresponding to the 10-kHz carrier frequency (~6 nm in this setup). This effect is not critical at 10 kHz because the distance between the three beams is small compared to the size of the bead. At much higher frequencies, however, this phenomenon can create multiple weak "ghost" traps in the sample plane.[12] Power spectra can be taken at various driving frequencies with trapped beads of several sizes to determine the range of conditions under which the trap, or traps, act as a single linear spring with a linear restoring force.

There are two alternatives for creating multiple traps in this system that eliminate the above concerns about rastering the trap quickly between two positions. First, if two separate RF inputs from two digital frequency synthesizers are mixed prior to amplification and are used to drive a single piezoelectric transducer on the AOD, two different, independently controllable first-order deflections of the incident beam are produced and can be used for dual-beam trapping experiments. Another alternative for multiple trap experiments is to use the zero-order, undeflected beam from the AOD as a fixed trap and perform feedback using the beam deflected in both X and Y. The dual-beam trapping experiments described later are compatible with this type of system, but still use force feedback on the movable trap to gain high-resolution step size information.

Feedback Techniques for Improved Resolution of Molecular Motor Steps and Forces

Force feedback mechanisms are commonly implemented in optical traps to maintain relatively constant force on the trapped bead by maintaining a constant force between the bead center and the center of the trap for accurate step size measurements that are independent of linkage stiffness. This technique has proven a very powerful tool for studies involving molecular motors, because it enables unprecedented accuracy in measuring the velocity and step size of a molecular motor at constant load under a variety of conditions. The use of force feedback has expanded the traditional range of molecular motor properties measured in optical trapping experiments to include studies that determine the effect of load on the ATPase kinetics and stepping velocity of molecular motors much more precisely than previous experiments, [2,5,20–22] reviewed in Ref. 23.

Some optical trapping experiments have also used position clamping, in which the bead position is maintained as constant as possible to attempt to create iso-metric conditions.[3,24] In experiments on molecular motors, position feedback has been used for isometric force measurements over short distances (i.e., with non-processive motors, which take one power stroke and then diffuse away from the filament). In practice, however, position feedback simply creates a stiffer trap, not an isometric one. Because the trap is not completely isometric (particularly on fast time scales) and because of compliances in the linkages of the system, force measurements on nonprocessive motors are still very difficult.[8]

If either force or position feedback is desired, the response time of the computer program must be very fast, updating the trap position on submillisecond time scales. In general, faster updating of the trap position is better because events on shorter time scales can be resolved and the constant force or position conditions imposed by the feedback are subject to smaller fluctuations. Because of these considerations, a compiled low-level software routine is often used to perform feedback calculations for adjusting the trap position in response to the bead position. A force feedback routine must always begin with a calibration of both trap stiffness and the voltage response of the QPD to known distance deflections of the bead. A proportional in-tegral derivative (PID) method is then used to calculate the proper adjustment of the

[21] M. J. Schnitzer, K. Visscher, and S. M. Block, *Nature Cell Biol.* **2**, 718 (2000).
[22] C. Veigel, F. Wang, M. Bartoo, J. R. Sellers, and J. E. Molloy, *Nat. Cell Biol.* **4**, 59 (2001).
[23] A. D. Mehta, *J. Cell Sci.* **114**, 1981 (2001).
[24] R. M. Simmons, J. T. Finer, S. Chu, and J. A. Spudich, *Biophys. J.* **70**, 1813 (1996).

bead position so that force remains constant as the motor steps along the filament. A helpful tutorial for describing how PID controllers work is available on the Internet from Carnegie Mellon University and the University of Michigan (http://www.engin.umich.edu/group/ctm/PID/PID.html), and a good reference is Franklin *et al.*[25] To maximize the working range of the QPD, the feedback system holds the bead in place, imaged on one edge of the QPD. The motor steps against the trap until the distance from the center of the bead to the trap multiplied by the trap stiffness is equal to the desired load. Feedback then maintains the separation between the bead and the trap to keep the force constant. The motor takes steps under constant load until it reaches the other end of the working range of the QPD, about 300 nm away in the setup described here. The feedback system has a "rail" such that when the bead begins to exit the linear range of the QPD, the trap stops moving, the force increases as the motor steps, and eventually the motor stalls and returns to the center. In practice, the techniques described above for making multiple traps and performing feedback on traps are used when they are needed, and the type of experiment depends heavily on the properties of the motor to be studied. The next section describes actual techniques for performing optical trapping experiments on several types of motors, focusing on myosin VI as a particular example.

Optical Trapping Experiments on Molecular Motors

Measurable Properties and Limitations of Optical Trapping on Processive and Nonprocessive Motors

Optical trapping experiments have been performed on both processive motors, which undergo multiple productive catalytic cycles and associated mechanical steps per diffusional encounter with their filament, and nonprocessive motors, which bind, undergo at most one productive catalytic cycle associated with a single mechanical step, and then diffuse away from the filament. There are differences in the experimental geometries that can be used for processive and nonprocessive motors as well as the information gained from them. With both processive and nonprocessive motors, the dwell time (the average time that the motor spends attached to the filament per step) can accurately be determined. For processive motors, the transitions between successive steps can be used to determine dwell times. For nonprocessive motors, dwell times are measured by observing the sudden drop in variance of the bead position when the motor binds to the filament, causing a sudden increase in the stiffness of the "spring" composed of the bead–motor–filament linkages confining the bead to the trap.[26] Under certain experimental

[25] G. F. Franklin, J. D. Powell, and A. Emami-Naeini, "Feedback Control of Dynamic Systems." Addison-Wesley, New York, 1994.
[26] J. E. Molloy, J. E. Burns, J. Kendrick-Jones, R. T. Tregear, and D. C. White, *Nature* **378,** 209 (1995).

setups, a second method for measuring dwell times of nonprocessive motors using correlated thermal diffusion was described by Mehta *et al.*[27] For measuring step sizes of processive motors, the compliances in the system (the linkages between the beads, motors, and filaments, as well as compliances within the motor) must be measured and corrected for in a fixed trap.[1,13] Techniques for oscillating the bead have been used to measure step sizes independent of system compliances.[4,22] The most accurate way of measuring step sizes is using an optical force clamp,[2,5,6] which gives an error in step size measurements of <1 nm. The broad step size distributions of some processive motors in an optical force clamp, for instance, myosin VI,[6] reflect the range of step sizes taken by these motors. With nonprocessive motors, only a single binding event and corresponding fraction of a single catalytic cycle is observed. Due to Brownian motion, the position of the bead is not well known before the binding event, leading to a large spread of values in the measured step size. By collecting a large amount of data, the step size of a nonprocessive motor can be found accurately, but the error in the step size distribution for these motors largely reflects the Brownian motion of the bead, not the distribution of step sizes for the motor. Processive motors can walk several steps out of the center of a fixed trap, which causes the separation between the trap and the bead and, therefore, the load, to increase until they stall, making it possible to measure the stall force of these motors. The force by single nonprocessive motors is difficult to measure unless the trap is extremely stiff because of the short distances that nonprocessive motors travel out of the trap. The optical trap must therefore be nearly isometric to measure forces by single nonprocessive motors. Furthermore, the system must be set up for an isometric force measurement so that the bead-motor linkage is very stiff to prevent the motor from simply pulling out its linkage to the bead rather than pulling against the nearly isometric trap. Reported forces of single nonprocessive myosins span a wide range (1–7 pN),[3,26] suggesting large systematic error in the experiment, possibly due to variability in the bead–motor linkage stiffness.[8]

Assays with processive motors can be performed using a "single bead" geometry (Fig. 4A). In this setup, the motor is attached to a polystyrene bead that is held in an optical trap, and the filaments are stuck to the glass coverslip surface. Although this experimental geometry is very simple, involving only one optical trap, it is also somewhat restrictive. Motors that must turn around their filament to walk will not work well in optical trapping experiments using the single bead geometry, although this geometry might allow a few mechanical advances to take place before the turning would become a problem. The single bead geometry is also difficult to use for most optical trapping experiments with nonprocessive motors, although it was used to measure single displacements of a mutant form of the kinesin family member, NCD (nonclaret disjunctional), which had displacements

[27] A. D. Mehta, J. T. Finer, and J. A. Spudich, *Proc. Natl. Acad. Sci. U.S.A.* **94,** 7927 (1997).

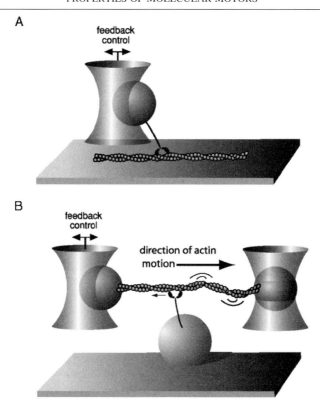

FIG. 4. Experimental setup for single bead and dumbbell experiments. (A) "Single bead" optical trap scheme for use with some processive motors. (B) "Dumbbell" optical trap scheme for use with nonprocessive motors, motors with short tethers between their motor domains and the bead linkage, and motors that must turn around the filament to walk. Force feedback measurements on only one bead in this orientation have been used to resolve step sizes of both myosins V and VI. [R. S. Rock, S. E. Rice, A. L. Wells, T. J. Purcell, J. A. Spudich, and H. L. Sweeney, *Proc. Natl. Acad. Sci. U.S.A.* **98,** 13655 (2001)]. In these assays, constant separation between the bead center and the trap center was maintained by force feedback on the left hand bead. The actin filament was slack, so that when the motors bound and advanced, tension built in the left-hand segment of the actin filament until the desired force was reached. The tension in the right-hand segment of the actin filament was zero, because the actin filament was slack and sufficiently flexible that the right-hand bead was not "pushed" out of its trap by the actin filament as the myosins pulled on the other bead.

in both directions along the microtubule.[28] In this case, very low ATP was used to increase the strongly bound state time for NCD to measure single displacements in a single bead assay.

For nonprocessive motors and motors that must twirl around their filament to walk, another experimental geometry is preferable. This setup, referred to

---

[28] S. A. Endow and H. Higuchi, *Nature* **406,** 913 (2000).

as the "dumbbell" geometry, involves trapping the filament track between two polystyrene beads that are held in two traps[3] (Fig. 4B). Alternatively, a dumbbell-type assay can be performed in which one of the two beads is permanently stuck to the surface. This "tethered particle" assay is useful for processive motors that turn around their filaments, in particular with RNA polymerase.[7] In a "dumbbell" experiment, the surface of the coverslip is coated sparsely with glass beads, which serve as surface platforms for the motors. Steps are measured as displacements of the filament "dumbbell" in the two optical traps. Although a dual trap setup is required in these assays, accurate step sizes can be measured using force feedback on only one of the two traps for actin-based motors.[6] A procedure for performing these assays on myosins V and VI is described below.

Single Bead Optical Trapping Assays on Processive Motors

Attaching Cytoskeletal Filaments to Glass. In the assays performed using the single bead geometry on myosin V, the coverslip is coated with neutravidin, and biotinylated actin filaments are then flowed in and attached to the surface. This procedure is described in detail elsewhere.[29] For assays on microtubule-based motors, polylysine-coated coverslips are used and microtubules adsorb nonspecifically to them.[2] Alternatively, axonemes, which are large bundles of 20 microtubules, can be adsorbed nonspecifically to glass for some assays.[30]

Coupling GFP-Motor Fusions to Polystyrene Beads in an Oriented Fashion. Fusions of many recombinantly expressed cytoskeletal motors with green fluorescent protein (GFP) have been created. The GFP is most often placed at the end of the coiled-coil region, where it should not interfere with the motile properties of the motor. GFP has been used successfully for a variety of both recombinant kinesins and myosins as a handle for coupling these motors to beads or to the coverslip surface in an oriented fashion.[6,30] If evidence of single molecule motility is seen by imaging the GFP by evanescent field or conventional microscopy, the same assay conditions and the same construct can be used for optical trapping on that motor.[6,30] A linkage designed to specifically couple a nonenzymatic portion of a motor protein to the bead has not been strictly needed to obtain results in optical trapping experiments. Squid axoplasm kinesin, for example, may bind preferentially to carboxylated beads with its coiled coil region stuck to the bead because of a positively charged region in its coiled coil. Squid kinesin nonspecifically adsorbed to beads in this manner has been successfully used in many experiments.[1,2] However, nonspecific adsorption of some motors in these assays can cause them to denature on the coverslip surface, or more dangerously, can adversely affect their activity without completely inactivating them. Therefore, a

[29] R. S. Rock, M. Rief, A. D. Mehta, and J. A. Spudich, *Methods* **22**, 373 (2000).
[30] K. S. Thorn, J. A. Ubersax, and R. D. Vale, *J. Cell Biol.* **151**, 1093 (2000).

specific linkage of a nonenzymatic portion of the motor to the bead or to the surface is preferred.

The protocol presented here involves nonspecifically attaching a monoclonal anti-GFP antibody to carboxylated 0.3- to 1-μm beads. It is also possible to covalently crosslink anti-GFP antibodies to carboxylated beads using an NHS ester and EDC to couple carboxylated beads to primary amine groups on polyclonal anti-GFP antibodies.[30] GFP-motor fusion proteins can then be adsorbed to the beads through the antibody in the proper orientation for motility. After attaching the anti-GFP antibody to the beads, they last $\sim$3 days at 4°. It is important, however, not to store the beads in reducing buffers as this degrades the antibody.

Attachment of Anti-GFP Antibodies to Microspheres

Buffers and Reagents

PBS: 100 mM phosphate-buffered saline

TMR–BSA: Bovine serum albumin labeled with tetramethylrhodamine isothiocyanate (Sigma, St. Louis, MO).

Beads: Carboxylated microspheres, 0.5–1 μm diameter (Polysciences, Warrington, PA)

α-GFP antibody: mouse monoclonal, 0.05 mg/ml in PBS (3E6 antibody, Qbiogene, Inc., Carlsbad, CA)

1. Dilute 10 μl of beads in 100 μl PBS and sediment at 20,000g for 1 min at 22°.
2. Resuspend beads in another 100 μl PBS and repeat step 1.
3. Resuspend in 20 μl α-GFP antibody +1 μl TMR-BSA. Incubate for 5 min at 22°.
4. Sediment beads at 20,000g for 1 min at 22°.
5. Resuspend in 20 μl PBS.
6. Repeat wash steps 4–5 three times.

Kinesins and myosins can be adsorbed onto these beads directly, and a 10 min incubation of motors and beads at 4° is sufficient.[6,30]

Bead Size Considerations

The size of beads is often important in optical trapping assays. Large beads ($\sim$1 μm) are easier to trap than smaller beads, and the trap stiffnesses achievable are higher for large beads. Smaller beads ($\sim$0.2–0.3 μm) can resolve faster events than large beads because their corner frequency is higher by a factor proportional to their radius. Small beads have a decreased stiffness in a 1064-nm optical trap relative to beads that are the same size as the wavelength of the trapping laser,[24] but this generally has not led to decreased resolution of steps relative to large beads.[2,5]

Experiments such as those measuring substeps of molecular motors, which can occur on submillisecond time scales, require either small beads,[31] or a very stiff trap since the corner frequency of the bead is also proportional to the trap stiffness. One consideration is that 0.5-μm beads are approximately the minimum size that can be sedimented in a tabletop centrifuge, and techniques for coupling beads to motors most often involve centrifugation. Beads smaller than 0.5 μm can be either column purified or sedimented in an ultracentrifuge. The protocols above are the same for 0.5-μm beads as for 1-μm beads because a large excess of antibody is used to coat beads. In attaching motors to small beads, however, a higher concentration of motor relative to the volume of the beads might be needed because the solvent accessible surface area will be larger relative to the volume of small beads than large beads.

"Dumbbell" Assay for Nonprocessive Motors or Motors That Must Turn Around Filaments

Unlike the single-bead trapping geometry, the dumbbell geometry shown in Fig. 4B can be used for both processive and nonprocessive motors, as well as for motors that turn around their filament. This geometry is also useful for motors with a short linkage between their motor domains and the bead attachment because the trapped beads are held about 1 μm farther from the surface, making it easier to connect active motors to their filaments without holding the bead too close to the surface. In the dumbbell geometry, the filament is held between two 1-μm polystyrene beads. The coverslip surface is coated with 1.5-μm glass beads, which serve as platforms for motors. Therefore, the trapped beads are approximately 1 μm away from the coverslip surface in this assay. As the motor steps, the filament slides and the motor pulls against the trapped bead at one end of the filament while pushing the filament toward the other trapped bead. In myosin-based assays, biotinylated actin filaments are coupled to neutravidin-coated polystyrene beads for creating the dumbbell.[29] A similar strategy was used to couple a biotinylated microtubule to streptavidin beads for optical trapping assays on the nonprocessive kinesin family member, NCD.[32] The assay described here was performed on recombinant fusions of GFP with both myosins V and VI.[6]

Buffers and Reagents

BEADED COVERSLIPS. Coverslips (18 mm^2) are coated with a monolayer of 1-μm polystyrene beads in 0.1% Triton X-100. Beads are sonicated for 30 sec prior to use in a bath sonicator. A good density of surface platforms on slides is achieved by using a solution of sonicated beads having an optical density of 0.15 at 595 nm. Coverslips are dipped in a 0.1% solution of nitrocellulose in isoamyl acetate and dried prior to use.

[31] M. Nishiyama, E. Muto, Y. Inoue, T. Yanagida, and H. Higuchi, Nat. Cell Biol. 3, 425 (2001).
[32] M. deCastro, R. Fondecave, L. Clarke, C. Schmidt, and R. Stewart, Nat. Cell Biol. 2, 724 (2000).

α-GFP antibody: mouse monoclonal 3E6 from Qbiogene, Inc., 0.05 mg/ml in PBS

AB: 25 mM imidazole hydrochloride, pH 7.4, 25 mM KCl, 5 μm calmodulin, 1 mM EGTA, 10 mM dithiothreitol (DTT), 4 mM MgCl$_2$, 2 mM ATP

ABSA: AB + 1 mg/ml BSA

O$_2$ Scavengers: 25 μg/ml glucose oxidase, 45 μg/ml catalase; this should be stored in PBS+ 50% (v/v) glycerol at $-20°$; 0.1% glucose is added to this mixture for the assay

NAV Beads: 1-μm biotinylated polystyrene beads, coated with neutravidin and thoroughly washed in ABSA[29]

BIOTIN–ACTIN FILAMENTS. Biotinylated actin[29] 50 μM was thawed, sonicated in a bath sonicator for 30 sec, and placed at 4° for 1 hr to depolymerize any actin complexes present in the solution. Then 50 mM KCl and 2 mM ATP was added, and the actin was polymerized for 1 hr at 4°. This results in slow polymerization of the actin into very long (many $\geq$10 μm) filaments. Stoichiometric ($\sim$50 μM) TMR–phalloidin (Molecular Probes, Eugene, OR) was added prior to use. Actin filaments can be used for up to 1 week, stored at 4°.

1. Make a flow cell using a beaded coverslip and double-stick tape.[29]
2. Coat the coverslip surface with α-GFP antibodies. Wait for 4 min.
3. Wash the slide with 10 μl ABSA buffer.
4. Flow in motor, at a dilution that should result in single molecule motility. For myosin VI, the concentration of motor used was $\leq$1 nM, but it varies depending on the motor and the type of linkage. This is discussed further below. Wait another 4 min.
5. Wash slide with 10 μl ABSA buffer.
6. Flow in a mixture of O$_2$ scavengers, ATP ($\sim$nM–mM concentrations), 1 : 300 NAV beads, and 1 : 4000 TMR-actin in AB.
7. Coat the edges of the flow cell with vacuum grease to prevent evaporation during the assay. The sample is ready to image.

Performing Force Feedback Measurements on Processive Motors in Dumbbell Geometry

In the geometry shown in Fig. 4B, force feedback can be performed on just one of the beads to make very accurate step size measurements if the filament track for the motor is relatively flexible, like actin. It is also possible to program force feedback for this type of system on both of the two beads, which is necessary for microtubule-based assays, in which the filament is very stiff. To perform force feedback measurements using two traps, the stiffness of both traps must be known, so that as the motor pulls one bead and pushes the other, constant force on both beads and a constant distance between them can be maintained. In actin-based

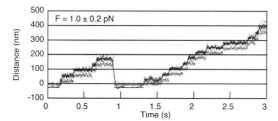 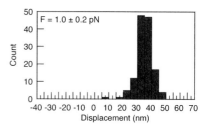

FIG. 5. Step size distributions and sample traces of GFP–myosin V stepping in the dumbbell trapping geometry. Data were taken at 2 mM ATP and 1 pN of force. Recombinant GFP–myosin V velocities were approximately 220 nm/sec with a step size of 35 ± 6 nm in dumbbell assays [R. S. Rock, S. E. Rice, A. L. Wells, T. J. Purcell, J. A. Spudich, and H. L. Sweeney, *Proc. Natl. Acad. Sci. U.S.A.* **98**, 13655 (2001)], consistent with previous results in single bead assays using tissue-purified myosin V [M. Rief, R. S. Rock, A. D. Mehta, M. S. Mooseker, R. E. Cheney, and J. A. Spudich, *Proc. Natl. Acad. Sci. U.S.A.* **97**, 9482 (2000)].

assays, the filament is quite flexible so the tension borne by the motor due to the bead that is being "pushed" (corresponding to the right bead in Fig. 4B) is negligible over short distances. While stepping, the myosin will pull itself along the actin toward one of the two beads (the bead in the left-hand trap in Fig. 4B) and push the actin out toward the other. The bead that the myosin is pulling against is used for force feedback. The actin filament is held with a slight amount of slack in it, so that the right-hand bead does not move throughout the entire feedback range of the left-hand bead (∼300 nm, in the setup described here). Therefore, the right-hand bead has no effect on the other components of the system, and force-feedback stepping measurements can be conducted using only the left-hand bead. Figure 5 shows the step size distribution and stepping traces using feedback on one trap for the dumbbell geometry using myosin V-GFP.[6] The step size and the distribution width (35 ± 6 nm) agree with previous measurements,[5,33] indicating that there are no systematic errors resulting from using this experimental geometry.

Movement by Single versus Multiple Motors

Recognizing single molecule motility is critically important when assessing the force, step size, and processivity of motors in an optical trap, because multiple motors creating the movement in an optical trap can lead to misinterpretation of the data and erroneous conclusions regarding all of these parameters. It can be somewhat difficult to determine what concentration of motors gives single molecule movement in the above assays because the number of motors per bead depends on a variety of factors such as the quality of the antibody and how much of the antibody adsorbs to the beads. The most reliable way to determine whether a new

[33] M. L. Walker, S. A. Burgess, J. R. Sellers, F. Wang, J. A. I. Hammer, J. Trinick, and P. Knight, *Nature* **405**, 804 (2000).

motor is processive is to begin at a high motor concentration and successively decrease to the single molecule level. Whenever possible, the motility of the motor should be characterized in both conventional microscope assays, which can give evidence of whether a motor is processive or nonprocessive, and in evanescent field microscope assays, which provide the most direct and stringent measure of processivity possible. These assays are summarized in Rock *et al.*[29] Briefly, the conventional microscope assays characterize the motion of motors at high concentration and gradually work down in concentration to the single molecule level. By using Poisson statistics, the number of molecules needed to sustain movement of filaments can be found. In combining all of these microscopy techniques to determine motor processivity, the GFP linkage to anti-GFP antibodies described here is beneficial because it can be used in conventional microscope-based assays to determine motor densities that should give single molecule motility. These conditions can then be duplicated in an optical trap, and the GFP can also be used under similar conditions without antibodies to image movements of processive motors directly in an evanescent field microscope.[6]

Conclusion

Optical trapping experiments very directly measure the velocity, force, step size, and processive run length of molecular motors. Force feedback techniques have enhanced the capabilities of optical traps to include extremely high-resolution determination of the step sizes of molecular motors, as well as the effect of load on their ATPase kinetics and stepping velocity. The ability to control the stiffness of the trap, positioning, making multiple traps, force feedback, and position feedback using AODs has greatly simplified the optical layout of the trap from previous designs and has added new capabilities to the instrument. Now, after nearly 10 years of using optical trapping to directly measure properties of molecular motors, it is difficult to imagine what the state of knowledge of these systems would be without these assays. The ever-increasing range of capabilities of optical traps and experiments that are performed using them continues to make optical trapping one of the most powerful techniques for elucidating the mechanisms of molecular motors and other biological systems in which force and movement are critical properties.

Acknowledgments

We thank D. Altman, C. Asbury, G. S. Lakshmikanth, E. Landahl, M. Lang, A. Mehta, Z. Oekten, R. Rock, and J. Shaevitz for helpful comments on the manuscript. R. Rock was instrumental in developing the assays described in the text. S. E. R. is supported by a Walter V. and Idun Berry fellowship. J.A.S. is supported by a grant from the National Institutes of Health. Portions of this manuscript (Fig. 4B and Fig. 5) are reproduced with permission from Rock *et al.*[6]

[7] Optical-Trap Force Transducer That Operates by Direct Measurement of Light Momentum

By STEVEN B. SMITH, YUJIA CUI, and CARLOS BUSTAMANTE

Introduction

"Optical tweezers" is a name given by Arthur Ashkin and colleagues to a device they invented, which uses light pressure to manipulate tiny objects. By focusing a laser beam through a microscope objective, they found that particles with high indexes of refraction, such as glass, plastic, or oil droplets, were attracted to intense regions in the beam and could be held permanently at a focal point.[1,2] Optical tweezers are useful in molecular and cell biology because several important forces are in an accessible piconewton (pN) range; e.g., ligand–receptor binding, DNA stretching, protein unfolding, and molecular motor stall forces.

Several good reviews cover the physics of optical traps,[3,4] but briefly their operation can be explained in either of two ways. In the first, light impinging on the particle is seen to be refracted or reflected by that particle. Because light photons carry a momentum $P = \hbar \mathbf{k}$ (where $\hbar$ is Plank's constant and $\mathbf{k}$ is the wave vector), the particle feels a reaction impulse that is equal but opposite to the change in the photon's momentum. If a particle acts as a positive lens and refracts light photons in a direction toward the object's center, then that object will become entrained or trapped in a light beam, especially if the beam has a narrow waist or focal point. The second explanation applies to particles that are much smaller than a wavelength of light, and can thus be treated as a Rayleigh scatterer possessing a polarizability, α. The electric field E from a light source induces a dipole moment αE in the particle, which experiences a force $\mathbf{F} = \alpha/2 \, \nabla(\mathbf{E} \cdot \mathbf{E}^*)$ attracting it to the focus of the light. Because α is proportional to the particle volume, the force holding the particle in the trap is proportional to the particle size, as well as the beam intensity gradient.

Methods to measure such forces in the optical trap have been under continuous development. Most force-measurement methods treat the optical trap as a harmonic potential well or "virtual spring" that pulls the bead toward the trap center. By measuring the displacement Δx of the particle within the trap and estimating a spring constant κ, the force is then given by $F = \kappa \, \Delta x$. The position of a spherical

[1] A. Ashkin, *Phys. Rev. Lett.* **24**, 156 (1970).

[2] A. Ashkin, J. Dziedzic, J. Bjorkholm, and S. Chu, *Opt. Lett.* **11**, 288 (1986).

[3] K. Svoboda and S. M. Block, *Annu. Rev. Biophys. Biomol. Struct.* **23**, 247 (1994).

[4] A. Ashkin, *in* "Methods in Cell Biology" (M. P. Sheetz, ed.), Vol. 55, p. 1. Academic Press, San Diego, CA, 1998.

bead inside a trap can be measured with subnanometer precision by one of several optical methods, and so this technique has proven very effective.[5-7]

Unfortunately, calibration of the virtual spring suffers from technical issues that complicate practical measurements. The particle-displacement sensor is often calibrated by translating a particle, fixed to a coverslip, by some known distance while measuring the response of a photodetector. However, such calibration applies only to other particles having the same size, shape, and orientation. Changes in the axial position (focal depth) of the particle can also affect the sensitivity of the position-detector. Calibration of trap stiffness κ is usually done with a test force, either an externally applied drag force or thermal Brownian force. However, both the Stokes law drag and thermal corner-frequency methods require knowledge of the object's size and shape as well as the local fluid viscosity in order to obtain the particle's drag coefficient.[3] If the trapped particle is near a stationary surface, like a coverslip or cell wall, the particle's drag coefficient increases.[8] Alternately, to estimate κ without using a drag coefficient, the equipartition theorem can be used to equate the particle's potential and thermal energies.

$$\tfrac{1}{2}\kappa < \Delta x^2 > = \tfrac{1}{2}k_B T \tag{1}$$

This method requires accurate distance measurement at high bandwidth along an axis corresponding to a single degree of freedom. If the particle's axial position or angular orientation couples to the transverse distance detector, then more than one degree of freedom will be sampled and κ will be underestimated. This method is also susceptible to instrumental noise and bandwidth limitations.[9]

Calibration problems can also arise when different particles are introduced into the trap or when optical conditions change. Then the shape of the potential well changes and κ must be re-calibrated for the new conditions. The trap stiffness is sensitive to the focal spot size and this, in turn, is affected by the spherical aberration of the objective lens. This aberration depends on the depth of the sample below the coverslip, so a calibration for κ performed 5 μm below the coverglass is invalid for experiments done at 10 μm depth (except for water-immersion lenses). Although early studies have shown that arbitrary biological objects could be manipulated inside cells,[10,11] it has been much more difficult to estimate the forces generated

[5] R. M. Simmons, J. T. Finer, H. M. Warrick, B. Kralik, S. Chu, and J. A. Spudich, in "The Mechanism of Myofilament Sliding in Muscle Contraction" (H. Sugi and G. H. Pollack, eds.), p. 331. Plenum Press, New York, 1993.

[6] K. Svoboda, C. F. Schmidt, B. J. Schnapp, and S. M. Block, *Nature* **365**, 721 (1993).

[7] H. Yin, M. D. Wang, K. Svoboda, R. Landick, S. Block, and J. Gelles, *Science* **270**, 1653 (1995).

[8] J. Happel and H. Brenner, "Low Reynolds Number Hydrodynamics," 2nd Ed. Kluwer Academic, Dordecht, The Netherlands, 1991.

[9] R. M. Simmons, J. T. Finer, S. Chu, and J. A. Spudich, *Biophys. J.* **70**, 1813 (1996).

[10] A. Ashkin, J. M. Dziedzic, and T. Yamane, *Nature* **330**, 769 (1987).

[11] M. W. Berns, W. H. Wright, B. J. Tromberg, G. A. Profeta, J. J. Andrews, and R. J. Walker, *Proc. Natl. Acad. Sci. U.S.A.* **86**, 4539 (1989).

by those objects because the stiffness of the virtual spring depends on the particle/ light field interaction. This interaction, in turn, depends on the size and shape of the organelle, its index of refraction, and the refractive index of the cell's cytoplasm. Calibration is straightforward for plastic beads of reproducible size in a simple environment, but there is no practical way to transfer those beads (or their calibration) into the complex environment of the cell's interior. Therefore most molecular motor experiments are currently being done *in vitro*.

Here we present a new[12] force-measurement method, based on an opposed-beam optical tweezers design, that overcomes some of these limitations. By measuring the angular intensity distribution of the laser light as it enters and leaves the trap, it is possible to determine the change in the momentum flux of the light beam, which, in turn, is strictly equal to the externally applied force on the particle. The force calibration now becomes independent of the particle's size, shape, or refractive index, its distance beyond the coverslip (spherical aberration), the viscosity or refractive index of the buffer, and variations in laser power.

A light-momentum force sensor calibrated from first principles (conservation of linear momentum) was thus constructed and has been used in the following structural and enzyme studies: Stretching single DNA molecules,[13] condensation of single DNA molecules,[14] unfolding and refolding single titin molecules,[15] mechanics of single RecA/DNA fibers,[16] pulling single chromatin fibers,[17] DNA polymerase activity versus template tension,[18] unfolding and refolding RNA structures,[19] and phage packaging motor activity versus force.[20]

Force Sensor Theory

A device to measure light-momentum force is outlined in Fig. 1. Here light is shown entering from the left and being focused by an objective lens into a spot where a particle is trapped. Light that exits the trap toward the right is collected by a similar objective lens. If an external force is applied to the particle, the light is refracted asymmetrically by the particle and exits the trap with a modified angular distribution. A position-sensitive photodetector (far right) measures the power and offset of this light to infer the external force.

---

[12] S. B. Smith, Doctoral Thesis, University of Twente, The Netherlands, 1998.

[13] S. B. Smith, Y. Cui, and C. Bustamante, *Science* **271**, 795 (1996).

[14] C. G. Baumann, V. Bloomfield, S. B. Smith, C. Bustamante, M. Wang, and S. Block, *Biophys. J.* **78**, 1965 (2000).

[15] M. S. Z. Kellermayer, S. B. Smith, H. L. Granzier, and C. Bustamante, *Science* **276**, 1112 (1997).

[16] M. Hegner, S. B. Smith, and C. Bustamante, *Proc. Natl. Acad. Sci. U.S.A.* **96**, 10109 (1999).

[17] Y. Cui and C. Bustamante, *Proc. Natl. Acad. Sci. U.S.A.* **97**, 127 (2000).

[18] G. Wuite, S. B. Smith, M. Young, D. Keller, and C. Bustamante, *Nature* **404**, 103 (2000).

[19] J. Liphardt, B. Onoa, S. B. Smith, I. Tinoco, Jr., and C. Bustamante, *Science* **292**, 733 (2001).

[20] D. E. Smith, S. Tans, S. B. Smith, S. Grimes, D. L. Anderson, and C. Bustamante, *Nature* **413**, 748 (2001).

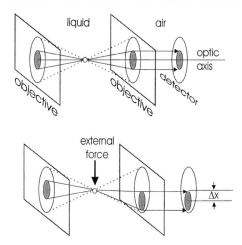

FIG. 1. (Top) Dashed lines indicate the outermost rays collected by an objective lens, defining the numerical aperture of the lens. The inner cone (solid lines) encloses the laser beam that enters from the left, passes through the bead, is imaged by the right objective, and exits. (Bottom) Application of an external force to the bead will cause it to equilibrate slightly off center in the trap so that light pressure from the deflected beam exactly balances the external force. The angular deflection, θ, of a ray leaving the trap is transformed by the right objective into an offset distance, ΔX, such that $\Delta X/R_L = n_1 \sin(\theta)$ where n_1 is the refractive index of the liquid and R_L is the focal length of the lens. The transverse light force F felt by a bead as it deflects a light ray of intensity W through an angle θ is given by $F = (n_1 W/c)\sin(\theta)$ where c is the speed of light. Therefore $F = (W/c)(\Delta X/R_L)$.

Consider light that is transmitted through a transparent liquid of refractive index n_1 and that interacts with an object immersed in that liquid. A light wave carries with it a momentum flux given by[21]

$$d(d\mathbf{P}/dt) = (n_1/c)\mathbf{S}\,dA \tag{2}$$

where $\mathbf{S}$ is Poynting's vector, c is the speed of light, and dA is an element of area normal to $\mathbf{S}$. The light force on the object is the difference in flux of the momentum entering ($\mathbf{P}_{\text{in}}$) and leaving ($\mathbf{P}_{\text{out}}$) the vicinity of the object. The force can be obtained by integrating the light intensity entering ($\mathbf{S}_{\text{in}}$) and leaving ($\mathbf{S}_{\text{out}}$) through a surface surrounding the object, provided elements of that surface (dA) are everywhere normal to $\mathbf{S}$

$$\mathbf{F} = d\mathbf{P}_{\text{in}}/dt - d\mathbf{P}_{\text{out}}/dt = (n_1/c) \iint (\mathbf{S}_{\text{in}} - \mathbf{S}_{\text{out}})dA \tag{3}$$

For an optical trap, integration can be performed over a distant ($R \gg \lambda$) spherical surface centered on the focal point. Here $\mathbf{S}_{\text{in}}$ is normal to the surface because

[21] J. P. Gordon, *Phys. Rev. A* **8**, 14 (1973).

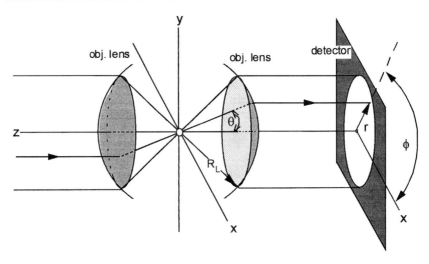

FIG. 2. Coordinate system for optical trap. Light enters from left and is focused to a spot where trapped bead is shown. A second lens, of focal length R_L, converts the exiting light to a parallel beam, which then falls on a photodetector.

the incoming wave is spherical, and $\mathbf{S}_{\text{out}}$ is normal because it emanates from a point (or object) at the trap. Because the radius of the sphere, R, is arbitrary, we can define an angular intensity distribution for light entering or leaving the focus, $I(\theta, \phi)\hat{\mathbf{r}}d\gamma = \mathbf{S}dA$ where θ and ϕ are angles shown in Fig. 2, $\hat{\mathbf{r}}$ is a unit vector from the focus, and $d\gamma = dA/R^2 = d\theta \sin\theta d\phi$ is an element of the solid angle.

Although it is difficult to *predict* $I(\theta, \phi)$ for light scattered from an arbitrary object trapped at a focus, it is not difficult to measure $I(\theta, \phi)$. Once this task is accomplished, the force exerted by light on the object can be computed as

$$\mathbf{F} = (n_1/4\pi c) \iint I(\theta, \phi)(\hat{\mathbf{i}} \sin\theta \cos\phi + \hat{\mathbf{j}} \sin\theta \sin\phi + \hat{\mathbf{k}} \cos\theta)d\gamma \qquad (4)$$

Here $I(\theta, \phi)$ is a radiant intensity, measured in watts/steradian, which is considered negative for rays entering the trap and positive for rays leaving it. If the trap is empty (no particle there to deflect the rays) then $I(\theta, \phi) = -I(-\theta, -\phi)$ and the integral over all angles equals zero.

A convenient way to measure $I(\theta, \phi)$ is afforded by a version of the Abbe sine condition,[22] which states that any ray emanating from the principal focus of a coma-free objective lens, inclined at an angle θ to the optic axis but still hitting

---

[22] M. Pluta, "Advanced Light Microscopy," Vol. 1, p. 163, Eq. 2.34c. Elsevier, Amsterdam, 1988.

the lens, will exit the image-side principal plane of that lens at a radial distance r from the optic axis given by

$$r = R_L n_1 \sin \theta_1 \tag{5}$$

Here n_1 is the refractive index of the liquid on the object side of the lens and R_L is an effective radius for the lens equal to its focal length. Although Fig. 2 fails to show the coverslip and air interfaces, which typically intervene along the path from focus to lens, the quantity $n_1 \sin \theta_1 = n_{Glass} \sin \theta_{Glass} = n_{Air} \sin \theta_{Air}$ is invariant (by Snell's law) for a ray traversing such flat boundaries so that Eq. (5) holds true regardless of such changes in media.

If the rays exiting the trap in a small element of solid angle $d\gamma/4\pi$ are projected without loss onto an area element $dA'(= r d\phi dr)$ on the image-side principal plane of the lens, then, by energy conservation, the irradiance E (in watts/m^2) on dA' is given by $E(r, \phi)dA' = I(\theta, \phi)d\gamma/4\pi$. If the lenses intercept all the light exiting the trap, then the expression for the force can be written as:

$$\mathbf{F} = \frac{1}{c} \iint E(r, \phi)\left(\hat{\mathbf{i}}\frac{r}{R_L}\cos\phi + \hat{\mathbf{j}}\frac{r}{R_L}\sin\phi + \hat{\mathbf{k}}\sqrt{n_1^2 - \frac{r^2}{R_L^2}}\right) r\,d\phi\,dr \tag{6}$$

where the integral is now taken over the surface of the image-side principal planes of the objectives.

The transverse ($\hat{\mathbf{i}}$ and $\hat{\mathbf{j}}$) components of the force can be integrated by placing position-sensitive photodetectors at those principal planes. A dual-axis detector of that type gives two difference signals, D_x and D_y, each proportional to the silicon detector's responsivity Ψ, and to the sum of local irradiances $E(x, y)$ weighted by their relative distances x/R_D or y/R_D from the detector center, where R_D is the detector's half-width. The light position signals are given by the following expressions:

$$D_x = \Psi \iint E(x, y)(x/R_D)dA' = \Psi \iint E(r, \phi)(r\cos\phi/R_D)dA'$$

$$D_y = \Psi \iint E(x, y)(y/R_D)dA' = \Psi \iint E(r, \phi)(r\sin\phi/R_D)dA' \tag{7}$$

where the integrals are over the surface of the detector. By combining Eqs. (6) and (7), expressions for the two components of force transverse to the optic axis, F_x and F_y can be obtained in terms of the detector signals and known constants.

$$F_x = \frac{D_x R_D}{c\Psi R_L} \qquad F_y = \frac{D_y R_D}{c\Psi R_L} \tag{8}$$

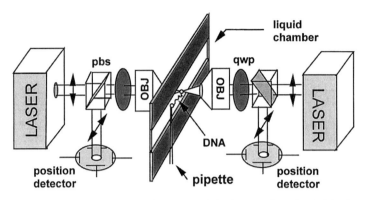

FIG. 3. Two diode laser beams with vertical linear polarization pass through polarizing beam splitters (pbs), quarter-wave plates (qwp), and microscope objective lenses (OBJ). A bead is trapped at their common foci in circular polarized light. The exiting beams are collected by the opposite objective, converted to horizontal-polarized light, and directed to position-sensitive photodetectors. The flow chamber comprises two coverslips selected for flatness, spaced 200 μm apart by Parafilm layers and sealed by heat. The pipette is drawn from 100-μm glass tubing down to a point with an opening of ~0.5 μm.

Instrument Design

Although several authors have used the deflection of the trapping laser beam to infer bead position, or empirically calibrate a test force,[23,24] it has not been possible to collect and analyze all of the light leaving a single-beam laser trap to determine its rate of momentum change. The measurement scheme depicted in Fig. 1 cannot actually be used in a single-beam trap because such a narrow cone of light (as depicted in Fig. 1) will not efficiently trap an object. The scattering force due to reflected light would overcome the axial gradient (trapping) force and the object would escape toward the right. If a high-NA beam were used instead, then the analysis lens shown in Fig. 1 could not collect the marginal-exiting rays. After interacting with the bead (subjected to an external force), marginal rays in a single-beam trap would be deflected further off axis and fall outside the NA of the collection lens. To avoid this dilemma, a counterpropagating dual-beam laser trap was constructed similar to those of Buican.[25] Here the use of low-NA beams inside high-NA objectives allows significant beam deflection while still collecting nearly all the light. Each objective is used twice, focusing one beam while collecting the other beam for analysis. The two beams are directed to different detectors by polarizing beam splitters as shown in Fig. 3. Use of quarter-wave plates before the

[23] L. P. Ghislain, N. A. Switz, and W. W. Webb, *Rev. Sci. Instrum.* **65**, 2762 (1994).
[24] M. Allersma, F. Gittes, M. J. deCastro, R. J. Stewart, and C. F. Schmidt, *Biophys. J.* **74**, 1074 (1998).
[25] T. N. Buican, in "Cell Separation Science and Technology" (D. S. Kompala and P. T. Todd, eds.), *Am. Chem. Soc. Symp.* **464**. American Chemical Society, Washington, D.C., 1991.

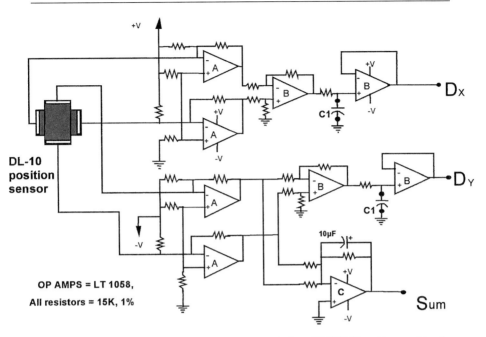

FIG. 4. Preamplifier to derive sum and difference signals from UDT DL10 position-sensitive detector. Current flows into the bottom layer of the detector (from upper op-amps) and out of the top layer (into lower op-amps). Capacitors C1 are chosen just large enough to average signal between data collection times, as set by the data acquisition system (not shown). Noise data were taken with capacitors removed. Then the rise time for a short laser pulse is ~10 μs, implying a bandwidth of ~100 kHz.

objective lenses ensures that light reflected from the particle is not returned to the lasers but instead is reflected back to photodetectors, where its momentum change is registered properly.

A position-sensitive photodetector can be thought of as a large-area PIN junction photodiode bonded to a planar resistor. A light ray falling onto the detector surface produces a localized electric current proportional to the ray's power. That current is injected into the resistor surface so that the signal is proportionately divided between the output resistor terminals, depending on the initial location of the ray. The difference between the two output currents is a signal representing the sum of the powers of all rays, weighted by their distance from the detector center, as per Eq. (7). A dual-axis detector (e.g., DL-10, United Detector Technology, Hawthorne, CA) has one planar photodiode sandwiched between two orthogonal planar resistors (see Fig. 4) so it outputs both x and y position signals. Such photodetectors cannot be placed at the output principal planes, which lie inside the objectives, so relay lenses ("L1" in Fig. 5) were used to reimage the principal

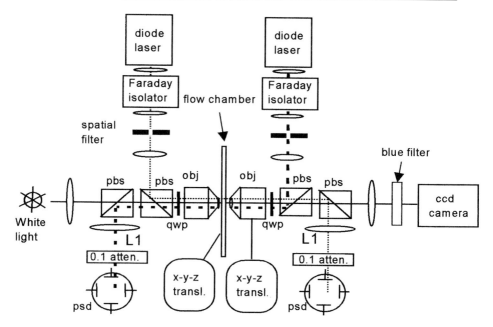

FIG. 5. Optical setup. Diode lasers (SDL 5431, 835 nm, 200 mW) are mounted in temperature-controlled mounts (Newport 700-C), beam circularized with anamorphic prisms (not shown), and protected from reflections by Faraday-effect optoisolators (OFR IO5-835-LP). A spatial filter (two 100-mm lenses, 40 μm pinhole) passes ~80% of the laser power. Polarizing beam splitters, (pbs) (Melles Griot 03-PBS-064) separate different polarizations for infrared beams but pass blue light in either polarization for ccd camera image. Quarter wave plates (qwp) (CVI QWPO-838-05-4) are used for circularly polarized beam at foci. Objectives are Nikon 60× plan-Apo-water NA 1.2 with correction collars.

planes onto the photodetectors. The detectors measure only the light exiting the trap, not entering it, so they perform only half of the integration required in Eq. (4). This problem is solved by aligning the detectors on the optic axis so that when no bead is present in the trap, the output beams are centered on the detectors and the difference signals vanish, corresponding to zero volts at outputs Dx and Dy in Fig. 4. The light entering the trap carries no transverse momentum in this frame of reference and need not be considered even after a particle has been introduced. Only the exiting light is affected by interaction with the particle.

In practice, alignment of the two laser beams must be corrected as the room temperature changes. If the two beam foci are coincident and a bead is introduced into the trap with no external force acting on it, then each detector registers zero change in the transverse force. If the two beams are offset, however, then one light beam pushes or pulls on the other light beam through their common interaction with the bead, and opposing force signals are registered at the two detectors. This difference signal can be used as feedback to correct misalignment by moving the

x–y–z translator connected to the right objective shown in Fig. 5. The external force signal, which is the sum of the two detector signals, remains zero despite alignment errors. If an external force (e.g., a tethered molecule or viscous drag) acts on the bead, the two detector signals act in concert, and their sum gives the force according to Eq. (8). System calibration is independent of laser power because the detectors measure the light power themselves; the signals Dx and Dy (Fig. 4) are proportional to the product of beam offset and power. Transmission losses in the objective lenses and associated optics are incorporated into an effective sensitivity Ψ for the detectors.

Momentum–Flux Calibration

The light momentum–flux force transducer is calibrated from first principles, i.e., conservation of light momentum, without use of test forces such as viscous drag or thermal motion. Approximate values for R_L, R_D, and Ψ in Eq. (8) can be obtained from manufacturers specifications but it is best to measure these quantities *in situ* to account for attenuation in lenses, cubes, or filters and also the magnification factor from lenses L1 acting on the position detectors. To find R_L, a test stand was constructed where the objective was held fixed while a pencil of light was directed backward through it (Fig. 6). The beam was offset various

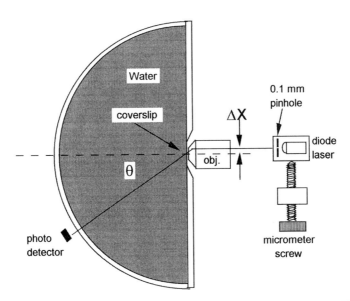

FIG. 6. Objective lens test jig with a diode laser (Melles-Griot 560LB108, 830 nm, 30 mW) carried on a translation stage. A semicircular trough of water is constructed of Plexiglas with a glass coverslip window. A split photodiode UDT SPOT-9 detects the output beam angle.

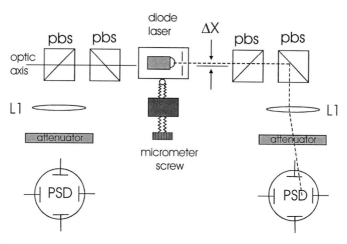

FIG. 7. Device to test for effective radius (R_D) of the photodetector. The objective lenses are removed from the trap setup and a test laser (830 nm, 30 mW with 100 μm pinhole) simulates light from an objective. The beam is translated a distance ΔX while the differential current from the position-sensitive detector is monitored.

distances ΔX while the output angle θ was measured. A value for R_L was obtained by fitting the ΔX versus θ data to Eq. (5). To find the effective detector radius, R_D, the objectives were removed and a movable source of parallel light (laser beam) was inserted in their place, as shown in Fig. 7. This beam was offset a variable distance ΔX as the voltage output Dx of the detector preamplifier (Fig. 4) was recorded. These data were fit to the function $Dx = Sum(\Delta X/R_D)$ where Sum is the total detector current (Fig. 4) and R_D is chosen for best fit. The effective detector sensitivity, Ψ relates the detector output current ("Sum") to the light intensity *at the trap focus*. Therefore the attenuation from one objective must be included, as well as that of the chamber, cubes, attenuators, and lenses. To measure Ψ a flow chamber was fitted as shown in Fig. 3, the objective foci were made coincident, but then only one laser was operated at a time. The power from this laser was monitored upstream and downstream from the pair of objectives using a Newport 840/818-ST meter. Then the power at the focus was estimated as the geometric mean of those two readings and Ψ was calculated as the detector output ("Sum") divided by the power at the focus.

To test the range of applicability of Eq. (5), seven objectives lenses were tested in the apparatus of Fig. 6: (1) Nikon E, 40× air, numerical aperture (NA) = 0.6; (2) Nikon E-plan 40× air, NA = 0.65; (3) Nikon CF plan-Achromat 60× air w/corr., NA = 0.85; (4) Nikon plan-Apo 60× water w/corr., NA = 1.20; (5) Zeiss plan-Neofluar 63× oil, NA = 1.25; (6) Nikon 100× oil, NA = 1.25; and (7) Zeiss plan-Neofluar 100× oil, NA = 1.3.

The input angles (θ) versus output offsets (ΔX) were tested at 15 to 20 positions across the lens and fitted to the function $\Delta X + m_1 = R_L n_1 \sin(\theta + m_2)$ where R_L, m_1, and m_2 were adjusted for best fit. All seven objectives matched Eq. (5) over their entire back-aperture width to within the measurement error, which was $\pm 0.05\%$ of an offset reading. A condenser lens was also tested, namely a large oil-immersion "Abbe" lens with NA = 1.25 and 1 mm working distance. It had a detectable error of 0.5% at its greatest offset. For visible light (670 nm) it was found that R_L always equaled 160 mm divided by the nominal magnification of the lens, to within $\pm 5\%$. The lens most often used in our tweezers, number 4 above, was tested again at 830 nm and found to have a slightly shortened effective radius of 2.50 mm at that wavelength. Because this lens was designed for a conjugate focal distance (tube length) of 160 mm, it was convenient to convert it to an infinite conjugate focus by placing a -150-mm lens immediately behind it (not shown in Fig. 5). This extra lens had the effect of reducing R_L to 2.25 mm. Also, R_L varied by $\pm 2.5\%$ across the range of that objective's correction collar settings. Therefore it is best to calibrate an objective at the trap wavelength, with its negative lens in place (if used) and with fixed correction-collar position.

Force Sensor Test Results

When a particle is trapped at a focus, most of the trapping light is refracted and collected by an objective (as shown in Fig. 1) but some light is scattered outside the collection angle (NA) of the objectives. This effect registers as a drop in the total power received by the photodetectors when a bead is introduced into the trap. To measure the efficiency of light collection, polystyrene beads of eight different sizes, ranging from 0.27 μm to 20 μm diameter, were introduced into the trap while the detector output ("*Sum*" in Fig. 4) was recorded. Table I lists the percentage of the light collected after introduction of a bead at the focus.

TABLE I
EFFECT OF BEAD SIZE ON LIGHT-FORCE CALIBRATION

Bead diameter (μm)	Light collected (%)	Light/Stokes	Light/thermal	Q_{trans}
0.27	99.8	0.85	0.90	0.005
0.54	98	0.90	0.95	0.03
0.76	96	0.88	0.96	0.06
0.82	95	0.90	1.03	0.11
2.03	98	0.88	1.04	0.25
5.10	99	0.94	1.02	0.39
10.00	99.5	0.96	a	0.40
20.30	99.5	0.95	a	0.40

[a] Values not accessible because corner frequencies are too low.

To test the effect of particle size on light–momentum force measurements, various sized polystyrene beads were trapped and exposed to fluid flow while the light–force sensor output was compared with the drag force, as calculated by Stokes law. Translating the fluid chamber back and forth past the laser focus should create a homogeneous flow field. The speed was varied by hand and monitored by an electronic micrometer attached to the chamber. Results over a range of speeds and bead sizes are plotted in Fig. 8. The vertical coordinate in Fig. 8 plots the light–momentum sensor output, converted to force by Eq. (8), where the constants R_L, R_D, and Ψ were calibrated as described above (Fig. 6 and Fig. 7). The points fall near a line with a slope of one, indicating agreement between the two methods. Individual best-fit slopes for eight different bead sizes are entered in Table I under "Light/Stokes."

Another way to check calibration of the force sensor is to measure the thermal forces acting on a trapped bead and compare the measured values with the theory. The spectral density of displacement noise for an overdamped particle in a harmonic potential is given by[26]

$$\langle \Delta x^2(\omega) \rangle_{eq} = \frac{2k_B T}{\xi \left(\omega_c^2 + \omega^2 \right)} \tag{9}$$

where ξ is the damping constant (drag coefficient) for the bead moving in a viscous medium, κ is the spring constant, and $\omega_c = \kappa/\xi$ is the corner frequency. Because the force $F = \kappa \Delta x$ in the trap, the spectral density of force fluctuations is given by

$$\langle \Delta F^2(\omega) \rangle_{eq} = 2\xi k_B T \frac{\omega_c^2}{\left(\omega^2 + \omega_c^2 \right)} \tag{10}$$

Far below the corner frequency, the force fluctuation spectral density is almost constant at a value given by

$$\langle \Delta F^2(\omega) \rangle_{eq} = 2\xi k_B T \tag{11}$$

Above the corner frequency, force fluctuations decay rapidly as $1/\omega^2$.

To calibrate with thermal noise, we recorded the force fluctuations for trapped beads using a fast A/D converter and obtained power spectral density (PSD) distributions by Fourier transform of the data. To make plots with units of piconewtons, the output from the preamplifier (in volts) was converted to force by using Eq. (8) with calibration factors obtained as per Figs. 6 and 7. If the light–momentum calibration is accurate, then the force PSDs should agree with theory and plateau at levels given by Eq. (11). Beads of five different sizes were tested and their PSDs are

---

[26] M. C. Wang, and G. E. Uhlenbeck, *Rev. Mod. Phys.* **17**, 323 (1945); reprinted in "Selected Papers on Noise and Stochastic Processes" (N. Wax, ed.), p. 113. Dover, New York.

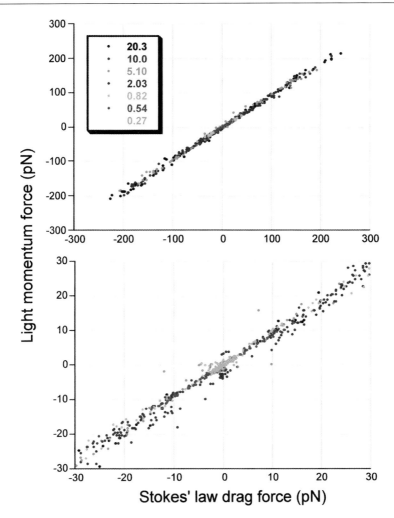

FIG. 8. Estimated viscous drag forces on eight different diameter polystyrene beads are compared to light–momentum sensor output, calibrated using light momentum [Eq. (8)]. All beads were Bangs Laboratories Estapor size standards, except one bead by PolySciences which was reported as 1.05 um in an earlier study (S. B. Smith, Doctoral Thesis, University of Twente, The Netherlands, 1998), but resized using dynamic light scattering (Particle Sizing Systems, Santa Barbara, CA) as 0.82 μm in the present study. For the largest (20 μm) beads, the walls of the fluid chamber were only 10 bead-radii distant, so Stokes law was corrected by 12% to account for a 6% effect of both walls [S. B. Smith, L. Finzi, and C. Bustamante, *Science* **258**, 1122 (1992)]. Bottom graph is 10× expansion of top graph, to show more small-bead data.

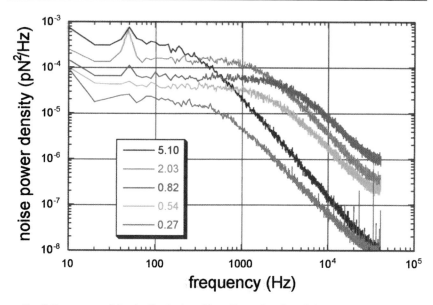

FIG. 9. Power spectral density distribution of force fluctuations from light–momentum force sensor using five different beads, sized 0.27, 0.54, 0.82, 2.0, and 5.1 μm in diameter. Five seconds of output was recorded at a 100 kHz rate using a National Instruments PCI-MIO-16XE-10 analog-to-digital converter with 16 bit accuracy. Data were converted to a frequency domain with a discrete FFT (Igor Pro Ver. 3, Wavemetrics).

plotted in Fig. 9. Note the plateau levels (force–axis intercepts) increase with bead size since the drag coefficient, ξ, for the bead also scales with size. Knowing the bead radius, r, and substituting the Stokes law drag coefficient, $\xi = 6\pi r \eta$ (where η is the viscosity), permits direct comparison between the light–momentum sensor output and the thermal force. Results are listed in Table I under "Light/Thermal" where the plateau levels for the PSDs were averaged over the frequency interval between 100 and 200 Hz, except for the largest bead, where 25 Hz is used due to its low corner frequency.

The maximum transverse holding force of the trap was measured by trapping polystyrene beads of various sizes and increasing the flow of water through the flow chamber until they escaped. The light–momentum sensor, calibrated by Eq. (8), measured the drag force up to the point of escape. A trapping efficiency was then calculated as per Wright et al.,[27] $F_{esc} = Q_{trans} n_1 W_{trap}/c$. This efficiency is labeled Q_{trans} in Table I, because the bead always escaped in a direction transverse to the optic axis. For large beads (>1 wavelength) the force sensor output becomes

[27] W. H. Wright, G. J. Sonek, and M. W. Berns, Appl. Opt. 33, 1735 (1994).

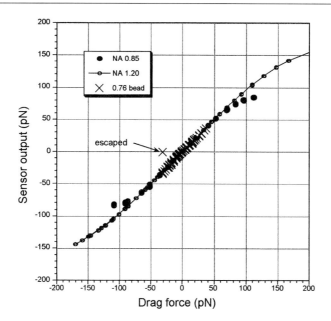

FIG. 10. Response of light–momentum sensor for beads that were pushed by fluid drag to the point of escape. Beads (2 μm diameter) were tested in two different trap setups, one with 0.85 NA objective lenses and 100 mW power, and the other with 1.20 NA lenses and 140 mW power. A smaller bead (diameter = 0.76 μm) was also tested in the trap with NA = 1.2 lenses.

nonlinear before F_{esc} is reached, as shown in Fig. 10, but smaller beads display linear output up to F_{esc}.

Finally, a different type of test force was used to determine the effect of various factors on light-sensor accuracy. When a dsDNA molecule is pulled above a certain force (65–70 pN) it undergoes a cooperative structural transition to a longer form.[13,28] Because the force remains nearly constant during that transition, it can be used as a standard to test force transducer accuracy. Therefore, individual molecules of lambda (λ) phage DNA were attached between the test bead and a pipette bead, as shown in Fig. 3, and extended in 500 mM NaCl buffer solution while the force sensor output was recorded. For a 7-μm polystyrene test bead, the light sensor reported a force of "68 pN" at the midpoint of the stretch transition. The sensor output remained constant for a 4-μm and a 2-μm test bead, as shown in Table II, but dropped to "64 pN" for a 1-μm test bead. Forces sufficient to cause the structural transition of DNA could not be generated using beads smaller than

[28] P. Cluzel, A. Lebrun, C. Heller, R. Lavery, J.-L. Viovy, D. Chatenay, and F. Caron, *Science* **271**, 792 (1996).

TABLE II

EFFECT OF BEAD SIZE AND TRAP POWER ON STANDARD FORCE MEASUREMENT

Bead diameter (μm)	Measured force (pN)	Power[a] (mW)	Measured force (pN)
		140	68
7	68	120	68
4	68	80	68
2	68	60	67.5
1	64	50	65

[a] Power calculated at trap from known objective transmissions. All power comparisons use same size (2 μm) bead.

1 μm. The effect of laser power on sensor calibration was tested in a similar way. Here a test bead of 2 μm diameter was pulled at the transition force of DNA while the laser power was varied. Comparing the total laser beam power at the focus ("Power" in Table II) with the sensor output ("force") gave a nearly constant result over a 2 : 1 power range. Next, the effect of changing the shape of the trapped object was explored by placing several test beads in the trap at the same time. Multiple beads align axially in a dual-beam trap and aggregate into chains. No change was seen in the transition–force reading when a single dsDNA molecule pulled a trap containing one, two or three beads (each 2 μm diameter). Finally adding sucrose to the 500 mM salt solution tested the effect of changing the refractive index of the buffer. In buffer containing only salt, the transition force output read "68 pN." Adding 20% sucrose (by weight) made the output drop to "67 pN" but increasing the sucrose to 33% restored the output somewhat to "67.5 pN." Here the refractive index was increased from 1.34 for salt buffer to 1.40 for high-sugar buffer, while the index for the polystyrene bead remained 1.57. Therefore the relative index (bead/buffer) decreased from 1.17 to 1.12 and the index change (bead minus buffer) dropped by ~30% without an appreciable change in force calibration.

Force Sensor Discussion

The counterpropagating beam optical trap described here analyzes the light–momentum flux leaving the vicinity of a trapped particle and reports the force that the flux has exerted on the particle. It is assumed that nearly all of the light leaving the trap is collected. As seen in Table I, this assumption seems justified because the measured collection efficiency for an objective with NA of 1.2 was 95–99%. Little scattering is expected for large diameter spherical beads with small displacements from the focus. Then, as sketched by Ashkin,[4] the output beam will be a cone of divergence similar to the input cone but deviated from the axis as shown in Fig. 1. For particles much smaller than one wavelength, the Gaussian waist at the focus bypasses the particle. Therefore the transmitted light is seen to increase for the two smallest beads in Table I. The worst collection efficiency (95%) occurs for beads about 1λ in diameter. There is currently no theory that successfully predicts the scattering pattern for such beads near a focus.[3,27] When an objective with NA of 0.85 is used, the collection efficiency drops to 90% for a 1-μm bead. Because this lens subtends only $\sim$11% of all (4π) solid angle as seen from the focus, the 1-μm bead must still scatter over a fairly narrow forward angle.

As seen in Table I, light–force measurements made on various sized beads exposed to fluid drag and Brownian forces agree rather well with Stokes law and thermal theory calibrations, but with slight systematic errors of two types. First, the average of all comparisons is low for the light–force sensor. This effect might result from errors in estimating R_L, R_D, and Ψ of Eq. (8), but then why does the light–force sensor agree better with thermal measurements than with drag force measurements? Alternately, the assumption of uniform velocity for all fluid in the chamber may be erroneous. Our beads are trapped far (100 μm) from any wall. If stage acceleration sets up inertial convection in our fluid cell (sloshing through the fluid ports), then it will register as a decrease in the relative flow past the bead. The "Light/Thermal" (Table I) results, where the chamber was not moved, are higher overall by 5%, in accordance with the sloshing hypothesis. A second type of systematic error occurs for beads which are $\sim$one wavelength in diameter. They show an additional 5% force deficit from the light–force sensor. This error probably reflects the uncompensated loss of 5% of the trapping light, scattered outside the lens NA. In future work, it should be possible to measure the light loss and compensate this force error.

It might seem odd that light–sensor accuracy improves again as the particles become smaller than the trap wavelength (e.g., 0.27 μm, Table I), since the tiny amount of light interacting with the particle is scattered over an increasingly large range of angles. To understand this effect, suppose the trapped particle was reduced in size until it became a Rayleigh scatterer with a perfectly symmetrical (about its own center) radiation pattern such that the irradiance is proportional to cos θ but omnidirectional in ϕ. Of the light that was scattered by the particle, only $\sim$60% would be collected by two objective lenses with NA of 1.2. Such loss seems to insure a force-measurement error. However, any transverse force on a trapped particle must be represented by a transverse asymmetry in the far-field scattered intensity pattern if momentum is conserved. Because the Rayleigh pattern itself is symmetric about ϕ, the requisite asymmetry must be caused by an interference between the Rayleigh waves (emanating from a source offset in the trap) and the spherical trap-beam waves emanating from the trap focus. Because the intense spherical waves are created by lenses and truncated to lie in a cone that is collected by those lenses, the asymmetrical interference pattern also lies within that cone, since there can be no interference where one set of waves (e.g., from the trap beam) is missing. The part of the Rayleigh pattern that misses the lenses is symmetrical and carries no net transverse momentum. Given that the transducer works properly for bead diameters much larger than a wavelength (ray optics) and much smaller than a wavelength (Rayleigh regime), it is perhaps not surprising that it also works fairly well for bead diameters near one wavelength.

Analysis of the transverse (x and y) components of light momentum is based on the Abbe sine condition [Eq. (5)]. Probably all objectives meet this criterion with negligible error since little error ($<0.05\%$) was detected in the seven objectives we tested. Three factors affect the choice of an objective: (1) It must have a long working distance to reach the middle of the fluid chamber. (2) Water immersion is preferable because the two foci remain coincident as the fluid chamber is translated sideways several millimeters. If the walls of the fluid chamber are not perfectly parallel, then variations in the chamber-water thickness cause beam misalignment for air-immersion lenses. Using water immersion, however, the total water thickness inside and outside the chamber remains constant. (3) The lens NA should be large. Although a counterpropagating dual-beam trap uses narrow beams which underfill the objectives, those beams deviate from the optic axis when an external force is applied to the trapped object (see Fig. 1). Extra aperture must be allowed for this deviation or light will be lost. If we call the radius of the trap beam r_{beam} and that of the back aperture of the objective r_{ba}, then the maximum distance the output beam can deviate before light signal is lost is the difference between these two radii $r_{ba} - r_{beam}$. This offset represents an angle θ given by $n_1 \sin\theta = (r_{ba} - r_{beam})/R_L$. The force that produces such a deviation is given by

$$F_{max} = (W/c)n_1 \sin\theta = (W/c)(r_{ba} - r_{beam})/R_L \qquad (12)$$

where F_{max} is a maximum force, above which the sensor loses accuracy. For instance, a Nikon CF plan Achromat $60\times$ air lens was used with the following specifications: NA $= 0.85$, $R_L = (160 \text{ mm}/60\times) = 2.667$ mm and $r_{ba} = R_L \times 0.85 = 2.267$ mm. The laser beam diameter was measured where it entered the lens and found to be $r_{beam} = 1.624$ mm. By Eq. (12), $F_{max} = 50$ pN for $W = 100$ mW. As seen in Fig. 10, the output from this force sensor does become nonlinear above 50 pN. When a $60\times$ lens with NA of 1.2 was used instead, F_{max} increased to $\sim$100 pN, as shown in Fig. 10.

Low values of Q_{trans} for bead diameters below one wavelength (see Table I) reflect a drop in the polarizability of small particles, proportional to the particle's volume. Rayleigh scattering arguments (above) suggest that force measurements remain accurate for any force, as long as the (small) particle remains trapped. Accordingly, Fig. 10 shows that a 0.76-μm bead escaped the trap without detectable nonlinearity, since $F_{esc} < F_{max}$.

Measuring Extension

The primary use of optical tweezers in biology has been to make real-time measurements of force versus molecular extension (or position). Thus, despite the fact that the light–momentum sensor allows us to infer the force on a particle without measuring its position, it is still necessary to measure particle positions quickly and accurately. One method to measure bead position is video microscopy. Figure 11 shows the image of two beads, one held on a pipette by suction and the other held in a laser trap. Köhler illumination is set for parallel rays at the object to enhance contrast. A bright spot forms at each center because the beads act as

A

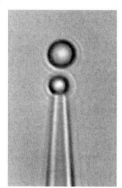

B

```
procedure fillNeighbors(level:byte; x,y:integer; fillByte:byte);
var brightness:integer;
begin
brightness:=videoPixelLevel(x,y);
if (brightness>level) and (globSize<globMax)   then
   begin
   xSum:=xSum+(x);
   ySum:=ySum+(y);
   putVideoPixel(x,y,fillByte);
   inc(globSize);
   fillNeighbors(level,x+1,y,fillByte);
   fillNeighbors(level,x-1,y,fillByte);
   fillNeighbors(level,x,y+1,fillByte);
   fillNeighbors(level,x,y-1,fillByte);
end;
```

FIG. 11. (A) Video image of 2-μm bead on pipette with 3-μm bead above in trap. (B) Recursive procedure "fillNeighbors" calls itself but returns when all interior pixels have been converted to dark (fillByte) level. Variables xSum and ySum divided by globSize give pixel-averaged x and y coordinates of bead.

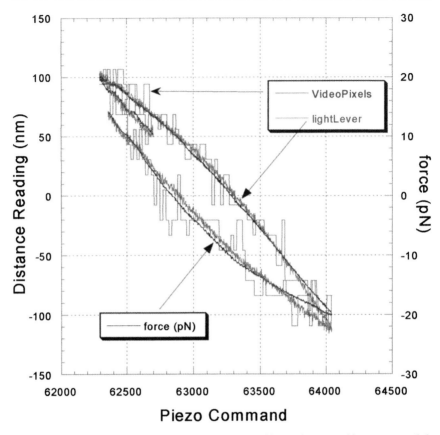

FIG. 12. Video-capture position (blue) and light lever position (red) measured in nanometers (left scale). Light–force sensor output (black) in piconewtons (right scale). All outputs plotted versus piezounits (positioner command from 16-bit A/D converter, open loop).

lenses. We locate the centroid of these spots by capturing the image and operating on it with an algorithm ("fillNeighbors," Fig. 11) that fills the interior of the dark rings while averaging the pixel positions. Using a $60\times$ magnification and a $1/3''$ CCD camera (Watec LCL-903HS), the pixel size in Fig. 11 is 150 nm. Pixel averaging refines this resolution by about 10 times, so that the RMS repeatability error ("jitter") between successive frames is 10–15 nm. Such jitter is evident in Fig. 12, which shows the video-detected position of a 2-um bead, held on the top of a pipette, which was moved up and down 100 nm by a piezoactuator.

Faster and finer position data can be recorded for the pipette bead by using a relatively inexpensive "light lever" device attached to the pipette stage (see

Supplementary Data in Ref. 9). Such a device comprises a single-mode optical fiber coupled to a diode laser (Thorlabs LPF-3224-635-FC), a PSD as in Fig. 4, and a lens attached to the pipette/chamber frame (Thorlabs C140TM-B). This lens collimates light from the fiber and directs the output bean onto the PSD. Because the focal length of the lens is short (1.45 mm) but the distance to the PSD is large (610 mm), a tiny movement in the relative position of the fiber translates into a 420-fold greater movement of the light spot on the PSD. When the PSD is coupled to a 16-bit A/D converter with 10 V input range (ComputerBoards PCI-DAS1602/16), the distance resolution becomes 1.5 nm. The light lever's increased resolution over video microscopy is evident in Fig. 12. Now the hysteresis of the piezoactuator stage (Thorlabs MDT-631) becomes clearly evident.

For ultimate resolution, the bead in the optical trap can act as sort of a light lever itself.[29] Although the output of the light–force sensor records force, that force can be related to a movement of the bead in the trap by dividing force by trap stiffness. The problem then becomes one of estimating trap stiffness, since it varies with a number of factors (see above). Figure 12 also shows the fine resolution and low noise of this method. Note, however, the force–signal scale was arbitrarily normalized to give the same heights to the trap-force and light-lever traces. Different scale factors would be needed for different sized beads or different laser powers.

Limiting Resolution

So what is the limit of spatial resolution for such a force-to-distance conversion? The single-bit resolution of the force detector circuit (0.016 pN), along with typical trap stiffness ($\sim$170 pN/μm for a 2-μm bead), suggests a limiting resolution of 0.1 nm. However, the actual resolution is limited by position noise between the pipette and the trap position, and the system bandwidth. To characterize such noise, the pipette was directly attached to the bead in the optical trap. That bead was then moved up and down by $\sim$1 nm using a piezoactuator stage. Bead position was monitored with both the light-lever and the light–force sensor at a sampling frequency of 200 Hz. Figure 13 shows such data, time averaged down to a bandwidth of 20 Hz. Here, position noise appears to exist mainly in separate time domains: slow drifts with >10 sec duration, and fast vibrations with <0.05 sec period. Its presence correlates with external factors such as room temperature change (slow), floor vibrations (fast), low-pitched voices (fast), fan motor noise (fast), dust crossing laser beams (slow), and general air currents. The instrument was on an air-supported table. To obtain Figure 13, we worked silently on a quiet weekend in a subbasement laboratory with room ventilation blocked. We allowed 1 hr (thermal) settling time after touching metal parts on the light lever or piezo

[29] W. Denk and W. W. Webb, *Appl. Optics* **29**, 2382 (1990).

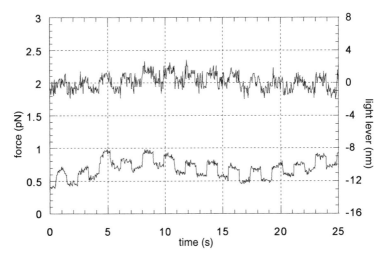

FIG. 13. Light–force sensor output for 2-μm bead in trap, which is also sucked onto pipette. Pipette is moved with step size of 10 piezounits = 1.3 nm. Upper trace is light lever, lower is light–force sensor.

stages with our hands. Some vibration noise in Fig. 13 could have been prevented, since a Fourier transform of the force–sensor noise (Fig. 14) shows a peak at 51 Hz due to acoustic noise from a (barely audible) fan motor on a nearby computer. The general rise in the spectrum at 100 Hz is due to room acoustic noise exciting a mechanical resonance of the pipette's piezo stage.

A real molecular experiment, however, usually involves a flexible coupling between the trap and the enzyme motor, such as a tether of DNA. Then thermal forces may set the limiting resolution of the experiment. The tether is elastic, so the bead undergoes Brownian motion in the trap, independent of any enzyme motor activity. To estimate such thermal motion, we integrate the noise power, Eq. (9), from $\omega = 0$ out to some bandwidth $B \ll \omega_c$ and get a value for noise displacement (Δx_{noise}) of a particle in a potential well.

$$\Delta x_{\text{noise}}^2 = (2 k_B T / \kappa)(B / \omega_c) = 2 k_B T B \xi / \kappa^2 \qquad (13)$$

where κ is the stiffness of a trapping potential. But κ for the bead is not due to the optical trap alone, but the sum of the trap and tether stiffness, since they act as two springs in parallel: $\kappa = \kappa_{\text{trap}} + \kappa_{\text{tether}}$. Thus the RMS (root mean square) distance noise for a trapped bead with tether is

$$\Delta x_{\text{noise}} = (2k_B T \xi B)^{1/2} / (\kappa_{\text{trap}} + \kappa_{\text{tether}}) \qquad (14)$$

Unlike the noise, the signal of interest, e.g., Δx_{step} of an enzyme motor, is reduced by using a tether with low stiffness. As pointed out by Yin et al.,[7] this signal is divided between two virtual springs in series, κ_{trap} and κ_{tether} and thus attenuated

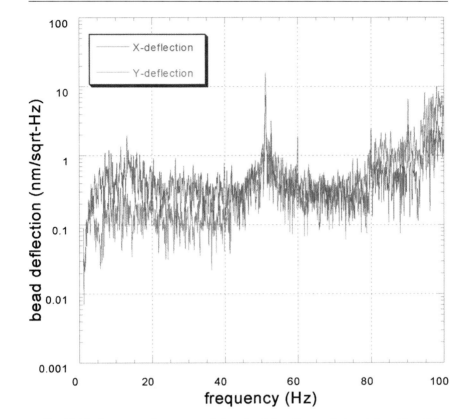

FIG. 14. Displacement–noise spectrum of pipette bead placed in an optical trap inferred from light–force signal with assumed trap stiffness of 10^{-4} N/m. X-axis noise spectrum in blue, y-axis noise in red.

at the trap where it produces a signal given by

$$\Delta x_{\text{signal}} = \Delta x_{\text{step}} \, \kappa_{\text{tether}} / (\kappa_{\text{trap}} + \kappa_{\text{tether}}) \tag{15}$$

To detect steps at the trap, the signal-to-noise ratio should be greater than one. Therefore Eqs. (14) and (15) give

$$\Delta x_{\text{step}} > (2 \, k_{\text{B}} T \, \xi_{\text{bead}} \, B)^{1/2} / \kappa_{\text{tether}} \tag{16}$$

Interestingly, the optical-trap stiffness does not affect limiting resolution.[3,30] Drag coefficients are fairly important. Thus smaller beads are better than big beads because they have smaller drag coefficients. To the extent this analysis may apply, an AFM cantilever should behave poorly because it has high drag. But the most important factor by far is the tether stiffness, as affected by average tether tension.

[30] C. Bustamante, C. Rivetti, and D. J. Keller, *Curr. Opin. Struct. Biol.* **7,** 709 (1997).

TABLE III
CALCULATED TETHER STIFFNESSES FOR 10-kb TETHERS OF ssDNA
AND dsDNA UNDER DIFFERENT TENSIONS[a]

Tension F (pN)	κ_{tether} dsDNA (pN/μm)	κ_{tether} ssDNA (pN/μm)
1	4	1.3
2	12	1.5
5	42	2.6
10	98	7.2
20	170	20
50	250	74

[a] Calculation assumes $P_{ds} = 50$ nm, $P_{ss} = 0.7$ nm, $L_{0,ds} = 3.4$ μm, $L_{0,ss} = 7$ μm, $S_{ds} = 1000$ pN, $S_{ss} = 800$ pN. Stiffness scales as $1/L$ for different length tethers.

Tether stiffness rises rapidly with tension due to the nonlinear entropic (bending) elasticity of a worm-like chain (WLC) with fixed contour length, roughly $F = (k_B T/4P)[(1 - x/L)^{-2} - 1 + 4x/L]$, where x is the end-to-end extension of the chain, P is its persistence length, and L is its contour length.[31] However, contour length is not strictly fixed, but increases with tension as $L = L_0(1 + F/S)$ where L_0 is the length at zero force and S is the enthalpic stretch modulus of DNA. Values of $\kappa_{tether} = dF/dx$ for the stretchable WLC using an improved power-series approximation[32] are given in Table III.

To test these predictions, we attempted to detect nanometer-sized steps through an 8-kbp tether of dsDNA. Figure 15 shows results where the pipette bead was moved up and down by $2\frac{1}{2}$ nm while a trap bead registered changes in the force transmitted to it by the tether. Compare the force traces when the average tether tensions were set at 2 and 20 pN. The signal is transmitted clearly through the tether under 20 pN average tension, but is barely discernible under 2 pN average tension. Indeed Eq. 16 and Table III predict that such a dsDNA tether, under 2 pN tension, requires a minimum step size of 2.6 nm to register above thermal noise in a bandwidth of 1 Hz, i.e., the frequency of our signal [in Eq. (16), B = 2π radians/sec].

Force-Extension Curves

Figure 16 shows seventeen superimposed stretch/release curves for a 5-kbp piece of dsDNA taken over a 5-min. period. The pipette position was measured using a light lever at a data rate of 100 Hz. Most stretch/relax curves superimpose,

[31] C. Bustamante, J. F. Marko, E. D. Siggia, and S. B. Smith, *Science* **265**, 1599 (1994).
[32] C. Bouchiat, M. D. Wang, J.-F. Allemand, T. Strick, S. M. Block, and V. Croquette, *Biophys. J.* **76**, 409 (1999).

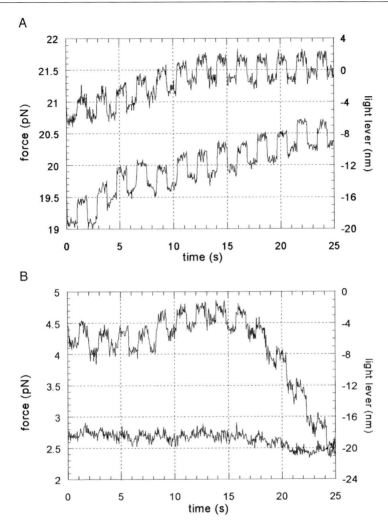

FIG. 15. (A) Transmission of distance steps through 8-kbp piece of dsDNA, in 100 mM NaCl buffer, at high (~20 pN) molecular tension; (B) same steps at low (~2.5 pN) tension. Upper traces are light lever output (pipette position), lower are light–force sensor. Piezo stage (Thorlabs MDT-631) moves pipette by 2.5 nm once every second. Trap bead diameter = 2 μm. Plot bandwidth = 20 Hz.

indicating good repeatability of force and distance measures. Fast kinetic effects are visible, such as torsionally constrained overstretching, nicked overstretching, strand melting, re-annealing, and final breakage. Three instrumental problems are visible in this plot: (1) The overstretching plateau reads 60 pN rather than 65 pN, as is appropriate for dsDNA in 100 mM NaCl and room temperature. Such errors

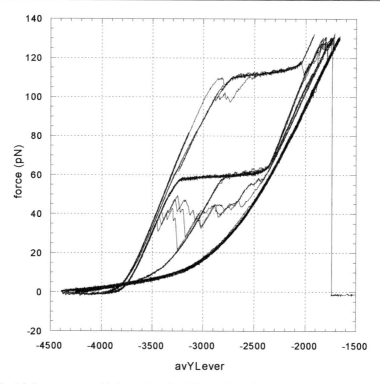

FIG. 16. Seventeen stretch/relax cycles of a 5-kbp section of plasmid pBACgusx11 (Novagen) pulled in 50 m*M* NaCl, 1 m*M* EDTA, 40 m*M* tris, pH 7.5. One strand melts off repeatedly but remains attached at an end. It often starts to reanneal when the force drops below 50 pN.

occur when dust collects on optics and light transmission is reduced below that measured when Ψ in Eq. (8) was calibrated. (2) When the molecule breaks, the force falls to -1 pN, not zero. Apparently the zero level has drifted since the last time the input light momenta were nulled (>5 min. previous). (3) The x-axis of this plot records only the pipette position, not the molecular end-to-end distance. Therefore the apparent molecular stiffness is softened by the trap compliance, and the curves lean toward the right.

If the trap compliance (inverse stiffness) were known, it would be possible to correct for trap-bead motion and obtain the true molecular extension. Fortunately video data give us the absolute (pixel) position of the trap bead at each force, albeit with 15 nm RMS jitter. That jitter averages out over many frames, giving a good average value for the trap bead movement versus force, i.e., the trap compliance. Thus subtracting the value (force* compliance) from each light-lever position gives the molecular extension. Force versus light-lever extension data, corrected for (video-detected) compliance, are shown in Fig. 17.

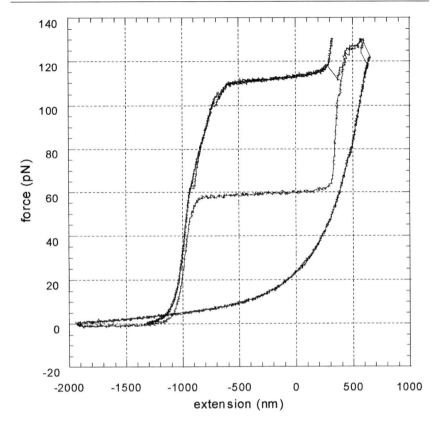

FIG. 17. First three stretch/relax cycles from Fig. 16 corrected for trap compliance. First cycle is reversible, staying on the top curve for stretch and relax. Plateau at 110 pN indicates transition of torsionally constrained molecule to composite S-P form (see Ref. 33). Second pull cycle follows upper curve, but a single nick occurs at 115 pN, when it jumps to the longer S form. Plateau at 60 pN indicates B–S transition (see Refs. 13 and 28). Further stretching causes the S form to melt at 130 pN (see Refs. 16 and 34). Bottom relaxation curve shows ssDNA characteristics (see Ref. 13). The end-attached strand reannealed only at zero force, since the third pull cycle followed the B form curve, then a B–S transition at 60 pN, then melted at 140 pN, then relaxed again as ssDNA.

Conclusion

The optical tweezers instrument described here is a light momentum flux sensor and actuator that makes it possible to directly exert and measure forces on objects in the range between 0.1 pN–200 pN. Because it is a double-beam instrument, its

[33] J. F. Leger, G. Romano, A. Sarkar, J. Robert, L. Bourdieu, D. Chatenay, and J. F. Marko, *Phys. Rev. Lett.* **83,** 1066 (1999).

[34] H. Clausen-Schaumann, M. Rief, C. Tolksdorf, and H. E. Gaub, *Biophys. J.* **78,** 1997 (2000).

optical alignment is more difficult and also more crucial for its operation. On the other hand, because this instrument operates on first principles, i.e., by determining the force from a direct measurement of the change of light momentum flux, its calibration is unaffected by bead size, shape, index of refraction, and location. This feature greatly facilitates its day-to-day operation, and may make it possible to measure forces in complex refractive media such as the cell interior. When coupled to an inexpensive light-lever system, this instrument can provide detailed force–extension data with a resolution of 1 nm at a bandwidth higher than 100 Hz. Subnanometer resolution should be possible using passive force measurement. Future improvements on vibration, sonic and thermal isolation should permit its use in the characterization of the individual steps of molecular motors such as DNA and RNA polymerases, as long as the motion is reported through a short, stiff DNA tether.

Acknowledgment

We thank Guoliang Yang for help in analyzing noise spectra and Fernando Moreno and Brian Gin for help in making the noise/step experiments. This work was supported by grants from the National Institutes of Health, GM-32543, and the National Science Foundation, MBC-9118482 and DBI-9732140.

[8] Application of Optical Traps *in Vivo*

By STEVEN P. GROSS

Introduction

A number of groups have combined sophisticated optical-trapping setups with bead assays to study the function of single molecules *in vitro* in exquisite detail. Such experimental setups have been well documented in a number of review articles.[1–3] However, much less has been written about using an optical trap to make quantitative measurements *in vivo*, the topic of this article.

What Is an Optical Trap?

An optical trap results when a high-numerical aperture lens is used to focus a laser beam to a diffraction-limited spot (see Fig. 1). When a small dielectric

[1] A. D. Mehta, J. T. Finer, and J. A. Spudich, *Methods Cell Biol.* **55,** 47 (1998).

[2] K. Visscher and S. M. Block, *Methods Enzymol.* **298,** 460 (1998).

[3] G. J. Wuite, R. J. Davenport, A. Rappaport, and C. Bustamante, *Biophys. J.* **79,** 1155 (2000).

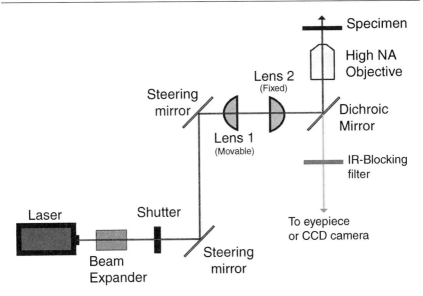

FIG. 1. Simple block diagram showing the path of the laser light from its origin at the laser, through its formation of an optical trap in the specimen. The beam expander (typically a $3\times$ or $5\times$ expander) is chosen so that the laser beam just fills the back aperture of the microscope objective: if the back aperture is overfilled, laser light is lost, and if underfilled, less of the lens is used, resulting in less of a gradient in the focused beam. The shutter is used to block the laser to minimize the sample's exposure to laser light when an object is not being actively trapped. The steering mirrors are used to control the orientation of the beam, to bring it into the objective. Lens 1 and 2 from a 1 : 1 telescope, and moving Lens 1 relative to Lens 2 allows steering and parfocalizing of the trap. Lens 1 and 2 are identical planoconvex lenses, placed so that the distance D between the two is twice their focal length f, i.e., $D = 2f$, so that parallel light entering the lenses remains parallel when it exits. This design was made popular by Block, and is discussed in a review article.[4] After passing through the two lenses, the laser is brought into the microscopes optical path by reflecting it off of a dichroic mirror, mounted below and in line with the microscope objective. The exact position of the dichroic mirror depends on whether one plans to simultaneously use the optical trap and fluorescence. If not, one can simply replace one of the fluorescent cubes with the mirror, a very easy operation. However, to build a system capable of simultaneous fluorescence and trapping, it is better to avoid trying to pass the laser through the fluorescence cube, since this leads to loss of laser power and also interferes with the choice of fluorescent filters. In this case, it is better to modify the microscope, to position the dichroic mirror below the lens, but *above* the filter cube. Such modification can be time consuming.

object such as a visicle, lipid droplet, or polystyrene or silica bead is placed in the beam, it feels a force pulling it toward the focus of the beam. The magnitude of the exerted force is a function of a number of parameters: the size and shape of the object, the difference between the index of refraction of the object and the surrounding medium, and the gradient in the intensity of the laser beam. The force, F, the trap exerts on an object is a linear function of the object's displacement, x, from the trap center, i.e., $F = -kx$, where k is the trap stiffness. Deviations from this linear relationship are observed at the very edges of the beam, typically about 200 nm from the center of the beam. The applied force increases linearly with laser power. The significance of object size and index of refraction differences are discussed below. Typically, a force of 10s of piconewtons can be applied to a small (0.5 μm) bead, although it is possible to exert more than 100 pN of force on a bead with a diameter of a few micrometers using a high-powered laser. In principle any laser can be used to make an optical trap; in practice infrared lasers are employed because they cause the least optical damage (see below).

Basic Trap Design

The basic optical trap is quite simple: a single-mode laser with Gaussian beam profile (TM_{00}) is passed through a beam expander (so that the beam fills the back aperture of the microscope objective, in order to use the entire lens, which generates the largest gradient in the focused beam), and then steered with a combination of mirrors and lenses into the microscope objective. This objective focuses the light in the sample plane of the microscope to a diffraction limited spot, forming the trap. The construction of such a basic optical trap is discussed in detail in a review article by Block.[4]

Typical Uses of Optical Traps in Vitro

In vitro optical traps are used to study polymers such as microtubules[5] or DNA,[6] and to measure properties of single molecules such as titin,[7] or molecular motors such as kinesin.[8,9] Most of these experiments are conceptually similar: attach a glass or polystyrene bead to the molecule or polymer of interest, and use this bead as a "handle" to manipulate the molecule. To be able to exert force on the molecule or polymer of interest, the molecule in question is also attached independently to an immobilized binding partner. As well as studying the strength of molecular interactions,[10] less traditional molecular motors, such as RNA polymerase, have also been studied.[11] Most of the optical traps used in these experiments are quite sophisticated. In particular, they involve quadrant-photodiode detection setups to

[4] S. M. Block, in "Cell Biology: A Laboratory Manual" (D. L. Spector, R. Goldman, and L. Leinwand, eds.). Cold Spring Harbor Press, Cold Spring Harbor, NY, 1998.
[5] H. Felgner, R. Frank, and M. Schliwa, J. Cell Sci. 109, 509 (1996).
[6] M. D. Wang, H. Yin, R. Landick, J. Gelles, and S. M. Block, Biophys. J. 72, 1335 (1997).
[7] L. Tskhovrebova, J. Trinick, J. A. Sleep, and R. M. Simmons, Nature 387, 308 (1997).
[8] C. M. Coppin, D. W. Pierce, L. Hsu, and R. D. Vale, Proc. Natl. Acad. Sci. U.S.A. 94, 8539 (1997).
[9] K. Visscher, M. J. Schnitzer, and S. M. Block, Nature 400, 184 (1999).
[10] A. L. Stout, Biophys. J. 80, 2976 (2001).
[11] M. D. Wang, M. J. Schnitzer, H. Yin, R. Landick, J. Gelles, and S. M. Block, Science 282, 902 (1998).

precisely measure the displacement of the trapped bead from the center of the trap to determine the force that the bead is applying on the molecule or polymer being studied. Further, many of these setups include a feedback loop, so that the force on the trapped object can be kept constant (either by modulating the laser power or the position of the laser relative to the trapped object).[1,2,11]

Typical Uses of Optical Traps *in Vivo*

Many *in vivo* applications of optical traps have also been discussed, although the majority of studies have employed traps in a nonquantitative manner. In this regard, optical traps have been used to manipulate the relative position of biological objects. For instance, cells have been sorted,[12] fibroblast cells have been brought into contact with a surface to investigate adhesion,[13] sperm have been brought into contact with eggs for *in vitro* fertilization, and chloroplasts and other internal organelles have been displaced intracellularly to investigate their mobility, and also the importance of the location of the organelle on its function.

There has also been progress in using optical traps to address quantitative questions *in vivo*. The majority of these studies have been conceptually similar to *in vitro* studies that use a bead as a handle to manipulate an attached protein or other object. In such cases, the beads are attached to proteins or antibodies that interact with the exterior of the cell to examine cellular response[14] or to investigate externally accessible properties such as the tension of the cell membrane[15] or the lateral mobility of membrane-bound proteins.[16] Responses of cells to externally applied forces have also been examined.[17] Because the manipulated bead is in the extracellular buffer and of known size and shape, these applications are effectively *in vitro* experiments with regard to the optical trap; the trapping setups described for *in vitro* work are all more than sufficient to make such measurements.

Finally, there have been some attempts to use optical traps to make quantitative measurements where the trapped object is itself in the cell. In these cases, one can directly trap and manipulate a cellular organelle such as a mitochondrion,[18] or lipid droplet,[19] or in principle introduce appropriately coated beads into the cell through endocytosis or microinjection. This article will discuss issues that arise as a result of attempting these measurements: local heating, optical damage, the

---

[12] S. C. Grover, A. G. Skirtach, R. C. Gauthier, and C. P. Grover, *J. Biomed. Opt.* **6,** 14 (2001).

[13] O. Thoumine, P. Kocian, A. Kottelat, and J. J. Meister, *Eur. Biophys. J.* **29,** 398 (2000).

[14] X. Wei, B. J. Tromberg, and M. D. Cahalan, *Proc. Natl. Acad. Sci. U.S.A.* **96,** 8471 (1999).

[15] D. Raucher and M. P. Sheetz, *Biophys. J.* **77,** 1992 (1999).

[16] K. Suzuki, R. E. Sterba, and M. P. Sheetz, *Biophys. J.* **79,** 448 (2000).

[17] L. M. Walker, A. Holm, L. Cooling, L. Maxwell, A. Oberg, T. Sundqvist, and A. J. El Haj, *FEBS Lett.* **459,** 39 (1999).

[18] A. Ashkin, K. Schutze, J. M. Dziedzic, U. Euteneuer, and M. Schliwa, *Nature* **348,** 346 (1990).

[19] M. A. Welte, S. P. Gross, M. Postner, S. M. Block, and E. F. Wieschaus, *Cell* **92,** 547 (1998).

importance of the index of refraction of the cytoplasm, and determination of the force applied to the trapped object.

Heating

When focusing an intense laser beam into a small volume of the cell, the first question to be addressed is how much the area will be heated, and, concurrently, how much optical damage the laser will cause. Both of these concerns are to some extent dependent on the exact details of the cellular environment, but there are a few general guidelines.

Typically, to minimize absorption of the laser light by either the proteins or the water in the cell, a laser in the near infrared (800–1100 nm) is used. Because water is relatively transparent to light in this part of the spectrum (see Ref. 20 for the absorption versus wavelength), heating is relatively mild. The localized heating due to an optical trap has been measured experimentally and the general rule for a laser operating at 1064 nm is to expect laser-induced heating of 1.45° for every 100 mW of laser power entering the sample.[21] The temperature rise is very rapid (about 1 ms). Only a small amount of the input laser power is absorbed; the rest passes through the sample. The absorption is not a linear function of the incident power, and can be calculated using Beer's law which is

$$A = \varepsilon bc = \log_{10}(P_{in}/P_{out})$$

where A is the absorption, ε is the Lambert absorption coefficient, b is the path length of the light, c is the concentration of the absorber (in this case set to 1 for water), P_{in} is the incident laser power, and P_{out} is the power in the beam after it exits the sample. From this law, we calculate that the power absorbed by the sample is

$$P_{abs} = (P_{in} - P_{out}) = P_{in}\left(1 - \frac{1}{10^A}\right)$$

As an example, using the experimental data from 1064 nm, let us calculate the expected heating from 830 or 980 nm lasers. From Palmer and Williams[20] we see that at a laser wavelength of 830 nm, $\varepsilon = 0.03$ cm^{-1}, for 980 nm $\varepsilon = 0.5$ cm^{-1}, and for a laser operating at 1064 nm, $\varepsilon = 0.14$ cm^{-1}. We assume that in the experimental measurements that determined a heating of $1.45°/100$ mW (at 1064 nm), the path length was 10^{-3} cm (10 μm). Then,

$$\frac{P_{abs,1064}}{P_{abs,830}} = \frac{[1 - (1/10^{0.14 \times 10^{-3}})]}{[1 - (1/10^{0.03 \times 10^{-3}})]} = 4.67$$

[20] K. F. Palmer and D. Williams, *J. Opt. Soc. Am.* **64,** 1107 (1974).
[21] Y. Liu, D. K. Cheng, G. J. Sonek, M. W. Berns, C. F. Chapman, and B. J. Tromberg, *Biophys. J.* **68,** 2137 (1995).

so the Beer–Lambert law suggests we should expect 4.67 times as much heating from a 1064-nm laser than from a 830-nm laser, i.e., we expect a temperature rise of roughly $0.31°/100$ mW at 830 nm. However, the absorbance of water is significantly higher at 980 nm than 1064, and we calculate $P_{abs,980}/P_{abs,1064} = 3.54$, implying that a 980-nm laser should cause roughly 3.5 times the heating of a 1064-nm laser, or approximately $5.1°/100$ mW. Note that the quoted laser power of 100 mW is the power in the sample, not the power at the laser itself. Since a good $100\times$ objective only transmits about 50–60% of the incident light, including other optical elements between the laser and the sample, typically approximately 45% of the laser light enters the sample. Thus, trapping with a (nominal) 200-mW laser functioning at 1064 nm in practice causes a rise of roughly $1.3°$ in the sample. The heating due to either a 1064-nm or 830-nm laser is probably irrelevant, but heating due to a 980-nm laser may be an issue, depending on how sensitive the process being studied is to the local temperature.

Optical Damage: Choice of Laser Wavelength

Although any laser can be used to make an optical trap, a few wavelengths are preferred due to minimized optical damage. A number of studies have investigated optical damage, primarily using two types of assays to measure cell health. In the first, the efficiency of cloning was measured as a function of exposure to the optical trap.[22] In the second, *Escherichia coli* bacteria were tethered to a surface by single flagella, and their attempt to swim resulted in regular rotations of the bacteria. The rate of rotation was used as an indicator of bacterial health.[23] Both studies came to the same conclusion: lasers at 980 nm do the least damage; 830 nm is second best, with the commonly used 1064 nm wavelength a distinct third. On a scale where the damage at 980 nm $= 1$, the damage at 830 nm $= 1.27$, and the damage at 1064 nm $= 2.5$.[22,23] Thus, an 830-nm laser does roughly 30% more damage that a 980-nm laser, whereas a 1064-nm laser does two and a half times as much damage.

It is surprising that these independent studies came to the same conclusion with regard to optical damage since they are presumably sensitive to different factors: in the cloning efficiency study the laser is focused on the nucleus, and damage presumably occurs inside the nucleus, whereas in the second the laser is focused on a part of the bacteria, and appears to affect metabolic processes. One possibility[23] is that the mechanism of damage is relatively generic, e.g., due to the generation of free oxygen radicals that damage whatever they encounter.

Additional studies by Liu *et al.*[24] used a number of different probes to investigate whether optically trapping cells (at 1064 nm) had effects on DNA structure,

[22] H. Liang, K. T. Vu, P. Krishnan, T. C. Trang, D. Shin, S. Kimel, and M. W. Berns, *Biophys. J.* **70**, 1529 (1996).

[23] K. C. Neuman, E. H. Chadd, G. F. Liou, K. Bergman, and S. M. Block, *Biophys. J.* **77**, 2856 (1999).

[24] Y. Liu, G. J. Sonek, M. W. Berns, and B. J. Tromberg, *Biophys. J.* **71**, 2158 (1996).

cell viability, and intracellular pH. No effect on either DNA structure or intracellular pH was observed at up to 400 mW. Some loss of viability for low-motility sperm was observed when they were held for extended periods (more than 2 mins) at 300 mW.

Ways to Decrease Optical Damage

Given that damage has been minimized to the extent possible by appropriate choice of laser wavelength, how can damage be decreased? If possible, the most practical solution is to increase the size of the trapped object—using the same power, a 0.5-μm bead feels significantly less force than a 1-μm silica bead (see below). The presence of oxygen is also a key factor in determining optical damage: in the bacterial assay,[23] eliminating oxygen using an oxygen scavenging system (introduced just before trapping) resulted in a 3- to 6-fold decrease in damage. Although the direct use of such a scavenging system is possible *in vitro* and when studying anaerobic organisms, it is of little utility when studying eukaryotic organisms. However, adding quenchers of singlet oxygen to the growth media[23] (e.g., antioxidants) might be helpful in this latter case, but to date there is no reported evidence supporting the efficacy of such additives.

Amount of Power to Be Used

The amount of power that can be used depends to some extent on the experiment. The bacterial assay suggests that optical damage is a gradual (rather than catastrophic) process, so minimizing the length of laser exposure is important. The assay using the efficiency of cell cloning as a readout for optical damage (trap focused on the nucleus) found that a 1-min exposure to 176 mW (in the sample) had no effect on cloning efficiency at 830 nm or 980 nm, and only reduced cloning efficiency by 10% at 1064 nm. A 3-min exposure of 176 mW had more effect, reducing the cloning efficiency by 50% at 830 nm, 20% at 980 nm, and 70% at 1064 nm. For manipulation of nuclei, it is probably reasonable to use up to approximately 200 mW power (in the sample) as long as a 830-nm or 980-nm laser is used, and exposure times are kept relatively short.

As far as manipulating mitochondria or other organelles, the amount of power that can be used again depends on wavelength. At 1064 nm, this has been measured directly: using up to 340 mW (measured after the objective, but before the sample) to manipulate and hold an organelle causes no obvious damage, even when the organelle is held for a number of minutes.[25] However, damage starts to appear at powers greater that 370 mW, though again only when the object is trapped for a number of minutes.[25] This result is consistent with the work of Ashkin *et al.*,[18] who

---

[25] M. W. Berns, J. R. Aist, W. H. Wright, and H. Liang, *Exp. Cell Res.* **198,** 375 (1992).

found that a 220-mW trap (the maximum power they required to stop all moving mitochondria, with the power of the 1064-nm laser measured at the sample plane) did no apparent damage to moving mitochondria—motion of the organelles was the same before and after brief periods of immobilization in the trap.

Because both the bacterial health and the cloning assays show that 830 nm is significantly preferable to 1064 nm (by roughly a factor of 2) and heating at 830 nm is less than at 1064 nm (by a factor of 4.67, see above), I estimate that at 830 nm one can probably use approximately double the maximum power possible at 1064 nm, i.e., 680 mW at the sample. Of course, the shorter the exposure, the better. For many experiments the manipulated object needs to be trapped only for seconds rather than the minutes of exposure used in the damage experiments.

Amount of Power Needed

In practice, forces of more than 8 pN on 0.5-μm-diameter lipid droplets[26] have been generated using a (nominal) 200-mW 830-nm laser, of which only approximately 35% made it through the lens (i.e., approximately 70 mW at the sample). Using 220 mW (at the sample), Ashkin *et al.*[18] were able to stop all moving mitochondria. The typical molecular motor exerts a force somewhere between 1 and 6 pN, and the limited experimental data *in vivo* suggest that at most approximately five motors function together on the same cargo.[18,19] Thus, the largest force one would estimate would be required for studies of cargo transport, to be able to stop all moving cargos, would be 30 pN. Because maximum stalling force is linearly proportional to laser power, assuming cargos similar to the lipid droplets, this would require approximately 300 mW at the sample, well within the range of what can be used at 830 nm. Because many motors require less than 6 pN to stall, this 300 mW is more or less an upper limit on the power required after the lens, for organelle-type studies. Thus, in general it should be possible to use enough power to stall any molecular motor-driven cargo *in vivo*. However, a priori it is not possible to make a definite estimate of the amount of power needed for a specific application because the force that can be applied *in vivo* will strongly depend on the optical properties of the trapped object, e.g., its size and refractive index. Thus, it may not be possible to effectively manipulate very small (200-nm) vesicles.

Choice of Laser

A number of factors must be considered in choosing which wavelength laser to use. If small laser powers will be used, it probably does not matter much. If large powers are likely, optical damage and heating become more of an issue. If

---

[26] S. Gross, M. Welte, S. Block, and E. Wieschaus, *J. Cell Biol.* **148,** 945 (2000).

the paramount concern is optical damage, 980 nm is best. However, if quantitative measurements are being made that could be affected by heating, it is probably better to use an 830-nm laser, trading a small increase in optical damage for a large decrease in heating. However, working at 830 nm potentially involves high cost: currently, the highest power single-mode (TM00) 830-nm diode is 200 mW, sold by Melles-Griot. The only way to get more power at 830 nm is to use a Ti : sapphire laser. Although a reasonable alternative, Ti : sapphire lasers tend to be more fussy to use than the other diode lasers available at 830, 980, and 1064 nm. At 980 nm, the choice of lasers is much greater; e.g., a 1-W MOPA laser (980 nm). Finally, there are a number of outstanding lasers at 1064 nm (Nd : YVO$_4$) with good beam stability and an excellent beam profile. One of the best of these lasers is made by Spectra-physics (Topaz), which has a 3-W output. Of course, at 1064 nm there are many less expensive alternatives as well. Even though 830 nm is probably preferable, for many studies the difference is insignificant—in the majority of cases one will end up using significantly less than the maximum 340 mW possible at 1064 nm.

Importance of Particle Size and Index of Refraction

For a trap based on a TM$_{00}$ laser beam (Gaussian beam profile), the axial force (along the beam path) is weaker than the lateral force (pulling the object in toward the beam center). However, for both axial and lateral forces, at a given displacement from the trap center, the force a trapped object feels is a function of its size: up to about 4 μm, the larger the object, the more force it feels. Theoretical calculations for objects whose size is small relative to the wavelength of the trap light (Rayleigh particles) suggest a cubic dependence ($Force_{max} \sim d^3$), where d is the size of the object.[27] However, for trapped particles of the same approximate size as the light's wavelength (e.g., 500-nm vesicles in an 830-nm trap), this dependence is somewhat weaker, in general somewhere between linear and quadratic. Empirically, for trapped latex beads, the maximum lateral force applied on a 1.02-μm bead was 7 times that on a 0.3-μm bead, and the maximum force on a 2.97-μm bead was 2.47 times more than on the 1.02-μm bead.[28]

The applied force is also a function of n, the relative index of refraction defined as $n = n_1/n_2$, where n_1 is the index of refraction of the trapped object and n_2 is the index of the surrounding medium. A good discussion of this can be found in Ashkin's review article.[29] For lateral forces, the larger n, the larger the applied force. Up to a point, this is also true for axial forces. By way of a typical example,

[27] A. Ashkin, J. M. Dziedzic, J. E. Bjorkholm, and S. Chu, *Opt. Lett.* **11,** 288 (1986).

[28] H. Felgner, O. Muller, and M. Schliwa, *Appl. Opt.* **34,** 977 (1995).

[29] A. Ashkin, *in* "Laser Tweezers in Cell Biology" (M. P. Sheetz, ed.), Vol. 55, p. 1. Academic Press, San Diego, CA, 1998.

consider how much the applied force changes when an object (e.g., a lipid droplet, $n_1 = 1.52$) is a water ($n_2 = 1.33$) versus in cytoplasm ($n_2 = 1.39$). Using Ashkin's published values (and a bit of linear interpolation) we find that going from $n = (1.52/1.33) = 1.14$ to $n = (1.52/1.39) = 1.09$ results in a decrease of the applied lateral force by approximately 20%. So, relatively small corrections are needed to enable calibrations to be done *in vitro,* and actual measurements to be done *in vivo.*

Particle–Coverslip Distance

While *in vitro* the trapped object is manipulated very close to the coverslip, controlling the object–coverslip distance is often not possible *in vivo.* High numerical aperture (NA) oil-immersion microscope lenses are corrected to decrease spherical aberration, but only close to the coverslip. As the lens is focused deeper into the sample, aberration increases, which results in a decrease in image quality, and more importantly, a decrease in trapping power. In practice, the extent of the change depends on the properties of the trapped object: *in vitro,* the effect is significantly more pronounced for polystyrene as opposed to silica beads. As a rough estimate, as the distance between the coverslip and the trapped object increases from 5 to 30 μm, there is a roughly 50% decrease in the applied trapping force for the same laser power (see Felgner *et al.*[28] for experimental data).

Although spherical aberration is important when working at depth, the extent of object–coverslip viscous coupling must be considered when the trapped object is relatively close to the coverslip. This effect does not actually alter the maximum force that the trap can apply, but does alter the way the trapped object moves, because its motion results from its exposure to both viscous and trapping forces. Thus, if one attempts to move the trapped object rapidly relative to the coverslip, the apparent trapping force is lower than might be expected if the object is close to the coverslip, because the viscous force opposing motion is high. Stokes law (discussed below) can be corrected to reflect this increased viscous drag; the functional form of the correction is given in Felgner *et al.*[28] Experimentally, given the same relative motion a 0.5-μm bead feels twice as much viscous drag at 1.5 μm from the coverslip than at 4 μm.[30]

Thus, for quantitative measurements, it is important to pay close attention to the distance of the trapped object from the coverslip. If the sample geometry is such that one needs to work at depth, it is worth considering using a high NA water-immersion lens, which is designed to minimize spherical aberration over a much larger range of focusing depths. However, unless absolutely necessary these objectives are not optimal for trapping, because they have a slightly lower

[30] J. Dai and M. P. Sheetz, *in* "Laser Tweezers in Cell Biology" (M. P. Sheetz, ed.), Vol. 55, p. 157. Academic Press, San Diego, CA, 1998.

NA (1.2), so that the gradient in the focused beam (and hence trapping strength) is decreased. They are also quite expensive.

Determination of Applied Force

In general, there are two ways that optical traps are used to determine stalling forces. In one approach, the position-determination method, a high laser power is used to stop all moving beads (or other objects), and then the force required to stop the bead is determined from the position of the bead in the trap, having previously calibrated the relationship between bead displacement and applied force. The second method instead looks at the escape force. In this approach, the initial calibration determines how much force it takes to remove the object entirely from the trap, as a function of laser power. Then, force measurements on a specific object are made by fixing the laser at a given power, and by scoring whether at that power (and hence applied force) the object is able to escape from the trap. This was the method used by Gross et al.[26] As a variation on this method, if the object's behavior is not rapidly changing, initially a high laser power can be used such that all the objects are trapped. Then, the laser power is gradually reduced to find the power at which the object escapes.

There are advantages to both methods. The position-determination method is typically used in vitro, and has the advantage of extremely high spatial and temporal resolution. However, it requires precise determination of the location of the trapped object relative to the trap center, usually to within a few nanometers. Such determination usually involves a sensitive detection setup that involves a quadrant photodiode and good electronic amplifiers, which can be somewhat involved. The quadrant photodiode can be placed either in the back focal plane of the microscope,[2] or used in an imaging mode.[1] When operating in the back focal plane, the deflection of the laser beam is measured after it passes through the sample. This method has the advantage that it does not measure the global position of the trapped object, but rather is sensitive only to the displacement of the object relative to the trapping beam. This often turns out to be impractical for in vivo work, where there can be numerous large scattering centers in the sample but after the trap's focus. These additional scattering centers cause additional (random) beam deflections, so that it is impossible to determine which contributions result from the trapped object alone. Thus, for most in vivo applications the quadrant photodiode should be imaged. One drawback of imaging the photodiode is that only a very small part of the field of view is imaged, so the trapped object to be measured must be carefully positioned. Thus, for such an approach to be practical in vivo requires an accurate computer controlled x–y translational stage to move the object to the appropriate location.

Instead of using a quadrant photodiode, particle tracking can be used to determine the particle's position so that the position-determination method can be employed without sophisticated electronics. With a good image, current software can determine the location of the center of a 0.5-μm bead to within 8 nm.[26] For a trap that produces a maximum of 10 pN of force, the trap stiffness constant is approximately $k = (10 \text{ pN}/200 \text{ nm}) = 0.05$ pN/nm, so an uncertainty of 8 nm corresponds to an uncertainty of 0.4 pN in determination of the stalling force. Relative to the use of a quadrant photodiode, particle tracking is slow (maximum of 30 Hz) and less precise, but is easier to set up.

The advantage of the escape-force approach is that the object can be trapped anywhere in the field of view, and it does not require detection electronics: the object is trapped, and whether or not it escapes at a given force can be scored by eye. A relatively precise measurement of the minimum force required to stall the object can be made by slowly decreasing the laser power, and noting at what point the object escaped. The disadvantage of this method is that it is slow, and it is impractical to slowly lower the laser force for objects whose behavior is rapidly changing. Thus, rather than measuring the stalling force for each object, one often ends up measuring a "population" stalling force: what was the average force required to stop the majority of moving objects, or what force was required to stop all moving objects (though many were stopped at lower powers). Both works describing *in vivo* measures of stalling forces[18,19] used variants of escape-force measurements.

Regardless of which method is used, the force the laser applies to the objects of interest must be calibrated. This is done *in vitro*. If beads are used as handles, the calibration is done on the appropriately sized bead. If a biological object (e.g., a vesicle) is the object being studied, these must be biochemically purified, so that calibration can be done on typical examples, again *in vitro*. All measurements are made at typical coverslip–bead working distances.

Given a single bead or vesicle trapped in buffer, the way the force is calibrated depends on whether one plans to do position-determination or escape-force measurements. If escape-force measurements are planned, the trapped object is moved relative to the buffer, until viscous drag forces pull it out of the trap. The force at that moment can then be calculated from Stokes law, which says that the viscous drag force, F, on a spherical object is given by

$$F = 6\pi\eta r v$$

where η is the coefficient of viscosity, r is the object's radius, and v is the velocity of the fluid relative to the object. The actual drag is modified by the particle's distance to the coverslip; see Felgner *et al.*[28] for details. The viscous drag is usually generated by keeping the trap fixed and using a computer-controlled x–y translational stage to move the stage relative to the trapped object.

If position-determination measurements are planned, the applied force can be calibrated in a variety of ways. As above, a constant force can be applied by moving the translational stage at a constant velocity. Then, as long as the applied force is below the escape force, the object stays in the trap, and its displacement relative to the trap center can by measured with whatever position detection scheme is being used. Using different velocities, one can experimentally determine the force-displacement curve. This is compatible with particle-detection approaches to position determination, as well as quadrant photodiode methods. Alternatively, one can use the Brownian motion of the object in the trap to determine the traps stiffness, as long as the position-detection equipment has sufficient bandwidth.[31]

Conclusions

The *in vivo* environment is determined by the organism rather than what is optimal for optical trapping. This leads to two sets of concerns: how to minimize optical damage and how to calibrate the force actually being applied. For *in vivo* work, lasers operating at 830 nm are particularly good, because they both minimize optical damage and cause relatively little local heating. Because optical damage is a gradual process, it is particularly important to minimize the duration that an object is exposed to the laser. Much of this article was devoted to optical damage, but the actual laser power required in many applications is low enough that it may not be an important design consideration.

To calibrate the applied force involves initial *in vitro* calibration, followed by corrections to compensate for the differences between working in buffer and in the relatively uncontrolled cytoplasm. The major concerns are the depth at which the measurements are made, the effective index of refraction of the cytoplasm, and the homogeneity and size of the trapped objects.

All of these difficulties notwithstanding, the laser trap is a tremendously powerful tool whose use can benefit many areas of research. With the recently completed genomes, we are starting to have a complete picture of all the proteins present in an organism, yet in many cases we still have little understanding of how these proteins function. By their very nature, processes in which the cytoskeleton plays an important role involve understanding forces and motion, questions for which *in vivo* optical traps are ideal tools. Although the field is still in its infancy, the use of optical traps is already allowing exciting advances in such diverse fields as immunology and cargo transport.

[31] K. Visscher, S. P. Gross, and S. M. Block, *IEEE J. Selected Top. Quant. Electron.* **2,** 1066 (1996).

[9] Cytomechanics Applications of Optical Sectioning Microscopy

By B. Christoffer Lagerholm, Steven Vanni, D. Lansing Taylor, and Frederick Lanni

Introduction

In the past half-century since the purification of actin, the fields of biochemistry and biophysics have produced a vast trove of information on the protein components and regulatory mechanisms of the cytoskeleton. Likewise, microscopy of both fixed and living cells has produced detailed information on the subcellular location of major cytoskeletal components, and to some degree, the kinematics. Nowhere has this information been more comprehensive than for the muscle fiber, in which the actomyosin contractile system is organized into a near-crystalline state. In the nonmuscle tissue cell, the cytoskeleton is much less ordered, but is much more dynamic and versatile. In addition to major changes that occur through the cell cycle, the nonmuscle cytoskeleton drives processes such as cell migration, morphogenesis of embryos and tissue, tissue maintenance and wound healing, axonal pathfinding, and metastasis of certain cancers. To fully understand the internal machinery that the cell assembles for each task, it is not enough to know structure. The action or mechanics of the cytoskeleton must be measured and matched to identifiable assemblies. In contrast to the muscle fiber, which undergoes uniaxial contraction, and for which detailed mechanics information can be obtained by use of a force transducer or strain gauge, a fibroblast applies complex tractions to the matrix in which it lives, and rearranges its internal machinery on a minute-to-minute time scale. To analyze this cellular machine, an elastic collagen-based model extracellular matrix (ECM) can be utilized as a strain gauge. The light microscope, when suitably automated and interfaced to a digital imaging system, then serves a dual purpose: (1) to image the configuration of the cytoskeleton with molecular specificity through fluorescence labeling, and (2) to image the deformations produced by the cell in the surrounding matrix. From deformation data, strain can be computed as a tensorial set of spatial derivatives and compared qualitatively and quantitatively to the cytoskeleton and the attachments between cell and matrix. If the elastic properties of the matrix are known, the analysis can be extended to the computation of stresses and traction forces on the cell–matrix boundary (Fig. 1).

Harris *et al.*,[1,2] provided a unique tool for cytomechanics analysis by developing transparent, elastic substrata that support the growth of cells. Contractile patterns

[1] A. Harris, K. P. Wild, and D. Stopak, *Science* **208**, 177 (1980).
[2] A. K. Harris, D. Stopak, and P. Wild, *Nature* **290**, 249 (1981).

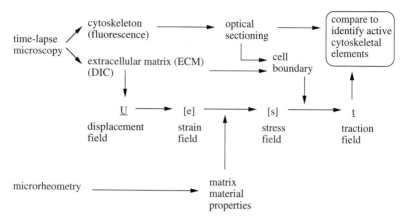

FIG. 1. Cytomechanics analysis via multimode light microscopy. In the constrained collagen gel cultures used in this study, most cells within the gel are oriented with major axis parallel to the cover glass. While serial-focus imaging is essential to follow individual cells over time, this configuration facilitates 2D image processing and plane–strain analysis. DIC images and FGI (fluorescence) optical sections of fibroblasts within the collagen-based matrix are acquired under a time-lapse, serial-focus, serial-field program of instrument control (see text). Single DIC images, which show the collagen matrix fibrils within the optical section, are processed through a 2D mechanics algorithm (DQA; http://dqa.web.cmu.edu) to derive the 2D vector field of material displacement, $U(x, y)$. From this field, the strain tensor field, $[e_{ij}(x, y)]$, is computed by taking spatial derivatives. This eliminates rigid-body movements (transverse drift and rotation) that do not contribute to strain. In the plane–strain approximation, $[e_{ij}]$ is a 2×2 symmetric matrix. The eigenvalues of $[e]$ are the 2D principal strains, and the corresponding orthonormal eigenvectors the principal directions of strain caused by the cell at each location in the gel. Patterns of traction application can be inferred from the directions of the principal strain vectors relative to the local boundary of the cell. However, if the elastic properties of the matrix are known through calibration or microrheometric measurements, the plane–stress tensor, $[s]$, can be derived from $[e]$. Along the cell–matrix boundary, traction (t) can be derived from stress and matched to fluorescence images showing the location of key cytoskeletal structures.

of forces applied by single cells produce characteristic patterns of wrinkles in Harris-type silicone membranes that are easily seen in phase-contrast microscopy and easily interpreted qualitatively. Additionally, the membranes permit reflection-interference microscopy (RIM, IRM) for identification of cell–substratum adhesions, and fluorescence microscopy. With careful characterization of the membranes by calibrated microneedle deflection, it is possible to use this system to estimate traction forces applied by locomoting cells.[3–5] However, because wrinkling is the result of a nonlinearity in the elastic response of the membrane, it is

[3] K. Burton and D. L. Taylor, *Nature* **385,** 450 (1997).
[4] K. Burton, J. H. Park, and D. L. Taylor, *Mol. Biol. Cell* **10,** 3745 (1999).
[5] B. A. Danowski, K. Imanaka-Yoshida, J. M. Sanger, and J. W. Sanger, *J. Cell Biol.* **118,** 1411 (1992).

notoriously difficult to make a general inverse analysis of wrinkle patterns. By reducing the compliance of the substratum to suppress wrinkling, and incorporating marker particles in a random array, Jacobson and co-workers[6–10] demonstrated that two-dimensional (2D) analyses of cell traction could be made by time-lapse microscopy of particle displacement, followed by digital image processing. Alternatives that circumvent complex mechanics include the use of microlithography and microcontact printing to prepare patterned glass or silicone surfaces incorporating arrays of independent spring elements or cantilevers that can be read out as strain gauges.[11,12] Polymeric alternatives to silicone have also been developed, most notably polyacrylamide (PAA) hydrogel thin films bonded to cover glasses.[13–19] PAA films are elastic, impervious to invasion, and inert to enzymatic degradation. They are highly transparent (requiring the use of marker particles or molded features for tracking deformation), can be made over a wide modulus range by varying the total acrylamide content or the acrylamide/bisacrylamide ratio, and can be modified readily by trapping or grafting biopolymers to "engineer" the gel surface to present biomolecular epitopes. Fibroblasts adhere poorly and fail to grow on 10% PAA, but will adhere and grow readily on PAA containing 0.1% gelatin or collagen.

   Although elastic substrata are now widely used, they differ in at least one respect from the tissue environment. Planar membrane or gel systems impose a nonnative geometric constraint on individual cells in that all adhesions are formed on one "side" of the cell, leaving an equal or greater membrane surface devoid of contact. A solution to this constraint is to grow cells at the interface between two contacted hydrogels, or within a hydrogel. Growth and patterning of fibroblasts have long been known to occur on the surface of collagen gels prepared by neutralization of acidic solutions of the protein.[20] Collagen model ECMs have

[6] J. Lee, M. Leonard, T. Oliver, A. Ishihara, and K. Jacobson, *J. Cell Biol.* **127**, 1957 (1994).
[7] T. Oliver, M. Dembo, and K. Jacobson, *Cell Motil. Cytoskeleton* **31**, 225 (1995).
[8] T. Oliver, K. Jacobson, and M. Dembo, *Methods Enzymol.* **298**, 497 (1998).
[9] M. Dembo, T. Oliver, A. Ishihara, and K. Jacobson, *Biophys. J.* **70**, 2008 (1999).
[10] T. Oliver, M. Dembo, and K. Jacobson, *J. Cell Biol.* **145**, 589 (1999).
[11] C. G. Galbraith and M. P. Sheetz, *Proc. Natl. Acad. Sci. U.S.A.* **94**, 9114 (1997).
[12] N. Q. Balaban, U. S. Schwarz, D. Riveline, P. Goichberg, G. Tzur, I. Sabanay, D. Mahalu, S. Safran, A. Bershadsky, L. Addadi, and B. Geiger, *Nat. Cell Biol.* **3**, 466 (2001).
[13] F. Lanni, "Development of Elastic Substrata for Optical Estimation of Cell Traction Forces." Science and Technology Center Internal Report (1993).
[14] T. A. Thomas, Ph.D. Thesis. Department of Chemical Engineering, Carnegie Mellon University, 1998.
[15] R. J. Pelham, Jr. and Y.-L. Wang, *Proc. Natl. Acad. Sci. U.S.A.* **94**, 13661 (1997).
[16] Y.-L. Wang and R. J. Pelham, Jr., *Methods Enzymol.* **298**, 489 (1998).
[17] R. J. Pelham, Jr. and Y.-L. Wang, *Mol. Biol. Cell* **10**, 935 (1999).
[18] M. Dembo and Y.-L. Wang, *Biophys. J.* **76**, 2307 (1999).
[19] S. Munevar, Y.-L. Wang, and M. Dembo, *Biophys. J.* **80**, 1744 (2001).
[20] T. Elsdale and J. Bard, *J. Cell Biol.* **54**, 626 (1972).

been developed for quantification of single-cell[21,22] or collective[23–26] contractile activity of fibroblasts and endothelial cells. Cells can be grown within the gel by casting a cell–gel composite or on its surface by inoculation of a formed gel. Much of our understanding of collagen contraction, both in terms of measuring net cell-generated tension and determining the regulating signaling pathways, has come from experiments in bulk format typically involving 10^5–10^7 cells per gel. Bulk type-I collagen gels in the 1–3 mg/ml range are hazy in appearance but have sufficient transparency for transmitted-light microscopy when cast with thickness less than 1 mm. The gel is elastic under small deformations,[27] is stable against proteolysis, and can easily be biochemically modified by the addition of other ECM polymers such as fibronectin. Because collagen fibrils in the gel network are directly visible in phase contrast or DIC, marker particles are not needed. Miniaturization of this preparation enables high-resolution light microscopy of single cells within the 3D collagen matrix, and the direct observation of deformations in the collagen matrix surrounding each cell.[28,29] In this case, the collagen matrix plays the role of an *in situ* strain gauge readable through the use of high-resolution light microscopy and digital image processing. At lower magnification, it is also possible to follow the movements and actions of a large number of cells within a cell–gel composite. This permits us to view directly and begin to analyze the emergent properties of a model tissue matrix.

Materials and Methods

Cell Culture

Swiss 3T3 fibroblasts (CCL-92, ATCC, Manassus, VA) are grown in bicarbonate buffered Dulbecco's modified Eagle's medium (DMEM; Gibco, Rockville, MD) supplemented with 10% calf serum, penicillin/streptomycin and, in the case of transfected cells, with 0.5 mg/ml geneticin (G418, Gibco). Fibroblasts are transfected to express yellow fluorescent protein (YFP)–actin using the LipofectAMINE Plus transfection kit (Life Technologies, Rockville, MD) and a commercially available yellow fluorescent protein actin vector (pEYFP-actin; Clontech, Palo Alto,

[21] P. Roy, W. M. Petroll, H. D. Cavanagh, C. J. Chuong, and J. V. Jester, *Exp. Cell Res.* **232**, 106 (1997).
[22] P. Roy, W. M. Petroll, H. D. Cavanagh, and J. V. Jester, *Cell Motil. Cytoskeleton* **43**, 23 (1999).
[23] M. S. Kolodney and R. B. Wysolmerski, *J. Cell Biol.* **117**, 73 (1992).
[24] M. S. Kolodney and E. L. Elson, *J. Biol. Chem.* **268**, 23850 (1993).
[25] Z. M. Goeckeler and R. B. Wysolmerski, *J. Cell Biol.* **130**, 613 (1995).
[26] F. Grinnell, *Trends Cell Biol.* **10**, 362 (2000).
[27] D. Velegol and F. Lanni, *Biophys. J.* **81**, 1786 (2001).
[28] S. Vanni, B. C. Lagerholm, C. A. Otey, D. Velegol, and F. Lanni, *Eur. Cells Materials* **2** (Suppl. 1), 21 (2001).
[29] S. Vanni, B. C. Lagerholm, C. A. Otey, D. L. Taylor, and F. Lanni, *Biophys. J.* **84**, in press.

CA). Swiss 3T3 fibroblasts expressing green fluorescent protein (GFP)–α-actinin were a generous gift of Dr. Carol Otey of the University of North Carolina at Chapel Hill.[30]

Fluorescence-Activated Cell Sorting

Stable transformants are enriched for highly expressing cells by fluorescence-activated cell sorting (FACS) on an EPICS Elite flow cytometer equipped with an argon laser tuned to the 488 nm line. Cells to be sorted are detached from tissue culture flasks with trypsin, pelleted at 500 rpm for 10 min, and resuspended in bicarbonate-buffered phenol red-free glutamine-deficient Ham's F12 medium (#9589, Irvine Scientific, Santa Ana, CA) supplemented with 3 mg/ml sterile bovine serum albumin (Sigma, St. Louis, MO) at a cell density of 10^6 cells/ml. Prior to sorting, the cell suspension is filtered to remove aggregates and stored on ice. Sorted cells are selected based on GFP or YFP fluorescence, as well as forward scatter and side scatter in which selected cells exhibit a fluorescence intensity greater than the typical maximum autofluorescence of untransfected Swiss 3T3 cells.

Cell-Collagen Suspension

Cell suspensions are prepared from type-I collagen (Vitrogen 100, Collagen Corporation, Fremont, CA) by following the manufacturer's suggested protocol. In brief, collagen is soluble in dilute acid, and gels on neutralization. A buffered stock solution is prepared by mixing cold Vitrogen 100 with $10\times$ phosphate-buffered saline (PBS: 0.2 M Na_2HPO_4, 1.3 M NaCl, pH 7.4) and 0.1 M NaOH at a ratio of $8:1:1$. This solution is stored on ice to inhibit gelation. Cells are detached from tissue culture flasks with trypsin, pelleted at 500 rpm in a clinical centrifuge for 10 min at $4°$, and resuspended in Ham's F12 medium supplemented with 10% calf serum and 25 mM HEPES (pH 7.3) for additional pH control in ambient atmosphere. Cells are kept on ice until use (typically <2 hr). Cell–gel composites are prepared by diluting the collagen stock solution with the cell suspension and additional media to give a final collagen concentration of 2.0 mg/ml. The cell suspension as well as all buffers (except $10\times$ PBS) and collagen solutions are stored on ice until use.

Supported Floating Cell–Collagen Gel Composites

Supported floating cell–collagen gel composite specimens are prepared from 65 μl of the cell–collagen stock solution pipetted into the center of a circular nylon mesh disc cut in the shape of a flat washer, and subsequent incubation for

[30] M. Edlund, M. A. Lotano, and C. A. Otey, *Cell Motil. Cytoskeleton* **48,** 190 (2001).

60 min in a closed 60-mm culture dish in a humidified, 37°, 5% CO_2 incubator. To avoid attachment of the gels to the dish, a sterile 40-mm glass coverslip is placed inside, and the mesh rings are placed on top of the coverslip. On gelation of the collagen, Ham's F12 medium is added to the dish to submerge the specimens. For imaging of single cells, samples are typically prepared with a final cell density of 250–500 cells/gel. The nylon mesh discs are cut from sheet stock (CMN-185, Small Parts, Inc., Miami Lakes, FL) by using 0.250 inch and 0.375 inch hole punches (#402 and #448; M.C. Mieth Mfg., Inc., Port Orange, FL) for the inside and outside diameters, respectively. In our sample configuration, gel contraction is initially limited to reduction in gel thickness while changes in gel area are restricted as long as the gel is attached to the nylon screen. To initiate transverse contraction, we make two parallel cuts in the gel thereby restricting transverse contraction to one axis (Fig. 2).

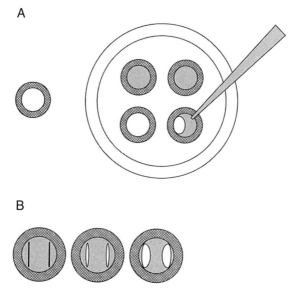

FIG. 2. Preparation of supported floating cell–collagen gel composites. (A) Stiff nylon mesh rings on a dry 40-mm cover glass in a sterile 60-mm dish are filled by pipette with 65 μl ice-cold cell suspension in neutralized collagen. The dish is then promptly covered and incubated 60 min under normal incubator conditions to set the gel. Evaporative loss averages 0.13 μg/min, accounting for less than 15% loss of gel weight during this period. This loss will become significant if the incubator is not well humidified. After gelation, culture medium is added, and the dish gently shaken to release the gels from the cover glass surface. (B) Lateral contraction of constrained collagen gel occurs after parallel cuts are made with a scalpel blade. In densely populated gels cultured 1–2 days, contraction occurs within 30 min. In sparsely populated gels, contraction occurs over 1–3 days. Nylon ring-supported gels are easily transferred between dishes or into environmental chambers for time-lapse microscopy.

Mounting and Maintenance of Specimens for High-Resolution Imaging

Supported cell-collagen composites are clamped onto a #1-1/2 glass coverslip (Fisher Scientific, Pittsburgh, PA) using two rectangles of fine stainless steel mesh (CX-200-B, Small Parts, Inc.) glued onto the coverslip with UV-setting cement (Norland Optical Adhesive, Edmund Scientific Co., Barrington, NJ). Clamping ensures that the specimen remains within the working distance of the microscope objective as much as possible. The coverslip is then mounted in a Focht perfusion chamber (FCS2, Bioptechs, Inc., Butler, PA) for long-term environmental control on the microscope (Fig. 3). Samples are typically imaged in bicarbonate-buffered, phenol red-free, low-riboflavin Ham's F12 medium supplemented with 10% calf

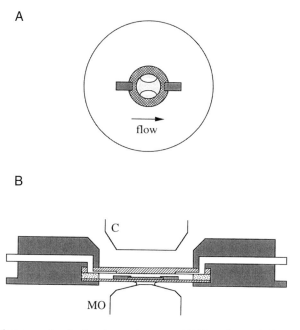

FIG. 3. Specimen mounting for time-lapse microscopy. (A) The nylon mesh ring-supported specimen is affixed to a 40-mm cover glass by insertion of the ring under a pair of screen tabs glued directly to the cover glass. (B) The affixed specimen is mounted in a perfusible temperature-controlled chamber (FCS2) with the gel cuts oriented to minimize disturbance by perfusive flow. Nearly the entire volume of the collagen gel lies within the working-distance range ($\sim$220 μm) of a high-NA immersion objective (MO). To obviate the problem of focus-dependent spherical aberration, an indirect water immersion objective (C-Apochromat 40$\times$ 1.2NA, Carl Zeiss) with dial-in cover glass thickness setting is used with distilled water as the immersion fluid. For long-term experiments (6 hr to 7 days), evaporation of the immersion water requires occasional replenishment from a small plastic syringe tube. Alternatively, we utilize perfluorotetradecahydrophenanthrene ($C_{14}F_{24}$), an inert clear liquid nearly isorefractive with water but having a lower evaporation rate.

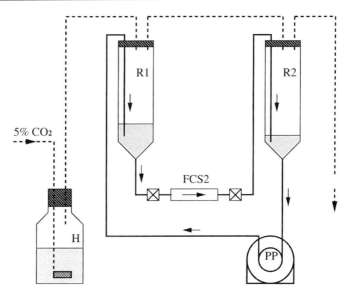

FIG. 4. Recirculating gravity perfusion system for long-term cell culture and time-lapse microscopy. Problems to be overcome include pulsation, gas exchange, evaporation, and medium conditioning. Temperature control is established by components in the FCS2 chamber unit. A peristaltic pump (PP) transfers medium from the drain reservoir (R2) to the supply reservoir (R1). Pulsation, which seriously affects focus plane stability, is eliminated by the large-diameter reservoirs that convert large flow rate variations into small changes in fluid column height. The resulting head difference causes smooth gravity-driven flow through the chamber, and this flow stabilizes to match average pump speed. Therefore, the steady-state perfusion rate is set simply by changing the pump speed. The reservoirs (50-ml syringe bodies) also provide a large fluid surface for gas exchange with the headspace. Both R1 and R2 are slowly ventilated at very low pressure with sterile, 5% CO_2 in air, which is first humidified by passage through a sterile water bubbler (H). The tubing length between R1 and the specimen is shortened to minimize CO_2 loss by permeation. To maintain medium conditioning by the cultured cells, the medium volume is minimized (5–10 ml) and recirculated. Reservoir, valve, and tubing components are cleaned and sterilized with USP hydrogen peroxide prior to use.

serum and 25 mM HEPES (pH 7.3) for additional pH control in ambient atmosphere. Use of pH indicator-free, low-riboflavin medium significantly reduces background in the fluorescence images. During time-lapse experiments, approximately 10 ml of medium is maintained in circulation over the specimen through a recycling gravity-flow system. This setup (Fig. 4) provides for gas equilibration and continuous perfusion of the FCS2 chamber with a limited volume of medium, thereby minimizing the problem of dilution of cell-produced growth factors.

Automated Time-Lapse Multimode Microscopy

Image data were obtained by use of the automated interactive microscope system (AIM) developed at the NSF Center for Light Microscope Imaging and

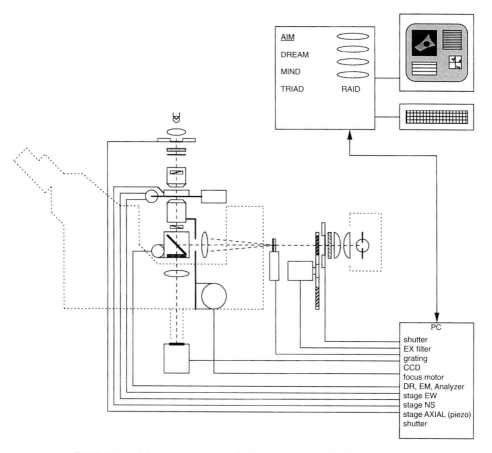

FIG. 5. Automated microscope system for time-lapse cytomechanics experiments.

Biotechnology.[31] The core components of the system consist of an inverted microscope (Zeiss Axiovert 135TV) with automation, a cooled charge-coupled device (CCD) camera (Hamamatsu C-4772-98), a perfusible thermostatted on-stage incubation chamber (Bioptechs FCS2), fluorescence grating imager optics,[32] and a computer system for instrument control, data acquisition scheduling, and image management (Fig. 5). AIM software provides a graphic user interface for setting multiple field coordinates in three dimensions through the coordinated control of a motorized stage and a motorized focus drive. Logged coordinates can range

[31] D. L. Taylor, K. Burton, R. L. DeBiasio, K. A. Giuliano, A. H. Gough, T. Leonardo, J. A. Pollock, and D. L. Farkas, *Ann. N.Y. Acad. Sci.* **820,** 208 (1997).
[32] F. Lanni and T. Wilson, *in* "Imaging Neurons—A Laboratory Manual" (R. Yuste, F. Lanni, and A. Konnerth, eds.), p. 8.1. Cold Spring Harbor Laboratory Press, Cold Spring Harbor, NY, 2000.

from independent single fields to complete 3D tilings of a specimen volume. For repetitive serial-focus image acquisition, a piezoelectric axial stage (Physik Instrumente) operates independently of the motorized focus drive. Independent (and multiple) image acquisition schedules can be set for differential interference contrast (DIC) microscopy and fluorescence to minimize photobleaching and phototoxicity. The microscope is equipped with a motorized six-position slider containing fluorescence filter sets and a Polaroid analyzer for DIC. Automated components and the CCD camera interface to a PC-based motion and acquisition processor pipelined to a fast dual-processor Unix-based computer (Onyx, Silicon Graphics, Inc.) equipped with a 20 GB RAID-type disk array. Because the AIM software was designed to utilize the SGI Irix 64-bit operating system, data address space is vastly larger than disk capacity. Therefore very large image sets can be directly addressed for efficient management and review. This is accomplished through software modules for multiimage navigation and display and for three-dimensional rendering of serial-focus image sets.

Optical Sectioning Microscopy

Optical sectioning refers to the formation of images, containing only in-focus information, that correspond to planar "slice" cross sections of a 3D object. Clearly, this has meaning only when the depth of field of the imaging device is much less than the depth of the object. A serial-focus stack of optical sections constitutes a 3D representation of the object. If the specimen is dynamic, as is generally the case with living cells in culture, speed in control of focus and in serial image acquisition is also essential for suppression of motion-related distortion in optical sectioning.

It should be stated at the outset that good optical sectioning depends on minimization of aberration in the microscope optical system. When looking deeply ($> 10 \mu$m) into low index specimens such as collagen gels, spherical aberration will become a limiting factor in oil immersion systems. For work in closed environmental culture chambers we recommend use of a high numerical aperture (NA) indirect water immersion objective with adjustable corrector for cover glass thickness.

In microscopy, both transverse and axial (depth) resolution sharpen with increasing NA.[33] This is most easily seen as a strong reduction of depth of field in high-NA optical systems. Because model ECMs and the cells within them are much greater in thickness than the depth of field in this case, any plane of focus within the specimen produces a direct image that contains a superposition of both in-focus and out-of-focus features. For both cytoskeletal imaging and tracking of deformation, this situation constitutes a serious impediment to extraction of quantitative information. Therefore, direct or indirect optical sectioning is a requirement in cytomechanical studies based on image processing and analysis. Although collagen

---

[33] F. Lanni and H. E. Keller, *in* "Imaging Neurons—A Laboratory Manual" (R. Yuste, F. Lanni, and A. Konnerth, eds.), p. 1.1. Cold Spring Harbor Laboratory Press, Cold Spring Harbor, NY, 2000.

gel fibril structures are easily visible in phase contrast, restriction of the condenser pupil by the annulus extends the depth of field and degrades optical sectioning performance in thick specimens. By comparison, DIC systems, which operate with both condenser and objective pupils fully unobstructed, provide a high degree of optical-sectioning performance in direct imaging. DIC performance can be degraded by the cumulative birefringence from oriented collagen fibrils in regions of the specimen in which cells have produced significant structural anisotropy, however, this has not been a limiting factor in minigels. Depth-of-field sharpness (optical sectioning) in DIC improves strongly with NA, therefore configuring the specimen for use with a moderate- to high-NA condenser and high-NA objective is important.

In fluorescence, quantification also requires optical sectioning, in this case selective elimination of background and out-of-focus contributions in each image. This is usually achieved directly by confocal scanning,[34] or indirectly by computational deconvolution of serial-focus direct-image data sets.[35] Although direct-imaging CCD camera-based instruments can be very fast, accurate deconvolution requires an extremely well-characterized optical system (in terms of accuracy and stability of the 3D point-spread function) and great attention to detail in image acquisition and processing. Single-point confocal scanning, on the other hand, inherently produces optical sections, but with a great disadvantage in speed because parallel data acquisition with a camera is replaced by a serial scanning process. Advances in multipoint scanning confocal design[36] and in direct-imaging systems have greatly increased the rate at which optical-sectioning image data can be obtained, and are particularly well-suited to live-cell studies. In place of multipoint confocal scanning, we have adapted to fluorescence imaging the original developments of Neil, Juskaitis, and Wilson[37–39] in encoded-field microscopy. In this type of instrument, which we refer to as a fluorescence grating imager (FGI), a movable Ronchi-grating mask is inserted into the illuminator at the location of the field iris (Fig. 6). When the grating is properly illuminated by a conventional, incoherent light source, it is projected into the specimen as a set of demagnified stripes that are sharply imaged on the in-focus plane of the specimen but are out of focus in the regions of the specimen "above" and "below" this plane. In the fluorescence image formed in the camera plane, in-focus features are therefore encoded by striping. If the stripes are shifted by moving the grating, in-focus features will "blink" whereas out-of-focus features will not. In

[34] J. B. Pawley, ed. "Handbook of Biological Confocal Microscopy," 2nd Ed. Plenum Press, New York, 1995.
[35] D. A. Agard, Y. Hiraoka, P. Shaw, and J. W. Sedat, *Methods Cell Biol.* **30,** 353 (1989).
[36] C. Genka, H. Ishida, K. Ichimori, Y. Hirota, T. Tanaami, and H. Nakazawa, *Cell Calcium* **25,** 199 (1999).
[37] M. A. A. Neil, R. Juskaitis, and T. Wilson, *Opt. Lett.* **22,** 1905 (1997).
[38] T. Wilson, M. A. A. Neil, and R. Juskaitis, *Proc. SPIE* **3261,** 4 (1998).
[39] T. Wilson, M. A. A. Neil, and R. Juskaitis, International Patent #WO 98/45745, 1998.

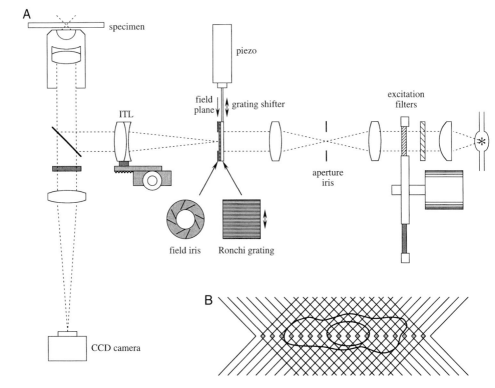

FIG. 6. Grating imager schematic. (A) A conventional inverted fluorescence microscope is modified by the replacement of the iris with a movable Ronchi grating in the field plane of the incident-light illuminator. In this configuration, the grating is projected into the specimen with a depth-of-focus set by the numerical aperture of the objective and the spatial frequency of the grating. A piezoelectric translator moves the grating in precise increments under the control of a computer. For practical reasons (see text), the excitation filters are relocated to a filter wheel that precedes the grating in the optical path. The illuminator tube lens (ITL) is adjusted axially to optimize grating focus. (B) Within the specimen, in-focus features are illuminated in the pattern of the grating, whereas features in the out-of-focus regions of the specimen are uniformly illuminated (see Refs. 37, 38, and 39). Shifting the grating between defined positions modulates in-focus features only, and generates a set of three or four images. Because the out-of-focus component in each image is the same, it is eliminated in the mean by digital subtraction. The in-focus features can then be demodulated by simple image processing algorithms (see text and Table I).

any given camera pixel, the received intensity will then consist of a "DC" out-of-focus component and a time-varying in-focus component. Following Neil *et al.,* this is done discretely by moving the grating in precise steps. The in-focus image component then can be recovered by simple algebraic operations on the resulting set of "input" images. This involves only digital subtraction to eliminate the common image component, followed by Pythagorean summation to demodulate. In

TABLE I
FGI OPTICAL SECTIONING FORMULAS

Grating shift sequence	Image processing formula
1/4-period, 3 images, 0–90–180	$(2^{-1/2})[(i_0 - i_{90})^2 + (i_{90} - i_{180})^2]^{1/2}$
1/3-period, 3 images, 0–120–240	$(2^{1/2}/3)[(i_0 - i_{120})^2 + (i_{120} - i_{240})^2 + (i_{240} - i_0)^2]^{1/2}$
1/4-period, 4 images, 0–90–180–270	$(1/2)[(i_0 - i_{180})^2 + (i_{90} - i_{270})^2]^{1/2}$

the simplest case, three images are sufficient, where the grating is shifted twice by 1/3- or 1/4-period (Table I). It is therefore relatively straightforward to estimate the light exposure needed for FGI operation. Because the Ronchi grating ideally transmits 50% of the incident light, the minimal 3-image set needed to produce one optical section requires a total light exposure only 1.5-fold greater than for a single conventional image. Advantages of the grating imager include simplicity, relatively low cost, use of standard light sources and filter sets, and speed. Additionally, no special optical elements are inserted into the image-forming light path within the microscope. Depth-of-focus of the pattern depends on both grating image spatial frequency and objective NA (Figs. 7 and 8).

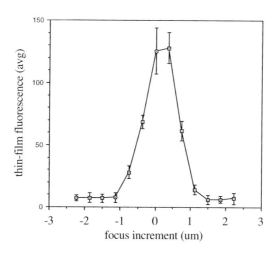

FIG. 7. Fluorescence grating imager axial response with the combination of a 1.30-NA objective and a projected grating period (L) equal to 2.0 μm. The specimen in this case was composed of a thin (0.10-μm) film of rhodamine-labeled poly(methyl methacrylate) applied to a cover glass and mounted with UV-setting optical cement. Image sets were acquired using the "0–120–240" shift sequence (Table I), and processed to generate optical sections. The focus drive was incremented by 0.375 μm between image sets. Thin film fluorescence was defined as the average pixel value in a central region of interest in each section. The Ronchi grating for this example was coarser by a factor of 5 than the theoretical best grating (see text), and produced optical sections with a full-width at half-maximum of 1.1 μm.

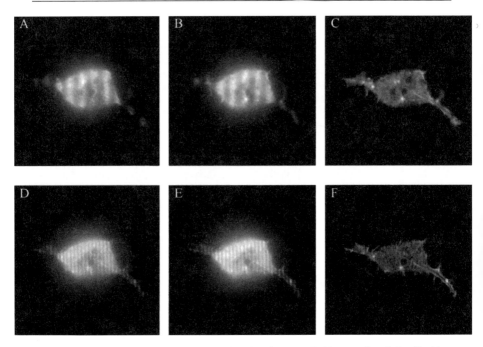

FIG. 8. Effect of grating frequency on optical section sharpness. In this example, a living fibroblast expressing GFP–α-actinin that was seeded in a collagen gel was imaged using the four-exposure 1/4-period shift sequence (Table I). Two images of the four are shown for a coarse grating (A and B; 5 μm projected period) and a fine grating (D and E; 1.5 μm projected period). Between images (A) and (B) or images (D) and (E), the striped pattern phase was shifted by 1/2 period [i_0 (A,D) and i_{180} (B,E)]. Optical sections (C and F) show considerable sharpening with increasing spatial frequency in agreement with Eq. (1). Under the operating conditions (wavelength $= 0.5$ μm, specimen refractive index $= 1.33$, NA $= 1.2$) we expect section thicknesses of 1.4 μm (C) and 0.6 μm (F), respectively.

A theoretical treatment shows that optical sectioning in the FGI is sharpest when the grating image period L equals λ/NA, twice Abbe's resolution limit:

$$\delta = (\lambda/2)/\{[n^2 - (NA - \lambda/L)^2]^{1/2} - [n^2 - NA^2]^{1/2}\}L \geq \lambda/2NA \qquad (1)$$

$$\delta_{\min} = (\lambda/2)/[n - (n^2 - NA^2)^{1/2}]L = \lambda/NA \qquad (2)$$

where δ is the approximate section thickness. In practice, a grating more coarse than the theoretical optimum appears to provide the best results, most likely because of contrast loss in very fine projected patterns (see below). The FGI is like a confocal scanner in that an optical section is produced for each plane of focus independently, rather than like a deconvolution system in which a serial-focus image set is required for the computation of any one optical section. If FGI image sets are obtained at serial planes of focus, such data can also be processed by a specialized deconvolution algorithm in place of the fast algebraic method. We have not yet experimented with this extension of the method.

There are several theoretical and practical limitations in FGI performance. Image processing as described is computationally simple, because it requires only single-pixel operations and is noniterative. However, the algebraic formulas are approximations because a real incoherently projected grating will contain weak harmonics of the fundamental sine wave that can be only partially compensated by the simplest shift sequence and processing algorithm. Such harmonics are seen as residual striping in the optical section. Chrome-on-glass square-wave gratings (Ronchi rulings) are the most practical and precise masks available for the FGI. Ideally, these contain only odd harmonics (3, 5, 7, . . .) of the fundamental pattern in a diminishing series. As pointed out by Wilson et al.,[39] a 1/3-period shift sequence and processing then produces an optical section in which the third harmonic is exactly compensated, making the much weaker fifth harmonic the first error term. In practice, we find the second harmonic to be the most significant source of error in our best optical sections. This is minimized when we utilize the 1/4-period, 4-image sequence and algorithm. Within limits, Fourier filtering can be used to remove residual striping from the optical section. When photobleaching is significant, the fundamental pattern also fails to be demodulated in processing. In this case, correction of the input images prior to computation of the optical section is necessary. In the simplest case, photobleaching in the culture medium and out-of-focus regions of the specimen uniformly decrease the background in sequential images, making it fairly easy to implement corrections for this type of artifact (Fig. 9). Patterned photobleaching in the plane of focus is a more complex problem for which compensation methods are in development. In our experiments, we find that background (autofluorescence) is significantly lower in the YFP band than in the GFP band, thereby reducing the overall photobleaching problem.

In principle, the axial resolution of the FGI matches a confocal scanner. However, because the FGI utilizes full-field illumination and detection, the out-of-focus fluorescence in any pixel will be proportionally greater than what would be seen in that same pixel in the confocal case. Therefore shot noise will also be proportionally greater. Even though the mean out-of-focus light level is removed by the digital subtraction steps in the optical-sectioning formula, the noise is not. In the FGI optical section, the resulting spatial noise level is the root mean square (RMS) accumulation from the input image set. In specimens with a high out-of-focus background, this becomes a limiting factor in the FGI relative to multipoint confocal scanning. Any factor that reduces projected grating contrast will also have the same effect. Because the FGI achieves optical-sectioning performance by projection of an *incoherently* illuminated grating into the specimen with finite depth of focus, it is limited by the incoherent transfer function of the microscope.[40,41] This means that grating contrast in the plane of focus will decrease with increasing

[40] M. Pluta, "Advanced Light Microscopy," Vol. 1. Elsevier, New York, 1988.
[41] S. Inoue and K. R. Spring, "Video Microscopy," 2nd Ed. Plenum Press, New York, 1997.

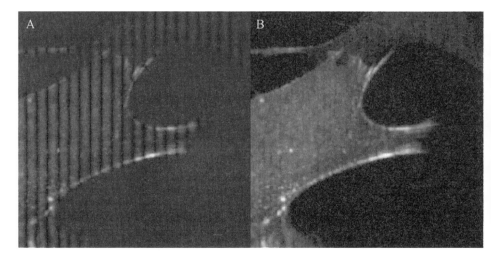

FIG. 9. Preprocessing correction for nonpatterned photobleaching. Grating imager optical sections using the four-exposure 1/4-period shift sequence (Table I) showing GFP–α-actinin in a living fibroblast. (A) Direct computation of the optical section produces a severe striping artifact caused mainly by uniform fading of background fluorescence. This can be estimated from the four-image set and corrected by addition of an increasing bias to images 2, 3, and 4. (B) Optical section computed from preprocessed corrected images.

grating spatial frequency. This lowers the signal-to-noise ratio, and therefore the optimum grating in practice may be coarser than the idealized case.

A number of instrumentation considerations significantly affect FGI performance. The basic instrument consists of a fluorescence microscope equipped with a cooled CCD camera system, programmable shutter, grating slider with drive mechanism, and desktop computer for control, image acquisition, and processing. The grating must be inserted into the fluorescence illuminator optics at the precise location of the field iris. The cooled CCD camera provides the high linearity, dynamic range, geometric precision, and low noise needed for digital processing. We recommend a camera with at least a 10-bit output, preferably 12 or 14 bits. As a fast drive mechanism for moving the grating, we have used low-cost piezo-electric translators with 70–200 μm ranges. Depending on the grating and shift sequence used, the translator must have sufficient range to move the grating more than $2 \times (1/3)$ or $3 \times (1/4)$ of its period. For a 10 line-pair/mm (LP/mm) grating (254 LP/inch), the period is 100 μm. To carry out a 1/4-period 4-image shift sequence, the grating must move from its initial position in three steps of 25 μm each. The necessary range of motion is therefore 75 μm. Best results are obtained if the piezo is not driven from "zero" extension to "maximum" extension, but rather within that range. In our instruments, we initiate piezo extension with a 2–3 μm offset, and occasionally cycle the piezo when not in active acquisition mode. Piezo nonlinearity, creep, and hysteresis are significant factors that can be minimized by filtering the control voltage step signals to the piezo driver through a Butterworth-type lowpass filter (Y. Y. Wang and F. Lanni, unpublished results, 2001), and by always stepping the piezo in the "loading" direction for a given image set. Alternatively, a feedback-stabilized piezo with sufficient range can be used. This is more expensive, but is extremely linear, accurate, and precise. Driven grating motion, vibration, or drift during image acquisition lowers contrast and will reduce the signal-to-noise ratio in the optical section.

The actual choice of grating depends upon the microscope in use and the objective magnification most useful to the experimenter. In most microscopes, the *demagnification factor* between the illuminator field plane and the specimen is *less* than the specified magnification of the objective; i.e., with a $100\times$ objective, the grating may only be demagnified by a factor of $50\times$. This difference is a result of the fact that a shorter focal length tube lens is usually used in the illuminator than in the imaging light path. This factor must be known or determined prior to investing in gratings or piezos. In this example, a 20 LP/mm grating will be projected down to a 1.0-μm period. Switching to a $40\times$ objective would increase the projected period to 2.5 μm. The shift *increment* is dependent only on the grating and shift sequence, and is not affected by switching the objective. Just as important as the projected period of the grating is the projected pixel array of the camera. In general, Nyquist sampling is a minimal requirement.[33,42] This is achieved when the camera pixel spacing, demagnified to the specimen, is less than or equal to $\lambda/4NA$, half the Abbe resolution limit. For a high-NA objective, the Nyquist limit is approximately 0.1 μm. For example, a 19-μm CCD array used with a $100\times$ objective and $2.5\times$ intermediate magnifier is effectively a 0.076-μm array in specimen coordinates, more than satisfying Nyquist. In practice, FGI performance improves with sampling over Nyquist, probably because of the difference between real camera arrays and idealized point sampling.

A number of other factors significantly affect projected grating contrast in addition to the transfer function and sampling; (1) Although the grating can be imaged into the field plane from a location outside the microscope, it is very difficult to do this without loss of contrast due to stray reflections and aberration. The best location for the grating is directly in the illuminator field plane, with the ruled features facing into the microscope. (2) The focus plane of the grating image must coincide with the focus plane of the microscope. Either the grating or the illuminator tube lens must be axially adjustable to set this focus. Adjustment may be necessary to compensate for the effect of longitudinal chromatic error when changing between filter sets. A fluorescent thin-film (<0.2 μm) slide is very useful for checking the coincidence of grating and camera focus. (3) The fluorescence excitation filter may aberrate the grating image. These filters generally have excellent

---

[42] F. Lanni and G. J. Baxter, *Proc. SPIE* **1660,** 140 (1992).

bandpass characteristics but often have poor wavefront-quality characteristics. This is of no consequence in conventional fluorescence microscopy, but matters in the FGI where the normal filter location follows the grating. This problem can be obviated in two ways: an equivalent imaging-grade filter can be substituted (emission filters are often selected for low wavefront error) or the excitation filter can simply be moved to a location where its wavefront quality is of no consequence—such as to a filter slider or filter wheel between lamp and grating (see Fig. 6). Likewise, relief of strain on the dichroic reflector due to its mounting fixture may also improve projected grating contrast. (4) Aberration due to improper use of an immersion objective or due to the illuminator tube lens will seriously degrade both grating image contrast and the sharpness of the grating depth of focus, and should be minimized. By taking these precautions, contrast up to 75% (7 : 1 peak/valley ratio) can be obtained in the projected grating.

Because optical sectioning in the FGI depends upon an input data set of three or four images, factors affecting image set consistency are very important. Light source variability over the time period required for image set acquisition will cause severe striping artifact in the optical section. Stability is generally very good with a 100-W mercury or 75-W xenon arc lamp used with a good-quality regulated DC power supply. AC arc lamps should be avoided. Image-to-image exposure time should be precisely matched. As always, photobleaching should be minimized through precise control of a programmable shutter, the best excitation and emission filters, and a camera with high detection quantum yield.

In addition to grating period and NA, the depth of focus of the projected grating image is strongly affected by the degree of spatial incoherence of the light across the grating. In general, the depth of focus will sharpen as the coherence is reduced. This is achieved by utilizing the full illumination numerical aperture (INA) of the objective. In practical terms, the light source should fill the back pupil of the objective (equivalent to Köhler illumination) with the grating not inserted, and should fill it uniformly. It may be possible to improve upon this limit for gratings close to the optimum by masking the light source to minimize regions of the zero order that do not superpose with the $\pm$ first orders in the back pupil of the objective.

Cytomechanics Analysis of Time-Lapse Image Data

Single DIC time-lapse image pairs, or longer image sequences, are processed in a software package that was developed in our laboratory and is available to researchers via the web: Deformation Quantification & Analysis (DQA; http://dqa.web.cmu.edu).

As schematized in Fig. 1 and shown in Fig. 10, the material displacement $U(x, y)$ is first derived at discrete locations in a grid pattern over the image by piecewise comparison of Image 1 with Image 2 (I1 and I2). At each grid point, a square test pattern from I1 is shifted over the corresponding region in I2. The grid point spacing (typically 10–30 pixels), the size of the test pattern (typically 20 × 20 pixels), and a limit on the search range (typically 10–20 pixels) are set by the user for each DQA run. The best match is defined by a maximum in the normalized cross-correlation coefficient computed between the two sets of pixel values. The vector shift (Δx, Δy) between the original grid point and the center of the test pattern at its best location in I2 is then recorded as $\mathbf{U}$ for that grid point. Because the interior of the cell does not constitute a region of the elastic external matrix, it is masked out for deformation quantification operations. The plasma membrane of the cell therefore represents the boundary across which traction is applied by cell to matrix. In the plane-strain approximation, this boundary is a closed curve represented by the border of the mask. For every image pair, a binary mask can be drawn using NIH Image for both I1 and I2, and is submitted along with the two images. The masks are generally very similar for consecutive images from a time-lapse sequence, and are combined into a single mask using a logical OR operation. The DQA process returns a map of $\mathbf{U}$ as a field of displacement vectors that has no entries in the mask region, and may have other missing entries due to a failure of the search to make a good match. By use of singular-value decomposition (SVD[43]) as a regularizer, the field $\mathbf{U}$ is smoothed, and missing entries are interpolated. From this field, two mechanics fields are then computed: (1) The 2D collagen density increment $\Delta d(x, y)$ is computed as the normalized surface area for every quadrilateral of grid points outside the mask. This is displayed qualitatively on a color scale from blue (density decrement) to red (density increment). (2) The plane-strain field is computed from the spatial derivatives of $\mathbf{U}$:

$$e_{ij}(x, y) = (1/2)(\partial U_i/\partial x_j + \partial U_j/\partial x_i) \qquad (3)$$

which removes all rigid-body components in the displacement (mainly transverse drift) and represents the gel deformation in its true tensor form. In the plane-strain approximation, $[e_{ij}]$ is a 2×2 symmetric matrix. At each grid point, the eigenvalues (λ_1, λ_2) and orthonormal eigenvectors ($\mathbf{v}_1$, $\mathbf{v}_2$) of $[e_{ij}]$ are computed, and the deformation is represented by the principal strains ($\lambda_1\mathbf{v}_1$, $\lambda_2\mathbf{v}_2$). These are shown qualitatively as a strain cross on each grid point in which gel extension (positive λ) and compaction (negative λ) are coded as blue and red, respectively. An accompanying text file provides numerical output at each grid point.

Our work currently is focused on mapping cytoskeletal structural features in relation to strains in the adjacent matrix. This is most conveniently done via side-by-side comparison of a fluorescence optical section with the corresponding DQA strain field (Fig. 11). In contrast to the lamellipodia of fibroblasts on a rigid planar substratum, pseudopods predominate in cells in a 3D collagen matrix. Along pseudopods the cortical actin cytoskeleton is prominent when traced with

[43] W. H. Press, S. A. Teukolsky, W. T. Vetterling, and B. P. Flannery, "Numerical Recipes in C," 2nd Ed. Cambridge University Press, New York, 1992.

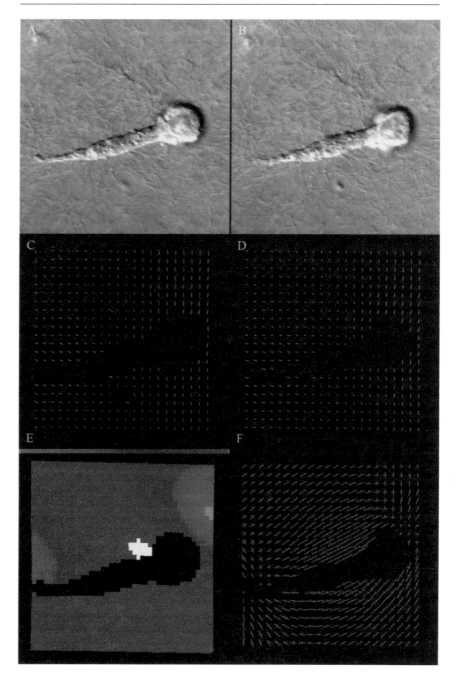

GFP–α-actinin or YFP–actin. Analysis of strain field patterns suggests that the pseudopod is the major contractile device for locomotion and for matrix contraction.[29] In this system, deformations caused by isolated cells over short time periods (minutes) appear to be reversible, whereas whole gel contraction requires either a longer time period (hours–days) or the collective action of multiple cells or both. In this respect, using collagen as a strain gauge is more complex than using a simple elastic continuum. Direct microrheometry shows that gels initially have average shear moduli (G) in the range of 55 Pa,[27] but also show a large point-to-point variation in this value. This intrinsic variation may explain some of the complexity seen in computed strain fields. Additionally, the average modulus is considerably lower than in 2D elastic substrata systems in current use.[18] The general similarity in motile behaviors (rate and range of movement, material displacement) between this system and other substrata provides an interesting clue to the mechanism by which cells regulate their internal machinery.

If the collagen gel is modeled as a Hookean elastic solid, the stress tensor $[s_{ij}]$ is related to the strain tensor by the continuum equation:

$$[s_{ij}] = \lambda tr([e_{ij}])\mathbf{I} + 2G[e_{ij}] \tag{4}$$

where $tr([e_{ij}])$ is the trace of the strain tensor, $\mathbf{I}$ is the identity matrix, and λ and G are the Lamé constants that characterize the elastic medium. G is the shear modulus. Both are expressible in terms of Young's modulus (E, also known as the bulk modulus) and Poisson's ratio, v; $\lambda = Ev/(1 + v)(1 - 2v)$ and $G = E/2(1 + v)$. Poisson's ratio is a measure of compressibility, ranging from $1/2$ for perfectly incompressible materials to zero for perfectly compressible materials. For slow deformations of the collagen gel network on the micron scale, we expect considerable compressibility due to the relative movement of solvent water and the gel network. Preliminary measurements in our laboratory put Poisson's ratio for these gels close to 0.23 prior to contraction. Other cytomechanically important quantities

---

FIG. 10. Deformation quantification and mechanics analysis. Panels show example input and output from the DQA website (http://dqa.web.cmu.edu). (A and B) Sequential DIC images (I1 and I2) from a time-lapse sequence, made at 14 min intervals, showing a fibroblast migrating through a collagen gel. (C) DQA output of the vector displacement field $\mathbf{U}(x, y)$ where red marks correspond to gridpoint locations in I1 and green segments point to the displaced material in I2. Most spurious vectors have been removed by statistical filtering. (D) Regularized and interpolated displacement field computed from (C) by use of singular-value decomposition (SVD). In this field, all vectors are computed from the fitted solution and missing gridpoints are replaced by interpolated vectors. (E) 2D gel density increment derived from (D). Density is mapped onto the color scale shown where red denotes regions of contracted gel and blue denotes regions of expanded gel. White and gray denote zones of maximal compaction and extension, respectively. (F) Incremental plane–strain field shown as principal strain crosses for each gridpoint in I1 (see text). Principal strains are color coded red for contraction and blue for extension. The length of each cross-segment is proportional to the magnitude of the corresponding principal strain. Field of view $= 150\ \mu$m.

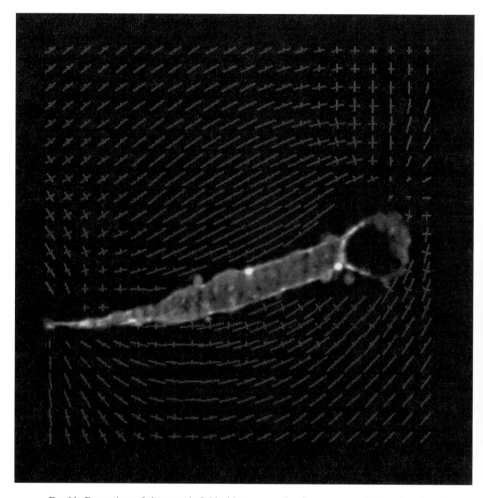

FIG. 11. Comparison of plane–strain field with corresponding fluorescence optical section showing GFP–α-actinin in a fibroblast migrating within a 3D collagen matrix. The monopodial configuration of this cell (see also Fig. 10) produced a deformation in which the major principal strains are contractions that parallel the pseudopod axis. This is most simply the result of inwardly directed tractions applied at opposite ends of the elongated cell. GFP–α-actinin shows a cortical concentration in the pseudopod and an absence of stress fibers. This is also seen with YFP–β-actin and suggests that the cortex may contain portions of the contractile machine.

can be derived from the stress field, such as the changes in stored elastic energy over time. On the boundary of the cell, tractions can be estimated from the relation: $\mathbf{t} = [s_{ij}] \cdot \mathbf{n}_{cell}$, where $\mathbf{n}$ is the unit vector normal to the cell boundary at each test point. In the plane–strain approximation, $[s_{ij}]$, like $[e_{ij}]$, is a 2 × 2 tensor, and the cell boundary is a closed curve in the image plane, rather than a closed surface.

Extension of cytomechanical analysis beyond the plane–strain approximation to 3D does not require fundamentally different experimental protocols. Our experiments already necessarily produce 3D image data in DIC through serial-focus image acquisition simply to make it possible to follow specimen shape changes and cell migration over time. The cross-correlation method for finding material displacements between sequential images can be implemented on 3D blocks of pixels in the same way as on 2D squares. Strain is computed no differently, except that $[e_{ij}]$ will generally be a 3×3 symmetric tensor rather than 2×2. Because the principal strains will then be represented as a triplet of scaled vectors at each grid point, and not as easily visualized graphically, new methods of display and quantification will be required. However, going to 3D analytical methods should greatly improve accuracy in identifying cytoskeletal structures and interpreting the observed pattern of strain in adjacent ECM.

Acknowledgments

We thank T. Brownlee and J. Airone for their experimental work in determination of Poisson's ratio; Dr. C. A. Otey for providing GFP–α-actinin expressing fibroblasts; D. Pane, M. Mantarro, and W. Galbraith for AIM software development; and A. Marciszyn for critical comments. This work was supported by Grants NIH AR-32461, NSF STC MCB-8920118, and NSF DBI-9987393.

[10] Measurements of Cell-Generated Deformations on Flexible Substrata Using Correlation-Based Optical Flow

By WILLIAM A. MARGANSKI, MICAH DEMBO, and YU-LI WANG

Introduction

The forces exerted by an adherent cell on the underlying substratum, the so-called cellular tractions, are important because of their potential involvement in cell motion, tissue morphogenesis, wound retraction, and in the transduction of information about the mechanical characteristics of the tissue.[1–4] Elastic substrate methods (ESMs) are the principal means for investigating the cellular tractions.

[1] C. G. Galbraith and M. P. Sheetz, *Curr. Opin. Cell Biol.* **10**, 566 (1998).
[2] D. P. Kiehart, C. G. Galbraith, K. A. Edwards, W. L. Rickoll, and R. A. Montague, *J. Cell Biol.* **149**, 471 (2000).
[3] A. Jacinto, A. Martinez-Arias, and P. Martin, *Nat. Cell Biol.* **3**, E117 (2001).
[4] C. Lo, H. Wang, M. Dembo, and Y.-L. Wang, *Biophys. J.* **79**, 144 (2000).

The essential idea is to put a cell on a flexible substratum of known mechanical properties and to use the way this material deforms as the basis for drawing conclusions. All ESMs have three main components. The first deals with the fabrication and characterization of an appropriate elastic substratum and the culturing of cells onto this material. The second focuses on the experimental determination of the precise manner by which the substratum deforms under the action of adherent cells. Finally, the last component deals with the computational problem of interpreting the substrate deformations in terms of forces exerted by different regions of the cell.

A number of ESMs have been developed over the past 20 years, each with its advantages and disadvantages.[5] This article will focus on the application of polyacrylamide (PA) substrata that we developed.[4,6,7] The most significant advantages of PA substrata are their nontoxicity, mechanical stability, and ease of preparation. Furthermore, the mechanical stiffness of PA substrata can be tuned by varying the concentration of acrylamide or the crosslinker, while the chemical properties of the surfaces are determined by extracellular matrix proteins that are covalently linked. Methods for the preparation and physical characterization of PA substrata and the collection of cell and substrate images have been described in several publications.[8,9] Likewise the theory for the deformation of elastic substrata is well understood and the methodology to interpret such deformations in terms of cellular forces has been described previously.[10,11] Therefore, this article will focus exclusively on recent advances in the analysis of the determination of substrate deformations.

PA substrata are transparent and deformations are detectable only with the aid of (fluorescent) marker beads embedded within the elastic medium. In early applications, the motion of the substrate was determined by simple visual inspections. In this approach a pair of fluorescent images of the marker beads is recorded, one while the cell is adhered to the substratum (referred to as the "strained" image, I_1) and the other after the cell is removed by enzymatic or physical means (referred to as the "unstrained" image, I_0). Corresponding beads in the two images are identified visually and their coordinates are used for constructing displacement vectors. Unfortunately this simple approach breaks down catastrophically when motions are large compared to the spacing of the markers. This difficulty, known as the "correspondence problem," occurs because the observer becomes confused about the identification of corresponding beads in the strained and unstrained images.

[5] K. A. Beningo and Y.-L. Wang, *Trends Cell Biol.* **12**, 79 (2002).
[6] K. A. Beningo, M. Dembo, I. Kaverina, J. V. Small, and Y.-L. Wang, *J. Cell Biol.* **153**, 881 (2001).
[7] S. Munevar, Y.-L. Wang, and M. Dembo, *Biophys. J.* **80**, 1744 (2001).
[8] Y.-L. Wang and R. J. Pelham, *Methods Enzymol.* **298**, 489 (1998).
[9] K. A. Beningo, C.-L. Lo, and Y.-L. Wang, *Methods Cell Biol.,* **69**, 325 (2002).
[10] M. Dembo and Y.-L. Wang, *Biophys. J.* **76**, 2307 (1999).
[11] M. Dembo, T. Oliver, A. Ishihara, and K. Jacobson, *Biophys. J.* **70**, 2008 (1996).

To overcome the correspondence problem we have adapted an approach that is well known in a number of image analysis problems, namely correlation-based optical flow.[12,13] Essentially one defines a small patch in I_0 that contains a number of markers and then searches in I_1 for patches with a similar characteristic pattern of pixel intensities. The correspondence problem is alleviated because instead of following the motion of a single marker, one follows the collective motion of uniquely recognizable marker groupings. The end result is a robust estimate of the substrate deformation, usually accurate to within a pixel. This estimate can be refined by automatic procedures for the correction of image registration errors, and by detection/correction of physically improbably modes of deformation. Finally interpolation methods can be used to obtain true subpixel accuracy. Software implementing this algorithm on Linux workstations is available from the authors.

Correlation-Based Optical Flow

We will start with the images I_0 and I_1 that show the distribution of fluorescent markers in the unstrained and strained substrate. These images are loaded into computers as large matrices of n_x columns by n_y rows:

$$I_k \equiv \begin{bmatrix} P_k(1,1) & P_k(2,1) & \cdots & P_k(n_x-1,1) & P_k(n_x,1) \\ P_k(1,2) & P_k(2,2) & \cdots & P_k(n_x-1,2) & P_k(n_x,2) \\ \vdots & \vdots & \vdots & \vdots & \vdots \\ P_k(1,n_y) & P_k(1,n_y) & \cdots & P_k(n_x-1,n_y) & P_k(n_x,n_y) \end{bmatrix} \quad (1)$$

where the subscript $k = 0$ or 1 and $P_k(x, y)$ is the intensity at the pixel (x, y). Before further processing of these arrays, a constant is subtracted from all the P_k so that the average pixel intensity in both I_1 and I_0 is equal to zero.

Suppose that (x, y) and (u, v) are the coordinates of certain pixels in I_0 and I_1 respectively, and that we wish to test the possibility that the motion of the substratum has carried (x, y) onto (u, v). This requires that we construct two regions (the "correlation windows") that extend for a distance of C pixels from the central locations of (x, y) and (u, v), respectively. If B_0 is the region surrounding (x, y) and B_1 the region surrounding (u, v), then we may say that these regions have similar "intensity patterns" if the pixel at $(x + \delta_x, y + \delta_y)$ in B_0 has an intensity that is in some sense "close" to the intensity of the homologous pixel located at $(u + \delta_x, v + \delta_y)$ in B_1. Mathematically the image similarity in B_0 and B_1 can be measured in a variety of ways and careful systematic comparisons of the cost and efficiency of different approaches have been reported.[14] In the case of typical

[12] P. Anandan, *Int. J. Comput. Vision* **2**, 283 (1989).
[13] B. Jähne, "Digital Image Processing: Concepts, Algorithms, and Scientific Applications," 4th Ed. Springer Press, Berlin, Germany, 1997.
[14] J. L. Barron, D. L. Fleet, and S. S. Beauchemin, *Int. J. Comput. Vision* **12**, 43 (1994).

fluorescent images of substrata, which have a very high contrast, our results indicate that the optimal similarity measure is the so-called normalized cross-correlation coefficient:

$$R(x, y, u, v, C)$$
$$\equiv \frac{\displaystyle\sum_{\delta_x}\sum_{\delta_y} P_0(x + \delta_x, y + \delta_y) P_1(u + \delta_x, v + \delta_y)}{\left(\displaystyle\sum_{\delta_x}\sum_{\delta_y} P_0^2(x + \delta_x, y + \delta_y)\right)^{1/2} \left(\displaystyle\sum_{\delta_x}\sum_{\delta_y} P_1^2(u + \delta_x, v + \delta_y)\right)^{1/2}} \quad (2)$$

The summations in this expression all range over the values δ_x and δ_y within the correlation windows of B_0 and B_1 (this means that δ_x and δ_y run between the limits of $-C$ and $+C$ except at locations close to the edges of an image).

This normalized cross-correlation has several properties that make it particularly suited as a measure of similarity in the current application. First, it can be shown that $R(x, y, u, v, C)$ falls in the range between -1 and $+1$ and that it approaches the upper limit only if $P_1(u + \delta_x, v + \delta_y)$ is equal to a positive constant times $P_0(x + \delta_x, y + \delta_y)$ for all choices of δ_x and δ_y. Similarly, $R(x, y, u, v, C)$ equals -1 only if $P_1(u + \delta_x, v + \delta_y)$ is equal to a negative constant times $P_0(x + \delta_x, y + \delta_y)$, i.e., if B_1 is the negative image of B_0. The function $R(x, y, u, v, C)$ is equal to 0 if the intensity patterns in B_0 and B_1 are completely uncorrelated. These properties mean that $R(x, y, u, v, C)$ is not affected by the linear rescaling of image intensity as might be caused by variations in the exposure time. Finally, constraining $R(x, y, u, v, C)$ to a range of -1 to $+1$ by normalization means that it has a comparable meaning regardless of the size and shape of the correlation windows. For these reasons the function $R(x, y, u, v, C)$ was clearly superior to the alternative approach of "sum of squared differences," and also slightly better than other measures of mutual variation within the regions B_0 and B_1.

Figure 1 shows typical examples of images I_0 and I_1. For aesthetic purposes each image is magnified so that the individual pixels can be discerned and in addition the intensities have been inverted so that the marker beads appear as black spots against a lighter background. Also illustrated is a correlation window in I_0 that is centered at the point (x, y) and that can be taken as a fixed reference. We also show two possible correlation windows in I_1. The first of these (shown with dashed outline) is centered at the same absolute image location as the reference window [i.e., $(u, v) = (x, y)$]. Because the substratum has moved, the features enclosed are quite dissimilar. The second window in I_1 (shown with solid outline) has been translated to a new location with center (u^*, v^*) such that $R(x, y, u, v, C)$ is maximized. The vector from the center of the dashed window to the center of the solid window gives a good estimate of the local substrate motion.

Among various parameters in correlation-based optical flow, the size of the correlation window C is of the utmost importance. Its value must be matched

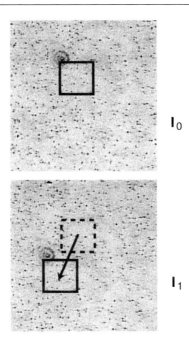

I_0

I_1

FIG. 1. Correlation-based optical flow. The basic premise is to compare pixel intensity patterns between two images, I_0 and I_1, to compute a motion field. The method begins by defining a reference window in I_0 (denoted by a solid outline). Then analogous patches within I_1 are examined, starting with a window (denoted by a dotted outline) centered at the exact pixel coordinates as the reference window. After finding the window (denoted by a solid outline) within I_1 that has an intensity pattern most similar to the reference window, a displacement vector is constructed that originates at the center of the dotted square and terminates at the center of the solid square. This vector gives an estimate of the local substrate motion.

by trial and error with the density of the marker beads. Too small a size would result in the inclusion of too few markers within the correlation window to form an unambiguous pattern, and the search for a matching pattern in I_1 will fail catastrophically. Too large a size would result in the loss of resolution since any differential movements within the correlation windows are neglected during the calculation of the cross-correlation coefficient.

Local Search Implementation

The ESM requires information about the substrate motion not at a single isolated point but throughout a large region surrounding a cell of interest. Generally it is neither necessary nor desirable to determine the movement at each pixel within this region, but only at nodes of a simple $m_x \times m_y$ lattice whose coordinates are expressed as (x_j, y_k) where $j = 1, 2, \ldots, m_x$ and $k = 1, 2, \ldots, m_y$. Once this lattice is established, the next task is to determine the substrate displacement at each lattice node to within integer precision. This means we have to form a correlation window B_0 surrounding each of the points (x_j, y_k) in I_0 and find corresponding integer pixel coordinates (u^*_{jk}, v^*_{jk}) in I_1 by testing correlation windows B_1 at various positions (u, v) in I_1 and determining the position where $R(x_j, y_k, u, v, C)$ is maximized. The maximum value of the cross-correlation at a given lattice site is denoted by R^*_{jk}.

A straightforward and efficient method for finding (u^*_{jk}, v^*_{jk}) is to perform an iterative search. In this approach, one starts with an initial estimate of the position in I_1 where $R(x_j, y_k, u, v, C)$ is maximized, $(u^{(0)}_{jk}, v^{(0)}_{jk})$. If no better guess is available then this is simply taken to be the point (x_j, y_k), (i.e., assume there was no displacement at this location). One then computes the values of $R(x_j, y_k, u, v, C)$ for all pixels within some distance S from $(u^{(0)}_{jk}, v^{(0)}_{jk})$. The pixel with the maximal $R(x_j, y_k, u, v, C)$ then becomes a new estimate of the target position, $(u^{(1)}_{jk}, v^{(1)}_{jk})$. If this new estimate remains the same as the old estimate, then $(u^*_{jk}, v^*_{jk}) = (u^{(1)}_{jk}, v^{(1)}_{jk})$ and the procedure terminates; otherwise we continue the search.

This iterative method converges to the correct answer in most cases but it may be subject to error if the search distance S is too small. The distance S therefore needs to be adjusted by trial and error. Although the chance of making an error can be minimized by using a large S, this is expensive since the computational work increases as S^2. Fortunately, this problem is generally avoided because the substrate deformations are being determined at a large number of locations and because these motions are known to be a continuous function of the position. An efficient search strategy is to first determine (u^*_{jk}, v^*_{jk}) on a sparse lattice and with a large value of S. The density of the lattice and the value of S are then progressively increased and decreased, respectively. With each refined lattice the displacement of the nearest lattice point on the previous sparse lattice is used as the starting estimate. Because the starting values derived in this way are already quite accurate, the algorithm will usually converge to the target point in a single iteration, even if the search radius is only a few pixels.

Subpixel Resolution

The behavior of $R(x_j, y_k, u, v, C)$ in regions near the global optimum (u^*_{jk}, v^*_{jk}) for an example optical flow computation is illustrated in Fig. 2. The function has a well-defined maximum, $R^*_{jk} \approx 0.93$, and is quite high at pixels immediately surrounding (u^*_{jk}, v^*_{jk}) (the range is between 0.84 and 0.90). Moreover, even at distances up to two pixels away from the maximum the cross-correlation coefficient remains significantly elevated above the background. This remarkable smoothness and continuity of $R(x_j, y_k, u, v, C)$ near a point of optimal correlation is a general behavior, reflecting the fact that the image of a marker particle has a diffraction limited diameter of 0.3–0.6 μm while pixels are generally 0.1–0.3 μm

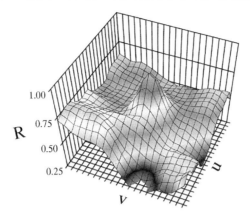

FIG. 2. Landscape of the cross-correlation coefficient around the global optimum. For fixed (x_j, y_k) in I_0, the cross-correlation coefficient $R(x_j, y_k, u, v, C)$ is plotted for various values of u and v in the neighborhood of the global optimum (u^*_{jk}, v^*_{jk}). As illustrated, $R(x_j, y_k, u, v, C)$ attains a well-defined maximum of $R^*_{jk} = 0.93$, but is still elevated well above the background even at distances ± 2 pixels away from the optimum. Thus, the cross-correlation coefficient is a smooth and continuous function of position and interpolation can be used to locate its optimum with subpixel accuracy.

in dimension. Therefore, the intensity of individual markers overlaps for a finite distance, within which the value of $R(x_j, y_k, u, v, C)$ changes only slightly.

We can exploit the smoothness and continuity of $R(x_j, y_k, u, v, C)$ to obtain a subpixel determination of the substrate motion using interpolation. After trying several schemes, we found that the simple five-point quadratic method is the most robust in real practice. This yields the following refined estimates for the coordinates of the maximum of $R(x_j, y_k, u, v, C)$:

$$u^{**}_{jk} = u^*_{jk} + \frac{0.5[R(x_j, y_k, u^*_{jk} + 1, v^*_{jk}, C) - R(x_j, y_k, u^*_{jk} - 1, v^*_{jk}, C)]}{2R(x_j, y_k, u^*_{jk}, v^*_{jk}, C) - R(x_j, y_k, u^*_{jk} - 1, v^*_{jk}, C) - R(x_j, y_k, u^*_{jk} + 1, v^*_{jk}, C)}$$

$$(3a)$$

$$v^{**}_{jk} = v^*_{jk} + \frac{0.5[R(x_j, y_k, u^*_{jk}, v^*_{jk} + 1, C) - R(x_j, y_k, u^*_{jk}, v^*_{jk} - 1, C)]}{2R(x_j, y_k, u^*_{jk}, v^*_{jk}, C) - R(x_j, y_k, u^*_{jk}, v^*_{jk} - 1, C) - R(x_j, y_k, u^*_{jk}, v^*_{jk} + 1, C)}$$

$$(3b)$$

This interpolation yields accuracy to approximately 0.1 pixel in ideal circumstances, where the correlation window undergoes motion with little internal deformation (see test results reported below).

If there are significant differential movements within the correlation window, then the displacement determined by interpolation may deviate slightly from the exact displacement at each marker particle. In this case a refined strategy has been developed. Instead of calculating displacements at nodes of a simple lattice in I_0, well-defined markers near the nodes are identified at $(x_{j'}, y_{k'})$ (again using a correlation-based method by searching for the pattern of a single marker). The displacement is then calculated with $(x_{j'}, y_{k'})$ as the center of the correlation window B_0. Once $(u^*_{j'k'}, v^*_{j'k'})$ is located, a single marker is again searched for within the radius of an Airy disk from $(u^*_{j'k'}, v^*_{j'k'})$ and the position of the identified marker is determined at subpixel precision and is used for defining the final displacement.

Detecting and Correcting Correspondence Failures

The optical flow algorithm as described above assumes the existence of unique pixel coordinates $(u^{**}_{jk}, v^{**}_{jk})$ in I_1 that maximizes the value of $R(x_j, y_k, u, v, C)$ for any given (x_j, y_k) in I_0. Generally, this is a fairly safe assumption because there will be many markers inside the correlation window and the chances of two correlation windows having an identical marker distribution is low. However, despite the best efforts there is still a finite chance that a wrong assignment can occur at a small number of lattice sites. This correspondence failure can happen if the radius of the correlation window is small, the density of markers in the substrate is low, the signal-to-noise (S/N) ratio of the images is poor, or if the substrate displacements are very large. The algorithm can also fail if there is some feature in I_0 that is completely absent in I_1, for example, when autofluorescence of the cell or specks of dust appear in the image of marker particles. These problems cause the local search algorithm to yield artifactual results at the sites in question.

The general approach to avoid these correspondence failures is to look for suspicious displacements and to remove or recalculate them. A simple screen for correspondence failure is to calculate the S/N within the correlation window B_0 for each (x_j, y_k) and to treat the node as suspicious if the S/N falls below a defined threshold. Alternatively, correspondence failure can be detected by comparing the value of $R(x_j, y_k, u, v, C)$ at each $(u^{**}_{jk}, v^{**}_{jk})$ against a defined threshold. Clearly if this maximal $R(x_j, y_k, u, v, C)$ is close to 0, the match between B_0 and B_1 must be treated as highly suspicious.

A more sophisticated screen takes into account the physical characteristics of the displacements of an elastic substrate. These displacements are continuous functions of position, whereas correspondence failures generally result in discontinuous, random displacements and can usually be detected by checking the relationship among the magnitude of neighboring displacements. To be specific, at a lattice node (x_j, y_k) we estimate the magnitudes of the so-called "in-plane strain components" as follows:

$$\epsilon^2_{xx} = 0.5 \left| \frac{u^{**}_{(j+1)k} - u^{**}_{jk}}{x_{j+1} - x_j} - 1 \right|^2 + 0.5 \left| \frac{u^{**}_{jk} - u^{**}_{(j-1)k}}{x_j - x_{j-1}} - 1 \right|^2 \tag{4a}$$

$$\epsilon^2_{yx} = 0.5 \left| \frac{v^{**}_{(j+1)k} - v^{**}_{jk}}{x_{j+1} - x_j} \right|^2 + 0.5 \left| \frac{v^{**}_{jk} - v^{**}_{(j-1)k}}{x_j - x_{j-1}} \right|^2 \tag{4b}$$

$$\epsilon_{xx}^2 = 0.5 \left| \frac{u_{j(k+1)}^{**} - u_{jk}^{**}}{y_{k+1} - y_k} \right|^2 + 0.5 \left| \frac{u_{jk}^{**} - u_{j(k-1)}^{**}}{y_k - y_{k-1}} \right|^2 \tag{4c}$$

$$\epsilon_{yy}^2 = 0.5 \left| \frac{v_{j(k+1)}^{**} - v_{jk}^{**}}{y_{k+1} - y_k} - 1 \right|^2 + 0.5 \left| \frac{v_{jk}^{**} - v_{j(k-1)}^{**}}{y_k - y_{k-1}} - 1 \right|^2 \tag{4d}$$

Then for each node we compute the norm of the strain tensor:

$$\| \epsilon_{jk} \| = \sqrt{ \epsilon_{xx}^2 + \epsilon_{yx}^2 + \epsilon_{xy}^2 + \epsilon_{yy}^2 } \tag{5}$$

If $\| \epsilon_{jk} \|$ is greater than some limiting value $\epsilon_{\max}$, then the values of $(u_{jk}^{**}, v_{jk}^{**})$ are regarded as suspect. The exact value of the cutoff limit, $\epsilon_{\max}$, is generally on the order of 1 but can be adjusted by trial and error depending on the nature of the data and the preference of the user. After all nodes have been checked, the displacement at suspect nodes is recalculated. In this recalculation, however, the radius C of the correlation window is increased so as to reduce the chances of a second correspondence failure. In addition, we use the value of (u_{jk}^*, v_{jk}^*) from the closest nonsuspect node as the starting estimate $(u_{jk}^{(0)}, v_{jk}^{(0)})$ for the local search at these suspect nodes.

Correcting Image Registration Artifacts

Small movements of the substrate or microscope stage that occur in between the acquisition of I_1 and I_0 cause systematic displacement of all marker particles by a constant vector (d_x, d_y). This so-called "registration artifact" is superimposed on the actual physical displacements and needs to be corrected before any calculation of the cellular forces is attempted. This correction is possible because as long as the lattice covers a substantial area outside the cell, many of the nodes (x_j, y_k) will be far away from the cell and will have a displacement that is due only to the registration artifact. Consequently, a simple approach to remove the registration error is to define a correlation window B_0 in a reference region far away from the cell and to calculate the substrate displacement at this location. This displacement (d_x, d_y) is treated as the registration artifact and is subtracted from the subsequent calculations. The major drawback of this approach is that sometimes it is difficult to identify such a reference region due to possible displacements caused by neighboring cells outside the field of observation.

A more robust, nonbiased strategy is built upon the fact that the displacement of the registration artifact should occur with the maximal frequency. Therefore if one constructs a histogram of the number of nodes versus the distance of displacement, the peak should be located at (d_x, d_y). We devised an efficient "nested histogram algorithm" to carry out this task. The scheme is initiated by sorting the uncorrected x and y displacements into a finite number of bins. This generates two crude histograms for the x and y displacements, respectively. Figure 3a shows such a histogram along the x direction for a typical experiment. As can be seen, the

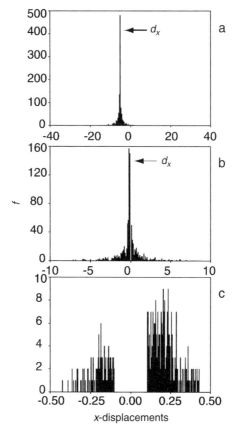

FIG. 3. Nested histogram algorithm. (a) A frequency distribution with a finite number of bins is created for the x displacements, $(u_{jk}^{**} - x_j)$. The center of the bin having the highest frequency (f) gives a first estimate of the most likely x displacement (d_x) of the substrate. The x coordinate of each node is then adjusted as $x_j = x_j + d_x$ so that the most frequent x displacement of the motion field will be the zero. (b) Finer correction of the registration artifact is performed by using the same number of bins while decreasing the interval size and then recomputing the histogram for the corrected x displacements, $[u_{jk}^{**} - (x_j + d_x)]$. (c) The process continues until the size of the interval or the maximum frequency becomes so small that a meaningful distribution cannot be generated. Note that the computation of the registration artifact in the y direction is completely analogous and, thus, is not shown here.

highest frequency occurs at $d_x \approx -4.0$ pixels. A second pair of histograms is then constructed, by centering at the most frequent displacement identified in the previous histogram and subdividing the bins to obtain a smaller interval and hence a higher resolution. As seen in Fig. 3b, the displacement with the highest frequency is now centered at $d_x \approx -0.5$ pixel. Combining the two corrections yield a net estimated registration artifact of $d_x \approx -4.5$ pixels. This process can be iterated, by doubling or quadrupling the resolution at each iteration and summing the corrections, to obtain the desired resolution. In the present example the process terminated with the histogram shown in Fig. 3c.

Overview of Optical Flow Algorithm

For easy reference the following flowchart summarizes the correlation-based optical flow algorithm in its most basic form. Slight modifications might be necessary in some instances.

1. Load images I_0 and I_1 into the computer as integer matrices.
2. Choose lattice nodes (x_j, y_k) and parameters $C, S, \epsilon_{\max}$.
3. Subtract average intensity from each image.
4. For each lattice node (x_j, y_k) perform steps 5–10.
5. Estimate $(u_{jk}^{(0)}, v_{jk}^{(0)})$ using existing results from a sparse lattice or by setting them equal to (x_j, y_k).
6. Use the local search algorithm to compute (u_{jk}^*, v_{jk}^*) and R_{jk}^*.
7. Use quadratic interpolation Eqs. (3a and 3b) to compute $(u_{jk}^{**}, v_{jk}^{**})$.
8. Compute $\|\epsilon_{jk}\|$ using Eq. (5).
9. Set $C = 2C$ and recompute $(u_{jk}^{**}, v_{jk}^{**})$ if $\|\epsilon_{jk}\| > \epsilon_{\max}$.
10. Recompute $\|\epsilon_{jk}\|$ using Eq. (5).
11. Calculate the most likely substrate displacement (d_x, d_y) caused by registration artifacts using the nested histogram algorithm.
12. For all (j, k), set $(x_j, y_k) = (x_j + d_x, y_k + d_y)$.
13. For all (j, k), write $x_j, y_k, u_{jk}^{**}, v_{jk}^{**}, R_{jk}^*$, and $\|\epsilon_{jk}\|$ to a data file.
14. Stop.

Illustrative Results

In Figs. 4 and 5, we illustrate the performance of our algorithm by determining the substrate displacements produced by a NIH 3T3 cell adhering to a PA substratum coated with type I collagen. The substratum in this experiment was fabricated using 5% (w/v) acrylamide and 1% (w/v) bisacrylamide and had an estimated Young's modulus of 2.8×10^4 Pa and Poisson's ratio of 0.30. The substratum contained 0.20 visible markers per micrometer square of the surface. The raw images I_0 and I_1 were initially stored as 8-bit tiff files with $[512 \times 512]$ pixels. The final magnification was 0.30 μm/pixel.

Substrate displacements were first determined by the standard method using a lattice of 32×32 nodes and with parameters $C = 5$, $S = 5$, and $\epsilon_{\max} = 0.316$ (Fig. 4a). A small correlation window was deliberately utilized in this calculation, so that there were a significant number of correspondence failures that were corrected using step 9 of the algorithm. To identify these corrections, the threshold

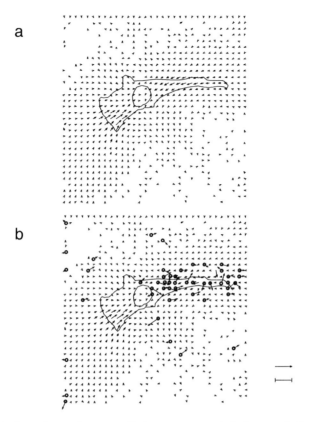

FIG. 4. Detecting and correcting suspect displacement vectors. A displacement field was computed on a lattice of 32×32 nodes with $C = S = 5$. The performance of the optical flow algorithm in detecting and correcting correspondence failures was studied by either setting $\epsilon_{max} = 0.316$ (a) or $\epsilon_{max} = 100$ (b). Lattice sites that were affected by changing ϵ_{max} in this way are indicated with small circles. The length scale represents 46.25 pixels and the displacement vectors are rendered at three times the actual motion for better visibility.

ϵ_{max} was reset to a very large value to bypass the corrections and the calculation was repeated (Fig. 4b). The lattice nodes in Fig. 4b where results were affected by the absence of any correspondence check are indicated by small circles. Clearly the check found all points where obvious problems occurred and there are very few if any false positives. We conclude that the check based on $\|\epsilon_{jk}\|$ and the recalculation with increased correlation radius is quite effective at detecting and correcting correspondence failures.

Figures 5 illustrates the removal of the registration artifact using our nested histogram algorithm. Using the same starting images described above, the substrate displacements were calculated on a lattice of 32×32 nodes with parameters $C = 10, S = 5$, and $\epsilon_{max} = 0.316$. In step 11 of the algorithm, the nested

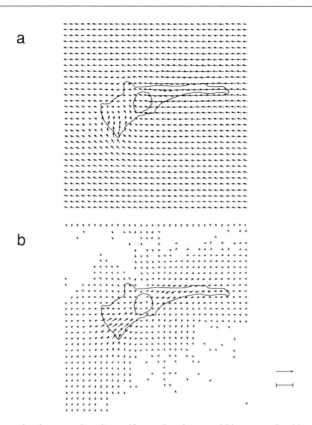

FIG. 5. Correcting image registration artifacts using the nested histogram algorithm. A displacement field was computed on a lattice of 32×32 nodes with $C = 10$, $S = 5$, and $\epsilon_{max} = 0.316$. The performance of the optical flow algorithm in correcting registration artifacts was tested by either omitting (a) or utilizing (b) step 12 of the algorithm. The length scale represents 46.25 pixels and the displacement vectors are rendered at three times the actual motion for better visibility.

histogram calculation indicated that the most likely substrate displacement was $\mathbf{d} = (-5.09, -0.83)$. Comparison of the results with (Fig. 5b) and without (Fig. 5a) the correction of registration artifact (step 12) indicates that the nested histogram method yields an excellent estimate of the registration artifact. The displacements observed give a much better representation of the action of the cell on the substratum.

As a simple empirical test of the overall accuracy of our optical flow algorithm, images of several standard PA substrata were recorded under conditions designed to simulate those of a typical experiment. After recording an initial image, the substratum was left on the microscope stage for 30 min before the second image was recorded. During this interval various sham procedures were implemented to replicate some standard procedures that might be involved in a real experiment

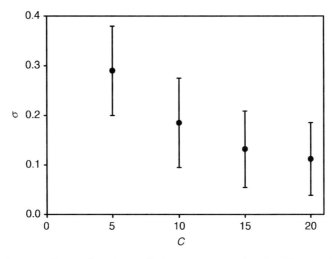

FIG. 6. Accuracy of computing substrate displacements vectors. A series of images of a cell-free, unstrained substrate were taken under conditions that simulated a real experiment. The motion between 30 image pairs was measured on a lattice of 100×100 nodes with $S = 5$, $\epsilon = 0.316$, and for values of C as indicated. The error $\sigma = \sqrt{\sigma_x^2 + \sigma_y^2}$ of the combined x and y motions was computed using all 10^4 lattice points. Each data point consists of the mean and standard deviation of σ. Results indicate that σ approaches ± 0.10 pixels when image quality is optimal.

(i.e., flushing and replacing the medium and purposely jostling the stage). The displacement field between two images was measured on a lattice of 100×100 nodes using parameters $S = 5$, $\epsilon_{max} = 0.316$, and various sizes C of the correlation window. The mean of the registration magnitude in these experiments was $(\sqrt{(d_x)^2 + (d_y)^2})_{ave} \approx 3$ pixels, which is typical of real experiments. The residual systematic motion after the correction of the registration artifacts varied between 0.06 pixels at $C = 5$ and 0.03 pixels at $C = 20$. For each value of C, the error $\sigma = \sqrt{\sigma_x^2 + \sigma_y^2}$ of the combined x and y motions was computed using all 10^4 lattice nodes and the results for 30 pairs of images are shown in Fig. 6. Note that the error is about ± 0.30 pixels when $C = 5$, decreases to ± 0.19 pixels when $C = 10$, and approaches ± 0.10 pixels at very large C.

These analyses demonstrate that a small C causes an increase in correspondence failure and error, due to the insufficient number of beads in many of the correlation windows. As C increases, the error σ approaches an asymptotic value greater than zero because factors other than the correspondence problem are limiting. These other factors include the inaccuracy in the registration error and the interpolation error. The asymptotic value of σ in this ideal limit is on the order of ± 0.10 pixels. Similar accuracy in the measurement of uniform motion fields using optical flow under ideal circumstances has also been reported by

Seitz.[15] Of course, these results should be regarded as lower bounds of the errors in more realistic circumstances. In particular, if motion of the substratum is nonuniform, then σ values may double or even triple those found in this simple test.

Summary

The optical flow algorithm presented here is a robust method that rapidly yields a high-density field of substrate displacement vectors based on two optical images. We found that one of the limiting factors, at least for inexperienced experimentalists, is the consistency of focusing or the drift in microscope focus. However, with properly collected images the standard error of the measurement was estimated to be on the order of ± 0.10 pixels. Finally, although the discussion has been focused on the displacement of flexible substrata, a similar method should be applicable for detecting movements on other types of images, as long as the movement involves a certain degree of local coordination.

Acknowledgments

This research was supported by the Computational Science Graduate Fellowship funded by the Department of Energy to W. A. Marganski, NIH Grant GM61806 to M. Dembo, and NIH Grant GM32476 and NASA Grant NAG2-1495 to Y.-L. Wang.

[15] P. Seitz, *Opt. Eng.* **27**, 535 (1988).

[11] Single-Molecule Imaging of Rotation of F₁-ATPase

By KENGO ADACHI, HIROYUKI NOJI, and KAZUHIKO KINOSITA, JR.

Introduction

A single molecule of F₁-ATPase has been shown to be a rotary motor, driven by adenosine triphosphate (ATP) hydrolysis, in which the central γ subunit rotates against a surrounding cylinder made of alternately arranged three α and three β subunits.[1-5] Together with another (yet putative) proton-driven rotary motor F₀, it constitutes the F₀F₁-ATP synthase that synthesizes ATP from adenosine

[1] P. D. Boyer, *Biochim. Biophys. Acta* **1140**, 215 (1993).
[2] P. D. Boyer, *Biochim. Biophys. Acta* **1458**, 252 (2000).
[3] K. Kinosita, Jr., R. Yasuda, H. Noji, and K. Adachi, *Philos. Trans. R. Soc. Lond. B* **355**, 473 (2000).
[4] K. Kinosita, Jr., R. Yasuda, and H. Noji, *Essays Biochem.* **35**, 3 (2000).
[5] H. Noji and M. Yoshida, *J. Biol. Chem.* **276**, 1665 (2001).

diphosphate (ADP) and inorganic phosphate using proton flow as the energy source. Isolated F_1 composed of $\alpha_3\beta_3\gamma_1\delta_1\varepsilon_1$ subunits only hydrolyzes ATP, and hence is called F_1-ATPase. Its subcomplex $\alpha_3\beta_3\gamma$ suffices for rotation driven by ATP hydrolysis. Single-molecule imaging of this subcomplex has revealed detailed mechanical and kinetic properties of the motor activity, and high-resolution atomic structures of F_1 are already available.[6,7] At present, F_1-ATPase is one of the best characterized molecular motors, or nucleotide-driven molecular machines. It is possible to learn a lot from this rotary machine about the molecular mechanism of chemomechanical energy transduction.

Because all molecular machines work stochastically, their operations can never be synchronized with each other in a rigorous sense. Thus, it is necessary to watch the individual behaviors closely. With F_1-ATPase, for example, we have been able to show that it rotates in a unique direction,[8] that it does so in discrete $120°$ steps,[9,10] and that $120°$ steps are resolved into $\sim90°$ and $\sim30°$ substeps at low ATP concentrations.[11] We have also been able to measure its rotary torque, and have shown that its energy conversion efficiency can reach $\sim100\%$.[9,12] We believe that it would be very difficult, if not impossible, to obtain these results without dealing with individual molecules. Here we describe in detail the techniques involved, hoping that they may also be applicable to other molecular machines, in particular to the detection of conformational changes underlying their function (note that a conformational change accompanies reorientation, or partial rotation, of one part against the other).

For the detection of rotation (or conformational changes), we recommend the complementary use of large and small probes. Here we describe two examples, an actin filament as a probe that is large compared to the rotary motor, and a single fluorophore as a small and less perturbing probe. We begin with the preparation of materials, and proceed to the setting of functional motor molecules on a glass surface and then to imaging and analysis.

Preparation of Proteins

The $\alpha_3\beta_3\gamma$ subcomplex of F_1 derived from thermophilic *Bacillus* PS3 is expressed in *Escherichia coli*.[13] To fix the subcomplex on a glass surface, 10 histidines

[6] J. P. Abrahams, A. G. W. Leslie, R. Lutter, and J. E. Walker, *Nature* **370**, 621 (1994).

[7] Y. Shirakihara, A. G. W. Leslie, J. P. Abrahams, J. E. Walker, T. Ueda, Y. Sekimoto, M. Kambara, K. Saika, Y. Kagawa, and M. Yoshida, *Structure* **5**, 825 (1997).

[8] H. Noji, R. Yasuda, M. Yoshida, and K. Kinosita, Jr., *Nature* **386**, 299 (1997).

[9] R. Yasuda, H. Noji, K. Kinosita, Jr., and M. Yoshida, *Cell* **93**, 1117 (1998).

[10] K. Adachi, R. Yasuda, H. Noji, H. Itoh, Y. Harada, M. Yoshida, and K. Kinosita, Jr., *Proc. Natl. Acad. Sci. U.S.A.* **97**, 7243 (2000).

[11] R. Yasuda, H. Noji, M. Yoshida, K. Kinosita, Jr., and H. Itoh, *Nature* **410**, 898 (2001).

[12] H. Noji, D. Bald, R. Yasuda, H. Itoh, M. Yoshida, and K. Kinosita, Jr., *J. Biol. Chem.* **276**, 25480 (2001).

[13] T. Matsui and M. Yoshida, *Biochim. Biophys. Acta* **1231**, 139 (1995).

(His tag) have been incorporated genetically at the N terminus of the β subunit. To label the stalk region of the γ subunit with a probe, γ-Ile-210* or γ-Ser-107* has been replaced with cysteine, and α-Cys-193, the only cysteine in the wild-type $\alpha_3\beta_3\gamma$ subcomplex, has been replaced with serine by site-directed mutagenesis. The mutant $\alpha_3\beta_3\gamma$ subcomplex is purified as follows.[12]

Purification of F₁

1. Incubate cell lysate at 65° for 15 min, and remove the denatured *E. coli* protein by centrifugation (e.g., 216,000g for 30 min at 4°).
2. Apply the supernatant containing F₁ on an Ni²⁺-NTA Superflow column (Qiagen) equilibrated with 50 m*M* imidazole, pH 7.0, and 100 m*M* NaCl. Wash the column with 100 m*M* imidazole, pH 7.0, and 100 m*M* NaCl, and then elute the enzyme with 500 m*M* imidazole, pH 7.0, and 100 m*M* NaCl.
3. Add ammonium sulfate to the fraction containing the enzyme to the final concentration of 10% saturation. Apply the solution to a butyl-Toyopearl column (Tosoh, Tokyo, Japan) equilibrated with 500 m*M* imidazole, pH 7.0, 100 m*M* NaCl, and 10% saturated ammonium sulfate. To remove endogenously bound nucleotides, wash the column with 10 column volumes of a solution containing 100 m*M* potassium phosphate, pH 7.0, 2 m*M* EDTA, and 10% saturated ammonium sulfate.
4. Elute the enzyme with 50 m*M* Tris–HCl, pH 8.0, and 2 m*M* EDTA, and store as precipitate in 70% saturated ammonium sulfate containing 2 m*M* dithiothreitol (DTT) at 4°.

Conjugation of F₁ with Streptavidin

For rotation assay with actin filaments, we use the cysteine mutant at γ-Ser-107; with the γ-Ile-210-Cys mutant, we rarely find rotating actin filaments.

1. Collect the mutant subcomplex (γS107C) stored as precipitate by centrifugation at low speed (e.g., 18,800g for 15 min at 4°), and dissolve in buffer A (20 m*M* MOPS–KOH, pH 7.0, 100 m*M* KCl) containing 5 m*M* dithiothreitol.
2. Incubate for 30 min at room temperature to fully reduce the sole cysteine (S107C) at the γ subunit.

---

* In thermophilic *Bacillus* PS3, the precursory polypeptide of γ subunit contains four amino acid residues at the N terminus to be truncated. The amino acid sequence of the γ subunit registered in the DNA databank starts from an Ala residue, which is the first amino acid residue after the truncation. In our expression vector, the precursory sequence of γ has been removed from the cloned DNA sequence and the residual DNA sequence starting from the Ala residue has been conjugated after the Met residue (start codon) to express the γ subunit. In this article, as well as our previous reports, the numbers of amino acids in the γ subunit have been determined from the DNA sequence including the Met residue in the expression vector, and are therefore larger by one than these in the DNA databank.

3. To remove dithiothreitol and possibly dissociated subunits, pass the solution through a size-exclusion column (Superdex 200 HR 10/30; Amersham Pharmacia Biotech, Piscataway, NJ) equilibrated with buffer A.
4. Incubate the enzyme with a two molar excess of 6-{N'-[2-(N-maleimide) ethyl]-N-piperazinylamido}hexyl-D-biotinamide(biotin-PEAC$_5$-maleimide, Dojindo, Kumamoto, Japan) in buffer A for 6 hr on ice.
5. Remove unbound biotin with a gel filtration column (PD-10, Amersham Pharmacia Biotech) equilibrated with buffer A. Specific biotinylation of the γ subunit is confirmed by Western blotting with horseradish peroxidase avidin D (Vector Laboratories, Burlingame, CA), and the capacity for streptavidin binding is confirmed by an assay using 4-hydroxyazobenzene-2-carboxylic acid (HABA).[14]
6. Add eight molar excess of streptavidin (Pierce, Rockford, IL) to the biotinylated enzyme, and incubate for 1 hr at 23°.
7. Purify the enzyme conjugated with streptavidin on a Superdex 200 column equilibrated with buffer A.

Fluorescent Biotinylated Actin Filament

Actin is extracted from acetone powder of rabbit skeletal muscles and purified as previously described.[15] Actin is covalently biotinylated at Cys-374 with biotin maleimide, and labeled with TMR–phalloidin.

1. Incubate 50 μM F-actin with 500 μM biotin-PEAC$_5$-maleimide in buffer B (10 mM MOPS–KOH, pH 7.0, 100 mM KCl, 1 mM MgCl$_2$) for 6 hr on ice.
2. Quench the reaction by adding 0.2% 2-mercaptoethanol, and collect F-actin by centrifugation (350,000g for 60 min at 4°).
3. Depolymerize the pellet in buffer C (2 mM Tris–HCl, pH 8.0, 0.2 mM CaCl$_2$, 0.2 mM ATP) overnight on ice, and clean the obtained G-actin by centrifugation (350,000g for 60 min at 4°).
4. Polymerize G-actin in buffer B overnight on ice. Then repeat the depolymerization protocol (step 3 above).
5. Remove unreacted biotin with a gel filtration column (PD-10) equilibrated with buffer C. Biotinylated G-actin is obtained.
6. Polymerize 5 μM biotinylated G-actin in buffer B containing 12.5 μM phalloidin–tetramethylrhodamine B isothiocyanate conjugate (TMR–phalloidin, Fluka, Ronkonkoma, NY). To remove free dye and nucleotides, precipitate F-actin by centrifugation (100,000g for 30 min at 4°), and resuspend in buffer D (10 mM MOPS–KOH, pH 7.0, 50 mM KCl, 2 mM

[14] Y. Kunioka and T. Ando, *J. Biochem. (Tokyo)* **119**, 1024 (1996).
[15] J. A. Spudich and S. Watt, *J. Biol. Chem.* **246**, 4866 (1971).

MgCl$_2$). This washing procedure is repeated twice. Estimate the final concentration of F-actin using $\varepsilon_{1\%}^{290\,nm} = 0.63$. Typically, about 70% of the actin is recovered through the washing. The amount of free ATP and ADP in 370 nM actin prepared in this way has been estimated by luciferin–luciferase assay[16] to be $\leq$50 nM after conversion of ADP to ATP by incubation with 0.2 mg/ml creatine kinase and 2.5 mM creatine phosphate for 30 min at room temperature, and does not increase for at least several weeks.

Labeling F$_1$ with Single Fluorophore

To detect the rotation of the γ subunit by the measurement of fluorophore orientation, the fluorophore should be firmly attached to γ. We have tested several combinations of γ-mutants (S107C, I210C, or S107C/I210C) and fluorescent dyes that are suitable for general single-fluorophore imaging (tetramethylrhodamine-5-maleimide, Molecular Probes, Eugene, OR; Cy3-maleimide or Cy3-bismaleimide, Amersham Pharmacia). We looked for the combination that would give the highest fluorescence anisotropy in solution, a measure of the wobble of the fluorophore on a protein molecule. Among those tested, the subcomplex $\alpha_3\beta_3\gamma$(I210C) labeled with Cy3-maleimide at the sole cysteine in γ gave the highest fluorescence anisotropy of 0.32 (λ_{ex} = 550 nm; λ_{em} = 590 nm; measured in a Hitachi F-4500 spectrofluorometer), indicating that the fluorophore wobble on the subcomplex was within a cone of having a semiangle $<25°$.[17] A bis-functionalized dye conjugated to two properly spaced cysteines would be the best approach for firm attachment, but bridging between the two reactive sites is not necessarily guaranteed. In our experience, a fluorophore, even without covalent attachment, often finds a (hydrophobic) site that embraces the fluorophore firmly enough. Trial and error is our preferred approach. The procedure for Cy3 labeling is given below.

1. Using the mutant subcomplex (γI210C), follow steps 1–3 of the procedure for the conjugation of F$_1$ with streptavidin above.
2. Mix the subcomplex in the column eluent with 1.1-fold molar excess of Cy3-maleimide in buffer A (20 mM MOPS–KOH, pH 7.0, 100 mM KCl) and incubate for 30 min at room temperature.
3. Remove unreacted Cy3-maleimide with a Superdex 200 column equilibrated with buffer A to terminate the reaction. Determine the labeling ratio by assuming $\varepsilon_{555}^{Cy3} = 150{,}000\,M^{-1}\,cm^{-1}$,[18] $\varepsilon_{280}^{Cy3} = 15{,}000\,M^{-1}\,cm^{-1}$, and $\varepsilon_{280}^{F1} = 154{,}000\,M^{-1}\,cm^{-1}$.[19]

[16] M. Deluca and W. D. McElroy, *Methods Enzymol.* **57**, 3 (1978).
[17] K. Kinosita, Jr., S. Kawato, and A. Ikegami, *Biophys. J.* **20**, 289 (1977).
[18] L. A. Ernst, R. K. Gupta, R. B. Mujumdar, and A. S. Waggoner, *Cytometry* **10**, 3 (1989).
[19] T. Matsui, E. Muneyuki, M. Honda, W. S. Allison, C. Dou, and M. Yoshida, *J. Biol. Chem.* **272**, 8215 (1997).

Preparation of Surfaces for F$_1$ Attachment

Ni^{2+}-NTA Polystyrene Beads for Observation of Actin Rotation

Amino polystyrene beads of 0.224 μm diameter (Polysciences, Warrington, PA) are covalently coated with amino-NTA [*N*-(5-amino-1-carboxypentyl) iminodiacetic acid, Dojindo] via ethylene glycol bis[succinimidylsuccinate] (EGS, Pierce) linkage as follows:

1. Dissolve EGS in *N,N*-dimethylformamide at 100 m*M* and add 1 volume to 9 volumes of 1% (w/v) bead suspension in 100 m*M* MOPS–KOH, pH 7.0. Incubate for 15 min at room temperature.
2. Add 100 m*M* amino-NTA to the mixture, and incubate for 1 hr at room temperature.
3. Wash the beads four times with 100 m*M* MOPS–KOH, pH 7.0, by centrifugation (e.g., 10,000*g* for 10 min).
4. Wash the beads four times with 10 m*M* NiCl$_2$ and 10 m*M* glycine to form Ni^{2+}-NTA. Remove unbound Ni^{2+} by washing four times with 100 m*M* MOPS–KOH, pH 7.0.

Ni^{2+}-NTA Glass Coverslip for Single-Fluorophore Imaging

Unless coverslips are properly cleaned, dust or impurities show up as numerous light spots under conditions for single-fluorophore imaging. Some dust particles, even bright ones, are apparently photobleached in one step, due possibly to detachment and more likely to actual photobleaching. These cannot be readily distinguished from genuine single fluorophores, and thus preparation of clean glass surfaces is crucial to single-fluorophore imaging.[20]

1. Immerse glass coverslips (Micro Cover Glass, No. 1, 24 × 36 mm^2 and 18 × 18 mm^2, Matsunami, Osaka, Japan) in 20 *N* KOH for 1 day to clean the surfaces. We use a ceramic container that holds each coverslip vertically and separately throughout the procedure.
2. Wash the coverslips extensively with ultrapure water.
3. Incubate the coverslips in 0.01% (v/v) acetic acid containing 2% (v/v) 3-glycidyloxypropyltrimethoxysilane (Fluka) for 3 hr at 90°. The pH should be adjusted to pH 5.
4. Wash the coverslips with ultrapure water.
5. Incubate the coverslips in 0.01 *M* NaHCO$_3$, pH 10.0, containing 10% (w/v) amino-NTA (Qiagen) for 16 hr at 60°. In this procedure, the coverslips are stacked in a small dish in a small amount of solution, because amino-NTA

[20] K. Adachi, K. Kinosita, Jr., and T. Ando, *J. Microsc. (Oxford)* **195**, 125 (1999).

is expensive. We reuse the reaction solution several times because of the stability of amino-NTA.

6. Wash the coverslips with ultrapure water.
7. Incubate the coverslips in 10 mM NiCl$_2$ (or NiSO$_4$) and 5 mM glycine, pH 8.0, for 2 hr at room temperature. The glycine is added to modify unreacted epoxy group.
8. Wash the coverslips with ultrapure water.
9. Store the coverslips in ultrapure water until use (within a few weeks).

Rotation Assay Using Actin Filaments

Flow Chamber

A flow chamber is constructed with two coverslips (bottom, 24 × 36 mm^2; top, 18 × 18 mm^2) separated by two greased spacers (Parafilm cover sheet) of ∼50 μm thickness.[8] The volume of the chamber is typically 10 μl. We fix a subcomplex of F_1 on surface-bound beads (0.224 μm in diameter) through 10 histidines (His tag) linked to the amino terminus of each β. The purpose of the bead pedestal is to let F_1 on the top rotate freely, without surface obstructions or impeding factors such as the higher than bulk friction near the glass surface.[21] A fluorescently labeled actin filament is attached to γ of F_1 through streptavidin–biotin linkage. Glucose oxidase, catalase, and glucose make up an oxygen-scavenger system[22] that serves to retard photobleaching of fluorescent dye. Creatine kinase and creatine phosphate are an ATP-regenerating system that maintains the initial ATP concentration.

1. Infuse 1 chamber volume of 0.1% (w/v) Ni^{2+}-NTA beads in buffer D (10 mM MOPS–KOH, pH 7.0, 50 mM KCl, 2 mM MgCl$_2$) containing 2 mM MgCl$_2$ into the flow chamber, and allow the beads to adhere to the glass surface for 15 min. The bead density will be several beads per 5 × 5 μm^2.
2. Wash the chamber twice with 3 volumes of buffer D′ [buffer D containing 10 mg/ml bovine serum albumin (BSA)].
3. Infuse 1 volume of 10 nM streptavidin-F_1 in buffer D′. After 2 min, wash the chamber twice with 3 volumes of buffer D′.
4. Dilute labeled actin filaments from the 370 nM stock 10 times in buffer D′ before use, and infuse 1.5 volumes of the 37 nM actin into the flow chamber. After 15 min, wash the chamber with 3 volumes of buffer D′.
5. Infuse 2 volumes of buffer D′ containing 0.5% 2-mercaptoethanol, 6 mg/ml glucose, 0.2 mg/ml glucose oxidase, 30 U/ml catalase, an ATP regenerating system consisting of 2.5 mM creatine phosphate and 0.2 mg/ml creatine kinase, and a desired amount of MgATP.

[21] A. J. Hunt, F. Gittes, and J. Howard, *Biophys. J.* **67**, 766 (1994).
[22] Y. Harada, K. Sakurada, T. Aoki, D. D. Thomas, and T. Yanagida, *J. Mol. Biol.* **216**, 49 (1990).

Observation

We observe the flow chamber on an inverted fluorescence microscope (TMD300, Nikon; excitation filter: HQ525/50, dichroic mirror: Q555, emission filter; HQ590/50, Chroma Technology) with an intensified (KS-1381, Videoscope) charge-coupled device (CCD) camera (Dage MTI) at 23°. Actin rotation can be observed for 30–40 min. In the absence of Ni^{2+}-NTA beads, or with untreated amino beads, few actin filaments are bound to the glass surface. Washing with 500 mM imidazole in buffer D' (10 mM MOPS–KOH, pH 7.0, 50 mM KCl, 2 mM $MgCl_2$, 10 mg/ml BSA) removes nearly 80% of actin filaments, whereas 500 mM KCl in buffer D' removes only ∼20%. In this way, we confirm that the actin filaments are attached to the F_1 subcomplex fixed on the Ni^{2+}-NTA beads through histidine tags.

In the presence of ATP, rotating fluorescent actin filaments are found on the surfaces of the chamber. On the bottom surface, the filaments rotate counterclockwise. On the inverted microscope where the light beam is reflected once before reaching a video camera or eye, an observed image corresponds to the view from above, or, for the F_1 on the bottom surface, the view from the F_0 side. At low (<1 μM) ATP concentrations, rotation is resolved into discrete 120° steps (Fig. 1).

A

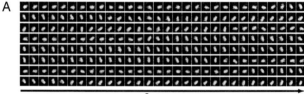

3 sec

B

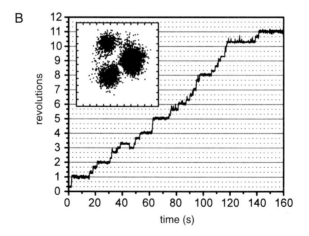

FIG. 1. Stepwise rotation of F_1-ATPase at 20 nM ATP. (A) Sequential images, at 0.1-sec intervals, of an actin filament of length 1.1 μm attached to the γ subunit of F_1-ATPase. (B) Time course of the rotation in (A). Inset shows the trace of centroid of the actin image.

Determination of Torque

When an actin filament is attached to the γ subunit of F_1, F_1 rotates the filament against hydrodynamic friction. The Reynolds number for an actin filament moving in water is extremely low, $<10^{-4}$. The torque generated by F_1-ATPase is therefore balanced with hydrodynamic friction. When a filament rotates around an axis at the middle of the filament, the friction is given by[21]:

$$\xi = (\pi/3)\eta\omega L^3/[\ln(L/2r) - 0.447]$$

where $\eta(= 10^{-3}$ N m^{-2} sec) is the viscosity of water at room temperature, ω is the angular velocity, L is the length of the filament, and $r(= 5$ nm) is the radius of the actin filament. Most filaments rotate around an axis away from the mid point. The friction in these cases is estimated as the sum of two parts:

$$\xi = (4\pi/3)\eta\omega\{L_1^3/[\ln(L_1/r) - 0.447] + L_2^3/[\ln(L_2/r) - 0.447]\}$$

where L_1 is the length from one edge to the axis and L_2 from the other edge. [In some of our earlier publications,[3,9,23] this equation for the case of $L_2 = 0$ was incorrectly cited as $(4\pi/3)\eta\omega L^3/[\ln(L/2r) - 0.447]$, where "2" in the denominator should have been omitted.] Note that these equations are applicable to filaments rotating well above a surface; filaments close to a surface experience much higher friction.[21] We try not to overestimate the torque, and thus we use these equations without introducing possibly ambiguous corrections.

We determine the position of rotation axis, the length of the filament, and the angular velocity as follows.

Axis. Find, in sequential images of a rotating filament, a fixed point. Placing a cursor on the monitor screen will help.

Length. Make sure that the video camera is not saturated, because, due to diffraction, an actin filament may appear longer (and thicker) by up to the wavelength of light, $\sim$550 nm for TMR fluorescence. We define the edges of a filament as the points at which the fluorescence intensity is half-way between the highest intensity and background. To obtain reliable torque values, we recommend analyzing long ($\gtrsim 2\ \mu$m) filaments.

Angular Velocity. The rotation angle can be determined, image by image, manually on a screen. If a digital image processor is available, the easiest method is to calculate the centroid of the filament image, which will move on a circular trajectory.[8] Alternatively, calculate the long and short axes of the filament from quadratic moments of the filament image. The latter method works even when the rotation axis is close to the middle of the filament.

At high ATP concentrations, F_1 bearing an actin filament shows continuous rotation (Fig. 2A and B). The angular velocity, however, is not strictly constant, presumably due to thermal fluctuations. We determine the angular velocity, and

---

[23] K. Kinosita, Jr., *FASEB J.* **13**, S201 (1999).

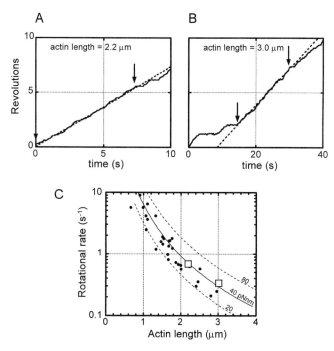

FIG. 2. Estimation of the torque of F_1. (A and B) Time courses of the rotation of an actin filament of length of 2.2 μm (A) or 3.0 μm (B) at 2 mM ATP. The average velocities are determined by averaging over five continuous revolutions (between arrows) as 0.69 and 0.33 revolution per second (dashed lines). (C) Rotational rate versus the length of the actin filament rotating around an axis at an edge. For actin filaments having a rotary axis away from the edge, $(L_1^3 + L_2^3)^{1/3}$ is plotted as the length because a torque value determined from the value is virtually as same as that determined from L_1 and L_2 within an experimental error. Large diamonds indicate the data in (A) and (B).

hence the torque, by averaging over at least five continuous revolutions that are not interrupted by unnatural pauses such as the last part of Fig. 2A and the first part of Fig. 2B. These pauses often occur at a certain angle(s), suggesting obstruction by nearby debris or glass surface (the filament may be oblique to the surface). Another reason might be that F_1 lapses at intervals into a catalytic inactive state called the MgADP-inhibited state. During this state, which lasts for ~30 sec in the wild-type $\alpha_3\beta_3\gamma$ subcomplex of thermophilic F_1, F_1 exhibits rotational fluctuation about a certain angle; continuous rotation resumes after a while.[24] The pauses by MgADP inhibition are readily distinguished, because they occur at angles 120° apart.[24]

Figure 2C shows rotational rate versus the length of the actin filament in the presence of 2 mM ATP. Most data are on or below the line representing the constant

[24] Y. Hirono-Hara, H. Noji, M. Nishiura, E. Muneyuki, K. Y. Hara, R. Yasuda, K. Kinosita, Jr., and M. Yoshida, *Proc. Natl. Acad. Sci. U.S.A.* **98**, 13649 (2001).

torque of 40 pN nm. We think that higher velocity values are more reliable, because any obstructions against rotation would reduce the velocity, and conclude that F_1 exerts a constant torque of about 40 pN nm regardless of the length of the filament.

Optics for Single-Fluorophore Imaging

The optical system is located in a clean room and mounted on a vibration isolation table. An optical layout for single-fluorophore polarization imaging is illustrated in Fig. 3. A 532-nm laser beam (DPSS 532-200, Coherent) is first focused onto a rotating diffuser disk (lemon-skin filter, Nikon, Tokyo, Japan) to average out speckle pattern and interference fringes. The linearly polarized laser beam is passed through a quarter-wave plate to make it circularly polarized. That the resulting excitation beam is isotropic in the image plane is confirmed by observing the beam through a polarizer placed on the sample stage. In experiments where the excitation

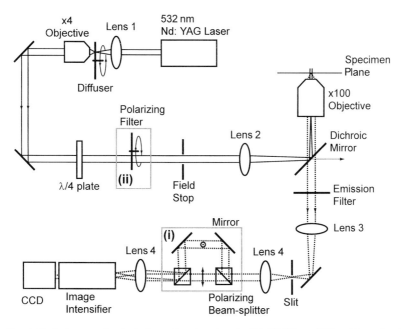

FIG. 3. Optical system for single fluorophore imaging. An Nd : YAG laser (532 nm) is used as the excitation light source. Lens 1 focuses the beam onto a rotating diffuser disk and the scattered beam is collimated with the ×4 objective. Lenses 4 produce a parallel beam in between. The focal lengths are lens 1, 50 mm; lens 2, 350 mm; lens 3, 180 mm; lens 4, 80 mm. Lens 3 is originally provided inside the microscope body, and the slit is positioned at the side camera port. Either (i) or (ii) is inserted, depending on the experimental demand. In the actual system, the excitation beam travels through the microscope from left to right.

polarization is rotated, a rotating sheet polarizer (HN32, Polaroid) is inserted after the quarter-wave plate [(ii) in Fig. 3]. The beam is introduced into a fluorescence microscope (IX70, Olympus) through a field diaphragm and a condenser lens (Lens 2, Fig. 3). A custom-made holder for a dichroic mirror (550DRLP, Chroma Technology; separation wavelength, 550 nm, Brattleboro, VT) allows entrance of the excitation beam from the right-hand side and exit of the unreflected portion to the left, preventing scattering in the body and thus reducing the background noise.[25] Lens 2 has a long focal length such that only the center of the objective aperture is illuminated; depolarization by the objective is thus avoided.

Fluorescence is collected through an oil-immersion objective (PlanApo 100×, NA 1.4, Olympus) and detected through a bandpass filter (BP590/50, Chroma Technology) with an image intensifier (VS4-1845, Videoscope, Sterling, VA) coupled to a CCD camera (CCD-300T-IFG, Dage MTI, Michigan City, IN). We have found that, whereas a water-immersion objective combined with quartz coverslips gives a lower background luminescence and usually a higher signal-to-noise ratio,[25] the signal itself from a single fluorophore is higher with the oil-immersion objective; for Cy3 and rhodamine fluorescence, therefore, we generally use the oil objective. When fluorescence emission is to be decomposed into vertically and horizontally polarized components, a dual-view apparatus[26] [(i) in Fig. 3] is inserted before the camera, onto which the two components are simultaneously projected. On the intermediary image plane after Lens 3 (Fig. 3), a rectangular slit is placed that defines the edges of the pair of images so that they can be positioned side by side without overlapping. We check whether the two components are detected with the same sensitivity by observing a thin layer of isotropic fluorescent solution sandwiched between coverslips. If disparity is found, we place coverslips in one of the split beams.

Rotation Assay Using Single-Fluorophore

Flow Chamber

Preparation of a flow chamber is carried out on a clean bench. The chamber is constructed of a bottom coverslip (24 × 36 mm^2) coated with Ni^{2+}-NTA and an uncoated top coverslip (18 × 18 mm^2) separated by two greased spacers. One chamber volume is typically ~10 μl. BSA is not used as blocking agent in the buffer for single-fluorophore imaging to avoid dust or impurities.

1. Infuse 50 pM Cy3-$\alpha_3\beta_3\gamma$ in buffer E (10 mM MOPS–KOH, pH 7.0, 50 mM KCl, 4 mM MgCl$_2$) into a flow chamber.
2. Allow 2 min for surface attachment, and wash the chamber with 5 chamber volumes of buffer E.

[25] I. Sase, H. Miyata, J. E. T. Corrie, J. S. Craik, and K. Kinosita, Jr., *Biophys. J.* **69**, 323 (1995).
[26] K. Kinosita, Jr., H. Itoh, S. Ishiwata, K. Hirano, T. Nishizaka, and T. Hayakawa, *J. Cell. Biol.* **115**, 67 (1991).

3. Infuse 5 volumes of degassed buffer E containing 0.5% 2-mercaptoethanol, 216 μg/ml glucose oxidase, 360 μg/ml catalase, 4.5 mg/ml glucose, 0.2 mg/ml creatine kinase, 2.5 mM creatine phosphate, and a desired amount of ATP.
4. To prevent evaporation, seal the open edges of the chamber with colorless nail lacquer.

Imaging

The orientations of individual fluorophores, and thus of γ, are assessed by two methods: (i) from the polarization of emitted fluorescence with the dual-view apparatus[27] and (ii) from the polarization dependence of the efficiency of light absorption by rotating the excitation polarization.[28] Method (i) has theoretically unlimited time resolution, and method (ii) is superior in angular resolution.

The flow chamber is observed using a fluorescent microscope in a clean room controlled at 23°. The laser power is decreased appropriately with neutral density filters; higher power increases emission from a single fluorophore but decreases the time to photobleaching. We set the excitation intensity at 1.1 mW over a sample area of 24 μm in diameter, Cy3 fluorophores can be observed for ~30 sec. We videotape dozens of fluorescent spots in a field of view until most of them are photobleached, and then move to a new field. A chamber is observed for a total of ~30 min. Usually, we cannot distinguish rotating fluorophores until we analyze the recorded images.

In method (i), a vertically oriented fluorophore will show up in the V (vertically polarized) image, and a horizontal one will appear in the H (horizontally polarized) image. Alternate appearance, as shown in Fig. 4A, indicates rotation of the γ-shaft. For method (ii), the polarization axis of excitation light is rotated continuously in the sample plane at 1 Hz, in the counterclockwise direction when viewed from above the microscope. A higher rotary rate may be employed when F_1 rotation is fast, at the expense of higher signal-to-noise ratio. Under rotating excitation, fluorophores are expected to fluorescence when the excitation polarization becomes parallel with their absorption transition moment. The fluorophore in Fig. 4B remains vertical from beginning to end, presumably being on an inactive F_1. In contrast, the fluorophore in Fig. 4C is initially at an 8 o'clock–2 o'clock orientation, turned into a 4 o'clock–10 o'clock orientation in the second row, and then turned through a vertical orientation to the 8 o'clock–2 o'clock orientation.

Analysis of Polarization

Images recorded on a Hi8 video recorder (EVO-9650, Sony) are analyzed with a digital image processor (C2000, Hamamatsu Photonics) and personal computer.

[27] I. Sase, H. Miyata, S. Ishiwata, and K. Kinosita, Jr., *Proc. Natl. Acad. Sci. U.S.A.* **94**, 5646 (1997).
[28] T. Ha, T. Enderle, D. S. Chemla, P. R. Selvin, and S. Weiss, *Phys. Rev. Lett.* **77**, 3979 (1996).

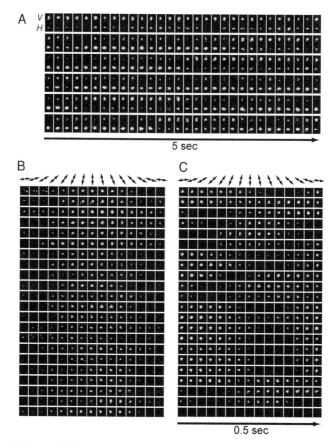

FIG. 4. (A) Sequential fluorescence images obtained with method (i). Images of a single Cy3-F_1 molecule rotating in the presence of 20 nM ATP are shown at 167-ms intervals. V, vertically polarized fluorescence; H, horizontally polarized fluorescence. Each image has been averaged spatially over 3×3 pixels; the size of the images shown is 15×30 pixels (1.5×3.0 μm^2). (B and C) Sequential fluorescence images obtained with method (ii). Single Cy3-F_1 at 20 nM ATP; 33-ms intervals. Each image has been averaged spatially over 3×3 pixels; the size of the images shown is 17×17 pixels. Arrows at the top indicate the direction of excitation polarization. In (C), the fluorophore was photo-bleached in the last row.

Fluorescence intensity of a spot is calculated as the integrated intensity over a square of 8×8 pixels (0.79×0.79 μm^2) enclosing the spot; background intensity calculated in the same way in the area after photobleaching is subtracted.

In method (i), time courses of spot intensities for V (vertically polarized fluorescence) and H (horizontally polarized fluorescence) are plotted as in Fig. 5A. Alternations of the two curves are the signature of rotation. Polarization ($P = [V - H]/[V + H]$) is calculated (Fig. 5C). At the low ATP concentration,

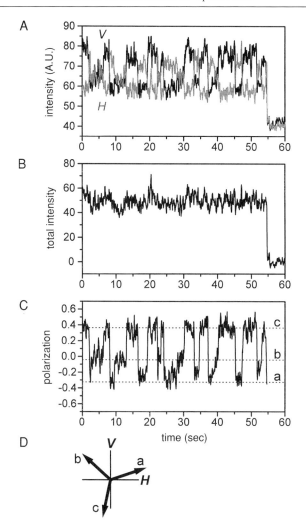

FIG. 5. (A) Time courses of spot intensities for V and H in Fig. 4A, median filtered over eight video frames (0.27 sec). The fluorophore was photobleached at $\sim$55 sec. (B) Time course of the total intensity, $V + H$, calculated from (A). (C) Time course of the polarization, $P = (V - H)/(V + H)$, calculated from (A) and (B). Dashed lines (a, b, and c) are calculated P for the three orientations in (D): $P = 0.4 \times [\sin^2(\theta + 18^\circ) - \cos^2(\theta + 18^\circ)]$, where $\theta = 0^\circ$, 120°, and 240°.

rotation of F_1 is expected to be stepwise, and indeed we see three levels of polarization, a, b, and c, in Fig. 5C. The three polarization values can be explained by the three orientations separated by 120° (Fig. 5D).

In method (ii), the fluorescence intensity oscillates with time as shown in Fig. 6A. When a fluorophore lies at an angle θ in the sample plane, its intensity is

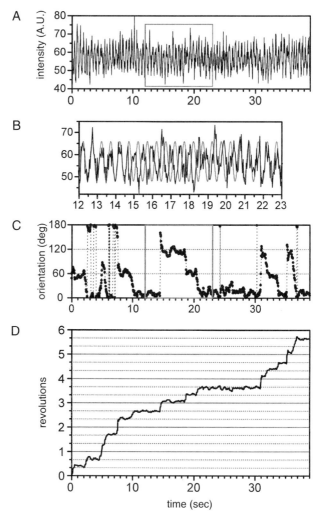

FIG. 6. (A) Time course of the fluorescence intensity in Fig. 4C. (B) Expanded view of the time course in (A) between 12 and 23 sec. (C) Time course of the fluorophore angle calculated from (A). (D) Accumulated rotation angle estimated from (C).

expected to flicker as a periodic function of time: $\cos^2[360°(t/T) - \theta] \propto \cos[360°(2t/T) - 2\theta]$, where t is time and $T(= 1 \text{ sec})$ is the period of excitation rotation. Thus, the orientation at time t, $\theta(t)$, is determined by fitting the observed intensity, $I(t)$, with this function over the period between t and $t + T/2$. Figure 6C shows $\theta(t)$ calculated from Fig. 6A; values between $0°$ and $180°$ are chosen. The orientation steps among three levels. If this fluorophore had remained at θ at $t = 12$ sec, $I(t)$ would have flickered as in the gray curve (Fig. 6B). In fact it changed its orientation at 14.7, 18.6, and 20.7 sec. The accumulated rotation angle, as shown in Fig. 6D, is obtained from Fig. 6C by assuming that all steps are counterclockwise; negative 60° steps (or positive 120° steps) in Fig. 6C are interpreted as counterclockwise 120° steps. The positive 60° step at ~5 sec is interpreted as rapid succession of two counterclockwise 120° steps within the 0.5-sec fitting window used for the angle analysis.

Conclusion

Single-molecule imaging with a large or small probe is a versatile tool for the investigation of the mechanisms of protein machines, revealing their conformational changes in real time (note that rotation we observe in F_1 is a conformational change in a large scale). Probes that are large compared to protein molecules are easily observed by conventional microscopy. Interpretation of the images is often straightforward, and intense signals from a large probe allows precise analysis. Asymmetric probes such as the actin filament are smart probes in that they amplify reorientational motions (= conformational changes) in a protein molecule without requiring a sophisticated instrument. In contrast, probes smaller than protein molecules, such as single fluorophores, are much less perturbing and enable observation under no load. By virtue of gene engineering, a single fluorophore can be attached at any desired site on a protein molecule, reporting the behavior of that particular site. We hope this article will encourage many laboratories to join the venture of single-molecule physiology, where one studies conformational dynamics of molecular machines in real time while they are at work.

Acknowledgments

We thank Masasuke Yoshida, Ryohei Yasuda, Hiroyasu Itoh, and the members of CREST (Core Research for Evolutional Science and Technology) Team 13 for suggestions and discussion, and Kerstin Steinert of Qiagen for the Ni^{2+}-NTA protocol. This work was supported in part by Grants-in-Aid from the Ministry of Education, Science, Sports, and Culture of Japan and CREST.

[12] Molecular Motors and Single-Molecule Enzymology

By Yoshiharu Ishii, Kazuo Kitamura, Hiroto Tanaka,
and Toshio Yanagida

Introduction

Molecular motors are the molecular machines that perform mechanical work using the energy released from the hydrolysis of adenosine triphosphate (ATP). Until very recently, mechanical and biochemical properties of the molecular motors were studied using myosin and actin in muscle fibers and in purified protein solutions. In these ensemble measurements, however, there were many problems in interpreting the data. First, these systems contained large numbers of molecules. Only averaged values were measured and many dynamic properties of the molecules were hidden. For example, the hydrolysis of ATP occurs in a series of reactions and individual molecules are in different phases. Thus, the average values obtained from the ensemble measurements were difficult to interpret. The entire reactions were usually broken down into individual steps and studied. Short-lived intermediate states, which were difficult to observe, could be studied by measuring the properties of analogous long-lived states instead. Second, it was difficult to measure both the biochemical and mechanical events using a single experimental system. Mechanical measurements in muscle fibers have been carried out extensively, whereas only a few biochemical measurements have been reported. In contrast to muscle fibers, the purified protein solutions have been used for biochemical measurements at the expense of the mechanical measurements. Thus, the biochemical and mechanical properties of actomyosin have been measured separately. The simplest idea to assume is that the mechanical events are tightly coupled with the chemical reactions. Of course, such assumptions must be tested.

Molecular motors such as myosin and kinesin are linear motors that slide along protein tracks, actin and microtubule, respectively. A sliding movement theory was first proposed for myosin and actin in muscle based on the observation of the changes in its striated pattern and physiological properties.[1] In the 1980s, 30 years after the proposal of the sliding movement theory, the sliding movement between purified myosin and actin was first demonstrated under a microscope.[2,3] The sliding movement of fluorescently labeled actin filaments moving over myosin molecules immobilized on the surface of a coverslip was visualized. These filaments moved with the same velocity as the shortening velocity recorded in intact muscles. This

---

[1] A. F. Huxley, *Prog. Biophys.* **7,** 255 (1957).

[2] S. J. Kron and J. A. Spudich, *Proc. Natl. Acad. Sci. U.S.A.* **83,** 6272 (1986).

[3] Y. Harada, K. Sakurada, T. Aoki, D. D. Thomas, and T. Yanagida, *J. Mol. Biol.* **216,** 49 (1990).

in vitro motility assay has offered an experimental system, in which both the mechanical and biochemical events could be monitored. In this method, however, it was not possible to detect where and when the ATP molecule was hydrolyzed and which myosin molecules were interacting with the actin filament. To elucidate how the myosin molecule works, it was necessary to measure the mechanical events generated by a single myosin and ATP molecule. A measuring system that could visualize and manipulate single molecules needed to be developed.[4]

In Vitro Motility Assay

The observation of the sliding movement of actin filaments over myosin molecules was essentially accomplished by visualizing the actin filaments.[5] The actin filaments were fluorescently labeled with tetramethylrhodamine–phalloidin. Phalloidin also served to stabilize actin in the filament form even at low concentration such as in the condition of this motility assay (monomer concentration of actin ∼10 nM) without interfering with the motile activity of actin. Actin filaments in solution show thermal Brownian motion in the absence of myosin. The addition of myosin and ATP increases this thermal motion, but the sliding movement of the actin filaments cannot be observed in solution. The sliding movement was observed when myosin was immobilized onto the glass surface (Fig. 1).

Procedures

Given that the *in vitro* motility assay forms a basis for the single molecule measurements described below, many techniques used for this assay are also used in other experiments. The surface of a coverslip (24 × 32 mm) is treated with silicone [5% Sigmacote in heptane (Sigma, St. Louis, MO)] at room temperature after being cleaned with 0.1 N KOH and ethanol. A myosin solution (∼ 0.5 mg/ml) is then applied by a micropipette to the silicone-treated coverslip and incubated for ∼1 min to allow the molecule to bind tightly to the glass surface. Immediately after the application of the myosin solution, a smaller coverslip (18 × 18 mm) with the diagonal edges turned slightly upward, is placed on top, which facilitates the solution change. Exchanging solutions can be achieved by placing the new solution at one upturned edge of the coverslip using a micropipette and simultaneously removing the solution from the other edge with absorbant tissue. Unbound myosin is removed by washing with a buffer. The actin–fluorescent phalloidin complex is prepared by mixing actin and phalloidin at a molar ratio of 1 : 1 and incubating overnight at 4°. The actin–fluorescent phalloidin complex at a concentration of 2.5 μM is then applied to the myosin-coated coverslip. Actin filaments are tightly

[4] Y. Ishii and T. Yanagida, *Single Mol.* **1,** 5 (2000).
[5] T. Yanagida, M. Nakase, K. Nishiyama, and F. Oosawa, *Nature* **307,** 58 (1984).

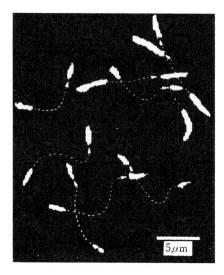

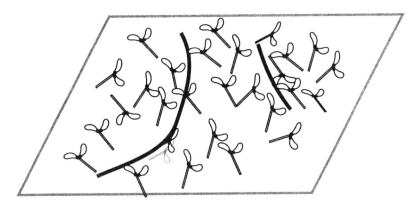

FIG. 1. The *in vitro* motility assay of the movement of actin filaments over the myosin molecules adsorbed on the glass surface. On the left are the images of the movement of actin filaments. Two images with the interval of 1.5 sec were taken on the same frame of a film by double exposure. On the right are actin filaments that move over myosin molecules adsorbed on the glass surface.

bound to the myosin immobilized on the coverslip. When ATP is added, the actin filaments start to move. To reduce photobleaching, 0.1 mg/ml glucose oxidase, 0.018 mg/ml catalase, 2.3 mg/ml glucose, and 1% 2-mercaptoethanol are included as an oxygen scavenging system in the assay buffer.

An inverted fluorescence microscope is placed on a vibration-free table. The sliding movement is recorded with a video recorder through a high-sensitivity SIT camera. The sliding velocity can be measured by analyzing the movement of the filaments on the videotape. The sliding velocity is expected to be ~6 μm/sec at 30°, corresponding to the maximum shortening velocity of muscle with no load.

Immobilization of Protein Molecules

In the *in vitro* measurements of biological function, it is critical to keep the proteins intact during the measurements. In the *in vitro* motility assay, immobilization of molecular motors on the glass surface is required. Immobilization of the proteins is also required for the imaging of single molecules (see below). However, the immobilization may affect the activities of these proteins. Myosin molecules are attached to the glass surface most likely through their C-terminal rod portion, which is not involved in the motile activities (Fig. 2a). Soluble fragments of the N-terminal head portion of myosin, single-headed subfragment-1 (S1) and two-headed heavy-meromyosin (HMM) have been successfully used instead of the whole myosin molecules, because the motor function of myosin is achieved by the head region.[6] The *in vitro* motility assays confirmed that the heads of myosin are sufficient to generate the sliding movement.[3,7] In this system, however, the direct interaction of the head portion of myosin with the artificial surface decreased the sliding velocity of the actin filaments (1.9 μm/sec) (on the left of Fig. 2b). Denatured myosin heads may bind to actin strongly even in the presence of ATP. The sliding velocity recovered to the same level as the whole myosin when myosin S1 was attached to the glass surface through the biotin–avidin bond (6.8 μm/sec).[8] The biotin–avidin bond links the myosin molecules and the glass as follows (on the right of Fig. 2b); Biotinylation occurs at the regulatory light chain (RLC) of myosin S1, which is located at the neck region (a region connecting the head and the rod portion) a considerable distance from the catalytic and actin-binding domain. The recombinant RLC is fused with biotin-dependent transcarboxylase (BDTC) and expressed in *Escherichia coli*. BDTC is a peptide out of the 1.3 S subunit of *Propionibacterium shermanii* transcarboxylase and is posttranslationally biotinylated in *E. coli*. BDTC–RLC is incorporated into myosin S1 by exchanging the light chain in myosin S1 (BDTC–S1). The surface of a silicone-coated coverslip is coated with biotinylated bovine serum albumin (BSA) (1 mg/ml). After washing out unbound BSA, a solution containing 1 mg/ml streptavidin is added to the glass coverslip. Finally myosin S1 containing BDTC–RLC is attached to the streptavidin–biotinylated BSA.

[6] S. S. Margossian and S. Lowey, *Methods Enzymol.* **85**, 55 (1982).
[7] Y. Y. Toyosima, S. J. Kron, E. M. McNally, K. R. Niebling, C. Toyoshima, and J. A. Spudich, *Nature* **328**, 536 (1987).
[8] A. H. Iwane, K. Kitamura, M. Tokunaga, and T. Yanagida, *Biochem. Biophys. Res. Commun.* **230**, 76 (1997).

a

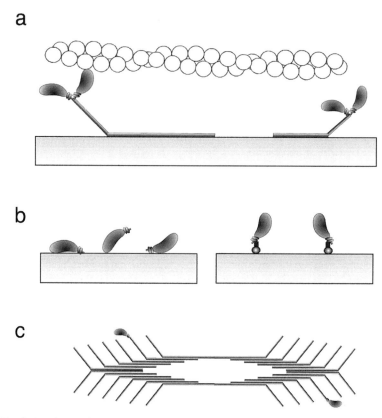

b

c

FIG. 2. Attachment of myosins onto the glass surface. (a) The whole myosin molecule is attached to the glass surface most probably through the rod portion. The head portion, which contains the catalytic domain and the actin-binding domain, does not interact directly with the glass surface. (b) When the myosin head fragment is attached to the glass surface, direct interaction with the glass surface occurs and interferes with the movement of actin filaments on the left. The myosin head fragment is attached through the biotin–avidin system on the right. (c) Single-headed myosin is mixed with myosin rod to form a synthetic myosin–myosin rod cofilament.

Myosin in Filaments

Myosin forms filaments through the rod portion at a physiological condition. Myosin filaments are useful to immobilize myosin molecules without any damage, because direct attachment of the head portion with the glass surface is avoided when myosin molecules are in the filaments on the glass surface (Fig. 2c). In addition, the use of myosin filaments has several additional advantages including (1) the position of the myosin head can be easily marked under a microscope, and

(2) the orientation of the heads is known in the filament. Regarding the effects of orientation on myosin, we have shown that the velocity and the displacement of the unitary step (step size) depend on the relative orientation between a myosin head and an actin filament.[9] Myosin–myosin rod cofilaments can be prepared by mixing myosin and myosin rod at a total concentration of 0.15 μM in a high salt solution (0.6 M KCl), in which myosin is in a monomeric form. The mixture of myosin and myosin rod is dialyzed against a low salt solution (0.12 M KCl) without stirring at 0°. The number of the heads on the filament can be controlled by changing the ratio of myosin and myosin rod. For experiments involving measurements from a single myosin head, the molar ratio of myosin (or single-headed myosin) to myosin rod is adjusted at 1 : 1000, which results in one or two myosin heads being included in the 5- to 8-μm-long filaments.

Imaging of Single Biomolecules

Visualizing Single Biomolecules

The visualization of single fluorophores is a technique required not only for visualizing the behavior of protein molecules but also for manipulation of single protein molecules. To visualize single proteins or other biomolecules, fluorescence probes are chemically attached to them.

In the early 1990s, single fluorescence dye molecules present on an air-dried surface could be visualized, using scanning near-field optical microscopy (SNOM).[10] With this method, however, it was difficult to trace the behavior of molecular motors in solution. For the visualization of single molecules, it was essential to reduce the background noise. Total internal reflection fluorescence microscopy (TIRFM) is the technique used to irradiate only the surface between the quartz glass and the solution to a depth of 100–200 nm.[11] Thus, only the fluorescent molecules positioned near the glass surface could be visualized. These single fluorophores can be observed as spots and the fluorescence disappears in a single stepwise manner as a result of photobleaching.

Objective lens-type TIRFM, in which the incident light is passed through the objective lens, is used for many other purposes such as the mechanical measurements as a working space above the specimen is created.[12] When the laser beam is introduced into the corner of the objective lens, the incident angle of the beam is larger than the critical angle of the glass–water interface (61.0°). The beam is

[9] H. Tanaka, A. Ishijima, M. Honda, K. Saito, and T. Yanagida, *Biophys. J.* **75**, 1886 (1998).

[10] E. Betzig and R. J. Chichester, *Science* **262**, 1422 (1993).

[11] T. Funatsu, Y. Harada, M. Tokunaga, K. Saito, and T. Yanagida, *Nature* **374**, 555 (1995).

[12] M. Tokunaga, K. Kitamura, K. Saito, A. H. Iwane, and T. Yanagida, *Biochem. Biophys. Res. Commun.* **235**, 47 (1997).

then totally internally reflected to produce an evanescent field at the glass–water interface (Fig. 3a). To demonstrate that the specimen is locally illuminated at the glass–water interface by an evanescent field, the fluorescence intensity of a fluorescent bead (diameter $0.2~\mu$m) attached to the tip of a scanning probe was examined as a function of the distance from the glass surface. Figure 3b shows the typical data of the fluorescence intensity, which could be well fitted to a single exponential curve with the $1/e$ penetration depth of 170 nm.

Visualizing Turnover of Single ATP Molecules

Molecular motors are also enzymes that hydrolyze ATP and fuel for mechanical work. We have visualized the turnover of single ATP molecules by a single motor molecule using fluorescent ATP.[11] The fluorescent ATP is chemically synthesized by attaching a fluorescent dye at the ribose of ATP.[12,13]

3′(2′)-O-[N-[2-(amino)ethyl]carbamoyl]-ATP is coupled with N-hydroxysuccinimide ester of Cy3.29 in 0.5 M triethanolamine, pH 8.5 at 25° for 3 hr. After the pH is neutralized, unreacted reagents can be removed by fast protein liquid chromatography (FPLC) reversed-phase chromatography on Mono Q. Purification after the synthesis is important, because many isoforms are present. The attachment of fluorophores may cause some alterations but this depends on the proteins being investigated. In the case of myosin, the fluorophores had no effect on the binding affinity of ATP and the ATPase rate.

To observe the ATP turnover by myosin, the Cy3-ATP solution is added to myosin immobilized on the glass surface. Myosin is labeled with different fluorescent dyes other than Cy3, for example, Cy5. The position of myosin molecules is marked and the fluorescence of Cy3-ATP is recorded at this position. Fluorescent ATP molecules can be detected as spots when they attach to enzyme molecules that are fixed to the glass surface (Fig. 4a). The fluorescent ATP molecules cannot be observed as spots when undergoing rapid Brownian motion in solution. Thus, at the positions where an enzyme molecule is located, a fluorescent spot can be observed when a fluorescent ATP binds to the enzyme molecule. The spot disappears when the nucleotide has dissociated, most likely after ATP is hydrolyzed to ADP. To confirm that the turning-on and turning-off of fluorescent spots is due to the turnover of ATP, the time interval that the nucleotide remained bound to the enzyme is measured for individual ATP molecules and then compared to the ATPase rate of the enzyme in solution. This process is stochastic, and the time interval distributes widely. The distribution could be well fitted to an exponential curve. In the case of myosin, the decay rate constant was consistent with the

[13] K. Oiwa, J. F. Eccleston, M. Ason, M. Kikumoto, C. T. Davis, G. P. Reid, M. A. Ferenczi, J. E. T. Corrie, A. Yamada, H. Nakayama, and D. T. Trenthum, *Biophys. J.* **78**, 3048 (2000).

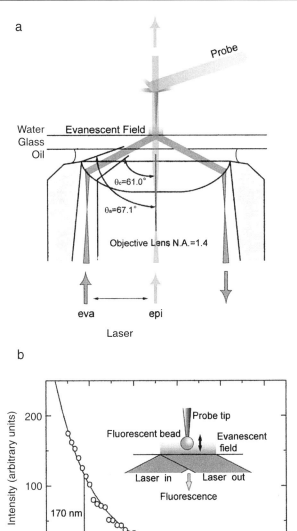

FIG. 3. Objective lens-type total internal reflection fluorescence microscopy (TIRFM). (a) Schematic drawing of the setup. Incident lights are switched between the TIRFM and the epifluorescence microscopy. The space above the specimen is used for the manipulation of a scanning probe. (b) The fluorescence intensity of a bead as a function of the depth from the surface of the glass surface and the solution.

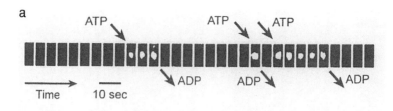

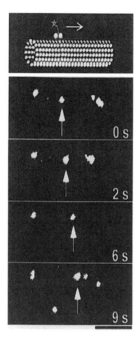

FIG. 4. Imaging of the function of single motor molecules. (a) The imaging of the turnover of single ATP molecules generated by a single myosin molecule. The fluorescence image of ATP was taken at the position where a myosin molecule is fixed. (b) The imaging of the movement of single kinesin molecules along single microtubules.

ATPase rate in solution. The rate is dependent on the concentration of ATP. An increase in the Cy3-ATP concentration accelerates the turnover rate and increases the rate for the binding. At the same time, high concentrations of Cy3-ATP increase the background noise, which in turn decreases the signal-to-noise (S/N) ratio. For optimum resolution a final concentration of Cy3-ATP is adjusted to be in the order of 10 nM.

Visualizing Sliding Movement of Single Motor Molecules

The sliding movement of a single motor molecule can be visualized for molecular motors that move along filaments of track proteins for long distances without dissociating. Kinesin is one such processive motor protein. Movement of single molecules of kinesin can be visualized after being fluorescently labeled (Fig. 4b).[14] The sliding movement of myosin, however, is difficult to observe with this method, because muscle myosin readily dissociates from actin filaments. Different types of unconventional myosin, myosin V and VI, have been found to move processively and the movement of the single motor molecules has been visualized.

It is crucial that the protein molecules are fluorescently labeled in a manner that does not interfere with the motor function. To achieve this, (1) the protein is fused to green fluorescent protein (GFP) or its mutants or (2) organic fluorophores are attached to a specific site on the protein or engineering sites. In the case of kinesin, a cysteine residue, of which the thiol group is highly and exclusively reactive for fluorescent dyes, is introduced at the end of the molecule using protein engineering. Fusion protein with GFP has several advantages; almost all protein molecules are labeled, and labeled proteins can be easily introduced into cells. However, the optical stability is less than that of some organic fluorophores for single molecule detection.

Manipulating Single Protein Molecules and Measuring Mechanical Events Generated by Single Motor Molecules

To measure the mechanical properties of single motor molecules, it is necessary to manipulate single protein molecules and allow the interaction to be monitored. A laser trap is a technique used to hold a bead utilizing the attractive forces exerted by a focused laser. By moving the focal point, it is possible to move a bead that attaches to a single protein molecule. A laser trap can also be used to measure the mechanical properties of molecular motors, because the trapped beads behave like a spring. When force is exerted on the protein, the bead becomes displaced and a restoring force is exerted by the trap toward the trap center to pull the bead back to the original position. Finally two forces are balanced.

All the instruments are set on a vibration-free table in a soundproof chamber to minimize the mechanical vibration and drift.

Manipulating Kinesin Using a Laser Trap and Measuring Its Mechanical Events

The processive movement of molecular motors can be observed by attaching the motors to beads and tracing the position of the beads (Fig. 5a). In these studies,

[14] R. D. Vale, T. Funatsu, D. W. Pierce, L. Romberg, Y. Harada, and T. Yanagida, *Nature* **380**, 451 (1996).

a

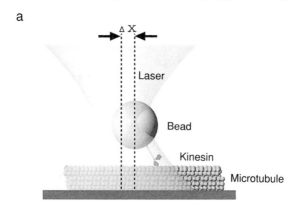

b

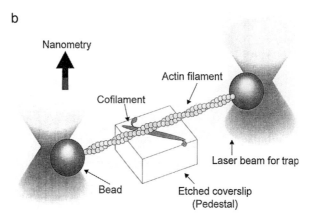

FIG. 5. Manipulation and nanometry of single motor molecules. (a) A single kinesin molecule is attached to a bead trapped by a laser and brought into contact with microtubules. The displacement of the bead was measured with nanometer accuracy. (b) A single actin filament was attached to two beads trapped by a laser at both ends and brought into contact with single myosin molecules in the myosin–myosin rod cofilament on a pedestal.

the resolution of the displacement measurement is limited because of the thermal motion of the beads. A laser trap decreases the thermal motion of beads, allowing the measurement of the displacement with a nanometer accuracy and the force generated by single kinesin molecules.[15,16]

[15] K. Svoboda, C. F. Schmidt, B. J. Schnapp, and S. M. Block, *Nature* **365,** 721 (1993).
[16] H. Kojima, E. Muto, H. Higuchi, and T. Yanagida, *Proc. Natl. Acad. Sci. U.S.A.* **91,** 12962 (1994).

Microtubules and beads are fluorescently labeled and their fluorescence images displayed on a TV monitor through an SIT camera. The stage of the microscope can be manipulated in the range from 0.1 μm to 2 mm by a hydraulic micromanipulator and a piezoactuator. The bead is trapped by a near-infrared laser (Nd:YAG) at $\lambda = 1064$ nm and laser power of 300 mW through convex and objective lenses ($\times 100$, NA 1.3). A bright image of a trapped bead is projected onto a quadrant photodiode sensor. The displacement of a bead in the x and y directions can be determined by measuring the difference in intensities on each pixel of a quadrant photodiode using difference amplifiers. The stiffness of the optical trap (K) can be determined from the variance of the thermal fluctuation of a trapped bead ($\langle x^2 \rangle$) based on the equipartition theorem, $K = k_B T / \langle x^2 \rangle$, where k_B is Boltzmann's constant, T absolute temperature, and x the displacement of the bead from the trap center. Variance is obtained by integrating the power spectrum density of the thermal fluctuation of the bead.

These studies have shown that kinesin moves along the microtubules in discrete steps of 8 nm. The time between the steps is dependent on ATP concentration, indicating that the mechanical event is tightly coupled with the biochemical reaction. Based on this finding, a "hand-over-hand model" has been proposed, in which a kinesin molecule walks on the tubulin heterodimers via its two heads in an alternating fashion.

Manipulating Actin Filaments Using a Laser Trap

The laser trap has also been used for the myosin–actin system. In contrast to processive molecular motors, conventional myosin readily dissociates from actin. To prevent an actin filament from diffusing away from the myosin during measurements, the actin filament is trapped at both ends (Fig. 5b).[17,18] The infrared laser (1064 nm, 500 mW) is split into two beams by an acousto-optical deflector (AOD) and incorporated into the microscope adapted for dual optical traps. Actin filaments are attached to beads via the avidin–biotin system. Streptavidin–biotinylated beads are attached to biotinylated actin in the filaments. Cy5-labeled actin monomers are also included in the actin filaments to visualize them and the filaments are stabilized by phalloidin.

To allow the actin filament to interact with the myosin molecules in the myosin–myosin rod cofilament, the cofilament is placed on a chemically etched pedestal on the glass surface.[9] This positions the actin filament higher than the radius of the bead above the glass surface. A chemical etched pedestal is prepared as follows. Cleaned coverslips are coated with chrome by vacuum evaporation and then

[17] J. T. Finer, R. M. Simmons, and J. A. Spudich, *Nature* **368**, 113 (1994).
[18] J. E. Molloy, J. E. Burns, J. Kendrick-Jones, R. T. Tregear, and D. S. C. White, *Nature* **378**, 209 (1995).

coated with photoresist. Square patterns are printed on the surface by irradiating with an ultraviolet (UV) light through a square pattern mask (15 × 15 μm). The resist film is developed and the irradiated area is removed. The chrome unmasked with resist is removed with chrome etching solution [7.5 wt% $HClO_4$, 12.1 wt% $(NH_4)_2Ce(NO_2)_6$, 80.4 wt% H_2O], and then the bare surface of the coverslip etched in a solution containing 3% NH_4F and 50% H_2SO_4 for 15 min. Etched coverslips are then rinsed with water, and the remaining chrome on the surface is removed with a chrome etching solution. The pedestals ∼5–10 μm × 5–10 μm in area and 5 μm in height are placed on the surface in intervals of ∼50 μm.

The position of the bead attached to the actin filament fluctuates due to thermal motion. Once myosin is bound, the thermal motion of the beads greatly decreases, reflecting an increase in the stiffness of the experiment system. We can determine when the actin filament interacted with myosin by estimating the variance of the Brownian motion. The starting position of the displacement, hence the size of the displacement, is not known for an individual event because of the thermal fluctuations. The displacement of the bead calculated relative to the mean position of the free beads had a distribution similar to that of the Brownian motion of free beads. Using this technique, only mean displacements could be obtained.[18]

Manipulation of Myosin S1 Using a Scanning Probe

Many of the problems encountered measuring the properties of the actin filaments using the laser trap were caused by compliance in the links between the bead and the myosin motors. Direct measurement of the displacement of the myosin head is advantageous over the measurement of actin filaments.[19] It is possible to manipulate a single myosin molecule with a microneedle. By attaching myosin S1 directly to the tip of a scanning probe as well as using bundles of actin filaments, the stiffness of the system increased to 1 pN/nm as compared with <0.2 pN/nm for the manipulation of the actin filament using the laser trap system. These numbers result in a root mean square displacement of 2.0 nm for the scanning probe method as compared with 4.5 nm for the laser trap system. This improvement increases the signal-to-noise (*S/N*) ratio.

A schematic drawing of the apparatus is shown in Fig. 6a. A scanning probe is prepared as follows. Highly sensitive fine glass microneedles 50–100 μm long and ∼0.3 μm in diameter are made from a glass rod (diameter ∼1 mm) using a glass pipette puller. The tip of a flexible microneedle is glued to a rigid glass rod ∼100 μm in diameter with epoxy resin, and then cut using a heated 5-μm-diameter platinum wire under a binocular microscope. The bending stiffness (spring constant) of the needle is calibrated by measuring the mean square of their thermal fluctuations in solution using the principle of energy equipartition, or by cross-calibration against

[19] K. Kitamura, M. Tokunaga, A. H. Iwane, and T. Yanagida, *Nature* **397**, 129 (1999).

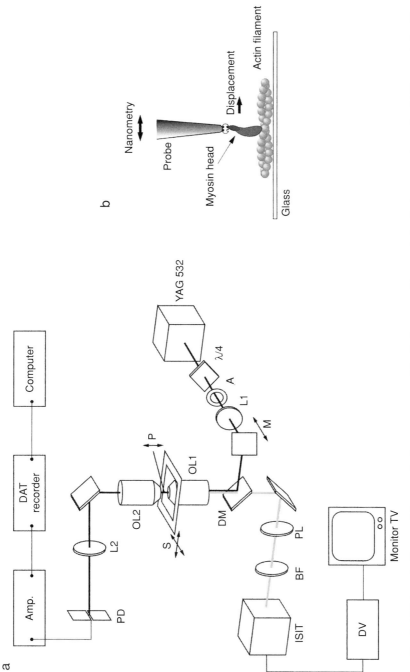

FIG. 6. (a) Schematic diagram of the setup for the measurement of the displacement of the myosin head using a scanning probe. (b) The myosin head is brought into contact with actin filaments and the displacement of the tip of the probe is measured.

stiffer calibrated needles. The needles with spring constants between 0.01 and 1 pN/nm are used. A ZnO whisker crystal is used for the tip of the probe. The crystal has a tetrapod-like structure with four 5- to 10-μm-long legs and the diameter of the tip is approximately 15 nm. ZnO whisker crystals are aminosilanized in ethanol containing 5% (v/v) 3-aminopropyltriethoxysilane, 2.9% water, and 1.2 mM HCl overnight at room temperature and then biotinylated in N,N-dimethylformamide containing 1 mM 5-[5-(N-succinimidyloxycarbonyl)pentylamido]hexyl D-biotin-amide [biotin-(AC$_5$)$_2$-Osu] overnight at room temperature. A biotinylated ZnO whisker is glued to the tip of a fine glass microneedle with epoxy resin under a binocular microscope.

Myosin S1 is then captured by a probe. BDTC-S1 and actin are labeled fluo-rescently (BDTC-S1* and actin*). The divergent beam of a frequency-doubled Nd:YAG laser (532 nm) is converted into a circularly polarized beam by a quarter-wave plate (λ/4) and focused by a lens (L1) on the back focal plane of the objective lens (OL1) (PlanApo 100× Oil, NA 1.4). Thus, specimens are illuminated by a collimated beam with Koehler illumination. Illumination can be switched between epifluorescence microscopy and objective-type TIRFM by shifting the position of the mirror (M) set before a dichroic mirror (DM1) (reflecting 532 nm laser line). Single Cy3-labeled S1 molecules on the glass surface are illuminated with the 532-nm laser by objective-type TIRFM. Fluorescence of single Cy3-labeled S1 molecules can be obtained by recording an image using an intensified SIT camera, and storing the image on digital video. Background luminescence should be rejected by the dichroic mirror (DM1), a holographic notch filter (NF), and an interference bandpass filter (BF). BODIPY FL-labeled actin bundles are also illuminated by the 532-nm laser and observed by switching the bandpass filter.

A solution containing the actin bundles is applied after the glass surface has been coated with α-actinin (0.1 mg/ml). Approximately 10^{-16} mol of BDTC–S1* in an assay buffer without ATP is then applied to the coverslip and allowed to react for 1 min. BDTC–S1* molecules bound to actin* bundles fixed on the glass surface can be observed briefly ($\sim$1 sec). The illumination should be minimized to avoid the photobleaching. The streptavidin-coated probe is then positioned above the surface of a coverslip and scanned over a 1 × 1-μm area to capture a single S1* molecule via the streptavidin–biotin bond. Contact of the probe with the surface is monitored by observing thermal fluctuations of the needle. The fluorescence of S1* at the tip of the probe is then observed to determine if a single S1* molecule had bound according to the fluorescence intensity and single-step photobleaching.

The scanning probe is illuminated with the 532-nm laser and the magnified ($\sim$300×) bright field image of the probe captured by the objective lens (OL2) (Plan 60× DIC) and a lens (L2) is projected onto a pair of photodiodes (PD). The displacements of the tip of the scanning probe can be determined from the differential output of the photodiodes. A custom-made sample stage was used.

The $x-y$ position of the sample stage is controlled by a piezoelectric scanner with subnanometer resolution. The scanning probe is also mounted on a piezoelectric scanner for the position of the z direction to be controlled. The displacement of S1 during the interaction with the actin bundle was measured by monitoring the displacement of the tip of the probe. This allowed the rising process of the displacement to be closely examined on an expanded time scale. The step we initially thought a single step caused by a single ATP molecule was found to contain several substeps of 5.5 nm.

Simultaneous Measurements of Mechanical and Chemical Events

Now that the *in vitro* motility assay has allowed both mechanical and biochemical measurements, it is possible to directly determine the coupling by simultaneously measuring the ATP turnover and the displacement of single myosin molecules.[20] The measurement system was set up by incorporating both the imaging and laser trap system on an inverted fluorescence microscope.

To ensure that single myosin heads generate the displacement, a myosin–myosin rod cofilament that has only a single myosin head must be used. The number of the myosin heads on the filament is estimated by observing the fluorescent spots of ATP*, when the cofilaments are fixed on the pedestal and ATP* is added in the absence of actin. A filament that had a single fluorescent spot that turned on and off was chosen. An actin* filament is then captured by a dual optical trap and the suspended actin filament brought into contact with a single one-headed myosin molecule in a myosin–myosin rod cofilament bound to the surface of a rectangular pedestal of $\sim$7 μm wide and $\sim$1 μm high. The interaction of actin with myosin is confirmed by monitoring the displacement of the actin filament. Use of a myosin filament facilitates the positioning of an avalanche photodiode (APD) for the measurement of the ATP fluorescence. Etched pedestals allow evanescent illumination of the myosin filaments on the glass surface. Individual ATP reactions of the single one-headed myosin molecule is observed by monitoring the fluorescence of ATP* with a high-sensitivity camera or the APD. Finally the fluorescence from ATP* is monitored by the APD to measure the binding of ATP and dissociation of ADP.

With low temporal resolution, good correlation between the biochemical event and mechanical event was obtained for actomyosin (Fig. 7): the binding of ATP to myosin causes the dissociation of myosin from actin and the dissociation of ADP causes a mechanical process. This coupling is consistent with the data of the ATP concentration dependence of the mechanical measurements. If the temporal resolution is increased to less than 10 ms, the timing between the two events becomes significant. This is especially evident if there is a wide distribution of

[20] A. Ishijima, H. Kojima, T. Funatsu, M. Tokunaga, H. Higuchi, and T. Yanagida, *Cell* **92,** 161 (1998).

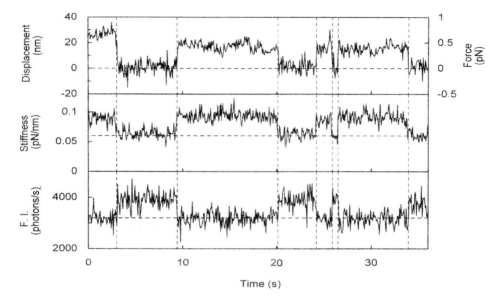

FIG. 7. Direct determination of the coupling between the mechanical and chemical event. *Top:* Displacement of the bead attached to an actin filament. *Middle:* The binding of myosin to actin was monitored by the stiffness estimated from the variance of the displacement. *Bottom:* The fluorescence changes from fluorescently labeled ATP at the position of a myosin molecule.

the delay time of the displacement after the dissociation of ADP. To examine the timing, however, it becomes essential to distinguish between the dissociation of ADP from the photobleaching for individual event. This is not possible, so the information must be derived statistically. In the case of myosin, the photobleaching of Cy3-ATP occurred with 0.008 s^{-1} for Cy3-ATP both directly attached on the glass and bound to myosin in the presence of vanadate under our experimental condition. This number should be compared with the rates of the mechanical events. The association rate constant of Cy3 ATP obtained from the durations of intervals was 0.13 s^{-1}. The dissociation of ADP was statistically much faster than the photobleaching. The probability that photobleaching occurred prior to nucleotide release was only $1 - \exp(-0.008\ s^{-1} 1/0.13\ s^{-1}) = \sim 6\%$. Thus, in some cases, the mechanical events are delayed after the release of ADP, indicating that the energy released from the breakdown of ATP has been stored.

Perspectives

The development of single molecule detection has allowed individual mechanical events to be measured. These results have deepened our understanding of the mechanisms underlying molecular motors. There are still many unanswered questions. We need further development of the technique. Different myosins have been discovered according to the development of molecular biology. Combination with protein engineering will be important in future research. The molecular motors are typical molecular machines and the conclusions obtained from the molecular motors will be shared with other molecular machines. The techniques we have developed as well as the knowledge we have acquired will be useful for future studies in the diverse field of bioscience.

We have learned that nanometer-sized protein molecules are influenced by thermal agitation. The input energy level released from the hydrolysis of ATP is the same order as the thermal energy. The measurements that are designed to monitor the behavior of single molecules are influenced by the effect of thermal disturbance. One question arises as to how the protein molecules work efficiently without being disturbed by the effects of thermal motion. Under these circumstances, the protein may harness thermal energy rather than work against thermal disturbance. The single molecule measurements will be tools to deal with thermal fluctuation, which, on the one hand, interferes with the measurements and, on the other hand, drives the function of the molecular machines.

[13] Visualization of Single Molecules of mRNA *in Situ*

By ANDREA M. FEMINO, KEVIN FOGARTY, LAWRENCE M. LIFSHITZ, WALTER CARRINGTON, and ROBERT H. SINGER

Introduction

The purpose of this article is to introduce methods and concepts that facilitate detection and identification of single molecules of mRNA *in situ* using fluorescence *in situ* hybridization (FISH). The methodology employs stringent imaging requirements that include a carefully calibrated quantitative epifluorescence digital imaging microscope, three-dimensional (3D) optical sectioning, constrained iterative deconvolution, and 3D interactive analysis software.[1–3]

[1] R. H. Singer and A. M. Femino, Patent Number 5,866,331. University of Massachusetts, Worcester, 1999.
[2] A. M. Femino, F. S. Fay, K. Fogarty, and R. H. Singer, *Science* **280**, 585 (1998).
[3] A. M. Femino, Doctoral Dissertation, Graduate School of Biomedical Sciences. University of Massachusetts Medical School, Worcester, 2001.

FISH is a very widely used technique in cell biology. Sensitivity of detection has been the major concern when implementing FISH technology. Previously, probes were designed to maximize signal in order to detect desired genes and transcripts. Large probes such as complete cDNA sequences were labeled with biotin and digoxigenin and detected using fluorochrome-labeled antibodies or streptavidin, respectively, to further amplify the signal.[4-10] Except in cytogenetics, which employed chromosome paints, high background due to nonspecific binding was a drawback and both hampered interpretation and compromised the reliability of information acquired with FISH. Improvements in specificity of FISH came with the use of short DNA oligonucleotide probes conjugated with biotin and digoxigenin.[11-13]

Fluorochrome-Conjugated Oligonucleotide Probes to Facilitate Analysis and Interpretation of FISH Images

Probes based on light emission afford the inherent advantage of accurate spatial detection. The detected signal can in principle be traced back to its source, unlike isotopic emissions that are detected via emulsions frequently at micrometer distances from the source. When a fluorescence signal arises from the fluorochrome molecules that are directly attached to the nucleic acid probe, the target molecule in the specimen can theoretically be located very precisely because the probe hybridizes within angstroms of the target nucleic acid.

The efficacy of using directly fluorochrome-labeled probes as FISH reagents was first tested in our laboratory (Fig. 1).[12,14] The immediate advantage is a one-step process for detection of hybridization sites. A one-step process is desirable because any number of spectrally distinct fluorochromes can be used simultaneously, limited only by the availability of dye molecules with distinguishable excitation and/or emission frequencies, affording detection of greater than two targets simultaneously.

The major disadvantage of fluorochrome-labeled probes is that they pose a detection problem due to their overall diminished intensity, compared to probes

---

[4] J. B. Lawrence, R. H. Singer, and L. M. Marselle, *Cell* **57**, 493 (1989).

[5] E. Viegas-Pequignot, B. Dutrillaux, H. Magdelenat, and M. Coppey-Moisan, *Proc. Natl. Acad. Sci. U.S.A.* **86**, 582 (1989).

[6] B. F. Brandriff, L. A. Gordon, and B. J. Trask, *Environ. Mol. Mutagen.* **18**, 259 (1991).

[7] B. J. Trask, *Methods Cell Biol.* **35**, 3 (1991).

[8] D. G. Albertson, R. M. Fishpool, and P. S. Birchall, *Methods Cell Biol.* **48**, 339 (1995).

[9] P. S. Birchall, R. M. Fishpool, and D. G. Albertson, *Nat. Genet.* **11**, 314 (1995).

[10] H. Yokota, G. van den Engh, J. E. Hearst, R. K. Sachs, and B. J. Trask, *J. Cell Biol.* **130**, 1239 (1995).

[11] K. L. Taneja and R. H. Singer, *J. Cell. Biochem.* **44**, 241 (1990).

[12] K. L. Taneja, L. M. Lifshitz, F. S. Fay, and R. H. Singer, *J. Cell Biol.* **119**, 1245 (1992).

[13] G. J. Bassell, H. Zhang, A. L. Byrd, A. M. Femino, R. H. Singer, K. L. Taneja, L. M. Lifshitz, I. M. Herman, and K. S. Kosik, *J. Neurosci.* **18**, 251 (1998).

[14] E. H. Kislauskis, Z. Li, R. H. Singer, and K. L. Taneja, *J. Cell Biol.* **123**, 165 (1993).

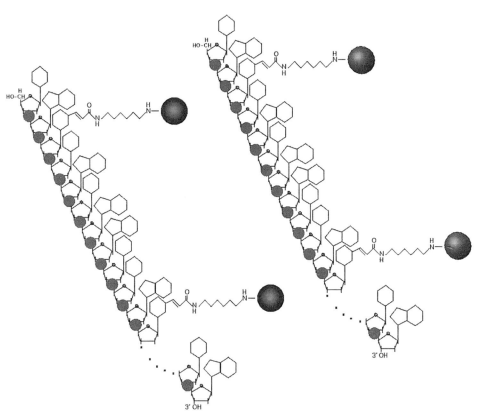

FIG. 1. Fluorochrome-labeled oligonucleotide probes. DNA oligonucleotide probes are specifically engineered to hybridize specific target sequences. A known number of fluorochrome molecules are covalently linked to the modified thymine analogs incorporated in the sequence. The intrinsic TFI of each probe can be calibrated. Many combinations of probes can be hybridized simultaneously.

using secondary detection. Initially species known to be present in high copy numbers such as poly(A) sequences, β-actin mRNA in chicken embryo fibroblasts, and CTG repeats in myotonic dystrophy were detected.[14–19]

An important attribute of the fluorochrome-labeled oligonucleotide probe is the predetermined number of fluorochromes to which it is covalently attached. The

[15] G. Zhang, K. L. Taneja, R. H. Singer, and M. R. Green, *Nature* **372,** 809 (1994).
[16] K. L. Taneja, M. McCurrach, M. Schalling, D. Housman, and R. H. Singer, *J. Cell Biol.* **128,** 995 (1995).
[17] D. A. Samarsky, M. J. Fournier, R. H. Singer, and E. Bertrand, *EMBO J.* **17,** 3747 (1998).
[18] J. C. Politz, R. A. Tuft, T. Pederson, and R. H. Singer, *Curr. Biol.* **9,** 285 (1999).
[19] H. L. Zhang, R. H. Singer, and G. J. Bassell, *J. Cell Biol.* **147,** 59 (1999).

probe provides a constant measurable signal and becomes amenable to calibration. Calibrated probes provide accurate intensity information and facilitate a definitive analysis of the fluorescence distribution in a FISH image in terms of number of probes hybridized.

Single Molecule Detection Feasible for Wide-Field Epifluorescence Microscopy

The feasibility of detecting and identifying specific single mRNA molecules with state of the art FISH and conventional light microscopy had not been entertained by those closely involved with *in situ* hybridization techniques even with the advent of digital imaging microscopy.[20] Previous detection of single mRNA molecules *in situ* was done by electron microscopy (EM).[21]

A very important finding was that with current digital imaging technology, there is adequate sensitivity to detect single probes *in situ* linked to as few as three to five fluorochromes using conventional epifluorescence microscopy.

Importance of Methodology

The methodology has made it possible to detect and identify a single molecule of mRNA in three-dimensional space, *in situ,* by applying a systematic approach to identify true hybridization signals.

The importance of having the ability to detect single molecules is highlighted by the interest and efforts of many investigators. In other laboratories implementation of single molecule detection has been accomplished under controlled experimental conditions that guarantee single molecules *in vitro* while employing highly specialized microscopic methods for detecting and studying those molecules.[22–26] In contrast, the methodology described here allows the identification of uncharacterized single molecules *in situ* in three-dimensional (3D) space through the analysis of the total fluorescent intensity (TFI) of hybridization sites.

Quantitative Epifluorescence Digital Imaging Microscope with Optical Sectioning

Development of Digital Imaging Microscope

3D digital imaging microscopy employing a conventional epifluorescence microscope with image deconvolution was a novel technology into the early 1990s

[20] D. M. Shotton, *Histochem. Cell Biol.* **104,** 97 (1995).
[21] G. J. Bassell, C. M. Powers, K. L. Taneja, and R. H. Singer, *J. Cell Biol.* **126,** 863 (1994).
[22] K. Peck, L. Stryer, A. N. Glazer, and R. A. Mathies, *Proc. Natl. Acad. Sci. U.S.A.* **86,** 4087 (1989).
[23] E. Betzig and R. J. Chichester, *Science* **262,** 1422 (1993).
[24] J. K. Trautman, J. J. Macklin, L. E. Brus, and E. Betzig, *Nature* **369,** 40 (1994).
[25] S. Nie, D. T. Chiu, and R. N. Zare, *Science* **266,** 1018 (1994).
[26] T. Funatsu, Y. Harada, M. Tokunaga, K. Saito, and T. Yanagida, *Nature* **374,** 555 (1995).

primarily limited to only a few laboratories. The 3D optical sectioning micro-scope was first described in the literature by Agard and Sedat.[27,28] Images were acquired as standard microphotographs that required digitization so that they could be deconvolved.

3D optical sectioning with deconvolution became more feasible with the advent of the charge-coupled device (CCD) cameras. The late Fredric S. Fay, with colleagues at the Biomedical Imaging Facility, located at the University of Massachusetts Medical School, developed a computerized digital imaging micro-scope (DIM).[29,30]

The availability of DIM for general research was a major contributor to the success of the work described in this article. The imaging station was optimized for low-light level and high-resolution imaging, which proved to be adequate for the detection and visualization of single molecules *in situ* as described in this work.

A Photometrics Series 180 back-thinned cooled CCD had high quantum ef-ficiency (QE), 80% at peak wavelength, that increased the signal for any given illumination compared to standard front acquisition CCDs. The higher sensitiv-ity reduced the requirement for longer exposures, which in turn reduced fading of fluorescent probes. The cooled CCD ($-40.5°$) allowed integration of signal without increasing the dark current significantly. Integration times were adjusted to improve the signal-to-noise (S/N) ratio in a read noise (50 e^- RMS at 50-kHz readout rate) limited system. The CCD camera was attached to a modified Nikon epifluorescence inverted microscope under computer control. The microscope was equipped with a stepping motor and eddy current sensor that provided accurate change of focus with 100 nm per step and a 100-W mercury lamp. The functions required to image a fluorescent specimen with optical sectioning were precisely synchronized under computer control.

The quantitative and spatial accuracy of fluorochrome-labeled oligonucleotide probes can best be used to advantage with a CCD detection system. The use of a CCD camera provides a sensitive, linear, detection method to capture emit-ted photons quantitatively onto a high-resolution pixel array. The photon energy, which is converted to a digital signal and further amplified quantitatively, becomes amenable to further quantitative processing and analysis.

General information about CCD cameras, including important considera-tions concerning their practical application to low-light imaging, can be found in "Methods in Cell Biology: Video Microscopy."[31]

[27] D. A. Agard and J. W. Sedat, *Nature* **302,** 676 (1983).
[28] D. A. Agard, *Annu. Rev. Biophys. Bioeng.* **13,** 191 (1984).
[29] F. S. Fay, K. E. Fogarty, and J. M. Coggins, in "Optical Methods in Cell Physiology" (P. D. Weer and B. Salzberg, eds.), p. 51. John Wiley & Sons, New York, 1986.
[30] F. S. Fay, W. Carrington, and K. E. Fogarty, *J. Microsc.* **153,** 133 (1989).
[31] G. Sluder and D. E. Wolf, eds., *Methods Cell Biol.* **56** (1998).

Quantitative Fluorescence Microscopy: An Important Tool in Cell Biology

With the advent of digital imaging, quantitative fluorescence microscopy became a sophisticated, high-performance technique by 1994, with broad applications.[32–36] The commercial availability of high-quality objective lenses and sensitive, cooled CCDs is credited for the major improvements in sensitivity, spatial resolution, and quantitative capability of conventional fluorescence microscopy.[37–39]

Three-dimensional applications using digital imaging microscopy have diversified considerably since the pioneering work of Fay and Agard, due to commercially available and affordable digital image microscopes and powerful computer processors.[40] Countless structures and biological constituents have been visualized for their spatial, temporal, and quantitative distribution *in situ*. Visualization of single genes, specific mRNAs, structural proteins, and transient pulses of chemical messengers such as calcium ions is described in the current literature.[41–44]

The importance of digital images stems from their potential to represent light emissions from biological specimens in terms of accurate numerical data. The numerical data can be further processed to extract quantitative information concerning the functions of cells and their cellular components. The majority of published works to date utilized relative fluorescence intensities emitted from an interrogated single cell to quantify and compare levels of cellular components or to define the spatial boundaries of objects of interest. Image analysis approaches have been highly customized to meet the specific needs of diverse experiments.

[32] G. R. Bright, *in* "Fluorescent Probes for Biology Function of Living Cells—A Practical Guide" (W. T. Mason and G. Rolf, eds.), p. 204. Academic Press, New York, 1993.
[33] D. L. Taylor and Y. L. Wang, "Fluorescence Microscopy of Living Cells in Culture." Academic Press, San Diego, CA, 1989.
[34] A. Waggoner, R. DeBiasio, P. Conrad, G. R. Bright, L. Ernst, K. Ryan, M. Nederlof, and D. Taylor, *Methods Cell Biol.* **30**, 449 (1989).
[35] B. Herman and J. J. Lemasters, "Optical Microscopy: Emerging Methods and Applications." Academic Press, San Diego, CA, 1993.
[36] C. V. Johnson, R. H. Singer, and J. B. Lawrence, *Methods Cell Biol.* **35**, 73 (1991).
[37] Y. Hiraoka, J. W. Sedat, and D. A. Agard, *Science* **238**, 36 (1987).
[38] R. S. Aikens, D. A. Agard, and J. W. Sedat, *Methods Cell Biol.* **29**, 291 (1989).
[39] M. Coppey-Moisan, J. Delic, H. Magdelenat, and J. Coppey, *Methods Mol. Biol.* **33**, 359 (1994).
[40] R. Rizzuto, W. Carrington, and R. A. Tuft, *Trends Cell Biol.* **8**, 288 (1998).
[41] R. M. Long, R. H. Singer, X. Meng, I. Gonzalez, K. Nasmyth, and R.-P. Jansen, *Science* **277**, 383 (1997).
[42] G. J. Bassell, H. Zhang, A. L. Byrd, A. M. Femino, R. H. Singer, K. L. Taneja, L. M. Lifshitz, I. M. Herman, and K. S. Kosik, *J. Neurosci.* **18**, 251 (1998).
[43] K. L. Taneja, L. M. Lifshitz, F. S. Fay, and R. H. Singer, *J. Cell Biol.* **119**, 1245 (1992).
[44] F. S. Fay, K. L. Taneja, S. Shenoy, L. Lifshitz, and R. H. Singer, *Exp. Cell Res.* **231**, 27 (1997).

Theoretical Basis of This Work

The following discussion is undertaken to provide an overview of some basic theoretical concepts in digital image processing that provide a theoretical foundation for quantitative imaging. Image formation theory that encompasses concepts relevant to single molecule detection will be highlighted.

The limitations of optical microscopy are reviewed to emphasize the need for optical sectioning and rigorous deconvolution algorithms to restore images to their true point source distributions. Deconvolution facilitates quantitative analysis of fluorescence distributions, especially in biological specimens with 3D distribution of multiple targets.

Formation of a Digital Image Described Mathematically

The basic principles of digital imaging were developed and described in great detail before the application to microscopy.[45,46]

The definition of an image takes on two forms. The true image, i.e., the input image, is comprised of point sources at specific locations in a sample. The output or actual observed image, however, is a distorted version of the input image, and is dependent on the optical properties of the imaging microscope and detector. A major source of distortion of the input image is the diffraction limited optics of a light microscope. For many purposes the observed image suffices to provide necessary information about the outcome of an experiment. It is possible, however, to solve for the true image using a mathematical approach. For the successful interpretation of 3D FISH images with single molecule accuracy it is necessary to solve for the true image.

A mathematical description for the generation of a fluorescence image by a quantitative fluorescence microscope begins with a test point, i.e., a point source with a spatial location, unit intensity, and no spatial extent. This point source is considered a unit impulse $\delta(x, y)$ and is defined such that its total brightness is one.[47]

$$\delta(x, y) = 0 \qquad \text{unless } x = y = 0$$

$$\int_{-\infty}^{\infty} \int_{-\infty}^{\infty} \delta(x, y) dx\, dy = 1 \tag{1}$$

The point source by definition is located at the origin of the x, y coordinate system, therefore only at coordinates $x = y = 0$ is the delta function nonzero. The

[45] K. R. Castleman, *in* "Digital Image Processing" (A. V. Oppenheim, ed.), p. 429. Prentice-Hall, Englewood Cliffs, NJ, 1979.

[46] K. R. Castleman, *in* "Digital Image Processing," p. 667. Prentice Hall, Upper Saddle River, NJ, 1996.

[47] I. T. Young, *Methods Cell Biol.* **30**, 1 (1989).

total brightness of the point source at $x = y = 0$ is 1. Therefore, integrated over all two-dimensional space, the total brightness of the point source is still 1, since the unit impulse contributes zero brightness to all other coordinates in 2D space.

Equation (1) states a basic and key premise in fluorescence digital imaging microscopy: light emanates from infinitely small point sources. Image processing can then also be based on the theory of image formation by point sources without spatial dimensions.

Images of cells interrogated with fluorescent probes are composed of point sources as well as extended sources. With the added concept of a linear, shift invariant system any light source can be thought of as a superposition of point sources. Extended sources can then also be modeled as a superposition of point sources. Shift invariance means that the intensity information from any point source in the input image must not be dependent on position. The objective lens and detector are the most critical components for this attribute. In a linear imaging system the relative brightness among the multitude of point sources does not change as the light traverses the optical system and is detected on a CCD array or other detector. The absolute intensity, however, of the point sources can be expected to change due to various sources of light loss, but that will not affect the relative weighting of the point spread functions.

Equations (2) and (3) describe the formation of an image.[47] A collection of unit impulses at different positions in a sample with weighting coefficients $a(u, v)$ can produce an arbitrary input image $a(x, y)$.

Input image:

$$a(x, y) = \int_{-\infty}^{\infty} \int_{-\infty}^{\infty} a(u, v)\delta(x - u, y - v)du\,dv \qquad (2)$$

The output image is dependent on the mode of detection of the input image. Each weighted input impulse, $a(u, v)\delta(x - u, y - v)$, when imaged through a microscope, generates a weighted point spread function $a(u, v)h(x - u, y - v)$. The sum of the weighted point spread functions is the resulting output image $b(x, y)$.[47]

Output image:

$$b(x, y) = \int_{-\infty}^{\infty} \int_{-\infty}^{\infty} a(u, v)h(x - u, y - v)du\,dv \qquad (3)$$

The input image $a(x, y)$ produces an output image described mathematically as $b(x, y)$ only if an imaging system is both linear and shift invariant.

Point Spread Function

In Eq. (3), $h(x, y)$ represents the output image produced by a single point source of light and is referred to as the point spread function (PSF). The PSF

contains all of the information necessary to describe the imaging system, which
leads to the convolution Eq. (4).[47] Equation (4) is frequently encountered but is
just another way of writing Eq. (3).

$$b(x, y) = a(x, y)^* h(x, y) \qquad (4)$$

The output image $b(x, y)$ is described as the product of convolution between
$a(x, y)$, the true distribution of input impulses, and the output image $h(x, y)$,
referred to as the PSF formed by a single input impulse or point of light. The
output image $h(x, y)$, is a distorted description of a true undistorted point source
with zero spatial dimensions and contains all the information necessary to describe
the imaging system.

The nature of the imaging process causes distortions in the input image as
it passes through the optical system. An image of a 3D PSF that is obtained
under identical imaging conditions as a FISH image (identical coverslip thickness,
mounting media, wavelength, optical setup) contains all the information necessary
to describe the optical setup to deconvolve the respective FISH image. Aberrations
and asymmetries along the optical axis as well as the diffraction pattern of in and
out of focus light all contribute to producing the 3D PSF. The 3D PSF completely
describes the behavior of a point source in 3D, assuming a linear, shift-invariant
system.

When an accurate description of $h(x, y)$ is available, for example, when an
empirical PSF for an imaging setup has been obtained, the output $b(x, y)$ for
any other input image $a(x, y)$ can be computed through the process of convolution [Eq. (4)]. Therefore one can simulate an output image, as it would appear
through a particular optical system from any model-input image. In fact, this same
process is used by deconvolution algorithms to solve for the input image $a(x, y)$
of an unknown fluorescence distribution. In essence, an input image $a(x, y)$ is
estimated and then convolved with the PSF. The resulting attempt at simulating
the output image $b(x, y)$ is compared to the actual empirical output image. When
the difference between the simulated output image and the empirical output image
converges, the image is said to be deconvolved or solved for the true input image
$a(x, y)$, because convolving the PSF with this estimated input image $a(x, y)$ in
fact produces the empirical output image $b(x, y)$.

Successful deconvolution is dependent on obtaining a PSF of high quality. A
desirable PSF is one that has good signal-to-noise (S/N) ratio at the plane of focus
and at out of focus planes. The in-focus plane should have equal numbers of out of
focus planes above and below. The number of optical sections in the PSF should
be equal to at least twice that of an image that will be deconvolved (Fig. 2).

A typical cultured cell 5 μm in thickness will require from 20 to 50 optical
sections to go out of focus above and below the cell using 0.250 or 0.100 μm z-steps,
respectively. The respective PSF must use the same z-step and have 41 to 101
planes, respectively. The most distant planes have expanded diffuse rings and loss

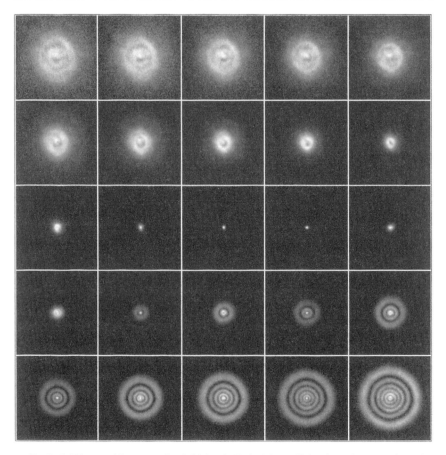

FIG. 2. A 199-nm red fluorescent bead (Molecular Probes) has sufficient intensity to acquire a PSF with good *S/N* ratio 50 planes above and 50 planes below the in-focus plane. The PSF is imaged with a Photometrics Series 200 back thinned camera on an Olympus AX70 upright epifluorescence microscope with a 60× infinity corrected objective, 1.4 NA, 4× camera eyepiece at wavelength 585 nm. The *xy* pixel size is 100 × 100 nm and the *z* step = 100 nm. Every fourth plane from a total 100 planes is depicted. The maximum intensity pixel in the in-focus plane (center panel) is 10,852 out of a possible 16,384 (66% of the dynamic range). The individual panels are autoscaled to view the dim rings.

of intensity per pixel, which reduces the *S/N*. A PSF acquired with a wide dynamic range 14-bit CCD camera that has a gray scale from 0 to 16,384 can use adequate exposure times to increase the *S/N* of the out of focus rings without saturating the pixels at the plane of focus. A PSF acquired with a 12-bit CCD camera that has a gray scale from 0 to 4096 will have less dynamic range to have good *S/N* for the out of focus rings without saturating the CCD with light in the plane of focus. A PSF taken with a variable exposure time, increasing exposure as one moves away from the focal plane, can overcome this limitation. The maximum intensity pixel in the plane of focus should not be greater than 85% of the full dynamic range of the camera: 3482 for a 12-bit vs 13,926 for a 14-bit CCD camera.

FISH Image

A FISH image is a record of the spatial coordinates of one or multiple copies of fluorescent probe molecules by virtue of their fluorescence emission. Their respective target molecules determine the exact locations of the probe molecules, in 3D space.

The fluorescence emitted from point sources is distorted due to resolution limitations of a diffraction-limited microscope. The resulting distorted image may appear to have "objects" with volume, shape, and surface boundaries. Objects will generally have poorly demarcated edges, especially when the signal is weak. The hazy appearance of objects and their boundaries masks the true collection of point sources that is producing the image.

To extract maximum information from an image, the technical limitations of lenses and S/N must be understood and optimized. The remainder of this section will highlight the limitations of optics and the need for deconvolution to improve resolution.

Diffraction-Limited Light Microscopy

The image of a point source of light formed in the focal plane of a converging lens is not a point. This is because a light microscope produces diffraction-limited images. The image is a circular diffraction pattern, which appears as a circular disk surrounded by progressively fainter secondary rings. This image of a point source is referred to as an Airy disk.[48] The image of a point source in an image plane is also referred to as a two-dimensional (2D) PSF. Any asymmetries in the ring pattern are related to illumination conditions and aberrations in the objective lens. A 3D PSF may include asymmetries in the cone rings that reflect mismatches in the refractive index between objective lens, coverslip, and mounting medium.

Two point sources whose images partially overlap but are located in the same focal plane can be resolved. Theoretically, the x, y positions of the respective point sources can be measured to 10 nm precision using the symmetry information in the diffraction pattern or using center of mass calculations.[49,50] Point

[48] D. Halliday and R. Resnick, in "Physics for Students of Science and Engineering," Combined Ed., p. 1020. John Wiley & Sons, New York, 1965.
[49] S. Inoue, Methods Cell Biol. **30,** 85 (1989).
[50] J. L. Harris, J. Opt. Soc. Am. **54,** 931 (1964).

sources in thin specimens (200 nm) are amenable to such precision measurements because out of focus light is minimized and image contrast can be sufficiently optimized.

Intensity Information in Focal Plane as Summation of In-Focus as Well as Out-of-Focus Signal

Thicker specimens such as cultured cells and thin tissue sections present a problem when using a conventional fluorescence microscope at high magnifications because of the small depth of focus.[39] Molecular targets detected by fluorescent probes will in all probability occur in 3D spatial distributions in thicker specimens (>200 nm). The formula for depth of focus, *DF*, is

$$DF = \frac{100}{7NAM_1} + \frac{\lambda}{2NA^2} \tag{5}$$

The parameters are the wavelength of illumination, λ, in micrometers, the total magnification, M_1, and the numerical aperture of the objective lens, *NA*. The resulting *DF* is expressed in micrometers. For a magnification of 300× and *NA* = 1.4 the depth of focus is 0.16 μm for a wavelength of 500 nm (Table I).

A typical cultured cell has a thickness equal to 5.0 μm. When viewing one optical section through a cell those objects that are in the focal plane or out of that focal plane by less than or equal to the *DF* are visualized with clarity. Structures just outside the *DF* are visible but blurred due to the 3D PSF that expands the total fluorescence intensity of a point source into peripheral rings thus diminishing the central intensity that demarcates the *x* and *y* coordinates of the point source. Structures at greater distances above and below the plane of focus are not

TABLE I
DEPTH OF FOCUS (DF) AS A FUNCTION OF TOTAL MAGNIFICATION, NUMERICAL
APERTURE (NA), AND EMISSION WAVELENGTH (λ)[a]

| Objective | Camera eyepiece | NA | DF (μm) | | | | μm/pixel[b] |
			$\lambda(0.450)$	$\lambda(0.520)$	$\lambda(0.585)$	$\lambda(0.667)$	
100×	2.5×	1.4	0.156	0.173	0.190	0.211	0.112
100×	5.0×	1.4	0.135	0.153	0.170	0.191	0.056
100×	2.5×	1.3	0.177	0.198	0.217	0.241	0.112
100×	5.0×	1.3	0.155	0.176	0.195	0.219	0.056
60×	2.5×	1.4	0.183	0.201	0.217	0.238	0.187
60×	5.0×	1.4	0.149	0.167	0.183	0.204	0.093

[a] The wavelength and DF are given in micrometers.
[b] The pixel dimensions at the specimen focal plane of the DIM optical setup used in this work are also listed for the record.

visible because the central intensity disappears but out of focus light from the same structures reaches the focal plane as diffuse rings and immerses objects in the focal plane in a haze or glow. Therefore, any one individual plane cannot provide a complete unambiguous interpretation of structures or events occurring in single cells or in thin tissue sections.

More importantly the accuracy of any measurement of TFI associated with low-light level objects such as single molecules is compromised. Hybridized molecular targets located throughout the spatial volume of cultured cells and tissue sections may contribute fluorescence from planes above and below the focal plane confounding the resolution of point sources in any one focal plane. Contributions from out of focus light are additive, decrease contrast, and preclude the unambiguous assignment of Airy disk patterns to individual point sources. This becomes problematic for determination of TFI attributed to any one point source and is incompatible with single molecule analysis.

Precise Optical Sectioning

The information in a 3D cell or tissue section can be more accurately determined if a number of different focal planes through the specimen are imaged and then examined jointly with the information reconstructed to display the total 3D view from the individual planes (Fig. 3).

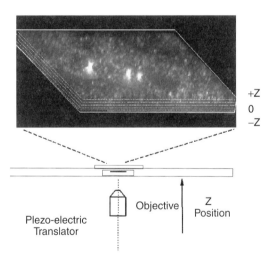

FIG. 3. Optical sectioning produces a stack of sequential two-dimensional images referred to as a 3D stack. Each 2D image contains both in focus and out of focus fluorescence intensity information. The image is viewed through the 2D stacks as a 3D rendering. A bright 100-nm fluorescent bead is at the left and two transcription sites are located near the center of the image. Single molecules of mRNA are seen as dim point sources immersed in the haze of out of focus light.

Optical sectioning provides a nondestructive process for viewing and documenting the fluorescence distribution of a specimen at different depths with incremental steps as small as 100 nm. The fluorescence signal is imaged and digitized at each step along the optical axis as a 2D image. The stack of sequential 2D images, referred to as a 3D stack, contains both in focus as well as out of focus information.

Previously an input image was described as a 2D distribution of weighted point sources. This model also holds true in the present case, but because the point sources are distributed throughout 3D space, an input image must be described at multiple optical planes to document the 3D spatial distribution of point sources and their respective intensity distributions.

The mathematical notation for optical sectioning based in a Cartesian coordinate system is described in Castleman and clarifies the notation that is used to describe deconvolution.[51] If two conditions are met, (1) the specimen function $f(x, y, z)$ is zero outside the field of view in x and y, i.e., the specimen is an isolated single cell and all point sources that contribute signal are located in the imaged field or within the cell boundaries, and (2) the function is zero outside the range $0 \leq z \leq T$, i.e., the optical sectioning went both above ($> T$) and below (< 0) the boundaries of the specimen, then the output image $g(x, y, z')$ can be described as a 3D convolution of the specimen function $f(x, y, z)$, the true fluorescence distribution, with a 3D PSF of the optical system.[51] There is adequate information in the 3D output image, $g(x, y, z')$, to determine the location of point sources of light, i.e., to solve for $f(x, y, z)$.

Deconvolution: An Indispensable Image Processing Step

What approaches can be used to recover the 3D specimen input function $f(x, y, z)$ from the series of output images $g(x, y, z')$ that are distorted by out of focus haze and other aberrations of the optical system?

The process of recovery of the unperturbed input image $f(x, y, z)$ is referred to as a deconvolution, restoration, or reconstruction.[51] Deconvolution will be the term used. The approach most important to this work is the constrained iterative deconvolution. A synthetic specimen function is deconvolved through an iterative process such that when this same function is blurred with the 3D PSF of the optical setup, the result yields approximately the recorded image of the actual biological specimen. The process of the constrained iterative deconvolution is to start with an initial approximation of the specimen function, $f_0(x, y, z)$, and determine the error, $e_i(x, y, z)$, that remains after the ith iteration,

$$e_i(x, y, z) = g(x, y, z) - [f_i(x, y, z)^* h(x, y, z)] \quad (6)$$

where $f_i(x, y, z)$ is the ith approximation of the specimen function, $g(x, y, z)$

[51] K. R. Castleman, *in* "Digital Image Processing," p. 566. Prentice-Hall, Upper Saddle River, NJ, 1996.

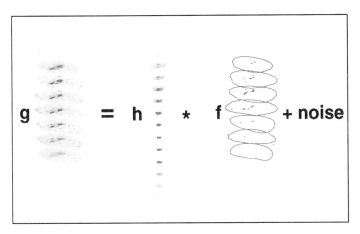

FIG. 4. Noise is a random component of a digital image. The empirical image (g) is formed by convolution of the input image (f) with the PSF (h) of the optical setup and includes random noise from the output electronics of the detector.

is the recorded image, $h(x, y, z)$ is a known 3D PSF, and * is convolution.[52] After each iteration, the ith approximation of the specimen function $f_i(x, y, z)$ is convolved with the 3D PSF. The product of the convolution is compared to the output image function $g(x, y, z)$ and the difference recorded as the error function $e_i(x, y, z)$. A correction is then made to $f_i(x, y, z)$ by some process based on the error function. A unique solution is arrived at when the system converges, a state where further changes in $f_i(x, y, z)$ do not improve the approximation evidenced by the error function. Accuracy of the solution is improved when more constraints are imposed on the solution. A common constraint is that the specimen function must be nonnegative. This is a very reasonable constraint for fluorescence images because fluorescent signal is intrinsically nonnegative.

The 3D imaging process is a blurring process with additional noise components introduced after blurring (Fig. 4). The noise comes primarily from the electronics. Such noise components do not correspond to any possible specimen components. Noise is a random contribution and will not form any consistent blurring pattern as would true point sources located in the specimen. Therefore noise in the final image $g(x, y, z)$ will confound any solution. The noise contribution must be remedied in order to converge to a unique and accurate solution.

A "well-posed" approximation is one that does have a unique solution and is not sensitive to the presence of noise.[53] Such a process is the constrained iterative

[52] K. R. Castleman, *in* "Digital Image Processing," p. 574. Prentice-Hall, Upper Saddle River, NJ, 1996.
[53] W. Carrington, *Proc. SPIE* **1205**, 72 (1990).

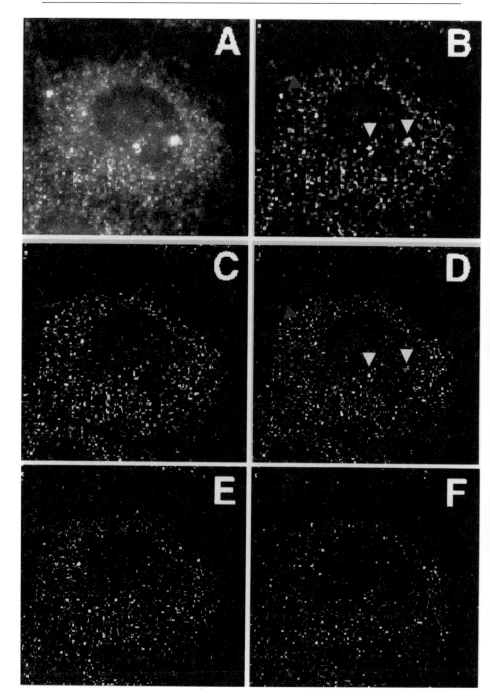

regularization procedure that seeks a solution that approaches the true input fluorescence distribution as the noise is reduced and as the image is sampled more finely and over a larger volume.[53] The regularization seeks a function $f(x, y, z)$ that minimizes the error function $e_i(x, y, z)$

$$e_i(x, y, z) = \sum_{i,j,k} \mid g_{ijk} - [f(x, y, z)^*h]_{ijk} \mid^2 + \alpha \iiint \mid f(x, y, z) \mid^2 dx\, dy\, dz$$

(7)

The function $f(x, y, z)$ is the nonnegative function that simulates the specimen function.[53] The second term on the right side of the equation enforces smoothness on $f(x, y, z)$ to prevent noise in g_{ijk} from introducing oscillations that would corrupt the solution for an accurate $f(x, y, z)$. When a large α value such as 10^{-3} is chosen, the solution is allowed to converge with a larger error function $e_i(x, y, z)$. When a small α value such as 5×10^{-6} is chosen, the soultion does not converge until a smaller error function $e_i(x, y, z)$ can be attained, which usually requires a greater number of iterations and increased processing time. The α parameter is adjusted to optimize the minimization process of Eq. (7). Optimization is here defined as a set of conditions that allows the iterative process to converge with a minimal error function $e_i(x, y, z)$ within a reasonable time without compromising the accuracy of the solution (Figs. 5 and 6). The selected α value is too small when it becomes incompatible with the level of random noise inherent in the empirical image $g(x, y, z)$ and the process cannot converge. Residuals are reported for a completed deconvolution and represent the percent of signal that could not be reassigned. The expected residuals for a given optical setup can be estimated.

Exhaustive photon reassignment (EPR) is the name given to the patented iterative numerical method for minimization of Eq. (7) published by Walter Carrington.[53] Work described in this article uses EPR exclusively. The EPR algorithm is broadly applicable to a wide range of image types.

FIG. 5. A deconvolution using EPR approaches the best solution when an image converges with the minimum of residuals that can be expected for the noise level of the system. (A) Raw image of a cell hybridized with a CY3 probe to the $3'$-UTR of β-actin RNA. The image is deconvolved using a variety of α-values (B–F). A solution is achieved using α-values between 1E-05 and 5E-06 (D and E). As the deconvolution approaches a solution the fiduciary 100-nm bead (red arrow) condenses to a point source. The transcription sites (yellow arrows) become less visible, which indicates that they are located deeper in the nucleus and we are observing their out of focus light in (B) The α-value, number of iterations achieved, convergence achieved (convergence is set at 0.001), and percent residuals are listed for (B)–(F): (B) 0.001, 64, 0.000915, 15.3; (C) 0.0001, 153, 0.000932, 11.0; (D) 1E-05, 321, 0.000974, 10.2; (E) 5E-06, 395, 0.0014, 10.2; (F) 1E-06, 400, 0.0303, 10.6. The α-value of 1E-06 (F) did not allow the image to converge within a reasonable number of iterations and therefore is assumed to be a limit to any practical improvement of the result. Pixel dimensions are $x = 187$ nm, $y = 187$ nm, $z = 250$ nm.

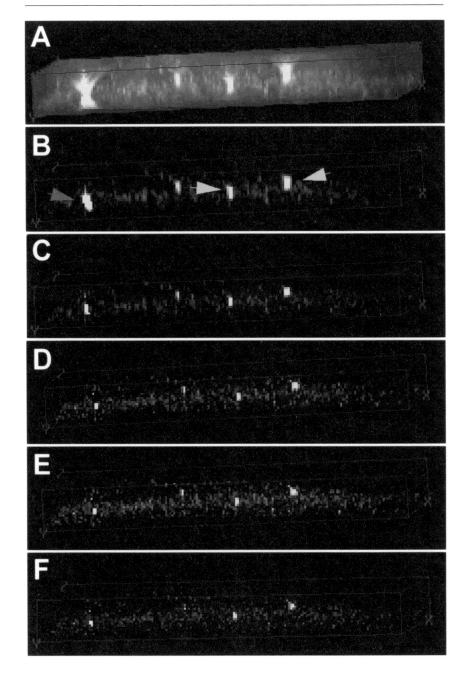

A very important aspect of the constrained iterative regularization algorithm, fundamental to the success of the work described in this article, is its quantitative accuracy. The EPR algorithm maintains the total intensity of the image as it approaches the best solution. The restoration process accounts for the total signal that is attributed to each point source. All the out of focus light, attributed to a particular point source, from above and below the plane of focus is returned to the plane of focus within the limits of resolution. The number of planes in the PSF determines the depth from which out of focus light is returned to each point source in the experimental image. To ensure that a point source, located in the highest or lowest optical plane of the experimental image, can be deconvolved accurately the number of planes on either side of the PSF focal plane must equal the number of planes in the experimental image.

The nonnegative constraint on the iterative method also leads to the possibility of resolution beyond the diffraction limit. Improved resolution is evidenced by a decrease in the full-width at half-maximum intensity of the image of a fluorescein-labeled 200-nm bead from 0.32 to 0.23 μm in the x and y dimension and 0.79 to 0.57 μm along the z axis.[54]

A version of EPR running on UNIX required an intermediate level of computer power that was significant but attainable. A Silicon Graphics workstation with a 100-MHz IP22 processor and 64 Mbytes of RAM and a 1-GB hard disk was sufficient and used until 1994 to process images used in this work. A 3D image stack of a 510×310 pixel array and 20 optical planes could take up to 14 hr to deconvolve depending on the choice of α value and the number of iterations required for convergence. Smaller segments could be restored within 2–3 hr.

After 1994 the deconvolution algorithm was adapted to run on four 8-parallel processor boards (Scanalytics Division, CSP, Inc., Billerica, MA) decreasing the restoration time for a $510 \times 310 \times 20$ image to 3 hr.

In June 1997, Signal Analytics Corporation Incorporated merged with Scanalytics, the biological imaging division of CSP, Inc. (Billerica, MA), to form a new

[54] E. D. W. Moore, E. F. Etter, K. D. Philipson, W. A. Carrington, K. E. Fogarty, L. M. Lifshitz, and F. S. Fay, *Nature* **365,** 657 (1993).

FIG. 6. A point source is deconvolved to a small volume compared to a transcription site that is deconvolved to a less dense and more asymmetric volume. Before deconvolution (A) and after deconvolution using a large α-value of 0.001 (B), point source (a 100-nm fluorescent latex bead, red arrow) and transcription sites (yellow arrows) appear very similar. (C–F) With decreasing α-values (0.0001, 1E-05, 5E-06, and 1E-06, respectively) the deconvolution brings the bead to a point source whereas the transcription sites comprise an irregular collection of voxels. The best solution is represented by (D) and (E). (D) and (E) converge to a minimum of residuals (Fig. 5) in a reasonable amount of processing time whereas longer processing with smaller α values does not appear to further improve the deconvolution.

corporation, Scanalytics Incorporated. Only a modified version of EPR, untested in this work, adapted to Microsoft Windows can be obtained from Scanalytics Inc., Fairfax, VA; info@scanalytics.com.

Relative Intensity Information Sufficient to Allow Accurate Image Processing but Not Interpretation

A fortuitous aspect of image formation theory is the abrogation of subjectivity. Relative intensity information, i.e., weighted point sources, is sufficient for accurate image processing and deconvolution. There is no need to interpret or make a value judgment about fluorescent sources, whether they represent a single point source or a superposition of multiple point sources.

Processing an image does not in itself provide any interpretation. The weighting factors provide relative intensity information, and after processing is completed, the interpretation of an image can only be in terms of relative intensities. This has been the major deficit and stumbling block for single molecule counting using fluorescence digital imaging.

Measurement of Intensity of One Fluorescent Probe to Provide Sufficient Information to Count Single Molecules Following Image Processing

What is the unit intensity of interest, that intensity or point source that represents a single fluorescent molecular probe? Image processing alone cannot a priori extract such information from a digital image. The a priori approach faces infinite possible solutions. An independent empirical measurement must be obtained to address this deficit. The unit intensity in question is not the same entity as the unit impulse $\delta(x, y)$. It is the unit point source intensity of the interrogation probe.

The availability of the cyanine dye series for labeling probes improved the signal-to-noise ratio of probes used for FISH due to the photo stability and large molar extinction coefficients of the dyes.[55] Fluorescein was also an acceptable dye because of its high quantum yield, although it was less photostable. Measurements indicated that interrogation probes could routinely be engineered to have total fluorescence intensities per probe that were greater than typical endogenous background signals in cultured cells. In such cases, the probe signal effectively became the primary contributor of fluorescence intensity to an image.

A conclusion, that will become more self-evident as data are examined, is that it is of utmost importance to know the unit intensity of one interrogation probe under imaging conditions. It is necessary in order to go beyond deconvolution and processing of images and progress to counting the number of fluorescent molecules. The goal to attain a more precise interpretation of the image data rests

---

[55] R. B. Mujumdar, L. A. Ernst, S. R. Mujumdar, and A. S. Waggoner, *Cytometry* **10,** 11 (1989).

with the ability to calibrate for the intensity expected of a single probe imaged through the optical setup.

Oligonucleotide Probes for Single Molecule Detection

Design of Fluorochrome-Labeled Oligonucleotide Probes

There are two major considerations in probe design to ensure the optimal interpretation of a fluorescence image. (1) To approach the theoretical potential to accurately describe the respective point source distribution in the true image, probes should be designed to conform as closely as possible to a point source. A single fluorochrome has limited spatial extent and therefore approaches a theoretical point source. Ideal probes would carry one fluorochrome. To realize adequate *S/N* three to five fluorochromes are more practical.[2] A DNA oligonucleotide probe with 50 bases has a linear extent of approximately 17 nm and therefore the attached fluorochromes approach the dimensions of a point source. (2) To ensure the accurate spatial localization of the probe relative to the target molecule, the fluorochrome must be tethered to the probe in a manner that maintains its spatial extent near to the probe and therefore also near to the target. Oligonucleotide probes hybridize within angstroms of their RNA or DNA targets. When they are conjugated to a fluorochrome via a relatively short linker arm, the fluorochrome is also constrained near the target (Fig. 1). The fluorescence emission of a hybridized probe then approximates a point source within angstroms of the target. Theoretically, maximum information can then be extracted about the spatial distribution of a probe and its respective target when the probe can be constrained close to the target and also treated as a fluorescent point source.

An appropriately designed probe will have four major attributes. (1) The probe is conjugated to fluorochromes that have adequate photostability to survive the continued exposure to excitation during optical sectioning. The presence of multiple fluorochrome molecules on one probe increases the probability that a target will continue to be detected through the entire stack of optical sections. The maximum intensity contributed by a probe is captured in the first optical section. Under our imaging conditions CY3-labeled probes faded at the rate of 2% per 2 sec exposure. After 20 optical sections the TFI of a CY3 probe would fade by approximately 40%. The probability that a probe will retain three out of five (60%) stable fluorochromes allows the continued detection of the target throughout the image stack. The total intensity contributed by CY3 probes to each optical section can then be corrected back to the total intensity of the first optical section to recover lost intensity information due to fading. (2) The probe has a defined size of 20 to 60 base pairs. The small size maximizes precision of hybridization to short sequences along the target nucleic acid. The probe may also consist of a collection of separate sequences iterated along a nucleic acid target, closely spaced, to produce a point source with increased intensity. (3) Probes are engineered with a GC content of approximately 50% to effect efficient and specific hybridization of diverse sequences under the same hybridization conditions. (4) The probe has a constant TFI at a specific excitation/emission wavelength. The modified base, a thymine analogue with an amino linker arm, is spaced adequately throughout the probe sequence to prevent steric hindrance during labeling and also to prevent quenching of fluorescence once the fluorochromes are conjugated. The ability to define a constant TFI for a fluorescent probe allows accurate quantitation.

Such a probe will possess the appropriate attributes to be identified as a point source *in situ* and will provide accurate quantitative and spatial information about a molecular target in a single cell. Fluorochrome-labeled DNA oligonucleotide probes used in this work meet all the requirements for specific, quantitative molecular probes that emit fluorescence signal from spatial extents approaching that of a point source.[2]

Probes Optimized for Signal-to-Noise Ratio in Biological Samples

The major contributors to the decrease in *S/N* ratio within a hybridized sample can be divided into three categories: photoelectronic noise, electronic noise, and autofluorescence. Each category is described.

1. Photoelectronic (photonic) noise is inherent in the quantum nature of light and is modeled as random using a Poisson distribution.[56] The standard deviation is equal to the square root of the mean number of photons emitted. Therefore, a point source emitting an average of 475 photons per second results in a standard deviation of $\sqrt{475} = \pm22$ photons of noise. The photon output of 10 fluorescence sources whose average photon output is 475 photons per second will show a statistical distribution curve (Fig. 7A).

2. Electronic noise contributes to the image after it is acquired. One example is CCD camera readout noise. Readout noise is random noise that is contributed once at the time of readout of the image. Older CCD cameras had higher readout noise. Readout noise was a major contributor to distortion of intensity information captured from low-light level fluorescence in the present work (Fig. 7B).

A typical *S/N* ratio for a scientific grade Photometrics Series 180 CCD camera similar to that used with our DIM is calculated using Eq. (8):

$$SNR = \frac{PQ_e t}{\sqrt{(P + B)Q_e t + Dt + N_r^2}} \qquad (8)$$

substituting the following values for each parameter: P, photon flux incident on CCD, 475 (photons/pixel/second); B, background photon flux incident on CCD,

[56] K. R. Castleman, "Digital Image Processing," p. 414. Prentice-Hall, Upper Saddle River, NJ, 1996.

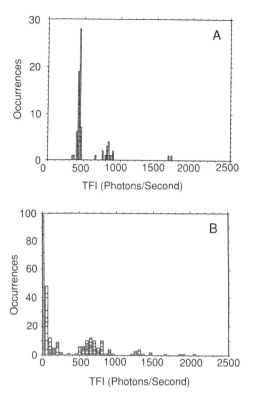

FIG. 7. The presence of random noise in an imaging system corrupts data. Histograms (A) and (B) show the theoretical extent of corruption of the TFI of point sources by random noise during the imaging process (contributed by Kevin E. Fogarty). A model, consisting of a collection of point sources of equal intensity scattered randomly onto a single plane that represents an array of pixels, was subjected to a convolution with an empirical PSF to simulate imaging through the optics of a light microscope. The intensity of the point source was chosen to be comparable to the intensity expected for one oligonucleotide probe labeled with five CY3 fluorochromes imaged through the DIM optical setup. (A) Photonic noise only. Photonic noise was added to 10 objects, each originally emitting 475 photons per second. The point sources were scattered randomly onto a single plane representing an array of pixels. The blurred image was deconvolved with the same empirical PSF from the DIM optical setup. Ten trials of 10 randomized objects were plotted as a histogram. The TFI of the restored objects was recovered as a distribution instead of a single value. (B) Photonic plus read noise. Photonic noise was added to 10 objects, each originally emitting 719 photons per second. The point sources were scattered randomly onto a single plane representing an array of pixels. The blurred image was then peppered with random noise as expected from the Photometrics Series 180 CCD camera readout. The new image was deconvolved with the same PSF as in (A). The deconvolved image was a Gaussian distribution of objects of varying intensity, with a mean close to the original intensity assigned to each point source. When the original signal was weak, i.e., from a single probe, the read noise had the greatest effect in corrupting an image signal.

0 (photons/pixel/second); Q_e, quantum efficiency of the CCD, 0.85; t, integration time, 2 (seconds); D, dark current, 1 (electrons/pixel/second); and N_r, read noise, 50 (electrons rms/pixel).

$$SNR = \frac{475 \times 0.85 \times 2}{\sqrt{(475 \times 0.85 \times 2) + (1 \times 2) + 50^2}} = 14$$

A graph of S/N vs. exposure time is plotted using this example (Fig. 8A).

The S/N ratio increases linearly with exposure time in the read-noise-limited region because an increase in exposure time accumulates more photons per pixel per unit time whereas the read noise, which is independent of time, remains the same. Therefore it is advantageous to increase exposure time when imaging low-level fluorescence in order to increase the S/N ratio. However, the stability of the fluorochrome must be evaluated and loss of signal due to fading must be weighed against increasing exposure time to provide an optimal increase in S/N ratio.

In the photon noise limited region there is less advantage to increasing the exposure time to improve the S/N ratio because the S/N ratio increases as the square root of the exposure time. Instead, acquisition of multiple shorter exposures may further improve the image through averaging of random noise.

As camera design improves, readout noise is decreasing. The Photometrics thermoelectrically cooled back-thinned CCD (model 180) camera used for this work had a readout noise of 50 electrons/pixel. Data acquired with the camera were read noise limited. A more recent Photometrics 200 Series CCD camera of comparable design had a readout noise of 12 electrons/pixel and comparable data acquired with the camera were photon noise limited (Fig. 8B). The lower read noise allowed an acceptable S/N ratio to be obtained using a 0.5 second exposure instead of a 2 second exposure.

Longer exposure times increase the number of photons emitted by a fluorochrome and therefore increase the S/N ratio. The brightest and most stable fluorochromes should be chosen in order to optimize the S/N ratio of probes against electronic noise. Especially where the target nucleic acid is known to occur in small copy number and/or single molecules, the fluorochromes of choice are CY3 and CY5, which emit in the red and far-red, respectively.

3. Autofluorescence is a major source of background inherent in biological samples that is contributed by biological components that have fluorescence absorption/emission in the same frequency range as the fluorescent probe of interest. The greatest amount of autofluorescence is exhibited in the blue and green emission spectrum of cells whereas the least is in the red through infrared. Autofluorescence is the major problematic factor remaining that dictates probe design. Longer exposure times cannot improve the S/N ratio of probe fluorescence versus autofluorescence. Using fluorochromes with high quantum efficiency and large extinction coefficients, linking multiple fluorochromes per probe (short probes, 24 to 60 nucleotides in length, can be linked with 3 to 6 fluorochromes, respectively, to

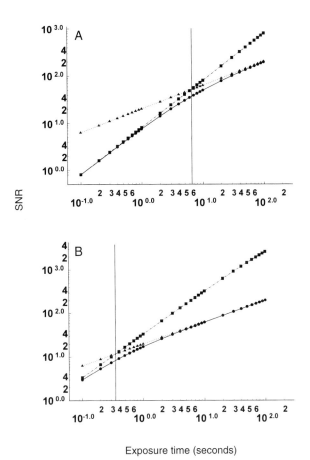

FIG. 8. Signal-to-noise ratios. Signal-to-noise ratios (SNR) are plotted as a function of exposure time (t) for two scientific grade CCD cameras using the equation $SNR = PQ_e t / \sqrt{(P + B)Q_e t + Dt + N_r^2}$. The plot area is divided into two regions at the intersection of the read-noise-limited equation $SNR = PQ_e t / N_r$ and the photon-noise-limited equation $SNR = PQ_e \sqrt{t} / \sqrt{(P + B)Q_e + D}$. The read-noise-limited region is designated to the left-hand side of the vertical line and the photon-noise-limited region to the right-hand side of vertical line. Photon noise ultimately limits the maximization of S/N as exposure time is increased (triangles). Read noise is the limiting factor for low-light level signal and short exposure time (squares). The camera read noise and the number of photons emitted per second from a source together determine the exposure time necessary to enter the photon-noise-limited region. (A) The Photometrics Series 180 CCD camera used in this work had the approximate S/N profile shown (circles) using read noise = 50 electrons rms/pixel and one fluorescent probe = 475 photons/pixel/sec. Any cytoplasmic signal originating from one fluorescent probe remained in the read-noise-limited region with exposure times of 2 to 3 sec per optical section. The respective signal-to-noise ratios were 14 and 20. (B) A Photometric Series 200 CCD camera had the approximate S/N profile shown (circles) using read noise = 12 electrons rms/pixel and one fluorescent probe = 475 photons/pixel/sec. The cytoplasmic signal for one fluorescent probe was in the photon-noise-limited region for exposure times of 1 to 2 sec per optical section. The respective signal-to-noise ratios were 17 and 26. The S/N was approximately two times higher compared to the Photometrics Series 180 CCD camera for the same exposure time. The advantage of lower read noise is reflected in the ability to use shorter exposure times that decrease fading while equal or better S/N is realized. The equations and plot format are taken from www.roperscientific.com.

ensure good *S/N* ratios), and iterating probes along the target sequence contribute the greatest advantage to improve the *S/N* ratio over autofluorescence. When a series of probes are iterated along a specific target sequence in such a manner that their closeness constitutes a fluorescence source that is small enough to be located in a single voxel, the signal emanating from that target will be amplified, increasing *S/N*. The total linear length of target should be restrained to about 140 nucleotides (48 nm of DNA at 3.4 Å/DNA base and 42 nm of RNA target at 3.0 Å/RNA base) or less in order to guarantee that the fluorescence source is as condensed as possible. Three 50-mer probes and up to five 25-mer probes could be iterated to a target and still remain a fluorescence source of approximately 50 nm in extended length.

Iteration of probes also helps to differentiate specific hybridization from nonspecific binding by their intensity differentials. There would be a decreased probability for random nonspecific targets to retain multiple probes compared to specific targets. Data will be presented in a later section that indicate nonspecific binding is negligible if appropriate conditions are used for FISH.

Some sources of autofluorescence can rival the intensity of the best probes. Efforts to discriminate autofluorescence from probe fluorescence through the use of spectral imaging is a potential solution to some of the current limitations of fluorescence imaging of single molecules at the higher emission frequencies. The emission spectra of a fluorochrome can be differentiated from the spectra of background autofluorescence and their respective contributions to pixel intensity can be calculated. An image representing only intensity information due to the fluorochrome can be generated.

A few practical measures that decrease autofluorescence include (1) the use of nutrient culture medium that does not contain phenol red, (2) fixation of samples with 4% *para*formaldehyde, and (3) using samples within 24 hr of fixation when the samples are stored in phosphate-buffered saline (PBS) at 4°.

DNA Oligonucleotide Probe Synthesis

The basic principles of synthesizing DNA generally, including oligonucleotide probes, are well known.[57] Those basic principles are applied to the design and synthesis of oligonucleotide probes used in this work.

A model 396 Applied Biosystems automated DNA synthesizer was used to synthesize probes used in this work. Typically, an oligonucleotide probe used in this work is obtained in a two-step process. The first step in the synthesis is to include a modified base at positions where a fluorochrome label is desired (Fig. 1). The preparation of amino-modified bases has been described in detail.[58,59] The

[57] M. H. Caruthers, *in* "Topics in Molecular and Structural Biology" (J. S. Cohen, ed.), p. 7. Macmillan Press, London, 1989.
[58] E. Jablonski, E. W. Moomaw, R. H. Tullis, and J. L. Ruth, *Nucleic Acids Res.* **14**, 6115 (1986).
[59] J. L. Ruth, *DNA* **3**, 123 (1984).

modified base used in this work is a thymine analogue that is available commercially as "Amino-Modifier C6 dT" (Glen Research, Sterling, VA), which is designed for use in conventional automated DNA synthesis. The synthesizer is run in the "Trityl On" mode to allow rapid purification of the modified oligonucleotide with an oligo purification cartridge (OPC). The second step is the covalent attachment of the fluorochrome label to each modified base. The purpose of the modified base used in the first step is to provide a functional group through which the fluorochrome label is covalently attached to the oligonucleotide in the second step. Various ligands can be conjugated to a DNA oligonucleotide probe through the amine derivatives.

The total number and spacing of the modified bases (and covalently attached fluorochrome labels) in the oligonucleotide can vary. Preferably, a modified base is incorporated within five bases from the 3' end of the oligonucleotide and thereafter at approximately every tenth base position in the nucleotide sequence of the oligonucleotide. Incorporation of modified bases, and thus fluorochrome labels, closer than every 10 bases may cause quenching of fluorescence and concomitant loss of visual signal strength.

Fluorochrome Labels

Various fluorochromes are useful in labeling probes for adequate signal strength. Standard methods for attaching fluorochromes onto amino groups have been described.[60] Preferably the fluorochrome is CY3 or CY5 (Biological Detection Systems, Pittsburgh, PA), and also fluorescein (Molecular Probes, Inc., Eugene, OR). The advantages of CY3 include (1) high molar extinction coefficient for absorption of light at excitation wavelength, (2) high quantum efficiency of emission, (3) pH insensitivity, (4) good water solubility, and (5) a very strong peak in the Hg lamp (a popular excitation source) spectrum at the excitation frequency. Water solubility reduces nonspecific adsorption to membranes, which results in lower background. More recently Alexa dyes (Molecular Probes) for excitation by specific laser frequencies have comparable or better attributes for brightness and stability.

Conjugation of Cyanine Dyes to Oligonucleotides

Our experience has been that commercial fluorescent nucleotide labeling kits do not optimally label DNA with CY3 and CY5 (Amersham Pharmacia Biotech, Piscataway, NJ) when the standard protocols are followed.

Accepted Commercial Protocol. Dissolve 30 nmol of dry sample in 0.5 ml of 0.1 *M* carbonate buffer (pH 8.5–9.0) and add to the dye vial. Cap the vial and mix

---

[60] S. Agrawal, C. Christodoulou, and M. J. Gait, *Nucleic Acids Res.* **14**, 6227 (1986).

thoroughly. Incubate the reaction at room temperatue for 60 min with additional mixing at 15-min intervals.

This protocol was intended for a DNA oligonucleotide with one reactive primary amine.[61] The final reaction solution contains hydrolyzed dye molecules (free dye), oligonucleotides conjugated to one dye molecule, and unlabeled oligonucleotides. The oligonucleotides are separated from free dye with a Sephadex G-50 column. Labeled oligonucleotides are separated from unconjugated oligonucleotides using reversed-phase high-performance liquid chromatography (RP-HPLC).

The labeling efficiency of this protocol is about 20%. More efficient labeling required an alternative protocol.

Alternative Protocol. To effect overall labeling efficiency greater than 80% for modified DNA oligonucleotides with multiple alkyl amino group (1) the concentration of dye (FluoroLink, Amersham Pharmacia Biotech) is increased by decreasing the reaction volume from 500 μl to 30–60 μl, (2) the ratio of dye to amino groups is increased further by reacting less oligonucleotide, and (3) the oligonucleotide is carried through three consecutive labeling reactions.

Oligonucleotide A (MW 15,774): Stock = 1.65 μg/μl or 0.133 nmol/μl (five alkyl amino groups per oligonucleotide) and concentration of amino linker = 0.665 nmol/μl.

Use 7.5 μl of Stock oligonucleotide A (5.0 nmol of amino linker) and evaporate. Dissolve oligonucleotide A in 30 μl of 0.1 M carbonate buffer (pH 9.3) and pipette the solution into a vial of FluoroLink Cy3 monofunctional dye (Amersham Pharmacia Biotech). Cap the vial and mix thoroughly. Incubate the reaction at room temperature for 60 min with additional mixing at 15-min intervals. Let it stand for an additional hour. Transfer the reaction mixture to a new vial of FluoroLink Cy3. Cap the vial and mix thoroughly. Incubate the reaction at room temperature for 60 min with additional mixing at 15-min intervals. Let it stand for an additional hour. Add an additional 30 μl of 0.1 M carbonate buffer to the reaction mixture and transfer the reaction mixture to a third vial of FluoroLink Cy3. Incubate it as above.

Separation of Labeled Oligonucleotide from Free Dye

For quantitative work the probe preparation must be free of unconjugated dye. The presence of free dye will give an erroneous calibration of the TFI per probe. A rapid method for purification of labeled oligonucleotide from free dye can be accomplished with a Sephadex G-50 (Sigma, St. Louis, MO) column. Sephadex G-50 is placed in excess 10 mM triethylammonium bicarbonate (TEAB) and deaerated overnight. A column is built to a bed volume of 32 ml in a disposable serological 25-ml pipette plugged loosely with cootton at the tip, and washed with

[61] L. M. Smith, M. W. Hunkapiller, T. J. Hunkapiller, and L. E. Hood, *Nucleic Acids Res.* **13**, 2399 (1985).

gravity flow for 10 min using 10 mM TEAB. The reaction mixture is applied to the top of the column and eluted with 10 mM TEAB. Fluorochrome-labeled probe elutes first, separating away from the free dye. Collect 1.0-ml aliquots immediately into 1.5-ml microcentrifuge tubes following application of sample. A well-built column will separate the probe from free dye by at least 2 ml. Detect the fractions that contain fluorescent probe using ultraviolet (UV) illumination. Dry and combine the fractions. Repeat the separation if necessary using a fresh column. The purity can be checked using HPLC.

UV/VIS Absorption Spectrum to Determine Concentration and Labeling Efficiency of an Oligonucleotide Probe

The concentration of the labeled oligonucleotide stock solution must be determined accurately in order to prepare accurate dilutions of labeled probe that will be used to generate a calibration curve to determine the TFI of one probe under imaging conditions. A Beckman DU 640 UV/VIS spectrophotometer was used.

The concentration of any labeled oligonucleotide can be determined by the DNA absorbance at 260 using Beer's law: $A = \varepsilon c l$ [A, absorbance; ε, extinction coefficient (cm^{-1} M^{-1}); c, concentration (M); l, 1-cm cuvette path length]. The extinction coefficient for a single-stranded oligonucleotide can be determined by summing the molar extinction coefficients for the individual bases multiplied by the number of each base A, C, G, and T (15,400A + 7,400C + 11,500G + 8,700T).

A more accurate method is to use a nearest neighbor calculation that includes nearest neighbor interactions of bases.[62]

$$2[\text{Sum}(E_{ab})] - [E2 + E3 + \cdots + E(n-2) + E(n-1)]$$

Where Sum(E_{ab}) is the sum of the pairwise extinction coefficients listed in the tabulation (http://www.genosys.com/technical) below and the E values are the normal single extinction coefficients for each individual base.

		3′Base		
5′ Base	A	C	G	T
A	13,700	10,600	12,500	11,400
C	10,600	7,300	9,000	7,600
G	12,600	8,800	10,800	10,000
T	11,700	8,100	9,500	8,400

Nearest neighbor calculations are given for ATGC and CGTA:

$$\varepsilon_{\text{ATGC}} = 2(11,400 + 9,500 + 8,800) - (8,700 + 11,500) = 39,200$$
$$\varepsilon_{\text{CGTA}} = 2(9,000 + 10,000 + 11,700) - (11,500 + 8,700) = 41,200$$

[62] C. R. Cantor and M. M. Warshaw, *Biopolymers* **9**, 1059 (1970).

The method of summation of molar extinction coefficients for individual bases does not differentiate between the sequences ATGC and CGTA:

$$\varepsilon_{\text{ATGC}} = \varepsilon_{\text{CGTA}} = (15{,}400^*1 + 7{,}400^*1 + 11{,}500^*1 + 8{,}700^*1) = 43{,}000$$

The fluorochrome labeling efficiency can be approximated by calculating the concentration of fluorochrome using the extinction coefficients (88,000 M^{-1} cm^{-1} at 490 nm, 150,000 M^{-1} cm^{-1} at 550 nm, 250,000 M^{-1} cm^{-1} at 649 nm) for fluorescein isothiocyanate (FITC) (pH 9.0), CY3 and CY5, respectively. An oligonucleotide with 5 amino linkers will be 80% or 100% labeled if the concentration of dye is four or five times that of the oligonucleotide, respectively. A precaution: to obtain valid information about labeling efficiency the probe must be pure with no free dye. Otherwise the DNA-to-dye ratio may erroneously show an apparently well-labeled probe.

Fluorescence *in Situ* Hybridization (FISH)

In situ hybridization with DNA oligonucleotide probes has evolved into a very straightforward process. Optimal hybridization conditions that provide high success rates have been published.[11,14] Most adjustments to the published techniques are a modification in the length of hybridization time and changes in concentration of formamide. Vanadyl adenosine complex and dextran sulfate were unnecessary components to obtain good hybridization results with DNA oligonucleotides.

The best results are obtained when care is taken to ensure similar melting temperatures, T_m, of all probes that are used for simultaneous hybridization. Oligo, Version 6.55 (Copyright 1989–2001, Wojciech Rychlik, Molecular Biology Insights, Inc.) software provides rapid assessment of T_m for any reasonable length sequence and GC content. This streamlines the choice of sequence and length of probe. In our laboratory sequences are chosen that have approximately 50% GC content. GC content has a dominant effect on T_m.

A typical hybridization protocol used in this work gives best results when cells are fixed for 10 min at room temperature in 4% *para*formaldehyde in phosphate-buffered saline (PBS) (2.7 mM KCl, 1.5 mM KH$_2$PO$_4$, 137 mM NaCl, 8 mM Na$_2$HPO$_4$). The cells are then washed and stored in PBS at 4° for less than 24 hr before hybridization.

Cells are gently permeabilized for 5 min in acetone at $-20°$ and then placed in PBS moments before hybridization. Alternatively cells could be permeabilized using 0.5% Triton in PBS for 10 min and washed once with PBS before hybridization. A mixture of fluorochrome-labeled oligonucleotide probes, 50-mer, each at a concentration of 10 ng per probe per coverslip, is hybridized for 3.5 hr at 37° in 50% formamide containing 2× SSC, 0.2% bovine serum albumin (BSA), and 1 mg/ml each of *Escherichia coli* tRNA and salmon sperm DNA. Following hybridization, the coverslips are washed successively for 30 min in 50% formamide/2× SSC at 37°, 30 min in 50% formamide/1× SSC at 37°, and 3 × 15 min in 1× SSC at room temperature.

The coverslips are mounted on slides with antifade mounting media [100 mg phenylenediamine in 10 ml PBS (adjusted to pH 8.2–8.5 with 0.5 M NaHCO$_3$) and 90 ml glycerol]. Multicolored (green plus red, 0.099-μm-diameter, custom synthesis, or TetraSpeck 0.1 μm-diameter Molecular Probes) latex beads are included in the mounting media and used as fiduciary markers to align images from cells interrogated with dual-labeled (FITC and CY3) probes. The coverslips are sealed with clear nail polish and allowed to dry at room temperature in the dark. The hybridized specimens are now ready for quantitative imaging. Slides are stored in the dark at $-20°$ up to 1 year with no detectable deterioration.

Calibration of Fluorescence with Epifluorescence Microscope

Calibration Procedure to Measure TFI of One Probe under Imaging Conditions[1]

The TFI per probe molecule is measured by first imaging a series of calibrated solutions of the probe and plotting the number of probe molecules versus the measured fluorescence intensity to generate a regression curve. The TFI per probe is determined from the corresponding slope. To successfully calibrate the probe it must be purified from unconjugated dye molecules and the concentration of probe must be determined accurately.

First, an aliquot of the original fluorescent probe stock (0.1–2 μg/μl) used in an *in situ* hybridization is set aside for calibration. The absorption at 260 nm is measured with a UV/VIS spectrometer. The concentration of oligonucleotide probe is calculated using the extinction coefficient expected for the specific oligonucleotide sequence.

Second, a range of concentrations of fluorescent probe is prepared. All dilutions are prepared gravimetrically because the concentrations become too low for independent verification with a UV/VIS spectrometer. Aqueous dilutions (A) of approximately 10 and 20 ng/μl for Cy3-labeled probes, 130 ng/μl for FITC, and 15 and 75 ng/μl for Cy5 are prepared. Final dilutions (B) are prepared with 2.5-, 5.0-, or 10.0-μl aliquots of (A) plus phenylenediamine/glycerol mounting medium containing fluorescent beads that have fluorescence emissions different from the probe emission frequencies. The final solutions are adjusted to have the same percent water content. The final probe concentrations are chosen to be compatible with the exposure times and optical setup conditions used for imaging the respective hybridized specimens in order to make a useful calibration curve that is in the linear range of the CCD camera. Useful preparations for our optical setup contained final concentrations of probe between 0 and 4.0 ng/μl for CY3 probes, 0 and 30 ng/μl for FITC probes, and 0 and 8.0 ng/μl for CY5 probes. Five microliters of solution is placed between a coverslip and slide and sealed with clear nail polish. Exposure times of 2000, 3000, and 15,000 milliseconds were used to image CY3, FITC, and CY5, respectively.

A key step to calibrate the TFI per probe is to obtain an accurate measure of the number of probe molecules that contribute to the resulting TFI imaged at a specific region of probe solution. The number of molecules imaged is in turn determined by the concentration of probe and the "imaged volume," which must be measured accurately. The imaged volume [imaged volume $= (x)(y)(z)\ \mu m^3$] is calculated using the effective pixel size at the specimen focal plane to calculate the area (x, y). The pixel size was calibrated for the optical setup of the microscope ($x = y = 93$ or 187 nm with a ×60 magnification, 1.4 numerical aperature (NA) objective, and ×5 or ×2.5 camera eyepiece, respectively). The z dimension is measured as the distance between the inner surfaces of a coverslip and slide using fluorescent beads adhered to both surfaces to delimit the respective inner surfaces at the imaged region (Fig. 9). There is considerable variation in z between slides

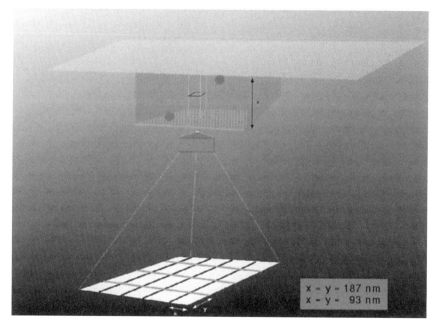

FIG. 9. Calibration of the TFI of one fluorescent probe molecule.[1] Each calibration solution, in succession, is placed between a coverslip and slide. Fluorescent beads (40–200 nm in diameter) adhere to the glass surfaces and delineate the distance between the inner surfaces of the coverslip and slide. The beads are brought into focus at each surface, while the stage position is monitored with an Eddy-current sensor, to measure the vertical distance (z) between the two beads. z is equal to the height of the volume element containing the imaged probe solution. The stage position was then set such that the focal plane was located equidistant from each bead. The dye solution was imaged at this focal plane. The image contained the total fluorescence contributed by all the fluorescent molecules in solution above and below the plane of focus in the imaged volume [imaged volume $= (x)(y)(z)$(number of pixels) μm^3]. The imaged volume included the full field of the CCD camera. The pixel size was calibrated for the optical setup of the microscope ($x = y = 0.093$ or 0.187 μm with a ×60 magnification, 1.4 numerical aperture objective, and ×5 or ×2.5 camera eyepiece, respectively).

TABLE II
TFI CALIBRATION DATA FOR FITC-LABELED PROBE SOLUTIONS[a]

Sample (ng/μl)	File	Pixels	TFI	$z1$ (μm)	$z2$ (μm)	$z2 - z1$ (μm)
7.96	afsha	158100	532104736	−0.05	13.79	13.84
7.96	afsha'	75990	256890720	−0.05	13.79	13.84
7.96	afshb	158100	617296832	1.44	17.54	16.10
7.96	afshc	158100	585917696	2.06	17.66	15.60
7.96	afshe	144330	561355456	0.04	15.52	15.48
14.68	afsga	158100	943511936	0.55	13.50	12.95
14.68	afsgb	158100	895207872	0.45	11.98	11.53
14.68	afsgc	142800	848046528	0.32	12.19	11.87
26.85	afsfa	158100	1247529856	−0.05	8.92	8.97
26.85	afsfb	126990	963209280	0.65	9.01	8.36
26.85	afsfc	158100	1188750976	0.06	8.37	8.31
26.85	afsfe	158100	1328312960	−0.04	10.14	10.18

[a] Two slides were prepared for each dye concentration. Two to three widely separated locations were imaged on each slide. The number of sampled sites for a concentration of 7.96 ng/μl (column 1) was five. The image of a site consisted of one optical section located at position $(z2 - z1)/2$ on the z axis. The image consisted of a maximum of 510×310 array of pixels (column 3). One optical section contained the total fluorescence intensity, TFI (column 4), of all probe molecules above and below the plane of focus in the imaged volume. The imaged volume was equal to the area of the pixel array at the sample focal plane multiplied by the distance between coverslip and slide (187 nm $\times$ 187 nm or 93 nm $\times$ 93 nm per pixel) $\times$ (number of pixels) $\times$ $(z2 - z1)$ μm.

(5–20 μm) and also at different locations under the same coverslip. The number of probe molecules in the imaged volume is then calculated from the concentration of probe in the solution. One optical plane, midway between a coverslip and slide containing the fluorescent probe solution, is imaged. The TFI of the single plane represents signal emitted by all probe molecules in the volume above and below the imaged plane designated the imaged volume. A sample calculation of the number of probes imaged and the TFI per probe is provided below.

Sample data from one imaging session are tabulated in Tables II and III for FITC and used to generate the respective calibration curve. Table II contains measurements taken during microscopy. The probe concentration of the calibration sample is tabulated in the first column. The file name of the imaged plane and the total number of pixels in the imaged plane are in the second and third columns of Table II, respectively. The cumulative TFI of all the pixels in the imaged plane is shown in column 4 (Table II). The position of the bead on the inner surface of the coverslip ($z1$), the position of the bead on the inner surface of the slide ($z2$), and finally the difference ($z2 - z1$) are tabulated in the fifth, sixth, and

TABLE III
SUMMARIZED CALCULATIONS FOR NUMBER OF FITC FLUOROCHROMES IN IMAGED VOLUME (nfiv)

Sample	imvol	Oligmol	nmiv	nfiv	tfipf
7.96 ng/μl					
Afsha	1.892491E$-$11	9.236778E$-$18	5562430	2.781215E+7	19.132097
Afsha'	9.096167E$-$12	4.439613E$-$18	2673555	1.336777E+7	19.217165
Afshb	2.201525E$-$11	1.074510E$-$17	6470746	3.235373E+7	19.079618
Afshc	2.133155E$-$11	1.041140E$-$17	6269791	3.134895E+7	18.690184
Afshe	1.932384E$-$11	9.431486E$-$18	5679684	2.839842E+7	19.767139
14.68 ng/μl					
Afsga	1.770792E$-$11	1.593088E$-$17	9593650	4.796825E+7	19.669510
Afsgb	1.576620E$-$11	1.418402E$-$17	8541682	4.270841E+7	20.960928
Afsgc	1.466036E$-$11	1.318916E$-$17	7942571	3.971286E+7	21.354458
26.85 ng/μl					
Afsfa	1.226564E$-$11	2.018533E$-$17	12155695	6.077847E+7	20.525850
Afsfb	9.182093E$-$12	1.511079E$-$17	9099789	4.549894E+7	21.169926
Afsfc	1.136315E$-$11	1.870012E$-$17	11261297	5.630648E+7	21.112150
Afsfe	1.392020E$-$11	2.290821E$-$17	13795426	6.897713E+7	19.257296

seventh columns (Table II), respectively. ($z2 - z1$) is the height, z, of the imaged volume.

Calculation of Number of Probe Molecules in Imaged Volume

Data and calculations are presented for an FITC-labeled probe.
$c = 0.13008$ is the concentration of stock solution in grams/liter and molwt $= 16,318$ is the molecular weight of an unmodified DNA olignucleotide comparable in sequence to the modified, fluorochrome-labeled probe. The molarity of the solution is the concentration (grams/liter) divided by the molecular weight:

$$M = c/\text{molwt} = \text{mol/liter}$$

$$M = 0.13008/16,318 = 7.9714\text{E}{-}06 \text{ mol/liter}$$

The area of one pixel of the CCD camera at the focal plane is determined by the optical setup and for $\times 60$ objective and $\times 5.0$ eyepiece is

$$\text{Pixel area} = 9.3\text{E}{-}08^{\wedge}2 \text{ m}^2$$

A representative measured z distance is 13.84 μm $= 1.384\text{E}{-}05$ m. (imvol) is the imaged volume in liters:

$$\text{imvol} = (z)(\text{unit pixel area})(\text{number of pixels})^*1000$$

$$\text{imvol} = (1.384\text{E}{-}05)(9.3\text{E}{-}08^{\wedge}2)(158100)(1000) = 1.892491\text{E}{-}11 \text{ liters}$$

Probe solution applied under a coverslip is a dilution of the stock solution:

Aliquot of stock solution $(V_a) = 3.07E{-}06$ liters

Final dilution volume $(V_f) = 5.0140E{-}05$ liters

Final molar concentration $(C_f) = M^* V_a / V_f$ mol/liter

$(C_f) = (7.9714E{-}06)(3.07E{-}06)/(5.0140E{-}05) = 4.8808E{-}07$ mol/liter

(oligmol) is the moles of oligonucleotide in the imaged volume and (oligmol) is the product of the imaged volume (imvol) and the molar concentration of the probe solution:

$\mathrm{oligmol} = \mathrm{imvol}^* C_f$

$\mathrm{oligmol} = (1.892491E{-}11)(4.8808E{-}07) = 9.23687E{-}18$ mol

Avogadro's number $= 6.022045E{+}23$ (molecules/mole). The number of oligonucleotide molecules in the imaged volume (nmiv) is

$\mathrm{nmiv} = \mathrm{oligmol}^* 6.022045E{+}23$

$\mathrm{nmiv} = (9.23687E{-}18)(6.022045E{+}23) = 5.56248E{+}6$ molecules

A representative total fluorescence intensity of the imaged volume is

$\mathrm{tfi} = 5.321047E{+}08$ arbitrary units

The number of fluorochrome molecules in the imaged volume is

$\mathrm{nfiv} = 5^* \mathrm{nmiv}$

The total fluorescence intensity per fluorochrome is

$\mathrm{tfipf} = \mathrm{tfi/nfiv}$

A summary of pertinent calculated quantities for FITC probes is tabulated in Table III. The TFI values for varying concentrations of probe are obtained from the fourth column (Table II). The corresponding number of fluorochromes in the imaged volume (nfiv) are obtained from the fifth column (Table III).

The TFI is plotted against the number of fluorochromes to generate a regression curve to calculate the TFI of one fluorochrome under imaging conditions. The regression curves are generated using Mathcad 7 Professional. The y axis intercept indicates the level of background. Point determinations of TFI per fluorochrome (tfipf) are not accurate due to the presence of background that contributes a high percentage of TFI especially at low probe concentrations. On the other hand, the background does not affect the slope.

The slope and correlation coefficient R are summarized for CY3, FITC, and CY5-labeled probes (Fig. 10). The slope of the regression curve is equal to the

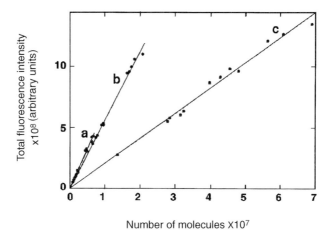

Number of molecules X10^7

FIG. 10. The amount of light emitted per probe was calibrated by measuring the TFI from a known number of probes in an imaged volume. The TFI was plotted against the number of fluorochrome molecules (five per probe) to generate the regression curve. The slope is equal to the TFI per fluorochrome (curve a = CY3, $m = 65 \pm 1.3$, $R = 0.992$; curve b = CY5, $m = 56.3 \pm 1.4$, $R = 0.992$; curve c = FITC, slope $m = 20.6 \pm 0.8$, correlation coefficient $R = 0.985$). Exposure times for CY3, CY5, and FITC were 3, 15, and 4 sec, respectively. A normalized TFI per probe was used to calculate the number of probes hybridized at discrete point sources of light in a restored image. Reprinted with permission from A. M. Femino, F. S. Fay, K. Fogarty, and R. H. Singer, *Science* **280**, 585 (1998). Copyright © 1998 American Association for the Advancement of Science.

intensity per probe or per fluorochrome, depending on the plotted variables. In this instance, the slope indicates the TFI of one fluorochrome molecule. The intensity associated with one probe is equal to five times that value.

Imaging Hybridized Specimen and Acquiring PSF

Hybridized specimens, PSF beads (200 nm), and calibration solutions are all prepared with the same mounting medium and identical coverslips to minimize variation of refraction, reflection, and light scattering. The probes are calibrated after the hybridized specimens and PSFs are imaged but during the same session. The optical setup, which affects the PSF and the light intensity at the specimen, may change between imaging sessions.

The hybridized specimens are optically sectioned using 100-nm steps with 50 optical planes or 250-nm steps with 20 optical planes. The corresponding PSF is imaged with either 100 or 250-nm steps and at least twice the number of optical sections as the specimen.

TFI of One Probe Calculated from the Regression Curve Is Equal to the TFI of a Probe Imaged in One Plane

The collective intensity of pixels that comprise the Airy disk of a point source imaged in the focal plane is equal to this value. A single probe that is optically sectioned and then deconvolved to a point source has a different value of TFI.

Calculated TFI for One Probe Is Adjusted to That for an Optically Sectioned and Deconvolved Probe

Exhaustive photon reassignment (EPR), the deconvolution process used in this work, integrates light over a depth (number of optical sections) defined by the PSF. Therefore, the TFI of a single probe is adjusted to that for a deconvolved image by a multiplication factor equal to the number of optical sections in the PSF used for deconvolution of the specimen image. The number of optical sections in the PSF used in this example is equal to 50.

The TFI for one probe molecule is adjusted to that for a deconvolved image CY3 (65 TFI units/CY3 fluorochrome) $\times$ (5 fluorochromes) $\times$ (50 PSF z-planes) = 16,250 $\pm$ 975; FITC (20.5 TFI units/FITC fluorochrome) $\times$ (5 fluorochromes) $\times$ (50 PSF z-planes) = 5125 $\pm$ 600; and CY5 (56.3 TFI units/CY5 fluorochrome) $\times$ (5 fluorochromes) $\times$ (50 PSF z-planes) = 14,075 $\pm$ 1050).

The TFI of a single probe, adjusted to an appropriate value for a deconvolved image, is used to interrogate a deconvolved image of a fluorescence *in situ* hybridization specimen.

The Adjusted TFI of One Probe Predicts the TFI of Immobilized Probes on a Coverslip

To test the accuracy of the calibration procedure an independent measure of the TFI of one probe was obtained by imaging single immobilized probes.

Fluorescent probes are diluted to a concentration of 2–4 pg/μl in mounting media. Five microliters is placed between a slide and coverslip and sealed with clear nail polish. The sample is left in the dark at room temperature for 1 hr and then placed in the freezer for 24 hr to allow probes to adsorb to the surface of the coverslip. A majority of probes settle out onto the coverslip far enough apart to be resolved as individual probes. There can be an uneven distribution where some areas of the coverslip are devoid of probe.

The probes were not detectable when examining the microscopic field without the aid of a CCD camera. However, as the sample was optically sectioned and imaged, points of light came into view essentially in a single focal plane. The deconvolved image consisted of discrete objects having a range of intensities.

A   B

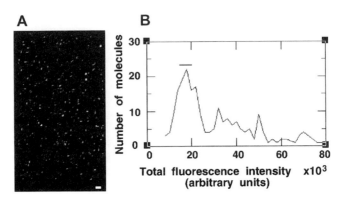

FIG. 11. Immobilized probes. (A) The accuracy of the calibration technique was tested by optically sectioning and restoring immobilized CY3-labeled probes adsorbed to glass coverslips. Pixel size = 93 nm. Bar: 1 μm. (B) Most frequent occurrences of TFI values per CY3 probe plotted as a histogram are comparable to the normalized value (solid bar) from the solution experiments. Reprinted with permission from A. M. Femino, F. S. Fay, K. Fogarty, and R. H. Singer, *Science* **280**, 585 (1998). Copyright © 1998 American Association for the Advancement of Science.

The mean TFI for an object was comparable to the calculated TFI for a deconvolved single probe (Fig. 11). The distribution of TFI of the immobilized probes mirrored the Gaussian TFI distribution of objects from the model system that included the random noise expected from the CCD camera used in these experiments (Fig. 7B).

Image Analysis

Interpretation of Digital Images

First, interrogation probes are small molecules. Small molecules provide an intrinsic level of precise spatial detail at molecular targets. The image, therefore, represents a collection of point sources of fluorescence that delineate the precise distribution and conformation of target molecules.

Second, the emitted light from point sources is detected on an array of pixels that has discrete boundaries. The pixels show a linear response to the intensity of light falling on the array. The intensity from one point source is distributed onto the pixel array according to the laws of diffraction.

Third, due to the Airy disk phenomenon, many pixels will detect some fraction of the light intensity from any single point source. Point sources, the basic unit in an image, appear as circular objects in the plane of focus due to the Airy disk pattern and as a cone in three dimensions. This is the PSF. A collection of spatially close point sources gives the appearance of a three-dimensional object due to the superposition of multiple, spatially distinct yet unresolved PSFs.

Fourth, an accurate model of the actual three-dimensional distribution of specific cellular molecules to which the fluorescent molecular probes have hybridized can in theory be reconstructed. Robust mathematical deconvolution algorithms applied to the fluorescence images can theoretically deconvolve the fluorescence information, using an empirical PSF, to estimate a true image.[63,64] The true image consists of point sources that are free of diffraction phenomena and other optical aberrations.

Fifth, the calibrated value for the TFI of one probe can potentially be used to identify sites in a restored image that represent one hybridized probe, the basic unit of fluorescence intensity in an FISH image. Therefore, interpretation of a fluorescent image should begin with the measurement of the TFI associated with objects in the deconvolved image.

Intensity Distributions in a FISH Image Are Hybridization Sites

It is important to establish statistical confidence in any new interpretation of FISH images and this is accomplished by analyzing the behavior of probes hybridized to specific targets. Analysis shows that probes bind to targets with predicted statistical probability, with predicted TFI, and with predicted spatial constraints.

First, the fluorescence intensity attributed to one fluorescent probe molecule is measured under imaging conditions.

Second, the amount of light emanating from a hybridization site is accurately measured. Optical sectioning and deconvolution are required. Optical sectioning provides complete spatial information necessary for successful deconvolution and reassignment of fluorescence signal to its point of origin in a 3D specimen. Reassignment of fluorescence signal to its point of origin facilitates the measurement of the total fluorescence intensity attributed to a hybridization site.

Third, the fluorescence intensity for one probe molecule, determined under imaging conditions, is corrected or normalized to be comparable to a hypothetical deconvolved single probe molecule. This allows a direct comparison of the intensity at a hybridization site in a deconvolved image with the predicted intensity of a probe molecule. Within statistical error the TFI representing a hybridization site occurs as multiples of the TFI attributed to one probe.

Fourth, to confirm a true hybridization site, three additional criteria are applied in addition to the TFI (Fig. 12). (1) The shape and (2) the size of a hybridization signal taken together establish the presence of a single hybridized probe, multiple probes, or noise. (3) Multispectral probes to a single target molecule must

[63] K. R. Castleman, in "Digital Image Processing" (A. V. Oppenheim, ed.), p. 293. Prentice-Hall, Englewood Cliffs, NJ, 1979.
[64] W. Carrington, *Proc. SPIE* **1205,** 72 (1990).

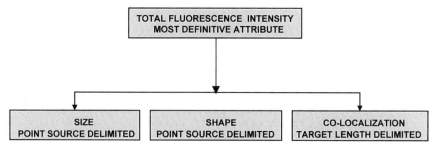

Fig. 12. The TFI is the most important criterion for identification of possible hybridization sites in a deconvolved image. A true hybridization site can be further confirmed through the requirement that the size and shape of the object conform to that expected for a deconvolved point source, especially if the TFI indicates the hybridization of a single probe (Fig. 17). More variable sizes and shapes are allowed when two or more probes are hybridized to different regions of a single target molecule, but are positioned close enough to remain unresolved. Two or more spectrally distinct probes hybridized to a single target molecule can further confirm a true hybridization site. The spatial constraints imposed by the selection and design of the probes will result in a predictable and distinctive colocalization pattern (Fig. 23).

show expected stoichiometry, colocalization, and predicted statistical probability distributions.

Fluorescence Signals Originating from Hybridization Sites Take on a Variety of Shapes and Sizes following Deconvolution

Deconvolution renders an image more amenable to quantitative analysis. Deconvolution aids the quantitative interpretation of hybridization sites in an image by placing the entire signal associated with a point source into a small concentrated volume in three-dimensional space. The deconvolution process also results in an increase in the definition of the boundaries of an isolated point source.

The history of the specimen to be analyzed is described briefly. Serum-starved, cultured normal rat kidney (NRK) cells were induced with serum, fixed, and hybridized with five probes targeted to the 3'-UTR of β-actin mRNA. The probes consisted of five different sequences, 50 nucleotides in length, each labeled with five CY3 fluorochromes. The five probes were iterated in close proximity to each other on the nucleic acid target (Fig. 13). The specimen was imaged with optical sectioning. A PSF was acquired. The probes were calibrated during the imaging session. The image stack was deconvolved using EPR with 400 iterations and an alpha value of 5×10^{-6}. The deconvolution converged minimizing the residuals to 10%.

A point source or single probe deconvolves to a 3D cluster of contiguous voxels and therefore the TFI of one probe is fractionally distributed among those voxels.

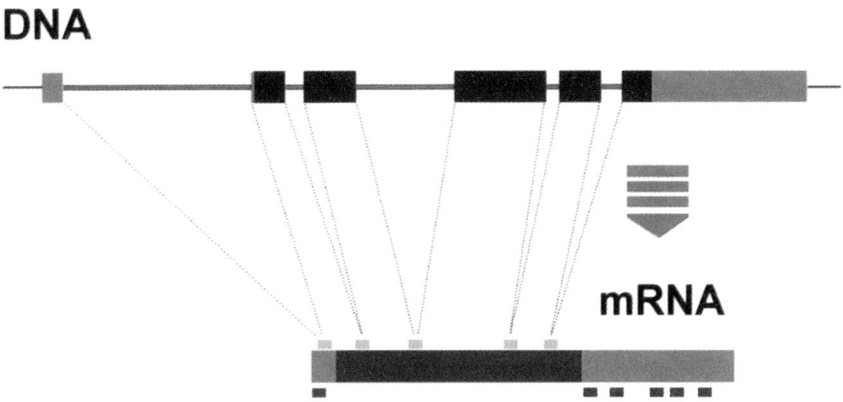

DNA

FIG. 13. A schematic depicts the rat β-actin gene and a typical processed β-actin messenger RNA. The positions of some representative oligonucleotide probes used in this work are shown hybridized to their target sequences (3′-UTR probes, red; splice junction probes, green; and one 5′-UTR probe, blue). The sequence of the β-actin gene is that published by U. Nudel, R. Zakut, M. Shani, S. Neuman, Z. Levy, and D. Yaffe, *Nucleic Acids Res.* **11,** 1759 (1983) (accession J00691). The nucleotide sequence for DNA is color coded for the 5′- and 3′-UTR (light blue), introns (pink), and exons (dark blue). The RNA nucleotide sequence is color coded for an open reading frame in the processed messenger RNA (blue) and the 5′- and 3′-UTR (light blue).

A deconvolved image also has many low intensity voxels that represent autofluorescence or background signal. A cell brought through a mock hybridization provides a representative intensity distribution of the autofluorescence and background signal. An appropriate threshold is applied to the images. All pixels with values less than the threshold value are set to zero generating a modified image. The threshold value should be low enough to minimize the loss of lower intensity voxels from the cluster that represents a single hybridized probe, but high enough to remove greater than 90% of the voxels attributed to autofluorescence signal within cells. The threshold was chosen to be less than 10% of the TFI expected for a single probe labeled with CY3. Setting a threshold facilitated the automated analysis of objects in the cell by creating discrete objects. A model, generated using the average value of the TFI of a single CY3 probe, was used to verify the process to establish a threshold, and showed that the criteria were valid for preserving the TFI of one probe (Fig. 14).

A binary mask was placed on the deconvolved image to exclude signal outside the boundary of the cell that was chosen for analysis. This was accomplished using an interactive computer program to trace along the outer edges of the cell membrane in each plane of the deconvolved image. There was enough autofluorescence to determine the cell boundary using the FITC image. The collection of boundary outlines from all the optical planes was then interpolated through the z axis to

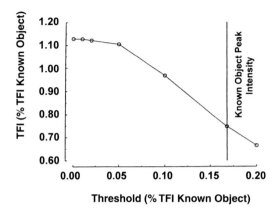

FIG. 14. A conservative calculated threshold minimized the adverse effects of background and noise on the measurement of the TFI of an object. Further analysis of the model system (Fig. 7B) indicated that setting a threshold at 10% of the total expected intensity of a single probe minimized the error for measuring the TFI of objects that represent single probes amid background noise (contributed by Kevin E. Fogarty). The average peak pixel intensity (maxima) within an object representing one probe falls at 17% of the TFI. This theoretical analysis supported the methodology that had been implemented for single molecule detection. A threshold that was too low did not separate objects adequately from background pixels and resulted in larger objects with TFI greater than 100% of that expected. A threshold that was too high removed pixels that were an integral part of the object leaving only the most intense pixels, those close to the object's peak intensity or maxima. Empirical data from immobilized fluorescent probes (Fig. 11) indicated that setting a threshold between 4.5 and 10% of the expected TFI minimizes the perturbation of the TFI of imaged objects. The data were for probes labeled with five CY3 fluorochromes and imaged with the described DIM system.

produce a shell delineating the volume of a single cell. The mask, when superimposed on the entire image, set to zero the value of all pixels outside of the shell boundary while leaving all pixel values inside the shell volume unchanged. This process facilitated an automated analysis of a specific region of interest (ROI) by segregating out the unwanted information, in this case every signal outside the perimeter of one cell. The remaining signal is a collection of discrete objects (Fig. 15A).

The experiment required two controls. The first control sample was hybridized with CY3-labeled probes specific to the 3'-UTR of α-actin mRNA. α-Actin is an isoform of actin found in muscle cells but not expressed in NRK cells. Such a control provided information on nonspecific binding of probes and was an indicator of good hybridization conditions. Optimal hybridization conditions provide adequate stringency for specific binding of β-actin probes and concomitantly reduce nonspecific binding to essentially zero (Fig. 15B). The second control cell was brought through a mock hybridization in which the labeled probes were omitted. The control provided information concerning the distribution profile of intensities

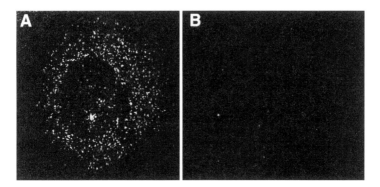

FIG. 15. Oligonucleotide probes only hybridize to their specific targets. Cultured NRK cells are induced for 15 min with serum and fixed. (A) A cell is hybridized with CY3-labeled probes targeted to the 3′-UTR of β-actin mRNA; 513 objects are detected, each having a TFI equaling one to five probes. Two transcription sites are also visible. (B) A cell is hybridized with CY3-labeled probes targeted to the 3′-UTR of α-actin mRNA; 20 objects are detected, each having a TFI equal to one probe hybridized. The α-actin gene is not induced by serum in NRK cells. The number of objects is equal to that found in a mock hybridization, and therefore can be considered background. The conclusion is that using carefully selected and controlled hybridization conditions, oligonucleotide probes do not contribute to background through nonspecific binding.

associated with autofluorescence. The control cells were deconvolved, processed, and analyzed with the exact steps used for the experimental cells.

The mock hybridization was analyzed to characterize the remaining autofluorescence contribution to a fluorescence image after applying a threshold (Fig. 16A). Autofluorescence originates from point sources that represent uncharacterized cellular components and has variable intensities. Analysis of an image hybridized with fluorescent probes to target sequences contains, in addition to the specific signal, a distribution of objects above threshold similar to that seen in the mock hybridization (Fig. 16B).

Predicted Shape and Size of Deconvolved Point Source

The discrete nature of the pixel array in the detector causes the raw signal to be detected as spatially discontinuous units of intensity rather than continuous variations of intensity. The perfect deconvolution would in principle place every point source at its respective spatial location. Ideally, one would expect the location to be in a single voxel. In practice this does not happen.

Through the use of the quantitative method described here that identifies true point sources by their TFI, empirical evidence indicates that a deconvolved point source that represents a single hybridized probe is present as a collection of contiguous voxels in 3D space. The exact number of contiguous voxels and their distribution in 3D space are dependent on several factors. The position of the point

A

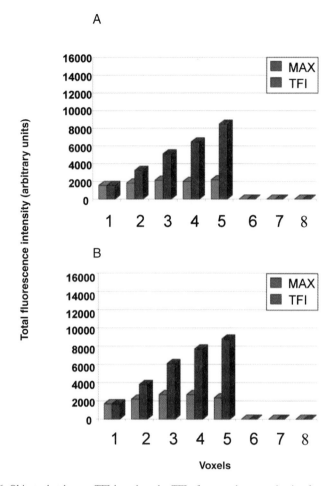

FIG. 16. Objects that have a TFI less than the TFI of one probe range in size from one to five voxels. The objects are considered background. The voxel with the maximum intensity in each object is not markedly higher than other voxels in the object. The increase in the TFI of objects is directly proportional to the increase in the size of the objects. (A) In a mock hybridization, the background objects are the major source of fluorescence. (B) In a cell hybridized with CY3-labeled probes to the 3′-UTR of β-actin, the background objects have a profile indistinguishable from the mock hybridization.

source in the specimen relative to the discrete pixel boundaries on the detector and the position relative to the focal plane in the series of optical sections will affect the shape of the cluster (Fig. 17). The sampling rate will additionally contribute to the size or number of voxels that makes up the object. Over sampling, using a voxel size smaller than the resolution limit will increase the size of a restored object.

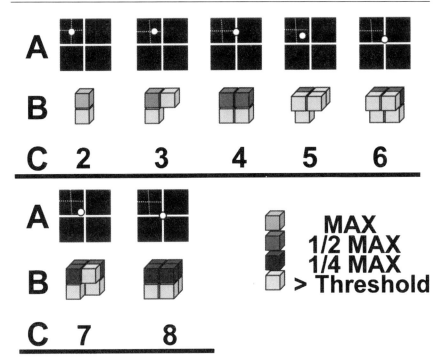

FIG. 17. The discrete nature of the pixel array detector and the discontinuous stepping associated with optical sectioning precludes the deconvolution of a point source to a single voxel, except in the one instance where the point source is exactly centered in a pixel and is in the focal plane. (A) A group of four pixels is shown. A point source is located at possible positions in one pixel. When the point source lies on an edge or at a vertex it, shares its TFI equally with two or four adjacent pixels. (B) The TFI of a point source imaged in the focal plane will be deconvolved to one maximal voxel (yellow) if it is centered in a pixel. If the point source is centered but out of the focal plane it will share a fraction of the TFI with an additional voxel along the optical axis (light blue). The TFI will be equally shared by two voxels (dark orange) if the point source is in the focal plane but located at an edge and four voxels (maroon) if it is in the focal plane and located at the vertex of four pixels. If the respective point sources are not exactly in the focal plane, their TFI will be shared with additional voxels along the optical axis (light blue) resulting in objects of four and eight voxels, respectively. Voxels that share a fraction of the TFI that falls below threshold are not shown and give the appearance of asymmetry to some objects. (C) The total number of voxels that can represent a deconvolved point source, theoretically, can range from one to eight as shown by this model. Empirical data acquired using probes that approach a theoretical point source support this model (Figs. 18 and 23).

Criteria to Identify Hybridization Sites to 3′-UTR of β-Actin mRNA in Deconvolved Image

A single CY3-labeled probe molecule, under imaging conditions, has the measured value, in this particular example, of 12,000 TFI units. Examination of the processed fluorescence image reveals objects varying in size and shape that conform to the expected TFI of one to five probes.

A

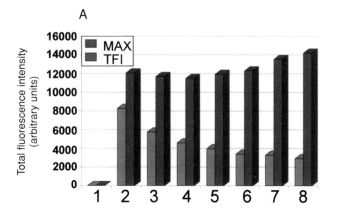

B

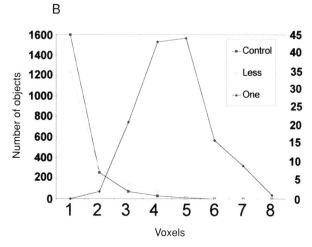

Voxels

FIG. 18. Objects that represent one hybridized probe are characterized. An NRK cell, hybridized with CY3-labeled probes targeted to the 3′-UTR of β-actin mRNA, is imaged and deconvolved. Objects in the deconvolved image whose TFI falls in the range of one probe hybridized (12,000–14,000 arbitrary units) are selected for further analysis. Their maximum pixel intensity, size (number of contiguous voxels), and frequency of occurrence are plotted. (A) Clusters ranging in size from two to eight contiguous voxels have a TFI expected of one hybridized probe. The maximum intensity voxel has the highest values in the smaller objects and the lowest values in the larger objects. This conforms to the subpixel placement of point sources (Fig. 17). (B) The most frequent sizes associated with objects representing one probe are three to six voxels, with four and five the most probable. Subpixel placement of point sources predicts that the least probable sites are those with the least redundancy such as the center of a pixel or the vertex formed by four pixels. The least probable sites result in one, two, or eight voxel objects. "Control" represents the objects in a mock hybridization (left ordinate). "Less" represents objects in a hybridized cell that have a TFI less than that expected for a single probe (left-hand ordinate). "One" is a histogram of the objects that represents one probe hybridized (right-hand ordinate).

How should we interpret the diverse presentation of objects? An object is herein defined as a discrete collection of contiguous voxels (i.e., sharing a face) remaining after application of a threshold. The object, in 3D space, is completely separate from all other objects by a surrounding shell of voxels that has been set to zero because it falls below the threshold intensity.

A true hybridization site was identified using several different criteria. The most definitive attribute, and the one used initially in this study, is the TFI. The TFI, calibrated for a single probe, is used to analyze the objects in the image. The objects are sorted according to the number of probes hybridized (number of probes hybridized = TFI/(TFI/probe).

Once an object is determined to have an intensity that is at least equal to the measured value of one probe under imaging conditions, it is further scrutinized for the attributes of size and shape. The size (number of voxels), TFI of each object, TFI of the maximum voxel associated with each object, and coordinates of the maximum voxel are tabulated using a computer program ("countobjs," Lawrence M. Lifshitz, UNIX operating system). A histogram shows all objects with a TFI value near the expected TFI of one hybridized probe sorted according to size, on the x axis. The TFI and frequency of occurrence are on the y axis (Fig. 18A, B) respectively. The most frequent size for one hybridized probe is three to six voxels for a voxel size of $187 \times 187 \times 250$ nm (x, y, z) after thresholding.

Quantitative Analysis of Hybridization Sites to Identify Single Molecules of β-Actin mRNA

Many hybridization sites are identified that have hybridized a single probe. Other sites show hybridization of multiple probe molecules (Table IV and Fig. 19). The original hypothesis was that a true hybridization site would be unequivocally identified through the use of iterated probes that hybridize to multiple sites along the 3'-UTR sequence, but that a single probe hybridized at a site would be indistinguishable from probes adhering nonspecifically to cellular components. However, it appears that there are many sites that hybridize a single probe. These sites cannot be discounted as nonspecific binding because the control cell that is hybridized with probes specific to α-actin mRNA shows very low numbers of objects that are identified as single hybridized probes similar to a mock hybridization (Fig. 15B, Table V, Fig. 20). Because the amount of nonspecific binding in a control probe is at best minimal, the interpretation is that greater than 90% of single probe sites represent a true target. The quantitation process combined with deconvolution facilitates a more accurate interpretation of fluorescence images. The conclusion is that quantitative analysis of hybridization sites identifies single molecules of β-actin mRNA.

Important information about the success of the deconvolution process emerges from the identification of single hybridized probe molecules in a 3D image. The

TABLE IV
FLUORESCENCE *in Situ* HYBRIDIZATION TO THE 3′-UTR OF β-ACTIN mRNA[a]

Number of probes hybridized	A (100% CY3)		B (100% CY3)	
	Number of objects	Percent of total	Number of objects	Percent of total
1	184	35.9	123	43.5
2	154	30.0	88	31.1
3	84	16.4	47	16.6
4	49	9.6	18	6.4
5	42	8.2	7	2.5
>5	10	—	18	—
Total objects (1–5 probes)	513	100.0	283	100.0

[a] Cells were hybridized with five different CY3-labeled probes and show objects hybridizing one to five probes. Objects from two cells (A and B) are sorted according to intensity. The objects are tabulated according to the number of probes hybridized.

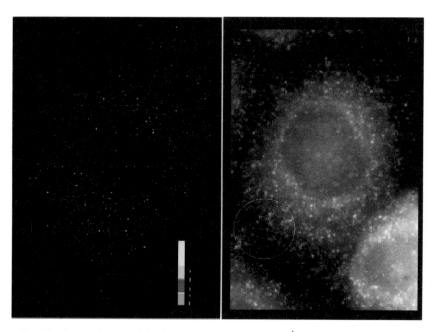

FIG. 19. Single molecules of β-actin mRNA are detected *in situ*.[1] (Right panel) A serum-starved NRK cell is hybridized with CY3-labeled probes to the 3′-UTR of β-actin mRNA and imaged. (Left panel) The TFI of each object in the deconvolved image is mapped to a single voxel at the coordinates of the respective maximum voxel. The single voxel objects are then color coded according to the number of probes hybridized. Most objects represent one to five probes hybridized, the number expected to hybridize a single β-actin mRNA. [1 = green, 2 = blue, 3 = pink, 4 = red, 5 = yellow, and >5 = white.]

TABLE V
CONTROLS IN HYBRIDIZATION[a]

	Mock hybridization (CY3)		α-Actin (CY3)	
Number of probes	Number of objects	Percent of total	Number of objects	Percent of total
1	22	73.3	20	100
2	6	20.0	0	0
3	2	6.7	0	0
4	0	0	0	0
5	0	0	0	0
>5	7	—	0	—
Total objects (1–5 probes)	30	100.0	20	100.0

[a] A mock hybridization and hybridization to α-actin, an mRNA not expressed in NRK cells, show about 20 objects that have the intensity of one hybridized probe.

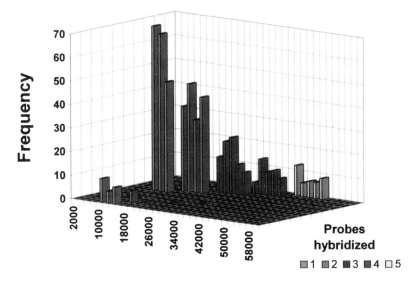

Total fluorescence intensity

FIG. 20. A histogram shows the frequency of objects detected in a cell according to their TFI. Each range of TFI is color coded for the number of probes hybridized. The front row represents objects in a cell carried through a mock hybridization. The second row represents objects in the cell depicted in Fig. 15A. The most frequent occurrences are objects representing one and two probes hybridized. Sites hybridizing three to five probes are less frequent.

observation that the TFI of a deconvolved point source is not contained in a single voxel, but is instead distributed in a contiguous cluster of voxels, in retrospect, seems appropriate (Fig. 17). The cluster of voxels, taken together and summed, contains the expected intensity of one hybridized probe and also has a finite volume. This observation has many consequences, as will be seen in the ensuing pages. This is the first time a fluorescent "object" has been interpreted *de novo in situ*.

The analysis of single probe hybridization sites indicates that the most common dimensions for a single probe acting as a point source is usually two to seven voxels after deconvolution and thresholding for a pixel size of $187 \times 187 \times 250$ nm. According to the model depicted in Fig. 17 the object size increases systematically as the point source location approaches an edge or a vertex of a pixel away from the center of a pixel. Therefore, the size of objects representing single probes in an image may reflect the subpixel position of the point source.[65]

The size, shape, and TFI of an object in the deconvolved image may also contain some contribution from random noise. The lowest intensity signals, such as those representing single probe hybridization sites, will be most affected by corruption due to random noise. The effect of random noise is to widen the range of values attributed to the TFI as well as to influence the shape and size of the deconvolved objects (Fig. 7B). The largest contributor to random noise in low-level fluorescence is readout noise from the CCD camera electronics (Fig. 8A).

Hybridization Behavior as Predicted by Statistics

Data indicate that single molecules of β-actin mRNA are detected. Can the results be rigorously tested or verified? One method is to hybridize competing probes labeled with spectrally distinct fluorochromes.

A set of five probes specific to the $3'$-UTR of β-actin is conjugated to CY3 and also to FITC. The two sets are then combined as a 50:50 mixture. Serum-starved NRK cells are induced with serum and fixed at 15 min. The cells are hybridized with the 50:50 mixture of CY3 and FITC-labeled probes. Other samples are separately hybridized with a set of 100% CY3-labeled probe or 100% FITC-labeled probe. The probes used for hybridization are calibrated for the optical setup. The samples are optically sectioned and deconvolved using 400 iterations and an α-value of 5×10^{-6}.

The objects in the deconvolved images are analyzed and categorized according to the number of probes hybridized. The objects detected in the CY3 channel from a cell hybridized with the 50:50 mixture are compared to a distribution of

[65] W. A. Carrington, R. M. Lynch, E. D. W. Moore, G. Isenberg, K. E. Fogarty, and F. S. Fay, *Science* **268,** 1483 (1995).

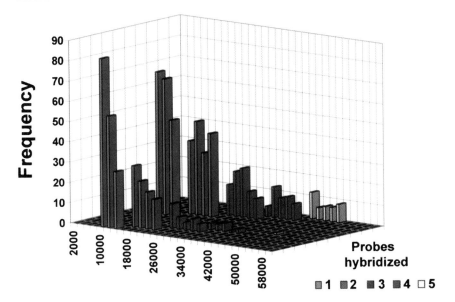

Total fluorescence intensity

FIG. 21. A histogram shows the frequency of objects detected in a cell according to their TFI. Each range of TFI is color coded for the number of probes hybridized. The front row represents CY3-labeled objects in a cell hybridized with a 50 : 50 mixture of CY3 and FITC-labeled probes to the 3′-UTR of β-actin mRNA. The back row represents a cell hybridized with 100% CY3-labeled probes. (Front) The CY3- and FITC-labeled probes compete with each other. The relative number of objects hybridizing one CY3-labeled probe increases above that found in the cell hybridized exclusively with CY3-labeled probes, while the number of objects hybridizing three, four, and five probes decreases dramatically.

objects from a cell hybridized with 100% CY3 label (Fig. 21). The CY3 signal becomes diminished in the cell hybridized with the 50 : 50 mixture of probes. This is expected when the hybridization is specific, because some of the specific binding due to the CY3 probe will be competed by the FITC probe. If the hybridization sites are in fact just nonspecific sticking, the signal could become additive with the result that the CY3 probe signal would remain largely unchanged.

There is a decline in the number of objects hybridizing three, four, and five probes, but surprisingly, an increase in the number of objects hybridizing one CY3 probe. The total number of objects can vary between cells. Therefore a better indication of a change between the 50 : 50 mixture and the 100% CY3 hybridization is a look at the relative distribution of objects within a single cell. The fraction of objects in a single cell that represents one hybridized CY3 probe

TABLE VI
A COMPETED HYBRIDIZATION

Number of probes	afsc2 (CY3)		afsd1 (CY3)	
	Number of objects	Percent of total	Number of objects	Percent of total
1	166	55.7	201	57.6
2	86	28.8	108	30.9
3	32	10.7	30	8.6
4	13	4.4	8	2.3
5	1	0.3	2	0.6
>5	16	—	5	—
Total objects (1–5 probes)	298	100.0	349	100.0

molecule increases from a range of 36–43% in a cell hybridized with 100% CY3 probes (Table IV) to approximately 56% in the competed hybridization (Table VI).

How can this be explained? The answer lies in the statistical probability distribution of hybridization to the $3'$-UTR when the access to some target sequences is limited. The profile of hybridization to the $3'$-UTR of β-actin mRNA indicates that one or two hybridization sites on a single mRNA molecule is the most frequent occurrence (Fig. 19 and Fig. 20). Target sites are not equally accessible at all times due to possible secondary structure or bound proteins.[66] A diagram depicts probable hybridization patterns that occur for a 50:50 mixture of CY3 and FITC probes targeted to the $3'$-UTR of β-actin mRNA (Fig. 22). The diagram clarifies the phenomenon of an increase in the number of objects that has hybridized one probe when hybridization is carried out with a 50:50 mixture of CY3- and FITC-labeled probes.

The diagram illustrates how sites that normally hybridize one probe will be hybridized with either a CY3 or FITC probe (Fig. 22A). Therefore, the number of objects hybridizing one CY3 probe should decrease by 50%. The empirical data show an increase in the relative number of objects that hybridize one CY3 probe (Fig. 21 and Table VI). The mRNA molecules that hybridize two probes have a 50% probability of hybridizing one CY3 and one FITC probe simultaneously, due to redundant permutations, but only a 25% chance of hybridizing two CY3 probes. Therefore the number of objects hybridized to two CY3 probes would be diminished by 75% of that expected from hybridization using 100% CY3 probes. Fifty percent of the possible objects hybridized to two probes will contribute to the

[66] K. Taneja and R. H. Singer, *Anal. Biochem.* **166**, 389 (1987).

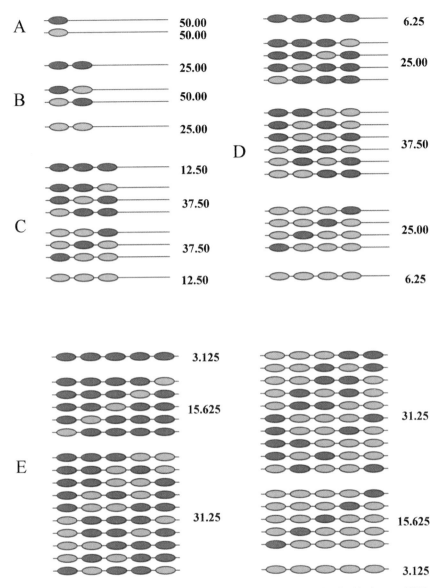

FIG. 22. A diagrammatic representation of the probability distribution of a 50 : 50 mixture of CY3-(red) and FITC- (green) labeled probes hybridized to the available sites on the 3'-UTR of β-actin mRNA. (A) The mRNA molecules that have only one site available will hybridize either a CY3- or a FITC-labeled probe. Therefore, there will be one-half as many objects in the respective image hybridized with one CY3 probe. (B) Molecules with two sites available have a 50% chance of hybridizing one CY3- and one FITC-labeled probe, but only a 25% chance of hybridizing two CY3 probes. Therefore, the number of objects hybridizing two CY3-labeled probes drops dramatically, whereas there is a concomitant contribution to the pool of observed objects hybridized to one CY3 probe. (C–E) Molecules with three, four, and five sites available will retain only 12.5, 6.25, and 3.125% of those sites exclusively hybridized to CY3. Therefore objects hybridized to three, four, and five CY3 probes become scarce.

objects that are categorized as hybridizing one CY3 probe. These will add to the apparent number of objects hybridizing one probe as detected in the CY3 channel. The other possible hybridization configurations, occurring for three, four, and five available sites, also contribute to the objects that are categorized as hybridizing one CY3 probe. The absolute number of objects contributed to the "one probe hybridized" category is small because there are fewer mRNA molecules that have three to five target sequences available. The overall result in the CY3 image is an apparent increase in the number of objects representing one hybridized probe molecule and an apparent diminishing of objects representing two to five probes hybridized.

A distribution of objects can be predicted quite accurately for a 50 : 50 mixture of probes. The respective probabilities of hybridization shown in the diagram can be used to calculate a hypothetical distribution based on the number of objects typically found in each category (one to five probes hybridized) of a cell hybridized with 100% CY3 probes (Table VII).

How do the statistics support single molecule detection? The TFI pattern of objects in the competed hybridization conforms to statistical prediction. Therefore the unit TFI that is used to identify an object hybridized to a single probe must indeed be the correct intensity for one probe. In turn, an object that hybridizes only one probe can only be one molecule. Objects that hybridize more than one

TABLE VII
STATISTICAL DISTRIBUTIONS OF HYBRIDIZED CY3 PROBES PREDICTED IN
COMPETED HYBRIDIZATION[a]

Number of probes hybridized	100% CY3 probe	50% CY3, 50% FITC	
		Predicted	Actual
1	184	218	201
2	154	100	108
3	84	35	30
4	49	9	8
5	42	1.3	2

[a] A typical distribution of objects in a cell hybridized with 100% CY3-labeled probes, column 2, provides a profile of the available sites on each $3'$-UTR of β-actin mRNA *in situ*. The probability distribution of a 50 : 50 mixture of CY3 and FITC-labeled probes (Fig. 22) is applied to the available sites and results in a "predicted" distribution of objects, column 3. A typical distribution of objects in a cell hybridized with a 50 : 50 mixture of CY3 and FITC probes, column 4, appears equal to the predicted distribution. A χ^2 test for the equality of distributions between two populations confirms that the "predicted" and "actual" distributions are equal. $\chi^2 = 1.301$, $\chi^2_{0.05} = 9.488$, df $= 4$.

probe, however, must also conform to the constrained spatial dimensions of one molecule to be unequivocally identified as one molecule.

Dual-Wavelength Hybridization to Resolve Intramolecular Distances as Close as 108 nm in Single Molecules of β-Actin mRNA

The objects detected in the FITC channel follow the same probability distribution pattern as those described for CY3 objects (Fig. 22). The competed hybridization diminishes the respective FITC signals compared to that of hybridization with 100% FITC-labeled probes.

Examination of the FITC image overlaid onto the CY3 image provides additional supporting evidence of single molecule detection of β-actin mRNA. The dual-wavelength image is zoomed in to focus on individual objects representing single molecules of β-actin mRNA (Fig. 23; Table VIII). Objects hybridized to two or three probes that also include both a CY3- and a FITC-labeled probe are selected. The probes show colocalization either through exact overlap of the respective maximal voxels associated with each spectrally distinct signal or through adjacent positioning of the respective maximal voxels. Both types of colocalization are expected if the object represents a single molecule of β-actin mRNA due to the fact that the set of five probes targeted to the 3'-UTR region is spaced relative to each other as close as 4 nucleotides and as distant as 360 nucleotides (Fig. 13).[2] The FITC and CY3 probes will show overlap colocalization when the hybridized

TABLE VIII
CYTOPLASMIC OBJECTS A, B, AND C AS SINGLE MOLECULES OF β-ACTIN mRNA[a]

Object	Fluorochrome	Object size (voxels)	TFI	Probes hybridized	Maximum voxel (TFI)	x, y, z
A	CY3	4	11630	1	4797	91, 199, 3
	FITC	5	3732	1	1191	91, 199, 3
B	CY3	3	16976	2	9727	98, 103, 11
	FITC	3	3066	1	1022	99, 103, 12
C	CY3	3	9747	1	4895	68, 136, 7
	FITC	3	3437	1	1866	69, 136, 8

[a] A 50 : 50 mixture of CY3 (red) and FITC (green)-labeled probes was hybridized to the available sites on the 3'-UTR of β-actin mRNA in situ. β-Actin mRNA may hybridize up to five 3'-UTR probes with one, two, or three the most probable number expected. Objects A, B, and C (from Fig. 23) have hybridized 2, 3, and 2 probes, respectively. The maximum voxels of the red and green components are within the expected linear constraints of the target 3'-UTR region. The TFI, the size of the respective red and green objects, and their distance apart confirm the hybridization to a single molecule of β-actin mRNA.

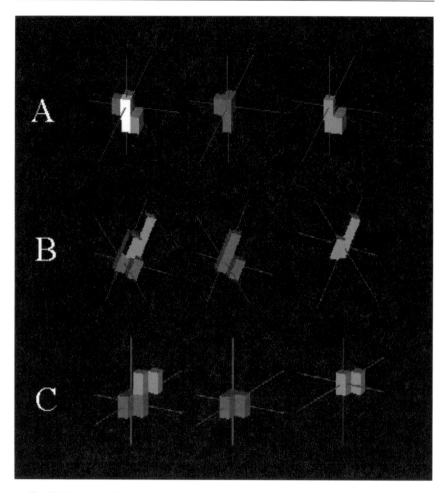

FIG. 23. Representative examples of β-actin mRNA that have simultaneously hybridized CY3 and FITC probes are shown zoomed in. The position of the maximum voxel for the respective CY3 and FITC objects is designated with a cross hair. The maximum voxel is the location of the point source (hybridized probe). (A) One CY3 (red) and one FITC (green) probe, same voxel colocalization (white). (B) Two CY3 probes and one FITC probe, adjacent voxel colocalization. (C) One CY3 and one FITC probe, adjacent voxel colocalization. Detailed data are provided for each object in Table VIII.

probes are linearly separated by 4 nucleotides, but will very likely be located in adjacent voxels when separated by 360 nucleotides (108 nm). Probes that are hybridized to separate molecules of mRNA would not be constrained to overlap or adjacent voxel colocalization (Fig. 24A). Therefore the colocalization as well as stoichiometry strongly supports single molecule detection of β-actin mRNA.

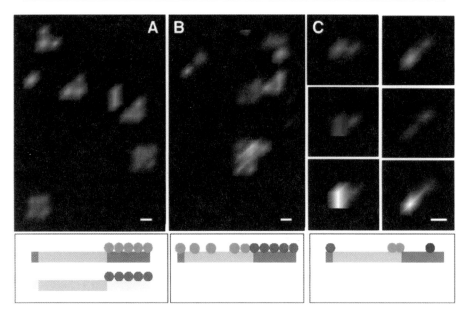

FIG. 24. Probing of intermolecular and intramolecular targets in the cytoplasm. (A) The detection of trans sequences. Both β- (green) and γ- (red) actin mRNA were detected simultaneously in the same cell with probes to their respective 3'-UTRs. Individual β- and γ-actin mRNAs segregate independently in the cytoplasm. [Statistical analysis of the frequency of measured separation distances in voxels (vo) (1 vo = 93 nm by 93 nm by 100 nm) between the red and the nearest green signal is expressed as cumulative percent: 0 vo = 0.1%, 1 vo = 1.5%, 2 vo = 6.4%, 3 vo = 13.7%, 4 vo = 25.1%, 5 vo = 36.5%; n (red) = 685, n (green) = 459.] (B) The detection of cis sequences. Two intramolecular regions of β-actin mRNA are resolved with five different probes targeted to the 3'-UTR (red) or the splice junction (SJ) sites (green). β-Actin mRNA molecules show nearest neighbor association of the 3'-UTR and the coding region. Frequency of separation distances in voxels: 0 vo = 2.4%, 1 vo = 15.8%, 2 vo = 32.9%, 3 vo = 59.1%, 4 vo = 72.6%, 5 vo = 87.2%; n (red) = 164, n (green) = 314. The 3'-UTR probes (red) are used to specifically detect β-actin mRNA. The SJ probes for β-actin (green) have homology with γ-actin, accounting for more green than red signal. Respective red and green pairs had TFI values consistent with single β-actin mRNA molecules. (C) Intramolecular measurement with three spectally distinct probes hybridized to two β-actin mRNAs. Two regions 631 bases (190 nm) apart were not resolved (CY5, blue; FITC, green); a third region, 1648 bases away (500 nm), was resolved (CY3, red). The colocalized voxels of the blue and green images are turned white. The distance between red and blue maxima is estimated to be 487 nm, consistent with a linear RNA. The schematic depicts the locations of the probes. Bar: 93 nm (A, B); 187 nm (C). Reprinted with permission from A. M. Femino, F. S. Fay, K. Fogarty, and R. H. Singer, *Science* **280**, 585 (1998). Copyright © 1998 American Association for the Advancement of Science.

Conclusion

Point sources of light emitting from fluorescent probes hybridized to molecular targets *in situ* are the basic units of information comprising a fluorescence digital image. Probes that conform as closely as possible to a point source maximize

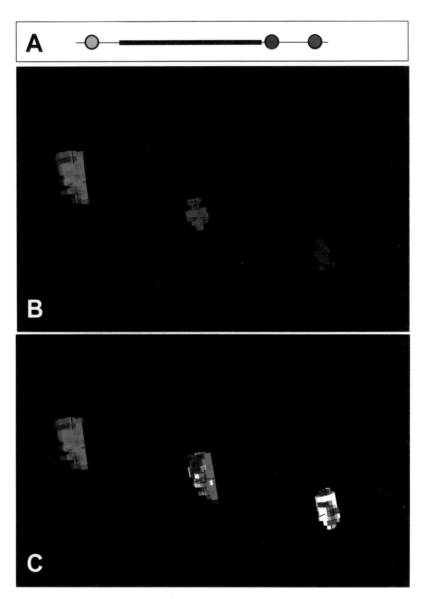

FIG. 25. A single β-actin transcription site is simultaneously detected with three probes targeted to the RNA, each labeled with a different fluorochrome. (A) A diagram of three spectrally distinct probes targeted to β-actin RNA. (B) The spatial distribution of signal from the probe targeted to the 5'-UTR (green) is expected to traverse the length of the gene during the peak of transcription at 15 min after serum induction. As expected, the signal appears to have the greatest spatial extent. The total number of green probes equals the total number of nascent RNA = 26. The spatial distribution of signal from both the proximal (14 probes) and the distal (8 probes) 3'-UTR is more condensed. (C) The signals from probes targeted to the proximal (red) and distal (blue) 3'-UTR are colocalized. Their respective spatial distributions are polarized toward one end of the 5'-UTR (green) signal. The interpretation is that the 5' end of a transcribing β-actin gene can be resolved from the 3' end.

analysis potential. Knowledge about image formation theory encourages a better appreciation for hybridization probe design.

The limitations of optical microscopy are reviewed to emphasize the necessity for optical sectioning and rigorous deconvolution. Deconvolution restores the TFI of an object to a delimited volume that is amenable to analysis for quantitative and spatial information. However, image formation theory, image processing, and deconvolution algorithms do not provide a solution to identify hybridization targets in an image. They do, however, provide a very important foundation for such a solution.

Definitive analysis of the fluorescence distribution in an image in terms of single molecules requires the identification of probe hybridization sites. Probes that approach the infinitely small spatial dimension of a point source and have a good S/N ratio are optimal for providing accurate spatial information about their respective targets and accurate intensity information to determine copy number following image deconvolution.

To accomplish a definitive analysis for an FISH experiment, a methodology was devised to calibrate oligonucleotide probes under imaging conditions. The calibration of the total fluorescence intensity attributed to a theoretical deconvolved single probe molecule led to the definitive identification of point sources that represented true hybridization sites in a deconvolved image. With adequate digital camera sensitivity and S/N ratio, it can be shown that many fluorescent objects represent single fluorescent probes hybridized to single target molecules

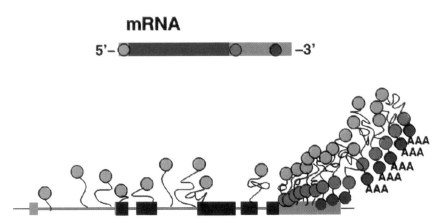

FIG. 26. A dynamic profile of transcription at one β-actin allele resolved with three spectrally distinct probes. The 5'-UTR probe detects all the nascent RNA (green). A second probe detects RNA that has progressed beyond the first 50 bases of the 3'-UTR (red), and the third probe detects RNA that has progressed to the distal region of the 3'-UTR (blue). The triple As indicate transcripts beyond the polyadenylation site. Reprinted with permission from A. M. Femino, F. S. Fay, K. Fogarty, and R. H. Singer, *Science* **280,** 585 (1998). Copyright © 1998 American Association for the Advancement of Science.

and, as a result, single copies of mRNA were detected and identified in the cytoplasm and nucleus of individual cultured cells.

Analysis of point sources representing one hybridized probe has provided additional information concerning the nature of the imaging process. The pixel does not suffice as the unit for analysis of a digital image involving fluorescence *in situ* hybridization. The unit for analysis is redefined to be the TFI of one probe molecule that has spatial dimensions, and is visualized as a three-dimensional cluster of contiguous voxels that varies in size and shape. The shape and size of a point source, representing one hybridized probe, contains information about the subpixel location of the point source.

The detailed characterization of biological molecules *in situ* is now possible using FISH. Unprecedented detail can be revealed from a hybridized cell that has been interrogated with carefully engineered, quantitatively accurate probes, followed by three-dimensional imaging and deconvolution.

The quantitative and spatial accuracy that results from the described analytical process provides a new and powerful visualization of biological events at the single molecule level. In addition to proving that single molecules could be visualized using epifluorescence microscopy, intramolecular spatial resolution, within one mRNA molecule, was demonstrated *in situ* (Figs. 23 and 24B and C). Additional experimental results demonstrated new detailed biological information concerning RNA processing at transcription sites as a consequence of accurate counting of the number of nascent RNA molecules at specific regions along the actin gene (Figs. 25 and 26).[2,3] The technology increases the potential of the FISH technique and dramatically extends the resolution limits of fluorescence light microscopy.

[14] Single Ion Channel Imaging

By ALOIS SONNLEITNER *and* EHUD ISACOFF

Introduction

Single-channel ionic current recording with the patch clamp, and by reconstitution into artificial lipid bilayers, greatly enhanced the understanding of ion channels by providing a direct view of the conductance and gating dynamics of channels. Here we describe an optical analog to the patch clamp that reveals protein motions on the single-channel level. Functional transitions that cannot be detected directly in ionic current recordings, because they do not open and close the channel gates, can be detected optically. We apply the method to a voltage-gated K^+ channel to detect the activation rearrangements of the voltage sensor that precede opening.

0076-6879/03 $35.00

The method extends voltage clamp fluorometry to the single-channel level by using total internal reflection (TIR) excitation. Changes in the fluorescence of a single dye molecule that is attached in a site-directed manner to a channel reveal local protein motion on a single-channel level. We describe the experimental setup and compare it to other approaches that have been proposed for the functional imaging of single ion channels.

Ion channels were the first complex molecules that could be studied in a biological context on the single-molecule level. This was achieved in the mid-1970s via patch clamp, developed by Neher and Sakmann in 1976,[1] and via incorporation of channels into lipid bilayers by Ehrenstein and colleagues.[2] The large ionic conductance of many channels ($\sim 10^4$ ions/ms) makes it possible to detect current flow through single channels, and thus reveals the dynamics of opening and closing of the channel's gates. The majority of functional transitions that ion channels make—which activate them, or transit them between inactivated or desensitized states of varying levels of stability—do not directly open or close the gates and are detected only indirectly as delays in gating. Some of these transitions, such as the activation transitions of voltage-gated channels, which precede opening, can be detected because they also move charge across the membrane via a structural rearrangement of a charged protein segment (this is the gating charge, and its movement generates the gating current).[3,4] However, the gating current is too small (~ 1–10 charges/ms) to be measured at the single-channel level. With the majority of functional transitions undetectable on the single-protein level a new approach for detecting single functional transitions in channels is required. We achieve this goal using fluorescence to monitor voltage sensing rearrangements at a single-channel level in the Shaker K^+ channel.[5]

Shaker K^+ Channel

The Shaker K^+ channel is a key component of action potential generation and propagation. It generates a transient outward current following membrane depolarization by its near cousin the voltage-gated Na^+ channel. Like the Na^+ channel, the Shaker current is transient because opening of the activation gate is followed by closure of one or more inactivation gates. The Shaker K^+ channel is a tetramer consisting of four identical subunits, each of which has six transmembrane segments (S1–S6) (Fig. 1A). The transmembrane segment S4 contains seven positively charged amino acids and acts as a voltage sensor. In response to membrane

[1] E. Neher and B. Sakmann, *Nature* **260**, 799 (1976).

[2] G. Ehrenstein, R. Blumenthal, R. Latorre, and H. Lecar, *J. Gen. Physiol.* **63**, 707 (1974).

[3] H. P. Larsson, O. S. Baker, D. S. Dhillon, and E. Y. Isacoff, *Neuron* **16**, 387 (1996).

[4] M. H. Wang, S. P. Yusaf, D. J. Elliott, D. Wray, and A. Sivaprasadarao, *J. Physiol.* **521**, 315 (1999).

[5] A. Sonnleitner, L. M. Mannuzzu, S. Terakawa, and E. Y. Isacoff, *Proc. Natl. Acad. Sci. U.S.A.* **99**, 12759 (2002).

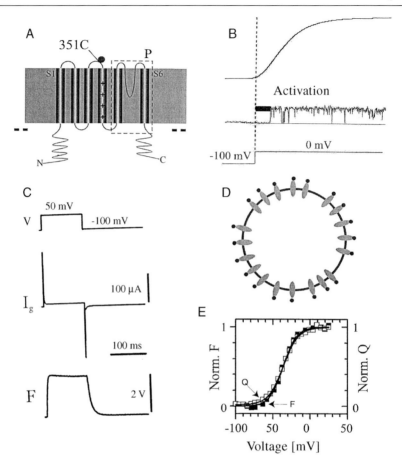

FIG. 1. Shaker K^+ channel topology, gating, and fluorescence measures of motion. (A) The Shaker K^+ channel consists of six transmembrane segments (S1–S6) with segments S5 the connecting loop and S6 forming the pore region (dashed rectangle, P) and S4, which contains positively charged amino acids, forming the main voltage sensor. A single cysteine is inserted at position 351 in the loop between S3 and S4 and labeled with tetramethylrhodamine maleimide (TMRM). (B) Ionic current induced by a voltage step from -100 mV to 0 mV for many channels from a whole cell recording (top trace) and a single channel (middle trace) in inside-out excised patches in patch clamp. This example shows a particularly long delay in onset of the single channel current following the voltage step (black bar), during which S4 undergoes its activation motion, leading to opening of the activation gate. (C) Gating current (I_g) from the whole cell and fluorescence (F) from the optical patch of membrane resting on the coverslip of a typical experiment where the oocyte is epiilluminated (penetrating excitation) and the oocyte expresses Shaker K^+ channels labeled to saturation with TMRM at site 351C. The voltage is stepped from a holding potential (-100 mV) first to a depolarized (50 mV) voltage and back to a hyperpolarized (-100 mV) voltage. I_g is transient because current stops being generated when all the S4s have moved, whereas F reaches a new stationary level when all of the S4 dyes are in the new environment. (D) Schematic representation of an oocyte expressing Shaker K^+ channels labeled with TMRM at an introduced cysteine. (E) Overlap of steady-state voltage dependence of fluorescence of TMRM attached to 351 (gray) and gating charge (black) both fit with single Boltzmann relations.

depolarization S4 undergoes a series of conformational rearrangements that together move five of the charged residues partway through the membrane-spanning portion of the protein, transporting the equivalent of three charges per S4 across the membrane electric field.[6–14] The motion of S4 triggers rearrangements in the pore domain of S5–S6, which open and close the activation and inactivation gates. In ionic current recordings, the phase of motion of the four S4s is seen as a delay in the onset of single-channel current (Fig. 1B, bottom trace, black bar). Variability in this delay and its multistep nature result in a sigmoidal rise of the macroscopic average current (Fig. 1B, top trace).

By interfering with a link between S4 and the pore via a mutation[15] one can decouple S4 from the inactivation gate in such a way as to shut the inactivation gate virtually permanently and render the channel nonconducting, leaving the motions of S4 and the activation gate intact. High-density expression of nonconducting channels ($\sim 10^{10}$ in a $Xenopus$ oocyte) enables the detection of the gating currents due to the motion of S4's positive amino acids across the membrane electric field. To study the Shaker K^+ channel in steady state we employ a mutant channel with its fast inactivation ball deleted.[16]

Fluorescence Report of Conformational Rearrangements

How can the "silent" conformational changes that do not directly open or close gates be detected in a channel? In 1996 we developed the method of voltage clamp fluorometry (VCF) in which a small environmentally sensitive fluorescent probe is attached to a specific location of a channel protein and changes in its fluorescence provide a report of local conformational changes in the channel during functional transitions that are measured at the same time by voltage clamping.[17] The use of environmentally sensitive probes in VCF is in contrast to the usual approach of using a fluorophore as a marker for the presence, location, and concentration of molecules in a complex environment, where insensitivity of the fluorophore to the specific environment is desired. As it turns out, many garden variety dyes are quite

[6] N. E. Schoppa, K. McCormack, M. A. Tanouye, and F. J. Sigworth, $Science$ **255,** 1712 (1992).
[7] S. Aggarwal and R. MacKinnon, $Neuron$ **16,** 1169 (1996).
[8] S. Seoh, D. Sigg, D. Papazian, and F. Bezanilla, $Neuron$ **16,** 1159 (1996).
[9] N. Yang, A. L. George, Jr., and R. Horn, $Neuron$ **16,** 113 (1996).
[10] N. Yang and R. Horn, $Neuron$ **15,** 213 (1995).
[11] S. P. Yusaf, D. Wray, and A. Sivaprasadarao, $Pflugers Arch.$ **433,** 91 (1996).
[12] D. M. Starace, E. Stefani, and F. Bezanilla, $Neuron$ **19,** 1319 (1997).
[13] O. S. Baker, H. P. Larsson, L. M. Mannuzzu, and E. Y. Isacoff, $Neuron$ **20,** 1283 (1998).
[14] H. P. Larsson, O. S. Baker, D. S. Dhillon, and E. Y. Isacoff, $Neuron$ **16,** 387 (1996).
[15] E. Perozo, R. MacKinnon, F. Bezanilla, and E. Stefani, $Neuron$ **11,** 353 (1993).
[16] T. Hoshi, W. N. Zagotta, and R. W. Aldrich, $Science$ **250,** 533 (1990).
[17] L. M. Mannuzzu, M. M. Moronne, and E. Isacoff, $Science$ **271,** 213 (1996).

sensitive to the environment. For example, tetramethylrhodamine maleimide (TMRM) changes its fluorescence with solvent polarity, sensing the difference between water ($\varepsilon = 78$) and ethanol ($\varepsilon = 24$) by an increase in peak fluorescence by 33% and shifting its emission peak from 575 nm to 567 nm.[17] We and others have used fluorescent probes attached to specific sites on membrane proteins to detect conformational changes that underlie their function.[17–25] So far, this work has been limited to macroscopic measurements from large populations of proteins.

When working with membrane proteins in living cells it is an advantage to employ membrane-impermeant fluorescent probes. These are attached to single cysteines that are introduced by mutagenesis into the external domain of the channel. Attachment is via very specific maleimide chemistry. Native cysteines are mutated away in these channels to ensure that the probe is attached only at the desired location. Channels are expressed at high densities in *Xenopus* oocytes and oocytes are labeled, resulting in labeling of native proteins and of Shaker channels (Fig. 1A and D). When the oocyte is subjected to two-electrode voltage clamping and the voltage is stepped from a resting potential (-100 mV) to an activating potential ($+50$mV) there is a large change in fluorescence intensity measured from a large number of channels that are labeled to saturation (each subunit has a dye at the same location, and the optical patch contains $\sim 10^9$ channels) (Fig. 1C, bottom). Responses to voltage are seen only when cysteines bearing Shaker channels are expressed: Oocytes expressing channels without introduced cysteines do not give a fluorescence report. When the dye is attached to S4, the fluorescence change follows the kinetics of the gating current (Fig. 1C, middle). The integral of the gating current is the gating charge. A plot of the steady-state voltage dependence of the gating charge is well described by a Boltzmann curve (Fig. 1E, black data points and curve). The voltage dependence of the simultaneously measured fluorescence change (Fig. 1E, black) closely follows the gating charge. In other words, the fluorescence intensity of the dye attached to a single cysteine on S4 directly reports on the gating charge carrying conformational rearrangements of S4. We now take VCF's ability to directly and selectively report on the functional rearrangements of a labeled domain of the channel and adapt it from population recordings to single-molecule recordings.

[18] A. Cha and F. Bezanilla, *Neuron* **19,** 1127 (1997).
[19] Y. Chang and D. S. Weiss, *Nat. Neurosci.* **5,** 1163 (2002).
[20] M. Li, R. A. Farley, and H. A. Lester, *J. Gen. Physiol.* **115,** 491 (2000).
[21] M. Li and H. A. Lester, *Biophys. J.* **83,** 206 (2002).
[22] B. K. Kobilka and U. Gether, *Methods Enzymol.* **343,** 170 (2002).
[23] F. J. Sharom, R. Liu, Q. Qu, and Y. Romsicki, *Semin. Cell Dev. Biol.* **12,** 257 (2001).
[24] P. L. Smith and G. Yellen, *J. Gen. Physiol.* **119,** 275 (2002).
[25] J. Zheng and W. N. Zagotta, *Neuron* **28,** 369 (2000).

Motivation and Prior Success of Single-Molecule Optical Measurements

Although ensemble experiments give important information, significant aspects of voltage sensing and gating remain uncharacterized in such an approach. Simultaneous single-channel optical and ionic recordings could fill in this gap. For instance, they could reveal whether the four S4s of a channel really move separately, whether each moves in a series of sequential steps, and whether all have to be activated for the activation gate to be open and the inactivation gate closed.

Addressing the above questions requires the ability to detect the motion of individual S4s. A single fluorophore can emit about 10^8 photons/sec under conditions of close to saturating excitation. With some photodetectors, such as avalanche photodiodes, that are able to detect a single photon, and through the use of a high numerical aperture objective one can approach a detection efficiency of the setup of about 5%, or 5×10^6 photons/sec. Because at least 50–100 counts per time point are needed to have a sufficient signal-to-noise ratio, this gives a maximal achievable time resolution of 10 μs/point. A time resolution of 10 μs might be sufficient to detect actual conformational transitions of S4. Further kinetic resolution can be gained by using fluorescence correlation spectroscopy, which should, in theory, be able to detect transitions down to the submicrosecond regime. This, however, requires detecting transitions of many molecules one by one in a homogeneous channel population in order to get sufficient numbers of photons per event at very short times. For the present, we do not use avalanche photodiodes, but instead work with an intensified charge-coupled device (CCD) camera, sacrificing orders of magnitude in sensitivity (and thus in time resolution), but gaining the ability to watch many single channels at a time. The success of the method should enable adaptation to avalanche photodiodes for monitoring one channel at a time at higher speed.

With the development of sensitive detection devices that detect single photons it has become possible to detect single fluorescent molecules in purified preparations and in living cells.[26] For example, single-molecule enzyme dynamics,[27] rotation of F_1-ATPase,[28] and individual ATP turnovers have been observed.[29] With single-fluorescent molecule experiments one can access different states of a protein by measuring parameters such as orientation,[30] lifetime, single-pair fluorescence resonance energy transfer (FRET),[31] or simply intensity of a single dye molecule.

[26] S. Weiss, *Science* **283,** 1676 (1999).

[27] H. P. Lu, L. Xun, and X. S. Xie, *Science* **282,** 1877 (1998).

[28] K. Adachi, R. Yasuda, H. Noji, H. Itoh, Y. Harada, M. Yoshida, and K. Kinosita, Jr., *Proc. Natl. Acad. Sci. U.S.A.* **97,** 7243 (2000).

[29] T. Funatsu, Y. Harada, M. Tokunaga, K. Saito, and T. Yanagida, *Nature* **374,** 555 (1995).

[30] G. S. Harms, M. Sonnleitner, G. J. Schütz, H. J. Gruber, and T. Schmidt, *Biophys. J.* **77,** 2864 (1999).

[31] T. Ha, T. Enderle, D. F. Ogletree, D. S. Chemla, P. R. Selvin, and S. Weiss, *Proc. Natl. Acad. Sci. U.S.A.* **93,** 6264 (1996).

Important steps toward a complete functional study of ion channels via fluorescence have been achieved. The diffusion of the K V1.3 K^+ channel in T-lymphocyte membrane has been measured in three dimensions by attaching a fluorescent-labeled toxin and following its trajectory.[32] Harms *et al.* used an EYFP fusion protein of the L-type Ca^{2+} channel to track its diffusion and stoichiometry on the equator of a cell.[33]

Experimental Approach: TIRFM

The most important step for observing single fluorescent molecules for an extended time in live cells is to reduce the background fluorescence from the autofluorescent cell content. To maximize the signal-to-background ratio it would be best to confine illumination to the membrane and to avoid illumination of the autofluorescent interior of the cell. Total internal reflection fluorescence microscopy (TIRFM) does exactly that. In TIRFM the laser beam is reflected off of the interface of glass and buffer when the angle of incidence is larger than a critical angle, producing total reflection of the beam and generating an evanescent field on the opposite side of the interface, where the cell is. The intensity of the evanescent field decays exponentially with distance from the coverslip, down to $1/e$ with a characteristic length $d = \lambda_0/4\pi[n_2{}^2\sin^2\theta - n_1{}^2]^{-1/2}$,[34] where n_1, n_2 are the refractive index of buffer and glass, respectively, θ is the angle of incidence, and λ_0 is the wavelength of the laser beam. The characteristic length for the decay is usually in the subwavelength regime for good TIRFM ($\sim$50 nm for our conditions: see Appendix: Materials and Methods). This provides a thin "illumination layer" directly above the glass surface in the inverted microscope. (For a detailed review of total internal reflection microscopy see Ref. 34).

The *Xenopus* oocyte, which we employ for its ability to express channels at high density, is surrounded by a glycoprotein layer and a vitelline membrane (Fig. 2A). For the plasma membrane of the oocyte to enter the evanescent field the vitelline membrane is removed mechanically and the 1-μm-thick glycosylation layer is removed by enzymatic digestion (see the Appendix: Material and Methods). This bares the plasma membrane (Fig. 2B), which is not flat, but rather folded in finger-like microvilli with diameters of $\sim$0.2 μm and a length of $\sim$1 μm (Fig. 3).[5] Because of the confined illumination in the z direction ($\sim$50 nm) only labeled Shaker channels at the tips of the microvilli will be excited.

For simultaneous two-electrode voltage clamp and fluorescence measurements with TIRFM we used the setup illustrated in Fig. 3. TIRFM is achieved by focusing

[32] G. J. Schütz, V. Ph. Pastushenko, H. J. Gruber, H.-G. Knaus, B. Pragl, and H. Schindler, *Single Mol.* **1,** 25 (2000).
[33] G. Harms, L. Cognet, P. H. M. Lommerse, G. A. Blab, H. Kahr, R. Gamsjäger, H. P. Spaink, N. M. Soldatov, C. Romanin, and T. Schmidt, *Biophys. J.* **81,** 2639 (2001).
[34] D. Axelrod, *Noninvasive Tech. Cell Biol.* **93,** 127 (1990).

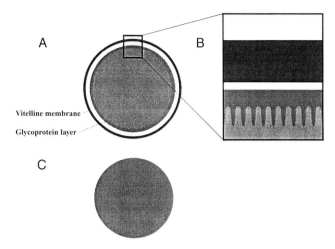

FIG. 2. Baring the oocyte membrane. (A) After isolation an oocyte has two protective layers: the vitelline membrane, which is a ~3-μm-thick clear membrane outside of the plasma membrane surrounding the oocyte. The plasma membrane is covered by an ~1-μm-thick glycoprotein layer. (B) Blowup of (A) showing that plasma membrane is made of fingerlike microvilli. (C) Mechanical removal of the vitelline membrane and enzymatic digestion of the protein glycosylation bare the oocyte membrane so that it can approach closely to the coverslip.

a laser beam onto the back focal plane of a high (1.65) numerical aperture (NA) objective off center, but parallel to the optical axis. This causes the laser beam to emerge from the objective at an angle and to be totally internally reflected off of the coverslip–buffer interface when the angle is larger than the critical angle given by $\theta_c = \sin^{-1}(n_1/n_2)^{34}$ ($\theta_c \approx 61°$ for our conditions), where n_1 is the refractive index of oocyte cytosol (~1.45) and n_2 is the coverslip refractive index (1.78, Olympus). (It is important to use an objective with an NA of 1.65, as we found the refractive index of the oocyte cytosol to be ~1.45.) The exponentially decaying evanescent field is produced at the point of reflection of the beam on the coverslip, exciting nearby fluorophores in the plasma membrane of the oocyte. The fluorescence is transmitted through a dichroic and a bandpass filter onto an intensified CCD camera. The gating currents from the two-electrode voltage clamp and the fluorescence change (detected by the intensified CCD) are recorded simultaneously.

The second important step in reducing the background is the labeling procedure. We use Shaker channel constructs with a single externally accessible cysteine. The fluorophore TMRM is covalently attached to this cysteine via a maleimide bond. To avoid labeling native proteins, we block the cysteines on the native proteins by labeling oocytes with a nonfluorescent probe tetraglycine maleimide before the arrival of channels on the cell surface.[17]

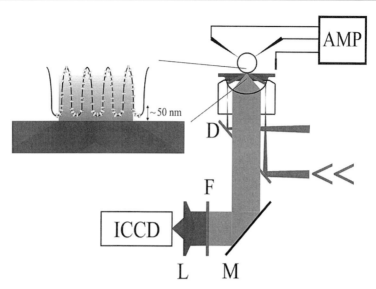

Fig. 3. TIRFM setup. Fluorescence detection and excitation were performed using an inverted microscope. The beam of a 100-mW Nd:YAG 532-nm laser is reflected off of a HQ545LP dichroic mirror (D) and aimed at the back focal plane of an Apo 100× 1.65 NA oil/∞ objective, so that it emerges into the immersion oil at an angle more shallow than the critical angle θ_c given by $\theta_c = \sin^{-1}(n_1/n_2) \approx 48°$, where n_1 is the water refractive index (1.33) and n_2 is the coverslip refractive index (1.78). (Refs. 29, 34, and 49). The maximum angle of emergence θ is given by $\theta = \sin^{-1}(NA/n_2) \approx 68°$, where NA is the numerical aperture of the objective (1.65). For good quality illumination (defined by a very shallow illumination depth), we adjust the laser beam to emerge close to the maximum angle θ, generating an evanescent field of excitation calculated to decay to $1/e$ in ∼50 nm. The diameter of the excitation spot is 15–25 μm. Single-fluorescent molecules are observed with excitation powers ranging from 500 μW to 4 mW. Fluorescent emission is filtered with an HQ572.5-647.5 bandpass filter (F). Detection is with a Pentamax GenIV intensified CCD camera. Two-electrode voltage clamping is performed with a Dagan CA-1B amplifier. Detail on left shows expanded view of the oocyte membrane at the site where the oocyte is closest to the coverslip. Because of the confined excitation by the evanescent wave, only the fluorescent labels attached to channels at the tips of the microvilli will be excited.

Ensemble Fluorescence: A Comparison of Penetrating vs TIR Illumination

In macroscopic optical recordings from channels labeled to saturation, TMRM reports on the channel's motions in a site-specific manner.[35] With TMRM attached to site 351, just outside S4, the fluorescence is bright at positive voltage and dim at negative voltage when measuring with a photomultiplier and exciting with conventional penetrating epiillumination (Fig. 4C). With epiillumination there is no need to remove the protective layers of the oocyte. With total internal reflection

[35] C. S. Gandhi, E. Loots, and E. Y. Isacoff, *Neuron* **27**, 585 (2000).

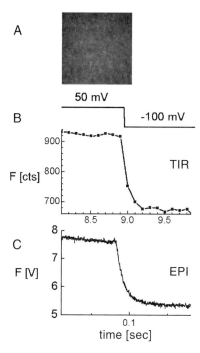

FIG. 4. Macroscopic fluorescence change from many channels labeled to saturation. (A) A frame from a movie of the fluorescence showing rather uniform labeling from channels labeled to saturation 25 ms per frame. Image size = $6 \times 6 \ \mu$m. (B, C) Fluorescence change in response to a step of voltage from $+50$ to -100 mV for TMRM attached to Shaker residue 351 illuminated with (B) TIR, with the plasma membrane within less than 50 nm of the glass surface, or (C) epiillumination with the vitelline membrane and glycoprotein layer intact.

illumination on oocytes stripped of vitelline membrane and protein glycosylation the fluorescence report detected with the CCD was the same as in the epiillumination (Fig. 4A and B). From the comparison of epiillumination vs. TIRFM several important conclusions can be made: (1) Channel function is not altered by mechanical and enzymatic stripping; (2) the close proximity of the Shaker channels to the glass surface does not interfere with their function; and (3) the fluorescence change in TIRFM is likely produced by the same mechanism as in epiillumination: due to changes in the local environment around the dye that accompanies local conformational change, rather than voltage-induced movement of the membrane in the evanescent field.

Detecting Protein Motion in a Single S4

To be able to image individual S4s we aim for a density of dye on the cell surface of less than $1/\mu$m^2. This requires that labeling be carried out with low

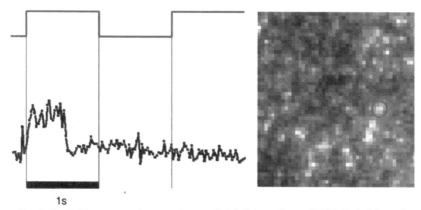

1s

FIG. 5. Single S4 movement from spot in sparsely labeled membrane. (Right) Typical frame from an experiment with TMRM attached to position 351C where labeling was sparse yielding isolated spots of brightness. (Left) Time trace of a fluorescent spot (circled in red in image on right). Note irreversible photobleaching midway through the first positive step eliminates any further modulation of fluorescence by voltage. The abrupt nature of the bleaching suggests that it is due to the photodestruction of a single dye, so that the trace that precedes it represents the report of protein motion in a single channel; 25 ms exposure time. (For additional examples see Ref. 5.)

concentrations of dye so that on average only a rare channel would be labeled, and then on only one of its four S4s. With our low labeling protocol (see the Appendix: Material and Methods) we estimate that we label about one channel in a thousand. Under these low labeling conditions, the CCD image loses its uniform intensity across the illumination area (Fig. 4A), but shows discrete spots of high intensity (Fig. 5, right). The intensity of these peaks changes in response to a step of membrane potential. A typical time trace for TMRM attached to site 351 at a single spot on the CCD (a 2×2 bin of pixels) can be seen in Fig. 5 (left). The fluorescence intensity starts off low at the negative voltage of -100 mV and gets brighter at the positive voltage of $+40$ mV, consistent with the macroscopic measurement (Fig. 4B and C). Midway during the 1-sec-long depolarizing step the fluorescence abruptly declines and then stays stable for an extended period. The repolarization to negative voltage produces no fluorescence change. And the next depolarization too has no effect. This indicates that the channel attached dye(s) in the spot experience irreversible bleaching. The time constant of bleaching of the optical patch was ~6 sec (not shown), indicating that the time of bleaching of different dyes is broadly distributed. The abruptness of the bleaching event (within one 25-ms frame) suggests that it was a "quantal bleach" due to photodestruction of a single dye molecule, and therefore that the fluorescence report preceding the photodestruction is due to the activation motion of a single S4. It is noteworthy that the dim state of the dye during the initial time at negative voltage is virtually at the same level as the background that remains after photodestruction (Fig. 5, left).

This means that the fractional fluorescence change due to S4 motion is very large. In the example that we show here TMRM attached to a site in S4 produces a very large fluorescence change. However, some sites show much smaller fluorescence changes,[35] presumably because the local protein motion does not change their environment as radically. For such sites polarization or FRET may help to amplify the signals.

Outlook

TIRFM-VCF extends single-molecule studies of ion channels from electrophysiological measurements of the dynamics of the opening and closing of gates to optical measurements of other functional transitions that do not directly gate the channel, but may control the gates. It may now be possible (1) to use polarized light measurements to determine if S4 rotates during activation, as has been proposed,[36,37] (2) to directly observe stochastic fluctuations of S4 across the membrane, which have so far been deduced from the bulk transport of protons across the membrane by histidines introduced into S4,[12] (3) to observe discrete serial substeps of S4 motion,[13] and (4) to observe a separate motions in the S4s in different subunits and relate them to the opening and closing of the channel gates.

The ultimate goal in single ion channel imaging is the simultaneous detection of conformational changes and function (single-channel current). Several groups have made attempts toward developing an approach that allows for the detection of ion channel current and fluorescence on the single-channel level. One attractive approach, at first glance, would be to use patch pipettes as a support for native and synthetic lipid membranes in which channels are fluorescently labeled.[38] This approach is limited by the fact that a patch is not formed on the rim of the patch pipette, but has a concave shape lining the walls of the patch pipette, making it impractical for fast imaging because of the nonplanar geometry. Progress has been made in lipid bilayers that can measure single-channel currents under conditions that enable visualization of single dyes.[39-44] Currently several groups are

[36] A. Cha, G. E. Snyder, P. R. Selvin, and F. Bezanilla, *Nature* **402**, 809 (1999).
[37] K. S. Glauner, L. M. Mannuzzu, C. S. Gandhi, and E. Y. Isacoff, *Nature* **402**, 813 (1999).
[38] J. Zheng and W. Zagotta, *Neuron* **28**, 369 (2000).
[39] K. Klemic, J. F. Klemic, M. A. Reed, and F. Sigworth, Planar Patch Clamp Electrodes. Patent WO 01/59447 A1 (2001).
[40] N. Fertig, Ch. Meyer, R. H. Blick, Ch. Trautmann, and J. C. Behrends, *Phys. Rev.* **E64**, 040901 (2001).
[41] A. Sonnleitner, G. Schütz, and T. Schmidt, *Biophys. J.* **77**, 2638 (1999).
[42] C. Schmidt, M. Mayer, and H. Vogel, *Angew. Chem. Int. Ed.* **39**, 3137 (2000).
[43] T. Ide and T. Yanagida, *Biochem. Biophys. Res. Commun.* **265**, 595 (1999).
[44] R. Pantoja, D. Sigg, R. Blunck, F. Bezanilla, and J. R. Heath, *Biophys. J.* **81**, 2389 (2001).

working on a simultaneous detection of fluorescence and ion channel current using gramicidin as a model system.[45,46] It is likely that these approaches will provide successful simultaneous optical and electrical measurements from single channels and enable fluctuations in structure to be related to fluctuations in functional state. Several of the questions that we listed above, however, should be approached effectively by the method we described here, where single-channel protein motions are related to the functional state of the population from whole cell current recordings. Moreover, because of its ability to work in live cells, the approach that we have described here may provide an important inroad into determining how (and whether) the activation of single membrane proteins alters the physiological state of the cell.

Appendix: Materials and Methods

Molecular Biology

Fluorescence experiments are performed on nonconducting (W434F),[15] ball-deleted (Δ6–46[47]) ShH4 Shaker channels,[48] after the removal of two native cysteines (C245V and C462A)[17] to ensure that membrane-impermeant fluorescent thiols would attach exclusively to a known position of cysteine addition. Site-directed mutagenesis, cRNA synthesis, and cRNA injection into *Xenopus* oocytes are as described previously,[17] leading to high-density channel expression in all of the experiments.

Oocyte Preparation

The vitelline membrane is found to be about 3 μm thick by scanning electron microscopy (not shown), and to bend or scatter the excitation light so that it penetrates into the cytoplasm, illuminating cortical granules. Mechanical removal of the vitelline membrane following 5 min exposure to an hypertonic solution (220 mM sodium aspartate, 2 mM MgCl$_2$, 10 mM EGTA, 10 mM HEPES, pH 7) reveals an approximately 1-μm-thick layer of extracellular matrix—too thick for penetration of the evanescent field. Therefore, following labeling (see below), the extracellular matrix is removed enzymatically by incubation in a mixture of 2 mg/ml hyaluronidase and 0.5 U/ml neuraminidase (Sigma, St. Louis, MO) for

[45] V. Bosisenko, T. Lougheed, J. Hesse, N. Fertig, J. C. Behrends, A. Woolley, and G. J. Schütz, *Biophys. J.* **82,** 46a (abstract) (2002).
[46] G. Harms, G. Orr, M. Montal, B. Thrall, S. Colson, and H. P. Lu, *Biophys. J.* **82,** 193a (abstract) (2002).
[47] T. Hoshi, W. N. Zagotta, and R. W. Aldrich, *Science* **250,** 533 (1990).
[48] A. Kamb, L. E. Iverson, and M. A. Tanouye, *Cell* **50,** 405 (1987).

12–20 min at room temperature. Oocytes are then mechanically devitellinized. Numerous microvilli that are approximately 1 μm in length and 0.1 μm in thickness are visible in scanning electron microscopy of oocytes following removal of the extracellular matrix and vitelline membrane (Fig. 1B). For scanning electron microscopy, oocytes are fixed in 2.5% glutaraldehyde and 1% OsO_4.

Fluorescent Labeling

Oocytes expressing Shaker channels with a single cysteine introduced are labeled with tetramethylrhodamine-5-maleimide (TMRM; Molecular Probes, Eugene, OR). The nanosecond excited-state lifetime of TMRM is much shorter than the microsecond or greater dwell times measured or estimated for many functional states of the channel, making the dye a good reporter of structural changes that take place around the dye during functional transitions. TMRM attachment to Shaker channels is maximized in several ways, described in more detail in Mannuzzu *et al.*[17] (1) by use of the charged TMRM fluorophore, which does not permeate the plasma membrane; (2) by use of the highly specific thiol-reactive maleimide chemistry to avoid attachment to other amino acids or to targets other than protein; (3) by blocking native membrane protein cysteines with a nonfluorescent maleimide reagent; and (4) by mutagenic elimination of two native Shaker cysteines, in S1 and S6. Oocyte incubation after RNA injection and the blocking and labeling procedures are as previously described[17] for oocytes expressing Shaker channels with a cysteine added at position 351C, except that lower TMRM concentrations (10–100 nM) are used, and better results are obtained with labeling at 10°, rather than on ice.

Total Internal Reflection Excitation and Detection

Fluorescence detection and excitation are performed using an Olympus IMT-2 microscope, after removal of the tube lens to make the microscope compatible with infinity-corrected lenses. Through the objective TIRFM[49] was done as follows. The beam of a 100-mW Nd:YAG 532-nm laser (Coherent, Santa Clara, CA) is reflected off of a HQ545LP dichroic (Chroma, Brattleboro, VT) and aimed at the back focal plane of an Apo 100×1.65 NA oil/∞ objective (Olympus, Melville, NY), so that it emerges into the immersion oil at an angle shallower than the critical angle θ_c given by $\theta_c = \sin^{-1}(n_1/n_2) \approx 61°$, where n_1 is the oocytes cytosol refractive index ($\sim$1.45) and n_2 is the coverslip refractive index (1.78, Olympus) (for review of TIRFM see Refs. 29 and 34). The maximum angle of emergence θ is given by $\theta = \sin^{-1}(NA/n_2) \approx 68°$, where NA is the numerical aperture of the objective (1.65). In these conditions, illumination generates an evanescent field of

[49] J. A. Steyer, H. Horstmann, and W. Almers, *Nature* **388**, 474 (1997).

excitation decaying exponentially from the glass–buffer interface into the buffer. The characteristic length of decay to $1/e$ in intensity can be calculated according to $d = \lambda_0/4\pi [n_2{}^2 \sin^2 \theta - n_1{}^2]^{-1/2}$ [34] which gives about 50 nm for our conditions assuming n_1 to be the refractive index of oocyte cytosol that we observed to be close to 1.45. The excitation spot is 15–25 μm in diameter. To minimize the rate of photodestruction, the power of the exciting light is attenuated with neutral density filters. Single-fluorescent molecules are observed with excitation powers from 500 μW to 4 mW. Fluorescent emission is filtered with an HQ572.5-647.5 band-pass filter (Chroma), and reflected laser light is excluded with a 532-nm notch filter (Kaiser, Optical, Ann Arbor, MI). Detection is with a Pentamax GenIV intensified camera (Roper Scientific, Trenton, NJ) at an intensifier setting of 80%. Images are usually recorded in the 2×2 hardware binned mode, which yields a pixel size in the image plane of 225 nm. This optical system enables the detection of single TMRM molecules that approach or adhere to the coverslip in dilute dye solutions. When oocytes are imaged, with the vitelline membrane and extracellular matrix surrounding the plasma membrane intact, evanescent field excitation yields fluorescence images consisting of some bright spots scattered against a diffuse background with considerable out-of-focus brightness. This can be explained by the high refractive index of the vitelline membrane, producing direct penetration of the exciting light into the cytoplasm. Following removal of the vitelline membrane and extracellular matrix, the fluorescence image becomes sharp and almost free of the out-of-focus haze. In contrast to dye in solution, where fluorescent spots are highly mobile from frame to frame (25–50 ms exposures), the fluorescence pattern of oocytes is stable for several seconds, with no lateral motion of the spots, consistent with limited diffusion of labeled membrane proteins.

TIR Excludes Autofluorescence from the Cell Interior

Oocytes have a special advantage not possessed by other cells: a pigment granule layer located just beneath the plasma membrane on the animal pole, which blocks excitation of, and emission from, the cell interior. A comparison of the autofluorescence of oocytes illuminated with epifluorescence excitation shows that the pigmented layer attenuates the autofluorescence by 7.6-fold in the average oocyte (pigmented animal pole fluorescence $= 0.66 \pm 0.03$ V PMT output; unpigmented vegetal pole $= 5.01 \pm 0.95$ V PMT output; $n = 10 \pm$SD). Evanescent field excitation eliminates the difference between the pigmented and nonpigmented poles, indicating that, as expected, the evanescent field does not penetrate into the cell interior. This indicates that evanescent wave excitation bypasses the need for a pigment layer, making other cell types equally suitable for both ensemble and single protein optical recording at the cell surface.

Electrophysiology and Solutions

Two-electrode voltage clamping is performed with a Dagan CA-1B amplifier (Dagan Corporation, Minneapolis, MN). The external solution contained 110 mM NaMES, 2 mM Ca(MES)$_2$, 10 mM HEPES, pH 7.5. TIR voltage clamping is performed in the presence of 30 mM of the triplet state quencher cysteamine (Sigma) to slow photodestruction, except where mentioned.

Fluorescence Analysis

Fluorescent spots that responded to voltage are identified by subtracting in WinView software (Princeton) the average of a pair of image frames acquired at the holding potential from an ensuing pair of frames taken during a depolarizing voltage step. Pixels with nonzero intensity in the difference image (spots where the voltage step evokes a ΔF) are analyzed by making a time trace of that spot's intensity across all frames in the movie using Metamorph (Universal Imaging, Downingtown, PA). The time traces are subsequently sorted for the single-molecule signature of abrupt disappearance (quantal bleaching), and these single-molecule fluorescence trajectories are analyzed up to the time of bleaching. Automated data processing is done with custom software written in Labview (National Instruments, Austin, TX).

Acknowledgments

We would like to thank Lidia Mannuzzu for ongoing support, Susumu Terakawa for development of the high N.A. objective, Shimon Weiss for critical comments to the experiments, Xavier Michalet for help and advice, Medah Pathak for stimulating discussions and Indian tea, all the people from the Isacoff lab for providing such a lively work place and Julia and Elias for providing distraction from life in the laser laboratory. Financial support for this project came from NIH Grant #R01NS35549. A.S. was supported in part by an Erwin Schroedinger Fellowship from the Austrian FWF.

[15] GFP–Fusion Proteins as Fluorescent Reporters to Study Organelle and Cytoskeleton Dynamics in Chemotaxis and Phagocytosis

By GÜNTHER GERISCH and ANNETTE MÜLLER-TAUBENBERGER

Green fluorescent protein (GFP) has become an invaluable tool to label proteins in living cells in order to investigate the redistribution of these proteins in response to external signals.[1] GFP is a globular protein of 27 kDa with a diameter of 2.4 × 4.2 nm along its short and long axis, respectively.[2] This means, by a steric argument, GFP may have a stronger or weaker influence on the folding, cellular localization, or activity of a protein to which it is bound. A flexible spacer (Gly-Gly-Ser)$_4$ inserted between the GFP moiety and the protein of interest may reduce the steric hindrance.[3,4] Normal localization of a GFP–fusion protein can be checked by comparison with antibody labeling of the endogenous protein in fixed cells. Normal function is established by rescue of the wild-type phenotype in mutants that lack the corresponding endogenous protein. However, a GFP–fusion protein may also exert a dominant-negative effect, or alter the phenotype of a cell simply by its abundance. In a living cell the activity of a protein may go through an optimum, such that a GFP–fusion protein might have a positive or negative effect depending on the degree of its overexpression.

There is a particularly high incidence of cases in which the addition of a GFP tag affects a protein that is translocated through a membrane, polymerizes or is prone to interact with a series of other proteins. To designate positions of the GFP tag at the N or C terminus of a protein X, we will use the terms GFP–(N)X and X(C)–GFP, respectively.

We discuss first the design of GFP fusions of cytoskeletal or organelle-specific proteins so as to minimize disadvantageous effects of the GFP tag. In addition, examples will be presented demonstrating that a GFP tag can block one particular function of a protein. In appropriate cases, this inhibitory effect of GFP can be used as an analytical tool to dissect the activities of a protein *in vivo* or to increase the specificity of its cellular localization.

In a second part, GFP constructs will be described that can be used to analyze responses to two external signals. In the chemotactic response the movement of

---

[1] K. F. Sullivan and S. A. Kay (eds.), *Methods Cell Biol.* **58,** (1999).

[2] M. Ormö, A. B. Cubitt, K. Kallio, L. A. Gross, R. Y. Tsien, and S. J. Remington, *Science* **273,** 1392 (1996).

[3] F. Hanakam, R. Albrecht, C. Eckerskorn, M. Matzner, and G. Gerisch, *EMBO J.* **15,** 2935 (1996).

[4] J. S. Weissman and P. S. Kim, *Cell* **71,** 841 (1992).

0076-6879/03 $35.00

cells is oriented by an external gradient of attractant; in phagocytosis the engulf-ment of a particle is initiated by its adhesion to a cell surface. These responses are related to each other by a common Gβ subunit in signal transduction and by similarities in reorganization of the actin system at the site of stimulation. In analyzing these responses, two complementary strategies can be employed. One is the targeted disruption of genes encoding constituents of the signal transduc-tion pathway and of cytoskeletal proteins involved. The second strategy focuses on the redistribution of GFP-tagged versions of the relevant proteins in living cells on stimulation by the chemoattractant cyclic AMP or by the attachment of particles.

In Fig. 1 an overview is given of organelles and cytoskeletal structures labeled in the motile cells of *Dictyostelium discoideum*. We will concentrate here primarily on proteins of this eukaryotic microorganism that illustrate different aspects of intracellular dynamics and responses to external signals, and provide information on proteins of other organisms in Tables I and II.

Organelle-Specific GFP Fusions

Positioning GFP Tag to Proteins of Organelle Membranes or Luminal Spaces

An overview of GFP–fusion proteins that have been used as organelle markers is presented in Table I. Preferably, GFP fusions of transmembrane proteins should be designed in a way that the GFP tag becomes located to the cytoplasmic face of the membrane. For example, a marker of endoplasmic reticulum (ER) membranes, calnexin, is correctly localized with a GFP tag placed at its cytoplasmic C terminus[5] (Fig. 1, panel 2).

For the correct sorting of luminal proteins of the ER usually two signals are required: an N-terminal hydrophobic leader sequence and a retention signal, for instance KDEL or HDEL, which prevents escape of the protein into the secre-tory pathway. As a marker of the ER luminal space, the Ca^{2+}-binding chaperone calreticulin can be used. Calreticulin is translocated through the membrane and retained in the ER if the GFP moiety is placed directly behind the N-terminal leader sequence to keep the C-terminal K(H)DEL retention signal accessible.[5] This signal is sufficient for retention, so that a minimal marker of the ER lumi-nal space comprises only a leader sequence followed by GFP and the K(H)DEL motif.[6]

For other organelles too, specific targeting motifs can be combined with GFP. The targeting into mitochondria requires an N-terminal presequence that must

[5] A. Müller-Taubenberger, A. N. Lupas, H. Li, M. Ecke, E. Simmeth, and G. Gerisch, *EMBO J.* **20**, 6772 (2001).
[6] J. Monnat, E. M. Neuhaus, M. S. Pop, D. M. Ferrari, B. Kramer, and T. Soldati, *Mol. Biol. Cell* **11**, 3469 (2000).

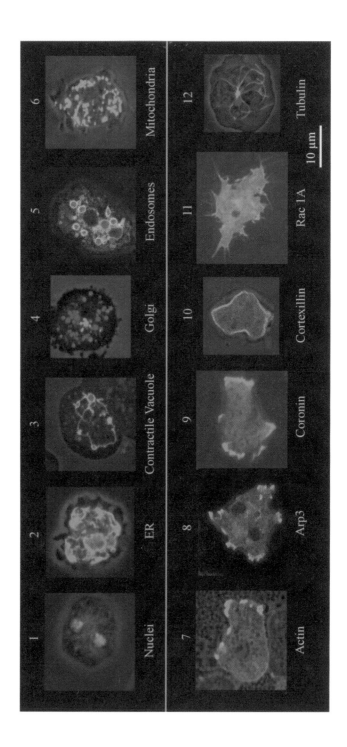

not be blocked, which means that GFP needs to be added to the C terminus.[7,8] For peroxisomal localization, the C-terminal tripeptide sequence S/A/C-K/R/H-L/M, called PTS1, is sufficient and can be linked directly to an N-terminal GFP moiety.[9,10] Nuclear targeting is achieved by fusing GFP to a nuclear localization signal (NLS). Commercially available GFP vectors (Clontech, Palo Alto, CA) make use of these organelle-specific targeting sequences.

Altered Localization or Function of GFP-Tagged Membrane Proteins

If the GFP moiety occupies a site required for trafficking of the protein, the intracellular localization of the fusion protein may be altered. The following examples illustrate how a GFP tag may impair the targeting or function of a membrane protein.

[7] M. Yano, M. Kanazawa, K. Terada, C. Namchai, M. Yamaizumi, B. Hanson, N. Hoogenraad, and M. Mori, *J. Biol. Chem.* **272**, 8459 (1997).
[8] L. Ni, T. S. Heard, and H. Weiner, *J. Biol. Chem.* **274**, 12685 (1999).
[9] E. A. C. Wiemer, T. Wenzel, T. J. Deerinck, M. H. Ellisman, and S. Subramani, *J. Cell Biol.* **136**, 71 (1997).
[10] D. Hoepfner, M. van den Berg, P. Philippsen, H. F. Tabak, and E. H. Hettema, *J. Cell Biol.* **155**, 979 (2001).

FIG. 1. Localization of GFP–fusion proteins to specific organelles (*top*) or to cytoskeletal structures (*bottom*). Cells of *Dictyostelium discoideum* are shown, most of them compressed by an agar overlay. Confocal images of the GFP fluorescence in green are superimposed to phase-contrast images in red or in blue (where a second label is colored in red). *Organelle markers:* (1) a binucleate cell with the nuclei labeled with histone 2–GFP (green). Microtubules emanating from centrosomes, one attached to each nucleus, are labeled with anti-α-tubulin antibodies (red) (this cell has been fixed; all other panels show living cells); (2) a cell expressing calnexin–GFP, showing labeling of ER membranes including the nuclear envelope in the left half of the cell; (3) the contractile vacuole complex consisting of bladders and a network of connecting tubules. The cell has been counterstained with TRITC–dextran (red) to label endosomes, which stay separate from the contractile vacuole compartment (Ref. 11); (4) golvesin(C)–GFP (green) provides a highly specific marker of the Golgi apparatus, showing no association with endosomes (red) (Ref. 12); (5) GFP–(N)golvesin is primarily associated with the membranes of endosomes, the lumen of which is labeled with TRITC–dextran (red) (Ref. 12); (6) mitochondria labeled with a polypeptide probe carrying an N-terminal targeting sequence and the GFP tag at its C terminus. *Cytoskeleton and associated proteins:* (7) GFP–actin in a cell growing on bacteria, showing actin accumulation in cell-surface projections and phagocytic cups (Ref. 15); (8) GFP–Arp3 representing localization of the Arp2/3 complex, which is recruited to cell-surface extensions and to late, neutral endosomes. The complex is displaced from early, acidic endosomes that are labeled with neutral red (Ref. 28); (9) GFP–coronin showing colocalization of coronin, a regulatory protein involved in phagocytosis, cell motility, and cytokinesis, with filamentous actin (Ref. 32); (10) GFP–cortexillin labeling the entire cell cortex (Ref. 33); (11) GFP–Rac1 enriched in the cell cortex and causing profuse filopod formation (Ref. 34); (12) GFP–α-tubulin visualizing the system of microtubules originating at the centrosome (Ref. 35).

TABLE I
REPRESENTATIVE GFP-TAGGED PROTEINS LOCALIZED TO SPECIFIC ORGANELLES

Protein	GFP position	Targeted organelle	Properties	Organism[a]
Annexin 2	C terminus	Secretory vesicles	Enrichment in actin tails propelling macropinosomes	Mammals[36]
Calnexin	C terminus	Endoplasmic reticulum	Excellent marker for ER membranes	Dd[5]
Calreticulin	Behind leader	Endoplasmic reticulum	Marker of the ER lumen	Dd[5]
Clathrin heavy chain	N terminus	Clathrin-coated vesicles	Rescues phenotypic defects of clathrin null cells	Dd[37]
β-COP	N terminus	Vesicles, Golgi	Enriched in the perinuclear region	Dd[38]
Dajumin	C terminus	Contractile vacuole	Dajumin without GFP localizes to plasma membranes	Dd[11]
Dynamin 2	N terminus	Vesicles, membranes	Functional link between dynamin and the actin cytoskeleton at podosomes shown	Mammals[39]
Dynamin 2	C terminus	Vesicles, membranes	Associates with clathrin-coated vesicles at plasma membranes, recruits to cortical ruffles, Golgi	Mammals[40]
Fibrillarin	N terminus	Nucleus	Found almost exclusively in nucleolus	Mammals[41]
Galactosyltransferase	C terminus	Golgi, ER, nuclear envelope	Constitutive cycling between ER and Golgi compartments	Mammals[42]
Golvesin	N terminus	Golgi, ER		Dd[12]
Golvesin	C terminus	Golgi, endosomes, ER, contractile vacuole		Dd[12]
Histone H1 or H2B	C terminus	Nucleus	Enables high-resolution imaging of chromosomes	Mammals[44] Sc[43]
HMG-17	N terminus	Nucleus	Colocalizes with endogenous counterpart	Mammals[4]
Lamin B1	N terminus	Nuclear envelope (NE)	Dynamics of NE visualized	Mammals[45]
Lamin B receptor	C terminus	Nuclear envelope	Localizes to nuclear envelope in interphase cells	Mammals[46]
Nup153	N terminus	Nuclear pore complex (NPC)	Dynamics of NPCs during interphase and mitosis	Mammals[4]
POM121	C terminus	Nuclear pore complex (NPC)	Dynamics of NPCs during interphase and mitosis	Mammals[4]
Protein disulfide isomerase	N terminus behind leader	Endoplasmic reticulum	57 residue C-terminal domain is sufficient for intracellular retention and ER localization localization	Dd[6]
Rab7	N terminus	Phagosome	Localizes to early and late phagosomes	Dd[47]

TABLE I (*continued*)

Protein	GFP position	Targeted organelle	Properties	Organism[a]
Rab11 (DdRab11)	N terminus	Contractile vacuole	Localizes exclusively to the contractile vacuole membrane system	Dd[48]
SEC24p	N terminus	COPII vesicles	Visualizes dynamics of COPII-coated transport complexes in ER-to-Golgi transport	Mammals[49]
SF2/ASF	N terminus	Nucleus	Functional in alternative splicing *in vivo*	Mammals[41]
TorA	C terminus	Mitochondria	Clustered at one end of mitochondria	Dd[50]
Vacuolin	N terminus	Endosomes	Localizes to postlysosomal endocytic compartment prior to exocytosis	Dd[51]
VSVG3	C terminus	Endoplasmic reticulum, trans-Golgi network (TGN)	Vesicular stomatitis virus G protein serves as marker to study TGN to plasma membrane traffic	Mammals[52]

[a] Dd, *Dictyostelium discoideum;* Sc, *Saccharomyces cerevisiae;* mammals, mostly mouse and human cell lines.

1. Dajumin–GFP, a marker of the contractile vacuole network, is prevented from cycling to the plasma membrane. Dajumin is a type 1 transmembrane protein that is found at the plasma membrane and on intracellular organelles. With GFP at its cytoplasmic tail, dajumin is strictly localized to the contractile vacuole system, an osmoregulatory organelle consisting of a network of ducts expanding into bladders (Fig. 1, panel 3). Even during discharge of the bladders through a pore to the cell surface, dajumin–GFP is not transferred to the plasma membrane.[11]

2. C-Terminal GFP turns golvesin, a post-Golgi protein, into a Golgi marker. Golvesin is a transmembrane protein of *D. discoideum* that normally distributes between the ER, the Golgi apparatus and two post-Golgi compartments, the contractile vacuole system and vesicles of the endosomal pathway, which are the most strongly labeled organelles.[12] GFP–(N)golvesin distributes similarly (Fig. 1, panel 5), whereas golvesin(C)–GFP differs from the untagged golvesin in not populating post-Golgi compartments. Instead, blocking of the C terminus by GFP causes golvesin to accumulate in the Golgi apparatus, providing a brilliant and highly specific label to visualize Golgi shape changes and the dynamics of tubule formation (Fig. 1, panel 4).[12]

[11] D. Gabriel, U. Hacker, J. Köhler, A. Müller-Taubenberger, J.-M. Schwartz, M. Westphal, and G. Gerisch, *J. Cell Sci.* **112**, 3995 (1999).

[12] N. Schneider, J.-M. Schwartz, J. Köhler, M. Becker, H. Schwarz, and G. Gerisch, *Biol. Cell* **92**, 495 (2000).

3. Caveolin 1 is involved in the uptake of virus particles through caveolae. Caveolin-1 localizes normally if tagged with GFP at either its C or N terminus. However, only C-terminally tagged caveolin-1 serves in mediating virus entry, whereas N-terminally tagged caveolin-1 exerts a dominant-negative effect, indicating that the N terminus of caveolin-1 is crucial for the uptake process.[13]

GFP Fusions of Cytoskeletal Proteins

GFP-fusions of actin, α-tubulin, or associated proteins have been used profusely to investigate microtubule movement and reorganization of the actin system in motile or dividing cells. A selection of suitable constructs is compiled in Table II. Elegant use can be made of single domains as probes for specific partners of interaction. For instance, a brilliant label for filamentous actin has been created by fusing GFP to the N-terminal domain of ABP120, an actin-binding domain of the α-actinin/spectrin type.[14]

Although in most instances, GFP fusions of cytoskeletal proteins reflect the localization and activity of the untagged proteins properly, there are a number of cases in which the GFP tag has a distinct inhibitory effect, as in the following examples.

Adverse Effects of GFP Tag on Cytoskeletal Protein Functions

GFP–Actin Fusion That Stops Myosin Movement

GFP–(N)actin has successfully been used in *Dictyostelium*,[15] in neurons, and in other mammalian cells[16,17] to study actin dynamics *in vivo*. With a molar ratio of GFP–actin to endogenous actin not higher than 1 to 10, adverse effects of a properly placed GFP tag on actin polymerization or myosin function are small enough to impair *Dictyostelium* cells only slightly. *In vitro,* the GFP–actin can polymerize, although less efficiently than untagged actin. In motility assays, actin filaments containing up to 30% of the GFP–actin are capable of moving on myosin subfragment-1 with only insignificant reduction in speed.[15] Contrary to these results, Aizawa et al.[18] reported that a GFP tag at the N terminus of actin

[13] L. Pelkmans, J. Kartenbeck, and A. Helenius, *Nature Cell Biol.* **3,** 473 (2001).
[14] K. M. Pang, E. Lee, and D. A. Knecht, *Curr. Biol.* **8,** 405 (1998).
[15] M. Westphal, A. Jungbluth, M. Heidecker, B. Mühlbauer, C. Heizer, J.-M. Schwartz, G. Marriott, and G. Gerisch, *Curr. Biol.* **7,** 176 (1997).
[16] A. Choidas, A. Jungbluth, A. Sechi, J. Murphy, A. Ullrich, and G. Marriott, *Eur. J. Cell Biol.* **77,** 81 (1998).
[17] M. Fischer, S. Kaech, D. Knutti, and A. Matus, *Neuron* **20,** 847 (1998).
[18] H. Aizawa, M. Sameshima, and I. Yahara, *Cell Struct. Funct.* **22,** 335 (1997).

TABLE II
SELECTED EXAMPLES OF GFP–FUSION PROTEINS OF THE ACTIN AND MICROTUBULE SYSTEMS

Protein	GFP position	Targeted structure	Functional properties	Organism[a]
Actin	N terminus	Actin filaments	Correct localization and function (if expression level is below 10%)	Dd,[15] Dm,[53,54] mammals[16,17,55]
β-Actin	C terminus	Actin filaments	Colocalizes with endogenous actin; only minor side effects on cell proliferation	Mammals[56,57]
α-Actinin	C terminus	Actin cytoskeleton	Localizes prominently at the leading edge in membrane protrusions	Mammals[58]
AIP1	N terminus	Actin cytoskeleton	Rescues mutant phenotype	Dd[20]
AIP1	C terminus	Actin cytoskeleton	Correct localization, but no rescue of mutant phenotype	Dd[20]
p41-Arc	N terminus	Actin cytoskeleton	Localizes to Arp2/3 complex	Dd[28]
Arp3	N terminus	Actin cytoskeleton	Localizes to and copurifies with Arp2/3 complex	Dd[28], Sc[59]
Calponin	N terminus	Actin cytoskeleton	Reduced motility in wound healing experiments; increased resistance to actin polymerization antagonists (only h1 isoform)	Mammals[60]
Capping protein	N terminus	Actin cytoskeleton	Concentrated in dynamic spots within lamella	Mammals[59]
Cdc42p	N terminus	Cytoskeleton	Targeted to medial ring contraction site during cytokinesis	Sp[61]
Citron-kinase	N terminus	Actin cytoskeleton	Localizes to cleavage furrow	Mammals[62]
Cofilin	N terminus	Actin cytoskeleton	Normal actin-binding activities *in vitro;* redistributes to actin-rich structures	Dd[63]
Coronin	N terminus	Actin cytoskeleton	Localizes to actin-rich structures	Dd,[32] Xl[64]
Coronin	C terminus	Actin cytoskeleton	Localizes to actin-rich structures	Dd[51]
Cortactin	C terminus	Actin cytoskeleton	Localized to lamellipodia of spreading cells	Mammals[65,66]
Cortexillin I	N terminus	Actin cytoskeleton	Localizes to cleavage furrow during cytokinesis	Dd[67]
DdCP224	C terminus	Centrosomes, microtubules	Cells exhibit supernumerary centrosomes and show a cytokinesis defect	Dd[68]
Dam1p	N terminus	Spindle	Localizes to intranuclear spindle MTs and spindle pole bodies	Sc[69]
DdLim	N terminus	Actin cytoskeleton	Localizes to the extreme rim of newly formed dorsal lamellipodia	Dd[70]

(continued)

TABLE II (*continued*)

Protein	GFP position	Targeted structure	Functional properties	Organism[a]
Duo1p	N terminus	Spindle	Localizes to intranuclear spindle MTs and spindle pole bodies	Sc[69]
Elfin (CLIM1)	N terminus	Actin cytoskeleton	Localizes to stress fibers of myoblasts	Mammals[71]
Fascin	N terminus	Actin cytoskeleton	Incorporated into actin bundles from the beginning of growth cone formation	Helisoma[72] (snail)
FHL2 (LIM protein)	C terminus	Cytoskeleton	Colocalizes with vinculin at focal adhesions	Mammals[73]
FHL3 (LIM protein)	C terminus	Cytoskeleton	Localizes to focal adhesions and nucleus	Mammals[74]
Kar9p	N terminus	Microtubules	Localizes to a single spot at the tip of the growing bud and the mating projection	Sc[75]
E-MAP-115 (ensconsin)	N terminus	Microtubules	Matches behavior of endogenous ensconsin	Mammals[76]
MHCK A	N terminus	Cytoskeleton	Recruited to actin-rich leading edge extensions	Dd[77]
Moesin	N terminus	Cortical actin cytoskeleton	Enriched in pseudopods, microvilli, axons, denticles, and other membrane projections	Dm[78]
Myosin II	N terminus	Cytoskeleton	Rescues myosin II-null defects	Dd[79]
Myosin V	N terminus	Cytoskeleton	Myosin Va head and head–neck domains codistribute with actin filaments	Mammals[80]
Myosin VI	N terminus	Cytoskeleton	Associates and colocalizes with clathrin-coated pits/vesicles	Mammals[81]
Myosin VII	N terminus	Cortical cytoskeleton	Rescues mutant phenotype	Dd[82]
Num1p	N terminus	Microtubules	Localizes at the bud tip and the distal mother pole	Sc[83]
α-Parvin	N terminus	Actin cytoskeleton	Localizes to membrane ruffles, focal contacts, tensin-rich fibers	Mammals[84]
Paxillin	C terminus	Actin cytoskeleton	Early visible in protrusive regions	Mammals[85]
Profilin IIB	N terminus	Actin cytoskeleton	Recruits to spindles and asters during mitosis	Mammals[21]
Rac1	N terminus	Cytoskeleton	Minimal effect on cellular morphologies and behavior; strong overexpression induces formation of filopodia	Dd[34,86]
RacE	N terminus	Cytoskeleton	Full rescue of mutant phenotype	Dd[87]
RacF1	N terminus	Cytoskeleton	Transiently associated with dynamic actin-dependent structures	Dd[88]

TABLE II (*continued*)

Protein	GFP position	Targeted structure	Functional properties	Organism[a]
α-Tubulin	N terminus	Microtubules	Visualizes brilliantly microtubule dynamics in different cell types; not recommended for studying assembly dynamics	Sc,[89] Dd,[90,91] Mammals[92]
γ-Tubulin	N terminus	Centrosome	Used to visualize centrosome movement	Mammals[93]
Twinfilin	N terminus	Actin cytoskeleton	Localizes primarily to cytoplasm, but also to cortical actin patches	Sc[94]
VASP	N terminus	Actin cytoskeleton	Localizes at focal adhesions and along stress fibers	Mammals[95]
Verprolin	C terminus	Actin cytoskeleton	Cell cycle-dependent distribution similar to cortical actin	Sc[96]
Zyxin	N terminus	Actin cytoskeleton	Incorporated into focal adhesions and stress fibers	Mammals[95]

[a] Dd, *Dictyostelium discoideum;* Dm, *Drosophila melanogaster;* Sc, *Saccharomyces cerevisiae;* Sp, *Schizosaccharomyces pombe;* Xl, *Xenopus laevis;* mammals, mostly mouse and human cell lines.

strongly impairs cytokinesis and motility of *Dictyostelium* cells. The strong effect observed *in vivo* is interpreted to mean that the GFP–actin forms a rigor complex with myosin II that acts as a brake, not allowing myosin II to move along actin/GFP–actin copolymers. The only obvious difference between the GFP–actin constructs that weakly or strongly impair cell behavior relies in the spacers inserted between the GFP and actin portions. These spacers comprise five amino acid residues in the former and two residues in the latter construct.

Myosin II (C)–GFP Does Not Polymerize

The tail of the conventional myosin II contains a region close to the C-terminus of the heavy chain that is important for bipolar filament formation. Accordingly, the linkage of GFP to the C-terminal end of the heavy chain blocks the assembly of *Dictyostelium* myosin II into filaments. In contrast, myosin II modified by GFP at its N terminus is not only capable of polymerizing but also of moving along actin filaments. The GFP–(N)myosin II is actually capable of rescuing the myosin II-null phenotype, providing evidence for its function *in vivo*.[19]

C-Terminally Tagged AIP1 Is Normally Targeted but Lacks Function

In the presence of cofilin, the actin-interacting protein 1 (AIP1) is an important regulator of actin filament organization. AIP1 fused to GFP at either side is

[19] S. L. Moores, J. H. Sabry, and J. A. Spudich, *Proc. Natl. Acad. Sci. U.S.A.* **93**, 443 (1996).

localized to F-actin-enriched structures like leading edges or phagocytic cups. However, only N-terminally tagged AIP1 rescues the wild-type phenotype in AIP1 null cells.[20] Thus, C-terminally tagged AIP1 is normally targeted but nonfunctional, indicating that a free C terminus is required for certain interactions.

Different Effects of GFP on Profilins I or II

Profilins I and II are small, ubiquitous regulators of actin polymerization. In mammalian cells, fusion proteins of profilin I or II carrying GFP at the N or C terminus are localized to areas of high actin dynamics, such as leading lamellae and ruffles.[21] *In vitro,* the affinity for actin is not decreased when profilins I or II are modified by GFP at the N or C terminus. However, GFP at the C terminus of profilin I abolishes the binding of this protein to polyproline or to the vasodilator-stimulated protein (VASP), whereas profilin II tagged at either side still binds to polyproline or VASP.[21a]

GFP–Fusion Proteins in Chemotactic Responses

Chemotaxis in *Dictyostelium* cells and neutrophils is based on signal transmission from cell-surface receptors to the actin system restructuring the cell cortex. In both cell types, the chemoattractant receptors are characterized by seven transmembrane domains and by their coupling to heterotrimeric G proteins. In *Dictyostelium,* an essential constituent of the signal transduction chain is CRAC (cytoplasmic regulator of adenylyl cyclase), which is recruited to the region of a cell exposed to the highest attractant concentration. A number of regulatory proteins of the actin system, including the Arp2/3 complex and coronin, are recruited to the site of stimulation, and are assembled together with polymerized actin into a leading edge.

Chemotactic Stimulation

For an efficient stimulation that makes rapid changes in the direction of a gradient possible, cells can be stimulated with a micropipette. This technique first applied to *Dictyostelium* cells[22] is also applicable to the stimulation of neutrophils with fMet-Leu-Phe[23] and of fibroblasts with epidermal growth factor.[24] It is crucial

[20] A. Konzok, I. Weber, E. Simmeth, U. Hacker, M. Maniak, and A. Müller-Taubenberger, *J. Cell Biol.* **146,** 453 (1999).

[21] A. Di Nardo, R. Gareus, D. Kwiatkowski, and W. Witke, *J. Cell Sci.* **113,** 3795 (2000).

[21a] N. Wittenmayer, M. Rothkegel, B. M. Jockusch, and K. Schlüter, *Eur. J. Biochem.* **267,** 5247 (2000).

[22] G. Gerisch, D. Hülser, D. Malchow, and U. Wick, *Phil. Trans. R. Soc. Lond. B.* **272,** 181 (1975).

[23] G. Gerisch and H. U. Keller, *J. Cell Sci.* **52,** 1 (1981).

[24] M. Bailly, J. Wyckoff, B. Bouzahzah, R. Hammerman, V. Sylvestre, M. Cammer, R. Pestell, and J. E. Segall, *Mol. Biol. Cell* **11,** 3873 (2000).

that the micropipette tip is narrowed to an inner diameter of 0.3–0.4 μm on a pipette puller. Most appropriate are borosilicate glass tubes of 1.0 mm outer and 0.5 mm inner diameter with a smaller tube inside, which allows filling with a 10^{-4} M cAMP solution through polyethylene tubing from behind. The solution is drawn into the tip of the micropipette by capillary forces, and air bubbles are easily removed by knocking against the glass tube.

Cells of *D. discoideum* need a starvation period of 5–6 hr before they are fully responsive to cAMP. For starvation, cells of the AX2 strain are harvested during exponential growth, washed, and shaken at a density of 1×10^7 cells/ml in 17 mM K/Na-phosphate buffer, pH 6.0, at 20–24°. Cells of other strains (AX3, AX4) may need pulsing with cAMP during the starvation period in order to reach the aggregation-competent stage. A concentrated cAMP solution is dropped into the shaken cell suspension every 5–7 min to obtain a final concentration in the cell suspension of 5×10^{-9} M at each pulse.

To examine cellular responses and redistribution of GFP-labeled proteins with oil immersion optics, an inverted microscope is used. A plastic ring (inner diameter 4 cm, height 4 mm) is sealed with silicon rubber on a 50×50-mm coverslip. The space within the ring is filled with a suspension of *D. discoideum* cells. The cells settled on the glass surface can be stimulated under these conditions by introducing the micropipette at a low angle. Using a micromanipulator the pipette should be moved with the tip kept several microns beyond the glass surface.

For confocal fluorescence recording it is important to realize that the leading edge is often detached and raised beyond the glass surface, so that it might be seen in phase-contrast images but not represented in confocal fluorescence images through the cell body. The tendency of tip rising can be lowered by coating the glass surface to increase adhesiveness. The coverslip is immersed for about 1 min in a solution of dimethyldichlorosilane in octamethylcyclotetrasiloxane (2%, w/v) and subsequently rinsed in ethanol.

GFP Constructs to Analyze Signal Responses

Red-shifted ($S^{65}T$) GFP excited with the 488-nm line of an argon-ion laser is appropriate to *Dictyostelium* cells, provided the light intensity is kept at a minimum necessary for high-resolution image series. Impairment of the cells immediately causes their rounding up. Because most of the relevant processes in *Dictyostelium* cells are fast, frame-to-frame intervals should be no longer than 10 sec.

Cyclic AMP receptors of the CAR1 type have been tagged at their cytoplasmic tail with GFP to demonstrate their uniform distribution.[25] Fluorescence resonance energy transfer (FRET) provides a measure for receptor-mediated activation of heterotrimeric G proteins. Enhanced yellow fluorescent protein (YFP) is fused to

[25] Z. Xiao, N. Zhang, D. B. Murphy, and P. N. Devreotes, *J. Cell Biol.* **139,** 365 (1997).

the N terminus of the Gβ subunit, and enhanced cyan fluorescent protein (CFP) is inserted into the first loop of the helical domain of the Gα_2 subunit (this GFP position has been guided by the crystal structure of heterotrimeric G proteins).[26] These fusion proteins are coexpressed to detect FRET from Gβ to Gα_2. The energy transfer declines if the cells are stimulated by cAMP, demonstrating a change of conformation and most likely dissociation of the G$\beta\gamma$ complex from Gα. Activation of the G protein level persists after adaptation of cells to the chemoattractant. These data indicate that the time course and spatial pattern of chemotactic responses are shaped by steps in the signaling pathway beyond or independent of the G protein cycle.

The redistribution of CRAC in response to chemoattractant represents a subsequent step in signal transduction. CRAC contains a pleckstrin homology (PH) domain, which recognizes changes in membrane lipid composition. Fusion of GFP to the C terminus of either full-length CRAC or its PH domain provides a tool to visualize local binding of the protein to the cytoplasmic face of the plasma membrane.[27] This binding occurs at sites where the cell is exposed to the highest chemoattractant concentration, and follows changing directions of the external gradient. Because the CRAC response is upstream of cytoskeletal changes, it can be observed in the presence of latrunculin A, an inhibitor of actin polymerization. The drug can be stored as a 1 mM solution in dimethyl sulfoxide (DMSO) at $-20°$, and is diluted to a final concentration of 10 μM in phosphate buffer. Cells round up under these conditions, so that the redistribution of CRAC can be examined without any interference caused by changes in the shape or position of motile cells.

Reorientation of cells in attractant gradients is primarily mediated by reorganization of the actin system, manifested in the assembly of polymerized actin in a complex with associated proteins. This assembly occurs within less than 10 sec upon stimulation, giving rise to a leading edge that directs the orientation of a cell toward the source of attractant. In cells with an intact actin system the site of response is more variable than the recruitment of CRAC in immobilized cells, which always occurs at the side of the cell pointing toward the source of attractant.[27] The chemotactic response in motile cells may even be initiated at a site opposite to the source of attractant, indicating that influences of cytoskeleton organization on the response are superimposed to the gradient of stimulation.[28]

As reporters for the chemoattractant-induced reorganization of the cytoskeleton, GFP–(N)Actin or GFP-tagged F-actin-binding proteins can be used (Table II). Because about half of the cellular actin exists as monomeric actin, considerable

[26] C. Janetopoulos, T. Jin, and P. Devreotes, *Science* **291**, 2408 (2001).

[27] C. A. Parent, B. J. Blacklock, W. M. Froehlich, D. B. Murphy, and P. N. Devreotes, *Cell* **95**, 81 (1998).

[28] R. Insall, A. Müller-Taubenberger, L. Machesky, J. Köhler, E. Simmeth, S. J. Atkinson, I. Weber, and G. Gerisch, *Cell Motil. Cytoskeleton* **50**, 115 (2001).

cytoplasmic background of GFP–actin is unavoidable. Brilliant images reflecting actin reorganization are obtained with the actin-binding proteins coronin[29] or AIP1[20] supplied with GFP either at their N- or C-terminal ends, or with an ABP120-GFP probe for filamentous actin.[14] Arp3, a constituent of the Arp2/3 complex, has been successfully tagged with GFP at its N terminus. The construct reflects the normal distribution of endogenous Arp3 as revealed by antibody staining.[28] Similar redistribution of p41-Arc, another constituent of the Arp2/3 complex, indicates that not only Arp3 but the entire complex is recruited to a nascent leading edge.

GFP–Fusion Proteins in Phagocytosis

The use of GFP for protein localization in phagocytes allows study of organelle dynamics in the endocytic pathway. *Dictyostelium* cells are professional phagocytes, which are easily transfected for the permanent expression of GFP–fusion proteins. Using GFP-tagged cytoskeletal proteins, the recruitment, enrichment, assembly, and dissociation of a specific protein can be observed by confocal microscopy during the uptake of a particle into the living cell. This technique can be complemented by quantitative fluorimetric measurements.

Improved Visualization of Uptake and Endosome Processing Using the Agar Overlay Technique

Cells gently flattened by this simple method[30] are optimally suited to observe endocytic uptake processes by conventional or confocal microscopy. The agar overlay technique has been applied to visualize association of the ER with phagocytic cups using calnexin–GFP[5] and to study endosome processing using GFP–(N)golvesin as a marker.[12] For combination with GFP, dextrans (molecular weight 10,000–70,000) labeled with TRITC or Alexa-Fluor 568 (Molecular Probes, Eugene, OR) can be used to study fluid-phase uptake, and similarly labeled yeast particles or *Escherichia coli B/r* to study phagocytosis. In *Dictyostelium* cells cultivated in liquid medium containing yeast extract and peptone, the endosomes are filled with fluorescent compounds. These can be removed by washing twice in a nonnutrient buffer and subsequent incubation in the buffer.

Preparation of Agarose Sheets. Agarose [e.g., Seakem ME, FMC, Rockland, ME, 2% (w/v)] is boiled in 17 mM K/Na-phosphate buffer, pH 6.0. On a cleaned slide, two 24 × 24-mm coverslips of 0.17 mm thickness are placed as spacers. About 0.5 ml hot agarose solution is pipetted between these spacers and pressed gently from above by the use of a slide until the agarose is solid. Excess agarose is cut off, and the sandwich is transferred to a petri dish containing phosphate buffer.

[29] G. Gerisch, R. Albrecht, C. Heizer, S. Hodgkinson, and M. Maniak, *Curr. Biol.* **5,** 1280 (1995).
[30] Y. Fukui, S. Yumura, T. K.-Yumura, and H. Mori, *Methods Enzymol.* **134,** 573 (1986).

The upper slide is carefully removed, so that the agarose sheet can be detached and cut with a razor blade into pieces of 5 × 5 mm. These pieces can be stored in cold phosphate buffer.

Microscopic Observation of Endocytosis. A suspension of D. *discoideum* cells is pipetted onto a cleaned coverslip. After settling the cells can by incubated with fluorescent dextrans to visualize macropinocytosis, or with yeast or bacterial cells to follow phagocytosis. Subsequently, a piece of agarose is placed onto the cells. Without moving the agarose, the buffer is carefully sucked off with a pipette and finally with a filter paper. The degree of suction is critical as too strongly compressed cells are impaired in phagocytosis. To observe macropinocytosis continually, the agarose piece is incubated before use for a few hours in a fluorescent dextran solution prepared with 17 mM K/Na-phosphate buffer, pH 6.0.

Quantitative Phagocytosis Assay

Fluorescent yeast particles are fed to *Dictyostelium* cells until, after various periods of incubation, a dye is added that quenches the fluorescence of the attached particles, leaving only the completely engulfed particles fluorescent. This fluorescence quenching (FQ) technique allows quantitation of ingested particles by fluorimetry.[31] Attached particles, i.e., quenched ones, can be counted on individual phagocytes and distinguished from the ingested, fluorescing particles by microscopic inspection. The method is easily adaptable to a variety of phagocytes.

[31] J. Hed, *Methods Enzymol.* **132**, 198 (1986).
[32] M. Maniak, R. Rauchenberger, R. Albrecht, J. Murphy, and G. Gerisch, *Cell* **83**, 915 (1995).
[33] I. Weber, G. Gerisch, C. Heizer, J. Murphy, K. Badelt, A. Stock, J.-M. Schwartz, and J. Faix, *EMBO J.* **18**, 586 (1999).
[34] M. Dumontier, P. Höcht, U. Mintert, and J. Faix, *J. Cell Sci.* **113**, 2253 (2000).
[35] R. Neujahr, R. Albrecht, J. Köhler, M. Matzner, J.-M. Schwartz, M. Westphal, and G. Gerisch, *J. Cell Sci.* **111**, 1227 (1998).
[36] C. J. Merrifield, U. Rescher, W. Almers, J. Proust, V. Gerke, A. S. Sechi, and S. E. Moss, *Curr. Biol.* **11**, 1136 (2001).
[37] C. K. Damer and T. J. O'Halloran, *Mol. Biol. Cell* **11**, 2151 (2000).
[38] M. R. Mohrs, K. P. Janssen, T. Kreis, A. A. Noegel, and M. Schleicher, *Eur. J. Cell Biol.* **79**, 350 (2000).
[39] G. C. Ochoa, V. I. Slepnev, L. Neff, N. Ringstad, K. Takei, L. Daniell, W. Kim, H. Cao, M. McNiven, R. Baron, and P. De Camilli, *J. Cell Biol.* **150**, 377 (2000).
[40] H. Cao, F. Garcia, and M. A. McNiven, *Mol. Biol. Cell* **9**, 2595 (1998).
[41] R. D. Phair and T. Misteli, *Nature* **404**, 604 (2000).
[42] K. J. M. Zaal, C. L. Smith, R. S. Polishchuk, N. Altan, N. B. Cole, J. Ellenberg, K. Hirschberg, J. F. Presley, T. H. Roberts, E. Siggia, R. D. Phair, and J. Lippincott-Schwartz, *Cell* **99**, 589 (1999).
[43] S. C. Ushinsky, H. Bussey, A. A. Ahmed, Y. Wang, J. Friesen, B. A. Williams, and R. K. Storms, *Yeast* **13**, 151 (1997).
[44] T. Kanda, K. F. Sullivan, and G. M. Wahl, *Curr. Biol.* **8**, 377 (1998).
[45] N. Daigle, J. Beaudouin, L. Hartnell, G. Imreh, E. Hallberg, J. Lippincott-Schwartz, and J. Ellenberg, *J. Cell Biol.* **154**, 71 (2001).

Silanization of Flasks. Erlenmeyer flasks (25 ml) are coated with repel solution (Repel silane ES, Pharmacia Biotech, Piscataway, NJ) in a hood for 20 min, dried for 10 min, washed once with absolute ethanol and once with 70% (v/v) ethanol, rinsed with distilled water, and dried.

Trypan Blue Solution. Trypan blue (2 mg/ml) is dissolved in 20 mM citrate buffer, pH 4.4, with 0.15 M NaCl. The solution is cleared through filter paper and subsequently sterilized through a 0.45-μm pore-size filter. Before the assay, 100 μl of trypan blue is pipetted to each of the required number of Eppendorf tubes.

Preparation of Fluorescent Yeast Particles. Yeast cells (5 g) (Sigma YSC-2) are suspended in 50 ml phosphate-buffered saline (PBS), pH 7.4, and subsequently boiled in a water bath for 30 min. The heat-killed yeast particles are washed five times in PBS, twice in phosphate buffer, pH 6.0, resuspended at 1×10^9 yeast cells per ml, and stored at $-20°$.

To label with TRITC, 2×10^{10} pelleted yeast particles are resuspended in 20 ml of 50 mM sodium phosphate buffer, pH 9.2, containing 2 mg TRITC (Sigma T3163, Isomer R) and incubated on a rotary shaker for 30 min at $37°$. The yeast particles are washed twice by centrifugation at $120g$ for 10 min in the phosphate buffer, pH 9.2, and four times in 17 mM K/Na-phosphate buffer, pH 6.0. The yeast particles (1×10^9/ml) are resuspended in this buffer and stored at $-20°$. Prior to

[46] J. Ellenberg, E. D. Siggia, J. E. Moreira, C. L. Smith, J. F. Presley, H. J. Worman, and J. Lippincott-Schwartz, *J. Cell Biol.* **138**, 1193 (1997).

[47] A. Rupper, B. Grove, and J. Cardelli, *J. Cell Sci.* **114**, 2449 (2001).

[48] E. Harris, K. Yoshida, J. Cardelli, and J. Bush, *J. Cell Sci.* **114**, 3035 (2001).

[49] D. J. Stephens, N. Lin-Marq, A. Pagano, R. Pepperkok, and J.-P. Paccaud, *J. Cell Sci.* **113**, 2177 (2000).

[50] S. van Es, D. Wessels, D. R. Soll, J. Borleis, and P. Devreotes, *J. Cell Biol.* **152**, 621 (2001).

[51] R. Rauchenberger, U. Hacker, J. Murphy, J. Niewöhner, and M. Maniak, *Curr. Biol.* **7**, 215 (1997).

[52] K. Hirschberg, C. M. Miller, J. Ellenberg, J. F. Presley, E. D. Siggia, R. D. Phair, and J. Lippincott-Schwartz, *J. Cell Biol.* **143**, 1485 (1998).

[53] V. V. Verkhusha, S. Tsukita, and H. Oda, *FEBS Lett.* **445**, 395 (1999).

[54] A. Jacinto, W. Wood, T. Balayo, M. Turmaine, A. Martinez-Arias, and P. Martin, *Curr. Biol.* **10**, 1420 (2000).

[55] J. R. Robbins, A. I. Barth, H. Marquis, E. L. de Hostos, W. J. Nelson, and J. A. Theriot, *J. Cell Biol.* **146**, 1333 (1999).

[56] L. Hodgson, W. Qiu, C. Dong, and A. J. Henderson, *Biotechnol. Prog.* **16**, 1106 (2000).

[57] M. L. C. Albuquerque and A. S. Flozak, *Exp. Cell Res.* **270**, 223 (2001).

[58] C. M. Laukaitis, D. J. Webb, K. Donais, and A. F. Horwitz, *J. Cell Biol.* **153**, 1427 (2001).

[59] D. A. Schafer, M. D. Welch, L. M. Machesky, P. C. Bridgman, S. M. Meyer, and J. A. Cooper, *J. Cell Biol.* **143**, 1919 (1998).

[60] C. Danninger and M. Gimona, *J. Cell Sci.* **113**, 3725 (2000).

[61] A. Merla and D. I. Johnson, *Eur. J. Cell Biol.* **79**, 469 (2000).

[62] M. Eda, S. Yonemura, T. Kato, N. Watanabe, T. Ishizaki, P. Madaule, and S. Narumiya, *J. Cell Sci.* **114**, 3273 (2001).

[63] H. Aizawa, Y. Fukui, and I. Yahara, *J. Cell Sci.* **110**, 2333 (1997).

[64] M. Mishima and E. Nishida, *J. Cell Sci.* **112**, 2833 (1999).

the assay, 1×10^9 particles are thawed, sonicated for 5–7 min, filtered through gauze of 8-μm mesh size, and put on a shaker. The absence of aggregated particles should be ascertained by microscopic inspection.

Fluorimetric Assay. Of a suspension containing 1.2×10^8 yeast particles, 120 μl is added to 10 ml of a suspension of 2×10^7 *Dictyostelium* cells in 17 mM K/Na-phosphate buffer, pH 6.0 (ratio 6:1 of yeast to *Dictyostelium* cells). The

[65] M. A. McNiven, L. Kim, E. W. Krueger, J. D. Orth, H. Cao, and T. W. Wong, *J. Cell Biol.* 151, 187 (2000).

[66] M. Kaksonen, H. B. Peng, and H. Rauvala, *J. Cell Sci.* 113, 4421 (2000).

[67] I. Weber, R. Neujahr, A. P. Du, J. Köhler, J. Faix, and G. Gerisch, *Curr. Biol.* 10, 501 (2000).

[68] R. Gräf, C. Daunderer, and M. Schliwa, *J. Cell Sci.* 113, 1747 (2000).

[69] C. Hofmann, I. M. Cheeseman, B. L. Goode, K. L. McDonald, G. Barnes, and D. G. Drubin, *J. Cell Biol.* 143, 1029 (1998).

[70] J. Prassler, A. Murr, S. Stocker, J. Faix, J. Murphy, and G. Marriott, *Mol. Biol. Cell* 9, 545 (1998).

[71] M. Kotaka, Y. M. Lau, K. K. Cheung, S. M. Y. Lee, H. Y. Li, W. Y. Chan, K. P. Fung, C. Y. Lee, M. M. Y. Waye, and S. K. W. Tsui, *J. Cell. Biochem.* 83, 463 (2001).

[72] C. S. Cohan, E. A. Welnhofer, L. Zhao, F. Matsumura, and S. Yamashiro, *Cell Motil. Cytoskeleton* 48, 109 (2001).

[73] H. Y. Li, M. Kotaka, S. Kostin, S. M. Y. Lee, L. D. S. Kok, K. K. Chan, S. K. W. Tsui, J. Schaper, R. Zimmermann, C. Y. Lee, K. P. Fung, and M. M. Y. Waye, *Cell Motil. Cytoskeleton* 48, 11 (2001).

[74] H. Y. Li, E. K. O. Ng, S. M. Y. Lee, M. Kotaka, S. K. W. Tsui, C. Y. Lee, K. P. Fung, and M. M. Y. Waye, *J. Cell. Biochem.* 80, 293 (2001).

[75] R. K. Miller, D. Matheos, and M. D. Rose, *J. Cell Biol.* 144, 963 (1999).

[76] J. C. Bulinski, D. Gruber, K. Faire, P. Prasad, and W. Chang, *Cell Struct. Funct.* 24, 313 (1999).

[77] P. A. Steimle, S. Yumura, G. P. Coté, A. G. Medley, M. V. Polyakov, B. Leppert, and T. T. Egelhoff, *Curr. Biol.* 11, 708 (2001).

[78] K. A. Edwards, M. Demsky, R. A. Montague, N. Weymouth, and D. P. Kiehart, *Dev. Biol.* 191, 103 (1997).

[79] S. L. Moores, J. H. Sabry, and J. A. Spudich, *Proc. Natl. Acad. Sci. U.S.A.* 93, 443 (1996).

[80] V. Tsakraklides, K. Krogh, L. Wang, J. C. S. Bizario, R. E. Larson, E. M. Espreafico, and J. S. Wolenski, *J. Cell Sci.* 112, 2853 (1999).

[81] F. Buss, S. D. Arden, M. Lindsay, J. P. Luzio, and J. Kendrick-Jones, *EMBO J.* 20, 3676 (2001).

[82] R. I. Tuxworth, I. Weber, D. Wessels, G. C. Addicks, D. R. Soll, G. Gerisch, and M. A. Titus, *Curr. Biol.* 11, 318 (2001).

[83] M. Farkasovsky and H. Küntzel, *J. Cell Biol.* 152, 251 (2001).

[84] T. M. Olski, A. A. Noegel, and E. Korenbaum, *J. Cell Sci.* 114, 525 (2001).

[85] C. M. Laukaitis, D. J. Webb, K. Donais, and A. F. Horwitz, *J. Cell Biol.* 153, 1427 (2001).

[86] S. J. Palmieri, T. Nebl, R. K. Pope, D. J. Seastone, E. Lee, E. H. Hinchcliffe, G. Sluder, D. Knecht, J. Cardelli, and E. J. Luna, *Cell Motil. Cytoskeleton* 46, 285 (2000).

[87] D. A. Larochelle, K. K. Vithalani, and A. De Lozanne, *Mol. Biol. Cell* 8, 935 (1997).

[88] F. Rivero, R. Albrecht, H. Dislich, E. Bracco, L. Graciotti, S. Bozzaro, and A. A. Noegel, *Mol. Biol. Cell* 10, 1205 (1999).

[89] A. F. Straight, W. F. Marshall, J. W. Sedat, and A. W. Murray, *Science* 277, 574 (1997).

[90] M. Kimble, C. Kuzmiak, K. N. McGovern, and E. L. de Hostos, *Cell Motil. Cytoskeleton* 47, 48 (2000).

[91] M. P. Koonce, J. Köhler, R. Neujahr, J. M. Schwartz, I. Tikhonenko, and G. Gerisch, *EMBO J.* 18, 6786 (1999).

[92] N. M. Rusan, C. J. Fagerstrom, A. M. C. Yvon, and P. Wadsworth, *Mol. Biol. Cell* 12, 971 (2001).

cells are shaken in 25-ml silanized Erlenmeyer flasks at 150 rpm. From this sus-pension, 1-ml aliquots are withdrawn at 5-min intervals, pipetted into 100 μl of trypan blue, incubated on a rotary shaker for 3 min, and centrifuged at 2500 rpm in an Eppendorf centrifuge. The pellet is resuspended in 1 ml phosphate buffer, vortexed, and immediately measured in a fluorimeter.

[93] B. A. Danowski, A. Khodjakov, and P. Wadsworth, *Cell Motil. Cytoskeleton* **50**, 59 (2001).
[94] B. L. Goode, D. G. Drubin, and P. Lappalainen, *J. Cell Biol.* **142**, 723 (1998).
[95] K. Rottner, M. Krause, M. Gimona, J. V. Small, and J. Wehland, *Mol. Biol. Cell* **12**, 3103 (2001).
[96] G. Vaduva, N. C. Martin, and A. K. Hopper, *J. Cell Biol.* **139**, 1821 (1997).

[16] Dynamic Imaging of Cell–Substrate Contacts

By AMIT K. BHATT and ANNA HUTTENLOCHER

Introduction

The coupling of fluorophores, such as green fluorescent protein (GFP), to focal adhesion proteins represents a powerful tool to study adhesive complex composi-tion and dynamics in live cells. On binding to extracellular matrix (ECM), integrins cluster in the membrane and recruit cytoskeletal and signaling proteins to form focal complexes or focal adhesions.[1,2] Focal complexes are early adhesive com-plexes that form at the cell periphery. Focal complexes contain small collections of integrins and associated cytoskeletal proteins. Focal adhesions, in contrast, contain larger aggregates of integrin and associated cytoskeletal proteins and are localized to the ends of stress fibers both at the center of the cell and the cell periphery. Ad-hesive contact sites may contain many different components including integrins; cytoskeletal proteins such as vinculin, talin, and α-actinin; and signal transduction proteins including focal adhesion kinase and protein kinase C (Fig. 1). Basic ques-tions regarding the sequence with which components enter the complex during complex formation and the mechanisms that regulate adhesive complex disas-sembly remain unanswered. Advances in microscopy methodologies and the use of fluorophore-tagged proteins have begun to provide insight into the extensive dynamics and molecular diversity of adhesive complex sites in live cells.

The fluorescence tagging of cell surface and intracellular proteins is a pow-erful method to visualize the dynamic temporal and spatial distribution of focal

[1] K. M. Yamada and S. Miyamoto, *Curr. Opin. Cell Biol.* **7**, 681 (1995).
[2] S. M. Schoenwaelder and K. Burridge, *Curr. Opin. Cell Biol.* **11**, 274 (1999).

0076-6879/03 $35.00

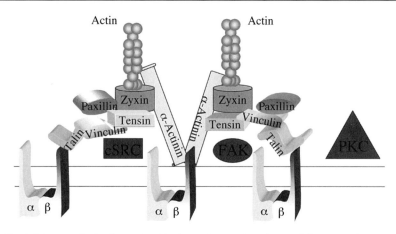

FIG. 1. Diagram of a focal complex: Focal complex consists of many different proteins, some of which are represented here.

adhesion components in live cells. Investigations have highlighted the importance of these advances in providing basic mechanistic insight into the regulation of adhesive complex sites *in vivo*. Studies by Kaverina *et al.*, for example, used GFP-tagged tubulin and GFP–zyxin to demonstrate the temporal and spatial relationship between microtubule targeting of zyxin-containing focal adhesions and their disassembly.[3,4] Critical to these studies was the ability to show a temporal and spatial relationship between microtubule targeting and focal adhesion disassembly in live cells, a relationship that would not have been elucidated by traditional immunofluorescence techniques, which require fixation for visualization of intracellular proteins. Investigations by Zamir *et al.* have demonstrated the existence of structurally and functionally distinct types of substrate contact sites in live cells. These studies describe the segregation of two different types of matrix adhesions as cells spread, including central fibrillar adhesions that contain the $\alpha_5\beta_1$-integrin and tensin, and peripheral focal contacts that contain $\alpha_v\beta_3$-integrin and paxillin.[5] In our studies we have employed dual imaging of GFP–α-actinin and red fluorescent protein (RFP)–zyxin to examine the temporal and spatial dynamics of these two focal adhesion components in live cells. We find that focal adhesions are dynamic structures that often contain both zyxin and α-actinin (Fig. 2A). In addition, we find that on colocalization of zyxin and α-actinin in focal adhesions the adhesive complex can undergo different fates including stabilization, disassembly,

[3] I. Kaverina, K. Rottner, and J. V. Small, *J. Cell Biol.* **142**, 181 (1998).
[4] I. Kaverina, O. Krylyshkina, and J. V. Small, *J. Cell Biol.* **146**, 1033 (1999).
[5] E. Zamir, M. Katz, Y. Posen, N. Erez, K. M. Yamada, B. Z. Katz, S. Lin, D. C. Lin, A. Bershadsky, Z. Kam, and B. Geiger, *Nat. Cell Biol.* **2**, 191 (2000).

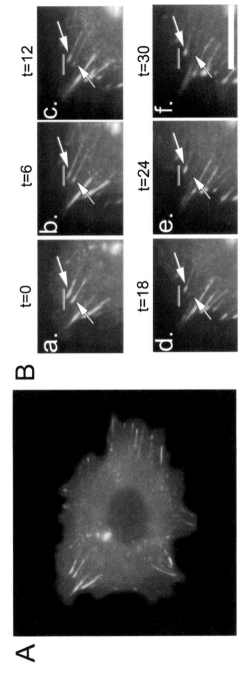

FIG. 2. Live fluorescent image of a cell cotransfected with zyxin–RFP and α-actinin–GFP. Both zyxin and α-actinin localize to focal complexes and actin filaments (A). Focal contacts that contain both α-actinin and zyxin can undergo at least two fates: translocation (upper arrow) and disassembly (lower arrow) (B). a–f show time lapse sequences at 6 min intervals. Bar, 10 microns.

or translocation. In this article we provide a practical discussion of the general methodology and principles used to study cell–substrate contact dynamics in live cells. Reviews have summarized the tools available for image quantification and analysis and are not the focus of the current discussion.[6–8]

Fluorescent Tags

The ability to attach fluorescent tags to proteins has revolutionized the visualization of proteins in live cells. Delivery of fluorescently tagged proteins into cells represents a major obstacle to the visualization of proteins in live cells. Early approaches used to visualize the dynamics of focal adhesion proteins in live cells involved the microinjection of fluorescently labeled purified protein. Unfortunately, this approach was time consuming and required protein purification and chemical labeling that may limit the activity of the protein. Furthermore, to visualize the intracellular distribution of the tagged protein in live cells required microinjection, which is feasible but limits its widespread application. An alternative approach used to examine the dynamics of cell surface focal adhesion proteins has included fluorescently conjugated antibodies against cell surface focal adhesion components. For example, integrin fate and dynamics have been examined in fibroblasts and leukocytes using fluorescently conjugated antiintegrin antibodies.[9–11] Limitations to these approaches include the potential effect that antibody binding may have on the normal function or distribution of the cell surface protein. A new, alternative approach to potentially visualize proteins in live cells is to use human immunodeficiency virus (HIV) Tat fusion proteins, which achieve high levels of transient expression without microinjection and may be conjugated to fluorophores to allow observation of intracellular distribution in live cells.[12,13]

Green fluorescent protein (GFP) isolated from the jellyfish *Aequorea victoria,* however, has become the preferred method for the visualization of proteins in live cells. Many GFP fusion proteins to focal adhesion components have been developed and characterized (Table I). The advantage of GFP–fusion proteins includes the ease with which the intracellular protein distribution may be visualized in live cells and the ability to use GFP–fusion proteins to observe the dynamics of proteins in live cells by time-lapse video microscopy. Further modification of GFP

[6] Z. Kam, E. Zamir, and B. Geiger, *Trends Cell Biol.* **11,** 329 (2001).

[7] J. White and E. Stelzer, *Trends Cell Biol.* **9,** 61 (1999).

[8] M. Whitaker, *Bioessays* **22,** 180 (2000).

[9] C. M. Regen and A. F. Horwitz, *J. Cell Biol.* **119,** 1347 (1992).

[10] S. P. Paleck, L. E. Schmidt, D. A. Lauffenburger, and A. F. Horwitz, *J. Cell Sci.* **109,** 941 (1996).

[11] J. D. Loike, J. el Khoury, L. Cao, C. P. Richards, H. Rascoff, J. T. Mandeville, F. R. Maxfield, and S. C. Silverstein, *J. Exp. Med.* **181,** 1763 (1995).

[12] M. Becker-Hapak, S. S. McAllister, and S. F. Dowdy, *Methods* **24,** 247 (2001).

[13] S. R. Schwarze, A. Ho, A. Vocero-Akbani, and S. F. Dowdy, *Science* **285,** 1569 (1999).

TABLE I

FUSION PROTEINS INVOLVED IN CELL MIGRATION

Fusion construct	Localization	Ref.
Paxillin	Adhesion complexes	5,14,15
α-Actinin	Actin stress fibers and focal adhesions	16
α_5-Integrin	Adhesion complexes	17
α_{11B}-integrin	Adhesion complexes	18
β_3-integrin	Adhesion complexes	18
ILK-3	Adhesion complexes	19
Tensin	Fibrillar adhesions	5,15
Parvin	Actin stress fibers and adhesion complexes	20
Zyxin	Focal complexes and actin stress fibers	21
Scar1/Wave1	Tips of protruding lamellipodia	22
Calponin	Central actin stress fibers	23
Arp2/3	Tips of actin stress fibers	24
Actin	Actin	25
Rack1	Early adhesion complexes	
Vinculin	Focal complexes	26
Tubulin	Microtubules	27
FRNK	Adhesion complex	28
FAK	Adhesion complex	28
E-Cadherin	Cell–cell adhesions	29
β_3-Endonexin S65T	Cytoplasmic/nuclear localization	30

has created variants of the protein that are less toxic (vitality GFP), brighter, and fluoresce at different wavelengths [GFP, cyan fluorescent protein (CFP), yellow fluorescent protein (YFP), and blue fluorescent protein (BFP)]. Vitality GFP has been useful in the creation of stable cell lines whereas the wavelength-shifted proteins have allowed dual imaging of proteins.

[14] R. Salgia, J. L. Li, D. S. Ewaniuk, Y. B. Wang, M. Sattler, W. C. Chen, W. Richards, E. Pisick, G. I. Shapiro, B. J. Rollins, L. B. Chen, J. D. Griffin, and D. J. Sugarbaker, *Oncogene* **18,** 67 (1999).

[15] E. Zamir, B. Z. Katz, S. Aota, K. M. Yamada, B. Geiger, and Z. Kam, *J. Cell Sci.* **112,** 1655 (1999).

[16] M. Edlund, M. A. Lotano, and C. A. Otey, *Cell Motil. Cytoskeleton* **48,** 190 (2001).

[17] C. M. Laukaitis, D. J. Webb, K. Donais, and A. F. Horwitz, *J. Cell Biol.* **153,** 1427 (2001).

[18] S. Plancon, M. C. Morel-Kopp, E. Schaffner-Reckinger, P. Chen, and N. Kieffer, *Biochem. J.* **357,** 529 (2001).

[19] S. N. Nikolopoulos and C. E. Turner, *J. Biol. Chem.* **5,** 5 (2001).

[20] T. M. Olski, A. A. Noegel, and E. Korenbaum, *J. Cell Sci.* **114,** 525 (2001).

[21] M. Reinhard, J. Zumbrunn, D. Jaquemar, M. Kuhn, U. Walter, and B. Trueb, *J. Biol. Chem.* **274,** 13410 (1999).

[22] P. Hahne, A. Sechi, S. Benesch, and J. V. Small, *FEBS Lett.* **492,** 215 (2001).

[23] C. Danninger and M. Gimona, *J. Cell Sci.* **113,** 3725 (2000).

[24] D. A. Schafer, M. D. Welch, L. M. Machesky, P. C. Bridgman, S. M. Meyer, and J. A. Cooper, *J. Cell Biol.* **143,** 1919 (1998).

Surprisingly, despite its large size (240 amino acids), fusing GFP to focal adhesion proteins, including the integrin cytoplasmic domain, does not perturb protein function. However, creating a functional GFP fusion protein can be challenging. Some of the factors to consider are amino acid composition of the linker sequence, length of linker sequence, and amino- vs. carboxy-terminal fusions. Some fusion proteins are only effective when fused to either the N or C terminus. For example, in muscle, α-actinin is functional only when fused to GFP on its C terminus. The length of the linker sequence and amino acids used may also be critical. In general, alanine is commonly used between GFP and the protein. If there is high background fluorescence and poor localization of a GFP–fusion protein increasing the linker length or alternatively the type of fusion protein (N vs. C) may reduce background fluorescence and promote localization. It is important to perform studies to determine that the focal adhesion–fusion protein is functional. At a minimum, it is important to determine that the fusion protein colocalizes with endogenous protein in fixed cells (methods described below), and expression of the GFP–fusion protein does not perturb the function of the endogenous protein. The ultimate proof of function is to show that the fusion protein can rescue function in a cell line that does not express endogenous protein.

Methods

Selection of Cell Type

To study focal adhesions it is necessary to select a cell type that forms focal adhesions that are easily visualized in live cells. We describe methods below that use Chinese hamster ovary (CHO)-K1 cells from the American Type Culture Collection (ATCC, Manassus, VA) to visualize focal adhesions. On a fibronectin substrate CHO-K1 cells generally form focal complexes within 1 hr and focal adhesions within 6 hr after adhesion. CHO-K1 cells adhere to fibronectin via the $\alpha_5\beta_1$ integrin and do not express endogenous $\alpha_v\beta_3$ integrin and, thus, represent a good system to study $\alpha_5\beta_1$ integrin-mediated substrate contact sites. Furthermore,

[25] M. Westphal, A. Jungbluth, M. Heidecker, B. Muhlbauer, C. Heizer, J. M. Schwartz, G. Marriott, and G. Gerisch, *Curr. Biol.* **7,** 176 (1997).
[26] N. Q. Balaban, U. S. Schwarz, D. Riveline, P. Goichberg, G. Tzur, I. Sabanay, D. Mahalu, S. Safran, A. Bershadsky, L. Addadi, and B. Geiger, *Nat. Cell Biol.* **3,** 466 (2001).
[27] M. Ueda, R. Graf, H. K. MacWilliams, M. Schliwa, and U. Euteneuer, *Proc. Natl. Acad. Sci. U.S.A.* **94,** 9674 (1997).
[28] D. Ilic, E. A. Almeida, D. D. Schlaepfer, P. Dazin, S. Aizawa, and C. H. Damsky, *J. Cell Biol.* **143,** 547 (1998).
[29] C. L. Adams, Y. T. Chen, S. J. Smith, and W. J. Nelson, *J. Cell Biol.* **142,** 1105 (1998).
[30] H. Kashiwagi, M. A. Schwartz, M. Eigenthaler, K. A. Davis, M. H. Ginsberg, and S. J. Shattil, *J. Cell Biol.* **137,** 1433 (1997).

CHO-K1 cells are easy to transfect with high efficiency, making it an ideal cell type to use in transient transfection experiments.

Expression of GFP–Fusion Proteins

The level of expression of the GFP–fusion protein is critical for the successful imaging of focal adhesions in live cells. Low expression of the GFP–fusion protein necessitates increased exposure times for visualization making it difficult to capture an informative image. However, overexpression of a GFP–fusion protein may also be problematic by increasing cytoplasmic levels of the protein and obscuring its localization. Consequently, it is important to control the expression level of the fusion protein to maximize image quality. There are many ways to introduce GFP–fusion proteins into cells, including lipid-based transfection methods and microinjection. Protein expression levels may be varied by changing the concentration of DNA used for the transfection or the duration of transfection reaction. The GFP–vinculin fusion protein demonstrates an example of the critical role that expression level plays in imaging of focal adhesion components. High expression levels of GFP–vinculin will obscure its localization to focal adhesions. For GFP–vinculin, a brief transfection using low concentrations of DNA is required. Specifically, for optimum expression of GFP–vinculin we recommend treating CHO-K1 cells with 1 μg DNA in LipofectAMINE (Invitrogen Carlsbad, CA) for an incubation of 2–3 hr. An alternative approach to optimize expression levels is to establish a stable cell line that expresses the GFP–fusion protein. In all experiments it is critical to include a control cell transfected with GFP alone, as GFP can localize to membrane protrusions or adhesion-like complexes at the cell periphery in some cell types. Methods used to transiently express GFP–fusion proteins are discussed below.

Transient Transfection Protocol

1. Plate CHO-K1 cells 24 hr prior to transfection at 3–4 $\times$ 10^5 cells on a 6-cm tissue culture plate in Dulbecco's modified Eagle's medium (DMEM) (Cellgro, Mediatech Inc., Herden, VA) supplemented with nonessential amino acids, 10% fetal bovine serum, 100 U/ml penicillin, and 100 μg/ml streptomycin so that cells are $\sim$70–80% confluent at the time of transfection. For a typical transfection use a mixture of 4 μg of DNA and 15 μl of LipofectAMINE. We prepare transfection reagent as described by the manufacturer. The concentration of DNA in transfection mixture (generally we start with 4 μg) may either be increased or decreased to control expression level. Allow the mixture to incubate for 45 min prior to treating cells.

2. Wash plates with phosphate-buffered saline (PBS) $\times$1 and add transfection reagent. Add 2 ml of serum-free media to the plate and incubate cells at 37° in 10% (v/v) CO$_2$ for 1–2 hr. Add 1 ml of 3$\times$ DMEM and allow the plate to incubate for

8–10 hr. If overexpression of the fluorescence protein is a problem, decreasing the total incubation time can help to optimize expression levels. Remove transfection reagent at the desired time by washing cells with at least 3 volumes of medium.

3. Divide cells 12 hr after the transfection reagent is added to the plate and allow them to recover for at least 24 hr or, if selecting for stable cell lines, replace media with selection medium. Maximal expression of the protein will generally occur 48–72 hr posttransfection.

Transient transfection does not allow for precise control of expression levels; however, in efficient transfections 60–90% of cells will express the fusion protein. In general, a spectrum of expression levels will be observed after transient transfection of the fusion protein and cells that express an optimum level for imaging may be used. With dual imaging the ratio of transfected DNA may have to be optimized to allow for imaging of the two fusion proteins simultaneously. An important consideration with transient transfections is that some lipid agents, LipofectAMINE in this case, may increase background fluorescence, making imaging of the GFP–fusion protein difficult. If this is a problem alternative approaches may be used including establishing stable cell lines or using an alternative method to express the cDNA including other lipid agents, such as Fugene (Roche Molecular Biochemicals, Berkeley, CA) or microinjection.

Microinjection. Microinjection is an alternative approach to introduce cDNAs or tagged proteins into cells as discussed above. We describe the protocol used in our laboratory for nuclear injections of cDNAs in CHO-K1 cells.

1. Calibrate stage as described in the section on stage calibration.
2. Plate cells on glass bottom plate at a density of 1×10^5 cells per dish.
3. Allow cells to adhere to the dish for 1 hr prior to microinjection. We use a DNA concentration of 50 μg/ml diluted into a solution of 5 mM potassium glutamate and 150 mM KC1. DNA concentration can be increased or decreased to adjust the expression level of the protein.
4. For microinjection use a 40× phase objective. For CHO cells nuclear injections may be done using an Eppendorf femto-jet (Eppendorf, Hamburg, Germany) at 95 Pa for 0.3 sec with an Eppendorf femto-tip needle. Do not layer on mineral oil until after microinjection is complete. Expression of the fusion protein will generally be observed 2–4 hr after microinjection, but this may vary depending on the conditions. Procedures for fluorescence imaging of GFP–fusion proteins may then be followed.

To optimize conditions for transfection or microinjection of a new fusion protein, initial characterization may be performed in fixed cells. This will ensure that the GFP–fusion protein localizes normally with endogenous protein. Below, we describe methods for preparation of fixed cells for imaging.

Imaging of GFP–Fusion Proteins in Fixed Cells

Coating Coverslips

1. Acid wash 22-mm coverslips by placing them in a solution of 20% (v/v) H_2SO_4 in a large flat bottom beaker. Periodically stir the beaker for the first 2 hr and then leave coverslips in acid overnight. Pour off acid and quench coverslips with a solution of 0.1 M NaOH for 5 min, then rinse twice with doubly distilled H_2O. Remove coverslips and place them individually between two sheets of Whatman (Clifton, NJ) paper to dry. When dry, coverslips can be stored between two sheets of Whatman paper in a petri dish.

2. Prior to use, sterilize coverslips by rinsing them once with ethanol and placing them in a sterile 24-well dish filled with 500 μl of PBS per well. Rinse once more with PBS to ensure ethanol, which can interfere with coating, has been removed.

3. Add desired concentration of ECM protein diluted in PBS (+Ca +Mg) (500 μl per well) and incubate at 37° for 60 min. Wash coverslip twice with PBS and block with 500 μl of 2% bovine serum albumin (BSA) in PBS for 30 min at 37°.

4. Wash coverslips two times with PBS and store at 4° until ready to use. Coated coverslips can be stored for a maximum of 1 week prior to use.

Plating and Staining Cells

1. Wash cells with a volume of PBS that is equal to the volume of medium on the plate. Aspirate PBS and add 0.02% EDTA in PBS. The volume of 0.02% EDTA varies depending on plate size. Our laboratory typically uses 500 μl per 6-cm plate and 1 ml per 10-cm plate. Place cells at 37° for 5 min. In 0.02% EDTA CHO-K1 cells will round up but may not completely detach from the plate. Check plate under a microscope to ensure that cells have detached.

2. After cells have detached, add an equal volume of medium, remove cells to a 15-ml conical tube, and spin at 1500g for 2 min. Wash cells twice by resuspending them in 2 ml of a serum-free medium such as CCM1 (Hyclone, Logan, UT). Plate 2 × 10^4 cells per coverslip.

3. Fix cells between 4 and 12 hr after plating depending on the phenotype being viewed. For optimum viewing of focal adhesions we recommend plating the cells for a minimum of 6 hr prior to fixation. There are a number of different methods for fixing cells. Generally, we use a formaldehyde fixation method. Wash cells gently with 500 μl of PBS (+Ca +Mg) and add 500 μl of 3% formaldehyde in PBS for 15 min.

4. Quench formaldehyde with 1 ml of a freshly made solution of 0.15 M glycine in PBS for 10 min and rinse twice with 500 μl PBS.

5. Permeabilize cells with 500 μl 0.2% Triton X-100 in PBS for 10 min and then rinse with 500 μl PBS twice and block coverslip with 500 μl of 5% goat serum

in PBS (filter sterilized through 0.22-μm filter) for 30 min at room temperature or overnight at 4°.

6. Coverslips are then moved and placed cell side up on a piece of Parafilm in a humidified chamber for staining. Dilute primary antibody to the appropriate concentration in 5% goat serum and incubate for 60 min. We use approximately 100 μl of diluted antibody per coverslip.

7. Rinse three times with PBS. Simultaneously aspirate and add PBS (allow about 100 μl to remain on the coverslip after last wash). Let coverslip incubate for 5 min. Repeat twice.

8. Add fluorescently conjugated secondary antibody diluted in 5% goat serum and incubate for 30 min (100 μl per coverslip).

9. Wash coverslip three times as in step 7 (increasing wash times may decrease background). Allow a fourth wash to remain on the coverslip to prevent drying. Mount cells using a mounting medium. There are many companies that sell mounting media. We prepare a mounting medium that consists of [6% (v/v) glycerol, 24 mM Tris–Cl, pH 8.5, 0.6 mg/ml phenylaminediamine, and 4.8% (w/v) polyvinyl alcohol]. This mounting medium is stored at −80° and is typically orange; if the mounting medium has turned a pale brown it should be discarded.

10. Mount on a glass slide that has been washed in 70% ethanol, dried, and labeled. Add 5 μl of mounting medium to the microscope slide. Pick up coverslip with forceps and rinse by dipping into a doubly distilled H_2O-filled beaker. (This removes PBS, which can leave behind crystals as the coverslip dries.) Aspirate excess H_2O and mount carefully with cell side down.

11. After mounting medium has solidified (about 1–3 hr) paint over the edges of the coverslip with clear nail polish. We find that clear nail polish has less background fluorescence than colored nail polish. Once the nail polish has dried slides can be stored at −20°. Wash coverslips with Sparkle (Wal-Mart) to remove excess mounting medium before viewing them on the microscope.

Live Imaging of GFP–Fusion Proteins

General Principles. The goal for live imaging is to maximize the signal obtained from the GFP–fusion protein while minimizing background fluorescence. Expression levels of the fusion protein are critical for image collection as described above. Minimizing background fluorescence is also an important aspect of image collection. Below we describe the hardware and methods used to maximize image quality using time-lapse video microscopy of GFP–fusion proteins.

Hardware. As is often the case, science is limited by the sensitivity of the tools available. Figure 3 diagrams the basic components necessary to perform both single and dual fluorophore live imaging. The basic design of the microscope must seek to achieve four general goals; maximize image resolution, maximize

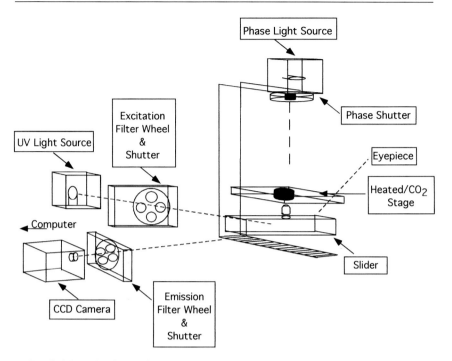

FIG. 3. Schematic of the basic components necessary for live fluorescence imaging. Filter wheels and shutters are placed in the light path between the UV source and cell for excitation and between the camera and image for emission. The shutter can also be placed between the phase source and cell for dual-phase/fluorescence imaging.

signal-to-noise ratio, provide user-friendly control of the hardware through the software, and maintain cell viability by controlling the extracellular environment.

The first of these goals, resolution, is the most critical when designing a system to image fluorescence. Two components that directly impact the resolution of a captured image are the objective and charge-coupled device (CCD) camera. When imaging focal contacts a 60× plan-apochromatic objective with a high numerical aperture (NA 1.4) is necessary to provide the best resolution and image quality. The CCD camera must also be able to capture a high-resolution image without increasing noise. Typically, the CCD camera is the limiting factor in the resolution of the obtained image. CCD cameras with high quantum efficiency are expensive; however, they are an integral part of an imaging system. Although the camera limits the resolution of the final image, it is important to design the microscope to optimize the signal that the CCD receives.

Using the correct filters can significantly increase the signal-to-noise ratio. For single emission fluorescence, narrow bandpass filters with sharp excitation

TABLE II
EXCITATION AND EMISSION WAVELENGTHS FOR
FLUORESCENT PROTEINS

Protein	Excitation	Emission
GFP	489	508
YFP	514	527
CFP	434	477
RFP	558	583
BFP	383	445

and emission cutoffs are used. These filters block excess excitation light that can increase background fluorescence. Dual fluorophore imaging presents unique problems because many of the fluorophores available have excitation–emission wavelengths that overlap (Table II). Filters that are commercially available to image two fluorophores are designed to allow for minimal overlap; however, in many instances overlap can cause bleed-through of the emission signal that can distort the image. Consequently, in addition to choosing the correct filters, one should select fluorophores that have the largest difference in their excitation–emission wavelengths. The dual-emission filters are set in filter wheels, as shown in Fig. 3, that can be controlled by software to switch excitation–emission wavelengths quickly while maintaining the optical position of the image. (*Note:* For imaging of a single fluorophore filter wheels are not necessary.) Photobleaching can also decrease the signal-to-noise ratio by requiring longer exposure times. Consequently, shutters are placed between the ultraviolet (UV) light source and the sample to minimize exposure times. Shutters can also be placed in the phase light path to allow for simultaneous differential interference contrast (DIC) and fluorescent imaging.

Many companies have developed software packages that can be used to control and integrate all aspects of the imaging system. Many of these companies, such as Universal Imaging (Downington, PA) and Inovision (Raleigh, NC), offer similar capabilities for varying prices. Although software is a significant investment, it is important that the software and the hardware work together relatively free of problems. However, because of the complexity of the system, inevitably there will be technical problems. Consequently, the most important consideration when choosing a software package is technical support. The software package should include on-site technical support as well as easy access to application support.

The system must also be able to maintain cell viability by controlling extra-cellular pH and temperature for the duration of the experiment. Many companies such as Ludl and Bioptechs make stage adapters that can maintain temperature and supply CO_2. One major pitfall of these stages is that one must continually check temperature settings and pH before and after the experiment to ensure that they have remained constant. These stages also can be quite expensive. To minimize expenses we have used a university machine shop to make a simple hollow brass stage through which heated water can flow. This stage can be attached to a circulating, heated water bath to maintain temperature and a homemade device may be used to supply CO_2 and maintain pH.

Background Fluorescence. When imaging focal contacts high background fluorescence can mask the localization of the GFP–fusion protein. Important considerations include the media and the type of plate used for the experiments.

Typical tissue culture media contain pH indicators such as phenol red that contribute to high background fluorescence. Many companies now carry phenol red-free tissue culture media, however, many of these still have high background fluorescence. We find that Ham's medium (Sigma, St. Louis, MO) provides the lowest background fluorescence of the media tested in our laboratory.

Resolving focal contacts through plastic plates traditionally used for tissue culture is difficult because of autofluorescence and thickness of plastic plates. Imaging focal contacts through glass bottom plates reduces working distance and background fluorescence. A cost-effective way to prepare plates is described below.

Preparation of Glass Bottom Plates

1. Acid wash 12-mm coverslips as described above.

2. Make 20-mm holes in the bottom of nontissue culture 35-mm petri dishes. (Size of the plate will depend on the size of the heated stage adaptor.) Typically, University machine shops can make these quickly and inexpensively.

3. Use optical adhesive to glue acid-washed coverslips to the petri dishes. Cure optical glue by exposure to long-wave-length UV light for 24 hr (12 hr per side). A typical tissue culture hood UV light is sufficient to cure optical glue. Using optical glue is important because it does not contain solvents that can enter the media during the experiment.

4. Coat plate with extracellular matrix protein as described in preparing coverslips section. Use approximately 2 ml of diluted substrate for coating the plate.

Ultraviolet Toxicity. Decreasing the background fluorescence will increase cell viability and decrease exposure times necessary to achieve a high quality image. Unfortunately, GFP–fusion proteins that are expressed at low levels may still require high exposure times and be more susceptible to photobleaching and UV toxicity. When imaging fixed samples, only one quality image is needed and therefore increasing exposure time is sufficient to improve image quality. However, with live imaging, increasing exposure times may improve image quality but will also lead to cell toxicity and photobleaching. An approach used to minimize toxicity while maximizing image quality includes obtaining long exposure times by decreasing the excitation intensity and integrating over a longer period of time. Reduction of excitation intensity can be accomplished by replacing a mercury light with a halogen or xenon light source or by placing a neutral density filter in the light path.

Neutral density filters are designed to decrease intensity by cutting all wavelengths of light equally. This method will decrease the temporal resolution of protein dynamics but can improve the quality of the image obtained.

Calibrating Stage before Live Fluorescent Imaging. Calibration of the stage prior to live fluorescence is critical to ensure that the cells remain viable through the duration of the experiment. Calibration of the stage should be started 1 hr before imaging to ensure that stage has stabilized before the experiment. Below, we describe the calibration of a Ludl heated stage.

1. Turn on circulating water bath and place a prewarmed glass bottom plate containing 2 ml of Ham's medium onto the stage. To prevent evaporation, layer 1–2 ml of mineral oil on top of the media. *Note:* Typically the temperature on the water bath is set higher than that desired on the microscope.

2. Turn on the microscope and use the appropriate objective (usually 60×). Focus objective to the plane of the cells. When the objective is in direct contact with the plate, it can conduct heat away from the plate. Heating the objective to 37° allows for a more accurate calibration.

3. Place a temperature probe into the medium such that it rests directly on the bottom of the plate.

4. Turn on 10% CO_2 balanced with air and allow the system to equilibrate for 30 min. After 30 min check the temperature and pH of the medium with pH paper. A temperature of 37° and a pH of 7.2 are typically used for time-lapse video microscopy. If temperature and pH are not correct adjust the water bath temperature and CO_2 level until the medium maintains a temperature of 37° and a pH of 7.2. The desired temperature and pH should be maintained for a minimum of 15 minutes before starting the experiment.

Our Ludl stage does not allow for the dynamic control of the temperature and CO_2 consequently, changes in room temperature and air circulation can have effects on the microscope calibration. As a result, it is important to check temperature and pH at the completion of the experiment to ensure that the stage is still properly calibrated. Using a HEPES-buffered solution can eliminate the need for CO_2. Manufacturers have started to develop systems that will enclose the microscope and are able to maintain constant temperature and pH, but these systems are expensive.

Fluorescent Imaging of Focal Contacts

1. Calibrate stage as described above.

2. Prewarm medium and reagents to 37° in a water bath. Ham's medium is pretreated at 37° in a 10% CO_2 incubator.

3. Wash cells once with 3 ml of PBS, add 0.02% EDTA, and incubate cells at 37° until detached. Avoid using trypsin to detach cells as trypsin may cleave cell surface receptors.

4. After the cells have detached, add an equal volume of medium and put cells in a 15-ml conical tube. Wash cells twice with 2 ml of serum-free media.

5. For CHO-K1 add 1×10^5 cells to 2 ml of prewarmed medium and add this to a matrix-coated glass bottom plate. Plate cells at low density in order to observe single cells that are not in contact. Allow cells to adhere for 1 hr at $37°$ and 10% CO_2.

6. Remove medium by washing plate once with PBS ($+Ca +Mg$) and replace with 2 ml of Ham's or equivalent low fluorescent medium. Put plate on the heated stage.

7. Layer mineral oil on top of the medium to prevent evaporation. Focus on the cells and allow the plate to adjust to the stage for a minimum of 30 min prior to imaging. (Focusing on the cells using DIC or phase will decrease exposure of the cells to UV light.)

8. For live imaging choose a representative cell that shows expression levels of the GFP–fusion protein optimum for imaging. Ideally, a minimum of two or three cells should be imaged per experiment. We recommend repeating the experiments 5–10 times to allow a total of 10–20 cells imaged per condition. When scanning the plate for a cell, reduce UV exposure by using a neutral density filter. When choosing a cell it is important to make sure that it is not in contact with another cell and that it expresses an optimum amount of the fusion protein for visualization. After choosing a cell, switch optics such that image can be captured by the CCD camera. It is beneficial for the CCD camera and the objective to be par a focal. If filter wheels are in place this may require placing a lens in the image path. In our experience, for most imaging applications, this lens does not significantly decrease the image quality.

9. Computer setups will differ depending on which software package is used; however, the CCD camera controls are similar between software platforms. Typical CCD camera software will allow you to control gain, exposure time, region imaged, and binning. Choose the settings so that they minimize exposure time while maintaining image quality. In our experiments we typically use gain $= 1$, exposure time $= 0.2$ to 1 sec, region imaged/binning $=$ center $640 \times 580/2$ (most older cameras allow you to choose only from specified regions). *Note:* Binning is a process that can decrease exposure times but will decrease the resolution of the image. We find that setting the camera bin to 2 does not significantly decrease image resolution.

10. Choose an interval with which to view cells. This interval will depend on the speed at which the cellular process is occurring. Protein localization can occur at varying speeds. When studying the dynamics of vinculin we capture an image every 3 min for 2 hr, whereas when visualizing RACK1 dynamics images are captured every 30 sec for 30 min. *Note:* It is also important to keep UV exposure to a minimum. Consequently, proteins may be visualized less frequently over larger time intervals.)

11. Begin capturing images with the software package. Monitor focus of the cell during imaging because as the cell changes shape the focal plane of the proteins may also switch. (*Note:* Microscope optics are sensitive to room temperature; thus, constant changes in focus may be due to fluctuations in room temperature.)

Variations. This protocol can be varied depending on the complexity of the imaging setup and software package used. For example, our microscope has a Ludl motorized X, Y, Z stage that allows us to capture images of more than one cell at a time. In addition, the Z control allows us to use digital deconvolution to improve image quality. Finally, it is also possible to collect dual DIC or phase and fluorescence movies by placing an additional shutter between the phase light source and the sample. (Phase shutter is necessary for the automation of the image capture.) It is important to compare the images obtained from the live imaging to samples that have been fixed and stained.

General Comments

When using a digital imaging system, displaying and archiving data can be complicated. Most computer software packages will come with software that can convert the captured images into animated movies. Depending on the length and resolution of these images movies can become quite large. Many journals only allow the submission of movies that are 6 MB or smaller; consequently, it may be necessary to shorten or crop images to decrease file size. In addition, data must be carefully archived, with duplicate copies in case the files get corrupted. Finally, and most importantly, when analyzing movies it is critical to remember to use a large sample size (10 or more cells).

Conclusion

The coming years promise to be exciting and informative in elucidating focal adhesion composition and function using live imaging approaches and GFP–fusion proteins. Advances in imaging technology and image analysis will continue to revolutionize our ability to understand focal adhesion structure and to identify the signaling proteins that regulate it. Critical for future investigation will be the ability to image multiple (three or four) components of the complex simultaneously to resolve the temporal and spatial distribution of these components and their regulation in live cells.

[17] Imaging Mitochondrial Function in Intact Cells

By MICHAEL R. DUCHEN, ALEXANDER SURIN, and JAKE JACOBSON

Introduction

Mitochondria play a central role in cell life and cell death. Not only do mitochondria generate adenosine triphosphate (ATP), the major currency used by cells in energy-requiring processes, but they also house a range of synthetic enzymes, including enzymes involved in heme and steroid biosynthesis, they are intimately involved in shaping the spatiotemporal characteristics of intracellular $[Ca^{2+}]_c$ signaling, and they house several proteins that play a key role in the regulation of programmed cell death or apoptosis.[1,2] In our attempts to understand these processes, it is essential to have tools that permit the study of mitochondrial function within cells and that provide unambiguous data to inform us about the fundamental mechanisms involved in processes so central to cell and tissue function and survival. The available methodologies used to follow mitochondrial function have developed dramatically in recent years along with the growing perception of the central role of mitochondria in the life of the cell. However, many of these approaches are open to misinterpretation, confusing our attempts to unravel the key events in the progression to cell death.[3,4] In this contribution, we will therefore focus on the principles and some of the problems associated with commonly used measurements and attempt to suggest solutions wherever possible.

The analyses of the biochemical and bioenergetic principles that underpin the standard biochemistry textbook description of the mitochondrion as the "powerhouse of the cell" were all based on studies of mitochondria isolated in bulk from tissues such as the liver, heart, or brain. Mitochondria were purified, and biochemical functions such as the rate of oxygen consumption and the rate of production of biochemical intermediates were assessed. Although such work still provides the mainstay for our understanding of mitochondrial bioenergetics, it also has its limitations. Most obviously, under these conditions, it is impossible to ask questions about the specialization of mitochondrial function in different cell types in which the yield of cells or of mitochondria might be limited. It is certainly not possible to ask questions about spatial and temporal changes in mitochondrial function in single cells during events associated with cell signaling or in response to pathophysiological conditions. It is these kinds of questions that have driven the

---

[1] J. C. Martinou and D. R. Green, *Cell Death Differ.* **2**, 63 (2001).
[2] J. Jacobson and M. R. Duchen, *Cell Death Differ.* **8**, 963 (2001).
[3] C. Fink, F. Morgan, and L. M. Loew, *Biophys. J.* **75**, 1648 (1998).
[4] M. W. Ward, A. C. Rego, B. G. Frenguelli, and D. G. Nicholls, *J. Neurosci.* **20**, 7208 (2000).

development of approaches that allow us to study mitochondrial function within individual cells. Although this contribution will deal specifically with fluorescence imaging approaches, it is important to bear in mind that valuable and important data may be accessible using the same principles but in populations of cells, using approaches such as fluorescence-activated cell sorting (FACS), in which information about changes in parameters in whole populations of cells may be accessible quickly and easily rather than employing the more arduous and laborious approach involved in imaging at the level of the single cell, although such studies may be confounded by the lack of synchrony among the population.

The discovery that proteins critically involved in the initiation and regulation of apoptotic cell death are housed within mitochondria, and may be released, either "accidentally" as a consequence of cell injury, or in a regulated fashion at the onset of programmed cell death, has set the mitochondrion center stage as a key player in major aspects of modern cell biology and medicine. Understanding the regulation of the distribution and redistribution of these proteins has become central in understanding the process of apoptosis, and imaging protein localization using fluorescent labels provides another important development in the application of imaging technology to the study of these organelles.

Another, perhaps less dramatic but also important, development has been the recognition that the Ca^{2+} uptake pathway in mitochondria may also play an important role both in the regulation of mitochondrial function and in the regulation of cellular calcium signaling.[5] Here again, the application of imaging technology to follow the spatiotemporal patterning of $[Ca^{2+}]_c$ signals and related changes in mitochondrial $[Ca^{2+}]$ or potential is central to the development of these ideas.

Basic Principles of Mitochondrial Bioenergetics

In this article, we will focus initially on approaches to the measurement of physiological functions of mitochondria. These involve measurements of mitochondrial membrane potential, mitochondrial redox state, and mitochondrial calcium handling. It is therefore necessary to review briefly some fundamental principles involved in the bioenergetics of these processes, as these principles define experimental design and underpin the mitochondrial response to biochemical agents and to changes in the cellular environment (see Fig. 1).

Figure 1 shows a diagram of a mitochondrion to illustrate the chemiosmotic principles that govern the process of oxidative phosphorylation. The supply of substrate, such as the product of glycolysis, pyruvate, to the citric acid cycle [the tricarboxylic acid (TCA) cycle or Krebs cycle] maintains the reduced state of the pyridine nucleotide (NADH) and the flavoprotein ($FADH_2$) pools. These reducing equivalents are supplied to the respiratory chain, NADH to complex I and $FADH_2$

[5] M. R. Duchen, *Cell Calcium* **28**, 339 (2000).

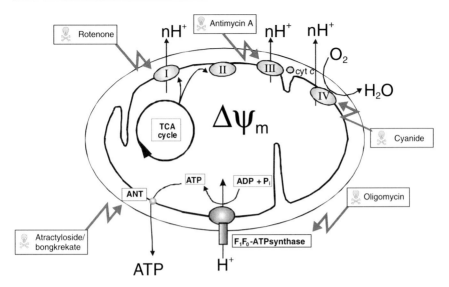

FIG. 1. Diagram to illustrate the chemiosmotic principles that govern mitochondrial oxidative phosphorylation. See text.

from succinate-linked substrates to complex II. Electrons are passed from either complex I or complex II via ubiquinone to complex III and are ultimately transferred to oxygen at cytochrome c oxidase, generating water. During the redox reactions of complexes I, III, and IV, protons are translocated across the inner mitochondrial membrane to the intermembrane space, generating an electrochemical proton gradient that is expressed largely as a membrane potential of the order of 150–200 mV negative to the cytosol. That potential, usually referred to as $\Delta\Psi_m$ ("delta psi"), provides the driving force for proton influx through the F_1F_0-ATP synthase. Rigorously, the proton motive force $\Delta P = \Delta\Psi_m + \Delta pH$ provides proton influx; δpH in mammalian cells is about 0.5–1, which corresponds to 30–60 mV and so $\Delta\Psi_m$ is the major component of driving force; most of the potentiometric probes discussed below respond only to $\Delta\Psi_m$.[6] The F_0 component of this enzyme system acts essentially as a proton conductance, which permits protons to move down their electrochemical potential gradient to drive the rotary motor provided by the F_1 complex, driving the phosphorylation of adenosine diphosphate (ADP). Ultimately, ATP is generated and transported to the cytosol. Some remarkable images of the operation of the enzyme were published by Noji et al.,[7] posted on the World Wide Web (seen at http://www.bmb.leeds.ac.uk/illingworth/oxphos/atpase.htm).

[6] D. G. Nicholls and S. J. Ferguson, "Bioenergetics 3," Academic Press, London, 2002.
[7] H. Noji, R. Yasuda, M. Yoshida, and K. Kinosita, Nature 386, 299 (1997).

The mitochondrial potential also provides the driving force for Ca^{2+} uptake into mitochondria. Ca^{2+} is taken up through the uniporter when the $[Ca^{2+}]_c$ is high; again, Ca^{2+} moves down an electrochemical potential gradient through the Ca^{2+} uniporter. This is a protein that has yet to be characterized, but it appears to behave like a channel, with an increase in open probability as $[Ca^{2+}]_c$ rises. Reequilibration of $[Ca^{2+}]_m$ is achieved through a Na^+/Ca^{2+} exchanger in the inner mitochondrial membrane (there is some uncertainty about the stoichiometry—either two or three Na^+ ions to one Ca^{2+}).[8]

The pharmacological or biochemical manipulation of these pathways is central to the investigation of mitochondrial function and so we have indicated on Fig. 1 the sites of action of some of the major reagents used. The use of these agents has also been described in detail.[9]

Measurement of Mitochondrial Membrane Potential

The mitochondrial potential, $\Delta\Psi_m$, lies at the heart of mitochondrial biology. It defines the transport of ions and (some) substrates, and of course provides the driving force for oxidative phosphorylation. The $\Delta\Psi_m$ has been assessed for many years by following the distribution of lipophilic cations, traditionally tetraphenylphosphonium (TPP$^+$). In a preparation of isolated mitochondria, TPP$^+$ will accumulate into mitochondria on the addition of substrate. Following the generation of a potential, the concentration of the ion in the extramitochondrial space falls and this can be followed using an ion-sensitive electrode manufactured with an appropriate semipermeable membrane. Fluorescence technology exploits the same principle, by following the (re)distribution of fluorescent lipophilic cations. Many fluorescent compounds, most notably rhodamine 123 (Rh123), tetramethylrhodamine ethyl and methyl esters (TMRE and TMRM, respectively), 5,5',6,6'-tetrachloro-1,1',3,3'-tetraethylbenzamidazolocarbocyanine (JC-1), DiOC$_6$, and DASPMI are both cationic and membrane permeant. They cross the cell membrane easily and partition between cellular compartments in response to the standing electrochemical potential gradients. The principle is illustrated in Fig. 2, in which a lawn of mitochondria on a glass coverslip was placed on a confocal microscope in the presence of the cationic dye tetramethylrhodamine methyl ester (TMRM, 50 nM). On addition of a substrate, succinate, the mitochondria accumulated dye and the fluorescence signal increased.

Considerable confusion has arisen in the literature because these dyes have been used in cells in different ways by different groups. It is essential to understand the

[8] T. E. Gunter, L. Buntinas, G. Sparagna, R. Eliseev, and K. Gunter, *Cell Calcium* **28**, 285 (2000).
[9] M. Mojet, J. Jacobson, J. Keelan, O. Vergun, and M. Duchen, *in* "Calcium Signalling" (A. Tepikin, ed.), p. 79. Oxford University Press, Oxford, 2001.

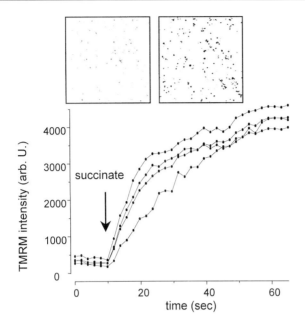

FIG. 2. Accumulation of potentiometric dyes in response to $\Delta\Psi_m$. In this experiment, isolated rat heart mitochondria were plated onto a glass coverslip bathed in a standard mitochondrial respiration buffer, but without substrate. TMRM (50 nM) was present throughout. Addition of 1 mM succinate (arrow) caused a rapid energization of mitochondria, which then accumulated the TMRM. The insets show the images before and after addition of succinate and the graphs show the intensity of signal over four individual mitochondria.

significance, application, and limitations of each approach if experimental design is to be accurately interpreted. Essentially, we will divide these approaches into two, which we shall call the redistribution approach, and the quench/dequench approach.

Measurement of Dye Distribution and Redistribution

For the rigorous measurement of mitochondrial potential, any of the dyes listed above can be used at a very low concentration—the lowest concentration consistent with a reasonable signal-to-noise ratio and that can be imaged with low light intensities. This is absolutely necessary because fluorescence from the dyes becomes a nonlinear function of dye concentration as the concentration increases, and also because all these indicators act as photosensitizing agents and are associated with photodynamic injury to cells, culminating in mitochondrial depolarization and ATP depletion. For TMRM or TMRE in our hands the optimal concentration

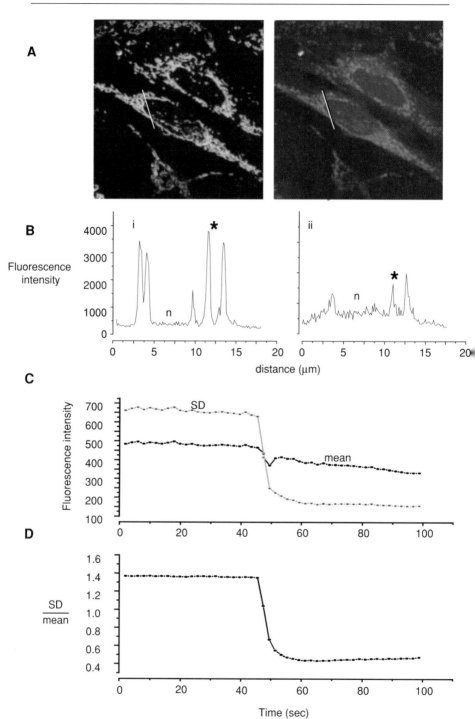

is $\sim$15–30 nM, but this may vary with cell type. The approach depends on the principle that at very low concentrations, the fluorescence signal shows a linear relationship with dye concentration and the dye concentration should simply reflect the Nernstian distribution of dye between compartments in response to local potential differences. According to this principle, one would expect an approximately 10-fold concentration of dye from saline to cytosol and then a concentration of some 400- to 800-fold from the cytosol into the mitochondria if $\Delta\Psi_m$ lies at about -150 to -180 mV. Thus, the fluorescence signals may range in intensity some 3000- to 4000-fold from the bathing saline to mitochondria, so that resolution of the signal in both cytosol and mitochondria by any imaging system demands digitization of the signal to at least 12 bits (4096 gray levels). Even then, this dynamic range is only just adequate to make accurate measurements, and so ideally one needs digitization to 14 or 16 bits, which are currently not available on confocal systems to our knowledge. The removal of contaminating out-of-focus signal is also necessary if measurements are to be made accurately. The images shown in Figs. 3 and 4 emphasize the need for confocal resolution if mitochondrial structures are to be identified clearly and if quantitative measurements are to be made. Even with reasonably high-resolution, good quality imaging on a cooled charge-coupled device (CCD) camera, the loss of spatial resolution caused by an out-of-focus signal makes it impossible to measure discrete mitochondrial signals accurately, except perhaps in very flat, very thin cells.[10] These issues have been discussed very carefully by Fink *et al.*[3]

For experiments in which dyes are used in the redistribution mode, cells are bathed in the dye at very low concentrations (10–30 nM) and allowed to reach equilibrium. This may take 30 min or more, and the time required must be established to ensure that a steady state has been reached before measurements are made. For all experimental work, the dye must be present in all bathing solutions.

---

[10] M. R. Duchen, J. Jacobson, J. Keelan, M. H. Mojet, and O. Vergun, *in* "Methods in Cellular Imaging" (A. Periasamy, ed), p. 88. Oxford University Press, Oxford, 2001.

FIG. 3. (A) Astrocytes in culture were incubated in 30 nM TMRM for 30 min to reach equilibrium, and imaged on a confocal imaging system (Zeiss 510CLSM). The concentration of dye into the mitochondria is evident (i). Addition of 1 μM FCCP from a locally positioned puffer pipette caused the rapid redistribution of dye (ii). The plots in (B) show the intensity profile along a selected line (shown on the images) before (i) and after (ii) FCCP. Note that in the control (i) the signal is intensely bright over mitochondria and very dark in the cytosol or nucleus, so that the signal varies from near 0 to 4000, giving a high standard deviation of pixel values within the cell. After application of the uncoupler (ii), the signal is much more uniformly distributed across the cell, so that the SD is much reduced. Note also that the specific mitochondrial signals (*) have decreased while the signal over the nucleus (n) and cytosol increased. In (C) we have plotted the mean intensity and its SD over a region of interest defined by the cell boundaries, whereas (D) shows the ratio of SD/mean signal.

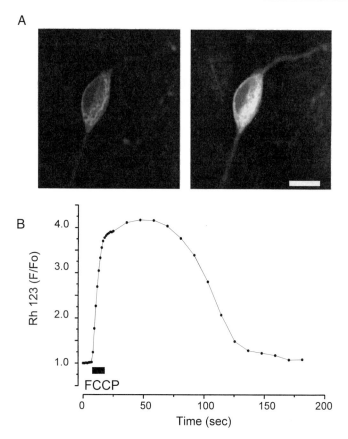

FIG. 4. (A) Hippocampal neurons grown in culture were loaded with rhodamine 123 at a concentration that will quench the fluorescence signal after loading and concentration by mitochondria (10 μg/ml for 10 min followed by washing) and images were acquired using a cooled CCD camera. (B) After application of FCCP, the fluorescence signal became much brighter, as shown in the plot of intensity with time. As the FCCP washed out, the signal recovered. *Note:* The change in signal is so great that a nonlinear "mapping" has been used to allow both signals to be displayed without either disappearing or saturating.

The approach is effectively used in two different types of experiment: (1) a dynamic measurement in which one asks how the potential will change in response to some change in the cell–cell signaling, or some pathophysiological condition, or (2) a comparison of populations of cells that have been previously exposed to different conditions.

When *dynamic measurements* are made, a fall in mitochondrial potential promotes the redistribution of indicator between compartments; in response to mitochondrial depolarization, dye will leave mitochondria and move into the cytosol.

In the short term, this can be resolved as a decrease in the specific mitochondrial signal and an increase in signal over cytosol and/or the nucleus (Fig. 3). In the short term (tens of seconds), this may not be associated with any change in signal at all over the whole cell (see Fig. 3C and below, Fig. 8D) and, if repolarization is rapid, the dye will simply be taken back up into mitochondria.[11] However, if the mitochondrial depolarization is sustained, then the loss of mitochondrial dye to the cytosol will be followed inevitably by a reequilibration of the cytosolic dye fraction, which will now move to the bathing saline, causing a loss of signal from the whole cell (see below, Figs. 6 and 8). Similarly, an increase in mitochondrial potential will initially cause an increase in specific mitochondrial signal and a decrease in the cytosolic signal with no change in net signal over the cell until the cytosol reequilibrates with the bathing saline, and the mean signal from the cell then increases.

In the second model, in which populations of cells are compared, the assumption is made that at equilibrium, the mitochondrial fluorescence signal will reflect the mitochondrial dye concentration, which is a direct function of the potential. Thus, if a manipulation has caused a loss of mitochondrial potential, the mitochondria will accumulate less dye and the mean signal from the mitochondria will be reduced at a steady state. Confocal imaging with fixed confocal optical thickness, laser power, and detector sensitivity should allow quantitative comparisons to be made using these nonratiometric indicators,[12] but with some important caveats, which we discuss below. It is also worth considering using cell sorting techniques for such measurements, but there are some real problems that we will highlight below, and so we would recommend a combination with confocal imaging to ensure the security of the data if at all possible.

Measurement of $\Delta\Psi_m$ Using Dyes in a Quench/Dequench Mode

It should be clear that although the approach outlined above is theoretically rigorous, it is also fraught with potential misinterpretations and errors. These problems have been discussed at some length by Ward *et al.*[4] and by Rottenberg and Wu.[13] When we started this kind of work in the late 1980s, we were obliged to use photomultiplier tubes to measure the averaged signal across a cell. In this case, if a dye simply redistributes from the mitochondria to the cytosol in response to a mitochondrial depolarization, the mean signal may not change. Several groups therefore adopted a strategy that may seem confusing: dyes such as TMRM, TMRE, and rhodamine 123 were loaded into cells at relatively high concentrations (~ 1–$20~\mu M$) for short periods, usually 10–15 min, followed by washing. As the dyes

[11] J. Huser and L. A. Blatter, *Biochem. J.* **343**, 311 (1999).

[12] B. Beltran, A. Mathur, M. R. Duchen, J. D. Erusalimsky, and S. Moncada, *Proc. Natl. Acad. Sci. U.S.A.* **97**, 14602 (2000).

[13] H. Rottenberg and S. Wu, *Biochim. Biophys. Acta* **1404**, 393 (1998).

accumulate into the mitochondria, they reach concentrations at which the signal shows a phenomenon called autoquenching; energy is transferred by collisions between monomeric dye molecules and the concentration of dye may promote formation of aggregates of dye molecules that may be nonfluorescent. Redistribution of the dye into the cytosol as the mitochondria depolarize relieves the quench and the net fluorescence signal increases (Fig. 4). Thus, in this mode, mitochondrial depolarization is associated with an increase in fluorescence. In our hands, these signals have always behaved exactly as predicted from chemiosmotic theory (see below and Fig. 9) and the signal provides an apparently reliable way to follow changes in $\Delta\Psi_m$ with time. Another important point is that measurements of the mean signal from a cell will give information about the average change in mitochondrial potential in that cell, and so useful information can be obtained without demanding high-resolution imaging. Indeed, we have routinely exploited this approach in low-power imaging of a field of cells in which we can bathe the cells with a given drug and follow changes in $\Delta\Psi_m$ in 20–30 neurons with time.[14]

It also turns out that rhodamine 123 is significantly less permeant across membranes than TMRM or TMRE.[14,15] This has the result that the redistribution across the plasma membrane in response to depolarization of $\Delta\Psi_p$ is much slower than in the case of TMRM, and so, even though the approach might seem less rigorous than the (re)distribution approach defined above, the dequench signals obtained with rhodamine 123 seem more reliable and unambiguous indicators of changing $\Delta\Psi_m$ and are less affected by changing $\Delta\Psi_p$ than any other that we have used. Of course, the approach is completely useless in attempting to compare mitochondrial potentials in populations of cells in response to manipulations, as the fluorescence signal is a nonlinear function of dye concentration, and so absolute intensity is now meaningless.

Problems, Caveats, and Solutions

ROLE OF PLASMA MEMBRANE POTENTIAL, $\Delta\Psi_p$. It is absolutely crucial to understand that the mean signal from the cell is not a unique function of changing mitochondrial potential. Thus, if the plasma membrane depolarizes, cytosolic dye will be lost. As the mitochondria are in equilibrium with the dye that has been concentrated into the cytosol in response to $\Delta\Psi_p$, the mitochondrial dye will reequilibrate and the signal will decrease, without any change in mitochondrial potential. The importance of this principle cannot be overemphasized. In attempts to study the role of mitochondrial membrane potential in glutamate neurotoxicity, for example, where exposure of neurons to glutamate causes massive changes in plasma membrane potential, it may be almost impossible to attribute the loss

[14] O. Vergun, J. Keelan, B. I. Khodorov, and M. R. Duchen, *J. Physiol.* **519,** 451 (1999).
[15] J. R. Bunting, *Photochem. Photobiol.* **55,** 81 (1992).

of signal from the cell specifically or separately to changes in either the plasma membrane or the mitochondrial potentials.[4] DiOC$_6$(3), often used as a probe of mitochondrial potential, has also been used as a probe of plasma membrane potential, and great care must be taken in interpreting signals with this dye.[13] We suggest two approaches to differentiate between effects of changing $\Delta\Psi_p$ and changing $\Delta\Psi_m$.

1. Control of $\Delta\Psi_p$—patch clamping and high [K$^+$]. The most rigorous way to examine the effect of plasma membrane depolarization is to record changes in fluorescence signal while changing the cell membrane potential under voltage clamp control. Using this approach, we showed that a voltage step from -70 mV to $+60$ mV (the potential at which the Ca^{2+} current reverses) caused no significant change in rhodamine 123 signal, whereas a step to 0 mV caused a small but significant ($\sim$20%) increase in signal.[16] This response was shown to be due to the mitochondrial response to a rise in [Ca^{2+}]$_c$ and mitochondrial Ca^{2+} uptake (which generates an inward current across the mitochondrial inner membrane and hence a transient depolarization). Unfortunately, use of the whole cell patch clamp technique as an approach to studying metabolic processes is always problematic, as [Ca^{2+}]$_c$ is usually buffered and ATP needs to be added to pipette solutions to avoid the washout of currents. An alternative is to use the amphotericin perforated patch technique,[17] but this has its own difficulties and limitations.

In practice, it is straightforward to test any procedure that causes an apparent change in mitochondrial signal against plasma membrane depolarization using high potassium, bearing in mind that changes in [Ca^{2+}]$_c$ can also cause changes in $\Delta\Psi_m$ and so the high K$^+$ (50 mM, with isotonic replacement of Na$^+$) containing solutions should also be Ca^{2+} free, at least in excitable cells.[18] Depolarization of the plasma membrane may cause changes in fluorescence signals from each of the dyes, and these must be understood if errors are to be avoided. We will discuss the behavior of the dyes in response to changes in $\Delta\Psi_m$ and $\Delta\Psi_p$, suggest a solution to differentiate between these physiological events, and illustrate some mathematical models that predict the behavior of signals.

2. Measurement of standard deviation and mean signals. One of the most elegant ways to quantify the distribution of a probe within a cell is to measure the standard deviation of the signal intensity throughout the cell. Thus, a signal that is highly compartmentalized within local structures but largely absent from the cytosol will show a high SD of intensity in the region of interest defined by the cell boundaries. Redistribution of the probe throughout the cell may cause no change at all in the mean signal, but the SD will fall as the signal becomes more uniformly

[16] M. R. Duchen, *Biochem. J.* **283**, 41 (1992).
[17] A. V. Nowicky and M. R. Duchen, *J. Physiol.* **507**, 131 (1998).
[18] J. Keelan, O. Vergun, and M. R. Duchen, *J. Physiol.* **520**, 797 (1999).

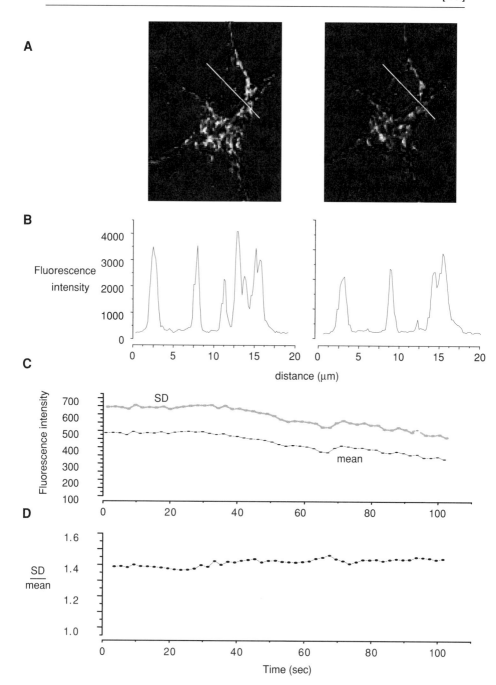

distributed throughout the cell (see Figs. 3, 5, and 6). This approach was used by Goldstein *et al.*[19] to quantify the redistribution of cytochrome *c*-tagged green fluorescent protein (GFP) during apoptosis and has also been suggested by Toescu and Verkhratsky[20] as a way to quantify the level of distribution of potentiometric dyes in mitochondria, although the approach suggested here is slightly different.

Figures 3 and 5 illustrate the principle employing TMRM. If $\Delta\Psi_p$ stays constant and $\Delta\Psi_m$ falls, dye is lost from the mitochondria to the cytosol. In the short term, the mean signal will stay constant (dye is simply moving from one compartment to another), but the SD will fall and the SD/mean ratio will fall (Fig. 3). Only later will the mean signal fall as the increased cytosolic dye concentration reequilibrates with the bathing solutions. In contrast, if $\Delta\Psi_m$ stays constant and $\Delta\Psi_p$ falls, dye will be lost from the cytosol, and the mitochondrial dye fraction will then reequilibrate more slowly with the cytosol. The mean fluorescence signal will inevitably fall, but the mitochondria will retain substantially more dye than the cytosol as long as they remain fully polarized, and so the SD of the signal will fall only slightly along with the decline in the mean signal, and the ratio of SD/mean should stay constant (Fig. 5). Thus, using the SD and mean signals with high-resolution imaging provides an excellent way to discriminate between changes in $\Delta\Psi_m$ and $\Delta\Psi_p$.

Some Mathematical Modeling

By now, it should be clear that these measurements are not simple, and that there are very significant chances of serious misinterpretation. In an attempt to clarify the behavior of the dyes, Ward *et al.*[4] generated a simple model that attempts to define the behavior of the dyes in response to changes in either plasma membrane potential or mitochondrial potential given the diffusion coefficient for the dye across membranes. The model is described in detail and was generously made widely available, and so we have adapted it and have also used it to try to resolve some of these issues. The following few paragraphs describe the behavior of signals according to the model.

The model takes into account the dependence of whole cell fluorescence on (1) transmembrane potentials, (2) the mitochondrial volume fraction in the cell,

---

[19] J. C. Goldstein, N. J. Waterhouse, P. Juin, G. I. Evan, and D. R. Green, *Nat. Cell Biol.* **2,** 156 (2000).
[20] E. C. Toescu and A. Verkhratsky, *Pflugers Arch.* **440,** 941 (2000).

FIG. 5. (A) Astrocytes in culture were allowed to equilibrate with 30 n*M* TMRM as described for Fig. 3. The cells were then exposed to a saline in which 50 m*M* KCl replaced 50 m*M* NaCl. (B) Note that the mean signal declined progressively but that the concentration of dye within the mitochondria was retained. (C) The mean signal and SD both fell, but the ratio of SD/mean shown in (D) remained flat.

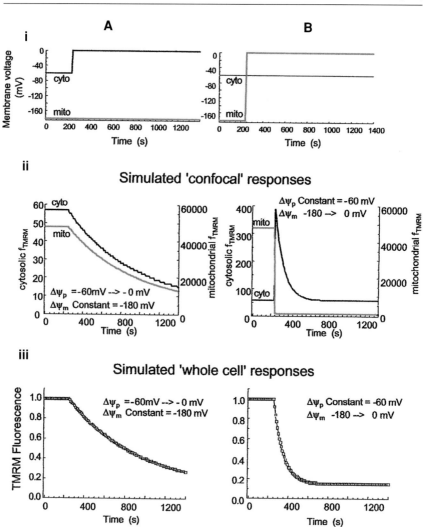

FIG. 6. A computational model was used to predict changes in the fluorescence signal from each compartment following either a step change in the plasma membrane potential (A) while the mitochondrial potential was held constant, or a step change in plasma membrane potential, while the mitochondrial potential was held constant (B). In this figure, the model was applied to changes in signal using TMRM at a concentration below that sufficient to reach the quench threshold, whereas the same principles were applied in Fig. 7 to explore the changes in signal using rhodamine 123 above the quench threshold. The protocols are represented by the plots shown in (i). (ii) The predicted change in signal arising from each compartment separately. (iii) The predicted change in whole cell fluorescence that results. See text for further details.

(3) the concentration at which the dye fluorescence undergoes quenching—the "quench limit," (4) the rate of probe diffusion through the cell membrane, and (5) the extracellular dye concentration. Application of the model reveals some features of potentiometric dye responses, which are not obvious and should be taken into account in the analysis of experimental data.[20a] We have applied the model to describe how the signals will change where (i) $\Delta\Psi_p$ changes at constant $\Delta\Psi_m$ and (ii) $\Delta\Psi_m$ changes with constant $\Delta\Psi_p$ and have compared TMRM in the redistribution mode at concentrations below the quench limit (Fig. 6) with rhodamine 123 in the dequench mode (Fig. 7). Essentially, in each we show the response to a step change in either $\Delta\Psi_p$ (Fig. 7A) or $\Delta\Psi_m$ (Fig. 7B); the changes in either transmembrane potential used for the calculations are illustrated in Fig. 7(i). In Fig. 7(ii) of each column, we show the computed changes in dye fluorescence signal over individual pixels of an image, i.e., equivalent perhaps to confocal resolution. *Depolarization of the plasma membrane* alone causes the gradual loss of dye from cytosol to saline, followed by the movement of dye from mitochondria to the cytosol (Fig. 7A, ii). *Mitochondrial depolarization alone* causes the rapid movement of dye from mitochondria to the cytosol followed by the slower loss of dye from the cytosol (Fig. 7B, ii). These changes in dye distribution are then reflected in changes in computed "whole cell" signal [Fig. 7(iii)], making assumptions about the volume fraction of the cell occupied by mitochondria. In both cases, we see a progressive loss of signal, but the change is faster with mitochondrial depolarization than with depolarization of the plasma membrane. Thus, the key findings are that, using TMRM, the signal will fall in response to both manipulations, reflecting the relatively high membrane permeability to this dye, and these two responses are essentially indistinguishable over a longer time frame. It is important to recognize that in the very short term (over tens of seconds), the mean signal changes very little with either manipulation—about 5% loss of signal at 20 sec.

In Fig. 7, we show the same computations but using rhodamine 123 at a concentration above the quench threshold. Rhodamine 123 crosses membranes far more slowly than TMRM, and so the responses are quite different. Now, in response to *plasma membrane depolarization alone,* the rhodamine 123 signal barely changes at all, and there is almost no redistribution of dye (Fig. 7A, ii) and no change in whole cell fluorescence signal (Fig. 7A, iii). However, on *mitochondrial depolarization alone,* there is a rapid shift in dye from mitochondria to the cytosol (Fig. 7B, ii) resulting in a robust increase in the "whole cell" signal (Fig. 7B, iii).

Calibration. We have also asked how these signals effectively calibrate, i.e., how the signals change with a series of step changes in $\Delta\Psi_m$ or $\Delta\Psi_p$, and this exercise especially reveals an important limitation in the use of rhodamine 123

[20a] M. R. Duchen and A. Surin, *Membr. Cell. Biol.* **19,** 4 (2002).

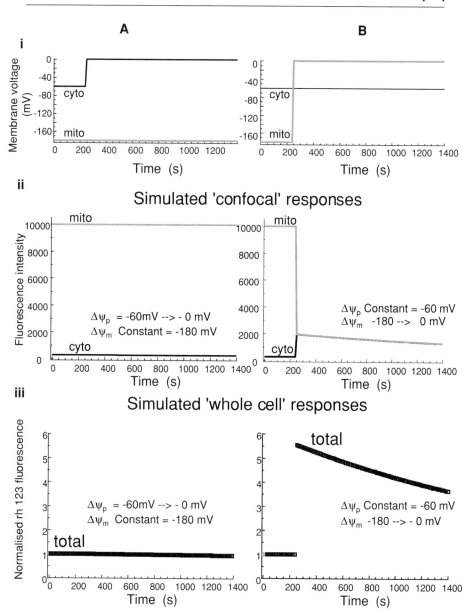

FIG. 7. The computational model (see Fig. 6) was applied to follow changes in signal in cells loaded with rhodamine 123 above the threshold to cause quenching after the concentration of dye into the mitochondria. Responses of individual pixels are shown following either a step depolarization of plasma membrane (A, i, ii) or mitochondrial membrane (B, i, ii), and the predicted change in whole cell fluorescence that results (A, iii and B, iii). See text for further details.

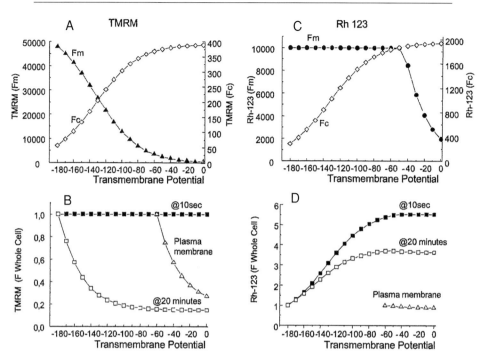

FIG. 8. We applied the computational model to explore the consequences of step changes in mitochondrial membrane potential for (A, B) TMRM (loaded at equilibrium with concentrations below the quench limit), and (C, D) rhodamine 123 above the quench threshold. Note that with TMRM, progressive mitochondrial depolarization causes the redistribution of dye from mitochondrial to cytosol. At 10 sec (filled squares), there is no change in whole cell fluorescence, but at 20 min, loss of dye from the cytosol now causes a decrease in whole cell fluorescence, which is fairly linear down to a potential of about −60 mV, but changes only slightly with further depolarization. Using rhodamine 123 at concentrations at which the intramitochondrial dye will quench, mitochondrial depolarization causes no change in mitochondrial signal and an increase in the cytosolic signal so that the total signal from the cell increases progressively. However, at about −80 mV, the quench limit in mitochondria is reached. A further depolarization fails to cause any further change in signal as this causes only a redistribution of dye from mitochondria to cytosol.

signals in the dequench mode. For TMRM at a low concentration the signal behaves quite simply (Fig. 8A): with progressive mitochondrial depolarization, the dye moves from the mitochondria (filled triangles) to the cytosol (diamonds). The computed change in whole cell signal is shown in Fig. 8B and is time dependent. Thus, at 10 sec, there is no significant change in the computed mean cell fluorescence (filled squares), but at 20 min there is a progressive loss of signal with progressive mitochondrial depolarization. Remarkably, according to the model, the signal falls to a minimum with a depolarization to just −80 mV or so, failing to fall further with further mitochondrial depolarization, as the dye in the cytosol remains in equilibrium with the dye in the bathing saline (Fig. 8B).

Figure 8C and D shows the results of the same computations for rhodamine 123 at quenching concentrations. This also reveals some surprises. Thus, with progressive mitochondrial depolarization, dye moves from the mitochondria (filled circles) to the cytosol (diamonds). The mean mitochondrial signal does not change at all until the intramitochondrial dye concentration reaches the quench threshold. As the membrane potential collapses further, the mitochondrial signal begins to fall. Initially, with modest changes in mitochondrial potential, the cytosol gains dye and so the whole cell signal (Fig. 8D) increases. However, once the mitochondrial quench threshold is reached, then further mitochondrial depolarization simply promotes the redistribution of dye from mitochondria to cytosol, and so the mean "whole cell" signal fails to change any further. Application of the model suggests that the total signal from the cell will saturate at a $\Delta\Psi_m$ of about $-60\,mV$ (Fig. 8D).

Although the model makes many assumptions and the numbers may not be strictly accurate, the principle is clear and has some important implications. Most particularly, it seems that it is never possible, using this approach, to define the complete collapse of $\Delta\Psi_m$. Before we recognized this limitation, we and others have routinely used the application of FCCP to give some kind of semiquantative scale against which we could calibrate changes in $\Delta\Psi_m$. Thus, at the end of many experiments, we applied FCCP to indicate the maximal signal available from rhodamine 123 dequench. We have then used that signal to scale the fluorescent traces between 0 (a resting, or polarized level) and unity, representing the response to FCCP, and assuming this to represent "complete" collapse of $\Delta\Psi_m$. The model predicts, however, that the change in signal with complete depolarization (e.g., with FCCP) is indistinguishable from the change in signal that would arise from depolarization to just $-60\,mV$.

In functional terms, in trying to understand the basis for mitochondrial pathophysiology, this limitation has important implications. For example, opening of the mitochondrial permeability transition pore (mPTP) would be expected to dissipate $\Delta\Psi_m$ completely. Inhibition of respiration through damage to the respiratory chain, for example, will allow the potential to dissipate, but reversal of the F_1F_0-ATP synthase maintains the potential to some degree and the rate of depolarization will depend on this and also on the leakiness of the mitochondrial membrane (both these are highly variable between cell types). Therefore, if measurement of a maximal rhodamine 123 signal cannot be attributed reliably to complete dissipation of $\Delta\Psi_m$ then opening of the mPTP becomes harder to identify as a mechanism of depolarization. If one needs to establish the scale of loss of $\Delta\Psi_m$, perhaps it is essential to use the redistribution mode of measurement, when complete redistribution (in effect the complete disappearance of any discrete identifiable mitochondrial signal) will occur only when there is effectively no remaining mitochondrial potential.

Figure 8 also emphasizes the difference in the behavior of the two dyes in response to changes in the plasma membrane potential (shown as open triangles)—essentially loss of the plasma membrane potential will cause the progressive loss of signal using TMRM (Fig. 8B) but almost no change in rhodamine 123 signal (Fig. 8D).

Role of the MDR. Another major problem in the use of these dyes is the export of the dyes by the multidrug resistance (MDR) carriers. All the dyes routinely used are good substrates for the MDR carriers, which will export the dyes from the cell and therefore limit their concentration by the mitochondria. This is particularly a problem in immortalized tumor-derived cell lines, and we now routinely check the effect of inhibitors of the MDR (e.g., verapamil, 10 μM) on dye loading to ensure that this is not a problem. Typically, in cells expressing the MDR, dye loading will be very slow to equilibrate but will be greatly enhanced by verapamil.

JC-1. JC-1 is a lipophilic cation with unique properties, which is now quite widely used, especially in FACS analysis of cell populations.[21] The principle of the indicator is that at high concentrations, the dye forms complexes ("J"-complexes) that show a red shift in the fluorescence emission spectrum—with excitation at ~490 nm, the peak emission of the monomer is in the green with a peak at ~539 nm whereas the J-complex emission has a peak at 597 nm, in the red.[22,23] The concentration of dye into mitochondria thus promotes aggregation, and so the greater the potential, the greater the red signal. In theory, the ratio of red-to-green fluorescence should increase with increasing $\Delta \Psi_m$.

It strikes us that the use of this dye has really not been carefully explored in terms of varying dye loading times or concentration. One striking feature of the dye is its rather slow distribution across membranes. In imaging experiments, the fluorescence shows curious crystalline structures within mitochondria (M. R. Duchen, personal observations),[24] which look quite different to the threadlike distribution of mitochondrial structures labeled with any other mitochondrial dyes. Further, fluorescence of the monomer is enhanced in a lipid environment[25] and therefore green fluorescence may reflect the accumulation of dye in membranes, whereas J-aggregates giving rise to red fluorescence appear in the aqueous phase. It is also not clear how specific the conditions are that cause J-aggregate formation, so that it is difficult to be sure that the aggregates arise only from mitochondria.

[21] A. Mathur, Y. Hong, B. K. Kemp, A. A. Barrientos, and J. D. Erusalimsky, *Cardiovasc. Res.* **46**, 126 (2000).

[22] S. T. Smiley, M. Reers, C. Mottola-Hartshorn, M. Lin, A. Chen, T. W. Smith, G. D. Steele, and L. B. Chen, *Proc. Natl. Acad. Sci. U.S.A.* **88**, 3671 (1991).

[23] M. Reers, T. W. Smith, and L. B. Chen, *Biochemistry* **30**, 4480 (1991).

[24] G. Diaz, A. Diana, A. M. Falchi, F. Gremo, A. Pani, B. Batetta, S. Dessi, and R. Isola, IUBMB Life **51**, 121 (2001).

[25] F. Di Lisa, P. S. Blank, R. Colonna, G. Gambassi, H. S. Silverman, M. D. Stern, and R. G. Hansford, *J. Physiol.* **486**, 1 (1995).

Dual-emission fluorimetry, with continuous measurements at $\sim$539 and $\sim$597 nm, might be expected to provide a ratio measurement of $\Delta\Psi_m$. Indeed, we found that an uncoupler does decrease the red signal and increases the green, although these typically change at different rates.[26] However, experience in different laboratories seems to show a substantial variability. Di Lisa et al.[25] also suggested that the red fluorescence signal appeared more sensitive to relatively small changes in mitochondrial potential than the green signal, which changed significantly only with larger excursions of $\Delta\Psi_m$.

The red fluorescence from the "J"-complexes also appears sensitive to factors other than $\Delta\Psi_m$. For example, Chinopoulos et al.[27] showed that H_2O_2 had profound effects on the red fluorescence independent of any evidence for a change in $\Delta\Psi_m$, and these authors therefore chose to measure only changes in the green fluorescence signal. Further, the equilibrium between monomers, dimers, and polymers is not solely due to membrane potential, as JC-1 is pH sensitive and its absorption spectrum may be affected by the osmolarity of its environment.[23]

Confocal imaging of JC-1 generates beautiful images, but it seems to us that the interpretation is uncertain. In some preparations, some mitochondria appear red and others green. Does this mean that there are populations of mitochondria with different potentials? We, and others, have even seen single mitochondria that are mostly green but have some red sections to them.[28] The problem lies in trying to address the significance of these measurements unambiguously. Careful confocal imaging of the same cell types with very low concentrations of TMRE have so far failed to reveal the same differences in population that appear with JC-1 (M. R. Duchen, unpublished observations), but these data still raise questions that we cannot answer.

Given all the considerations outlined above, we remain very cautious about the interpretation of JC-1 signals, especially in trying to follow changes in signal with time, although others have clearly used the indicator with great effect.[29]

Validation of Signals. Imaging of cells stained with most of the fluorophores listed above reveals obvious selective loading into mitochondria. Using JC-1 we have seen red fluorescent objects that are not mitochondria (including the nuclei of dead cells!). There is a tendency for reviewers to ask for validation using MitoTracker dyes, but there is a circularity of argument here, as the MitoTracker indicators typically load preferentially into mitochondria in exactly the same way as the potentiometric probes such as TMRM, and so of course the indicators colocalize. There are also concerns that the MitoTracker dyes themselves may impair

[26] M. R. Duchen, O. McGuinness, L. A. Brown, and M. Crompton, *Cardiovasc. Res.* **27**, 1790 (1993).
[27] C. Chinopoulos, L. Tretter, and V. Adam-Vizi, *J. Neurochem.* **73**, 220 (1999).
[28] S. T. Smiley, M. Reers, C. Mottola-Hartshorn, M. Lin, A. Chen, T. W. Smith, G. D. J. Steele, and L. B. Chen, *Proc. Natl. Acad. Sci. U.S.A.* **88**, 3671 (1991).
[29] G. Szalai, R. Krishnamurthy, and G. Hajnoczky, *Embo J.* **18**, 6349 (1999).

mitochondrial function and so if these dyes are used in functional studies, appropriate controls must be carried out to ensure that function has not been altered. It is necessary to be careful to address several concerns, using each dye and in any cell type: Does the signal change reflect only changes in mitochondrial potential or is there a substantial change with depolarization of the plasma membrane? How quickly and reliably does the dye follow changes in $\Delta\Psi_m$ in response to a range of manipulations that is expected alter $\Delta\Psi_m$? Is there any evidence of toxicity of the dye in terms of cell function, mitochondrial function, etc.?

Experimental Manipulation of Mitochondrial Membrane Potential. Most authors employ (at most) a mitochondrial uncoupler to demonstrate the change of fluorescence signal with mitochondrial depolarization. Although this is not unreasonable as a starting point, it hardly suffices to validate the use of the dye in full. We have also suggested the following scheme to test the behavior of an indicator (illustrated in part in Fig. 9)

1. Application of rotenone inhibits respiration at complex I and usually causes a slow depolarization or none at all (varying in rate with cell type).

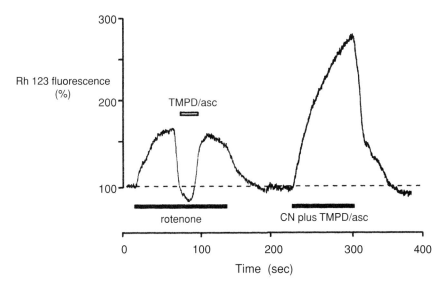

FIG. 9. This trace illustrates the application of a protocol to test dye behavior with imposed changes in $\Delta\Psi_m$. The trace shows changes in rhodamine 123 fluorescence under dequench conditions recorded from a mouse sensory neuron in culture. Rotenone was applied first, blocking respiration at complex I. This causes a slow modest loss of mitochondrial potential that can be reversed by application of TMPD (20 μg/ml) with ascorbate (1 mM), supplying electrons to complex IV and so bypassing the rotenone inhibition. In contrast, CN, which inhibits at complex IV, causes a mitochondrial depolarization despite the presence of the TMPD, which cannot bypass the CN block.

2. After washout of the rotenone (which recovers only slowly), addition of oligomycin may cause a small increase in $\Delta\Psi_m$ or no change at all (the hyperpolarization suggests some resting proton flux through the F_1F_0-ATPase, which is blocked by oligomycin). Application of rotenone will now cause a faster and larger depolarization, as the F_1F_0-ATPase cannot operate in "reverse mode" to maintain the potential.

3. Now application of tetramethylphenylenediamine (TMPD, $\sim$10 μg/ml) with ascorbate (to reoxidize the TMPD, $\sim$1 mM) restores the potential, as this bypasses the inhibition by rotenone and delivers electrons directly to cytochrome oxidase.

4. A further addition of cyanide will depolarize the potential again, despite the TMPD/ascorbate, as CN^- blocks downstream of the TMPD site.

The procedure generates predictable and reliable responses using rhodamine 123 and TMRE/TMRM in dequench mode. Using TMRM at very low concentration and high-resolution confocal microscopy it is possible to monitor the movement of dye between cytosol and mitochondria with this manipulation (M. R. Duchen, unpublished observations).

A Miscellany. Rhodamine 123 has always appeared satisfactory as an indicator of $\Delta\Psi_m$ in all cell types that we have tried except in cardiomyocytes. In these cells, the distribution of dye is clearly mitochondrial but uncouplers failed to cause a change in signal (but see Griffiths *et al.*,[30] although these authors used a very high concentration of the dye). TMRM and TMRE seem to behave in cardiomyocytes exactly as in all other cells. We do not pretend to understand the basis for these differences.

It is also extremely important to note that rhodamine 123 fluoresces red when excited at 530 nm or greater (M. R. Duchen, unpublished observations), although it is usually used with fluorescein isothiocyanate (FITC) filter configurations. This is not widely reported, but means that the dye cannot be used in combination with rhodamine-type dyes either for colocalization with rhodamine-conjugated antibodies or in physiological experiments (e.g., with rhod-2). This has given rise to serious errors.

Toxicity

Direct Toxic Effects

Most indicators have direct toxic effects on mitochondrial function that are usually ignored. Thus, carbocyanine dyes such as $DiOC_6(3)$ and JC-1 have long been known to inhibit complex I (NADPH dehydrogenase) and at concentrations of 40–100 nM, concentrations often used in flow cytometric studies, $DiOC_6$ inhibits

---

[30] E. J. Griffiths, M. D. Stern, and H. S. Silverman, *Am. J. Physiol.* **273**, C37 (1997).

mitochondrial respiration by $\sim$90%, equivalent to rotenone! Rhodamine 123 inhibits the F_1F_0-ATPase at high concentrations,[31] and Scaduto and Grotyohann have shown that the rhodamine-based dyes all tend to inhibit mitochondrial respiration, especially when loaded at concentrations in the micromolar range.[32]

Phototoxicity. Photobleaching of fluorescent dyes, whereby light-induced oxidation of the fluorophore results in a gradual loss of signal, may occur if the intensity of illumination by the excitation light is too great. This is particularly relevant if dyes are loaded at low concentrations when the relatively low fluorescence signals may prompt users to increase the excitation intensity. If this is unavoidable, limiting the duration of exposure by closing a shutter or scanning intermittently may reduce bleaching.

Light-induced oxidative damage is a major problem with all mitochondrially localized probes. Illumination of fluorophores increases the production of reactive oxygen species, and the consequent oxidative stress can have wide-ranging effects, including induction of local calcium release, inhibition of the electron transport chain, and induction of the mitochondrial permeability transition. As with photobleaching, limiting the intensity of the excitation light by attenuating the excitation output or reducing the period of illumination will reduce the oxidative stress. The toxic effects have been used by some to explore the consequences of a mitochondrially generated oxidative stress.[11,33,34]

NADH and Flavoprotein Autofluorescence

The term autofluorescence refers to the fluorescence that arises from endogenous compounds intrinsic to the cell, and is used to distinguish it from the fluorescence of indicators that are artificially introduced. Our understanding of these signals and their properties owes much to the pioneering work of Britton Chance in the 1950s and 1960s, who showed that the bulk of intrinsic fluorescence in cells arises from the compounds that act as hydrogen carriers, which ferry the protons and electrons from the citric acid cycle substrates to the electron transport chain. These signals can provide valuable indicators of changes in mitochondrial metabolism, as their properties change with the redox state of the carriers. Thus the fluorescence of the pyridine nucleotide, NADH, is excited in the UV (peak excitation at about 350 nm) and emits blue fluorescence with a peak at about 450 nm.[35,36] The oxidized form, NAD^+, is not fluorescent. An increase in UV-induced blue fluorescence therefore indicates an increase in the ratio of NADH

[31] R. K. Emaus, R. Grunwald, and J. J. Lemasters, *Biochim. Biophys. Acta* **850,** 436 (1986).
[32] R. C. Scaduto and L. W. Grotyohann, *Biophys. J.* **76,** 469 (1999).
[33] D. B. Zorov, C. R. Filburn, L. O. Klotz, J. L. Zweier, and S. J. Sollott, *J. Exp. Med.* **192,** 1001 (2000).
[34] D. Jacobson and M. R. Duchen, *J. Cell Sci.* **115,** 1175 (2002).
[35] B. Chance, B. Schoener, R. Oshino, F. Itshak, and Y. Nakase, *J. Biol. Chem.* **254,** 4764 (1979).
[36] J. Eng, R. M. Lynch, and R. S. Balaban, *Biophys. J.* **55,** 621 (1989).

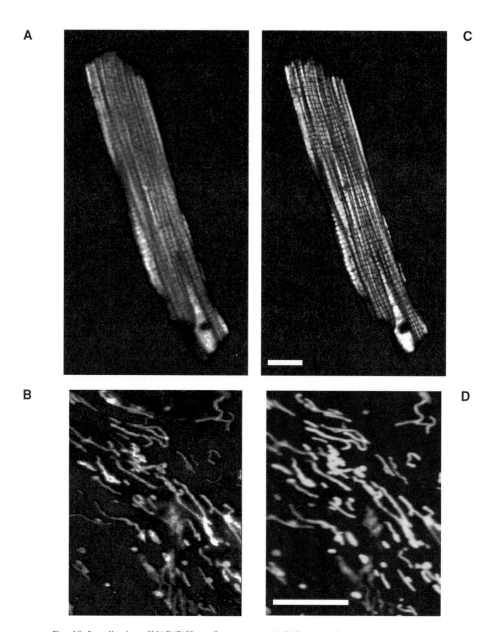

FIG. 10. Localization of NAD(P)H autofluorescence. (A, B) Images of a cardiomyocyte loaded with TMRM to show the colocalization of the NAD(P)H autofluorescence (A) with the TMRM staining (B). Images were acquired on a confocal imaging system (Zeiss 510 CLSM) and the images were acquired on separate channels; autofluorescence was elicited by excitation at 351 nm and the signal measured between 430 and 470 nm, and the TMRM signal was elicited by excitation at 543 nm and measured using a long pass filter at 585 nm. Note the mitochondria run in bands along the long axis of the cell. (C, D) Similar images are shown at higher magnification from a field of astrocyte cytosol. Again, the colocalization of NAD(P)H (C) and TMRM (D) stained structures is evident. Bars: 10 μm.

to NAD^+—a net shift in the pyridine nucleotide pool to the reduced state. Figure 10 illustrates the localization of NADH fluorescence to mitochondria in cardiomyocytes and astrocytes that were also loaded with TMRM.

It is important to emphasize that these changes in signal do not indicate net changes in the absolute size of the total pool but rather a change in the balance of reduced to oxidized forms. NADPH is also fluorescent with very similar spectral properties, but in most cell types is present at much lower concentrations than NADH. NADH and NADPH are both present in both mitochondrial and cytosolic compartments: however, several properties tend to mitigate in favor of the mitochondrial signal—there is more of it, and the binding of NADH to membranes enhances the fluorescence whereas enzymatic binding tends to quench the cytosolic fraction.[37]

Flavoproteins ferry electrons using a flavin or FAD molecule. Flavoprotein fluorescence is excited in the blue (with a peak at about 450 nm) and fluorescence emission is maximal in the yellow/green, with a peak at about 550 nm. In contrast to NADH, flavoprotein fluorescence *decreases* when the carrier binds electrons. A decrease in flavoprotein autofluorescence reflects an increase in the ratio of reduced to oxidized flavoprotein—the inverse of the response of the pyridine nucleotides.

The redox state of the pyridine nucleotide and flavoprotein pools reflects the balance between the rate of reduction by substrate processing and the rate of oxidation by mitochondrial respiration. Thus, both upregulation of substrate processing and inhibition of respiration will favor a net balance toward a reduced state. Conversely, an increase in respiratory rate favors net oxidation of the pool. Thus, upon inhibition of respiration by CN^- or by anoxia, the respiratory chain cannot oxidize the reduced forms, which will accumulate to a new steady state. NADH autofluorescence increases and flavoprotein autofluorescence falls. Mitochondrial respiration responds to collapse of $\Delta\Psi_m$ by an uncoupler with an increase in respiratory rate. This promotes maximal oxidation of NADH to NAD^+ and $FADH_2$ to FAD, decreasing the autofluorescence from NADH and increasing that from FAD (Fig. 11).

Clearly, measurement of these signals alone does not provide useful or meaningful measurements, but changes in signal may give information about changing substrate utilization or oxygen consumption, which is otherwise inaccessible, especially if calibrated against the maximal reduced level (achieved by complete inhibition of respiration with cyanide) and the maximally oxidized level (achieved by maximal stimulation of respiration with an uncoupler such as FCCP).

Protocols

Pyridine Nucleotides – NADH. Because the transmission of microscope optics usually falls off rapidly below 350 nm, quartz optics should be used if available with

---

[37] B. Chance and H. Baltscheffsky, *J. Biol. Chem.* **233**, 736 (1958).

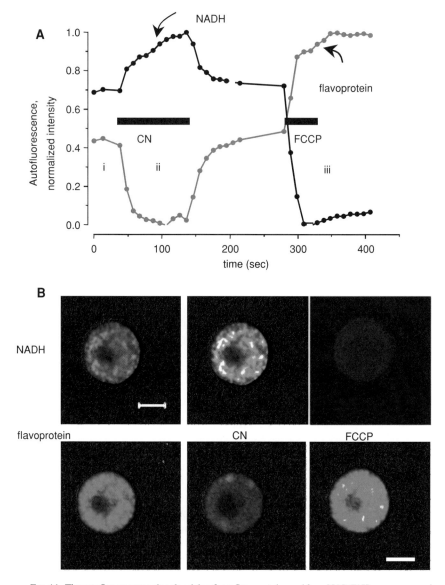

FIG. 11. The autofluorescence signals arising from flavoproteins and from NAD(P)H were measured from adipocytes grown in culture. Images were acquired on a confocal imaging system switching between excitation at 351 nm with the signal measured between 430 and 470 nm (NAD(P)H, and excitation at 458 nm with the signal measured through a long pass filter at 505 nm (flavoprotein). Cyanide (2 mM) and FCCP (1 μM) were applied as indicated from a nearby puffer pipette to induce fully reduced (CN) and fully oxidized (FCCP) states of the intermediates. The changes in signal with time are illustrated in (A) whereas (B) images sampled from the sequences at the times indicated (i, ii, and iii) are shown. The signals have been normalized with repair to the maximal and minimal signals (see p. 379).

a 350 ± 20-nm bandpass filter. Any light above 390 nm may excite flavoprotein fluorescence, which will not only contaminate the NADH signal, but also result in an underestimate of the NADH signal, as flavoprotein fluorescence changes inversely compared to the NADH signals. The emission may be measured using a wide bandpass filter with a peak at 450 nm and a bandwidth from ± 20 to 40 nm. The fluorescence tends to bleach, and excessive illumination can cause photodamage to the cell, and so, as in every case, it is best to keep the illumination intensity to a minimum consistent with reasonable signal to noise. This is always maximized in any fluorescence system by maximizing the efficiency of the light collection pathway. Use lenses with a high numerical aperture (NA) and sensitive low noise detection systems. On the Zeiss 510 confocal system, we have obtained excellent images by using a 40× oil immersion objective (NA 1.4) and carefully centering and focusing the laser and pinhole alignment, and by opening the pinhole to aid light collection (Figs. 10 and 11).

Flavoprotein autofluorescence is best elicited using excitation at 450 nm ± 20 nm and is measured between 500 and 600 nm (we have used a bandpass filter at 550 nm ± 40 nm, or simply image with a long pass filter at >510 nm). It is worth noting that on a conventional argon laser-driven confocal system, significant autofluorescence may be excited by a 488-nm argon laser and will contaminate many fluorescence signals if the dye emission is low and if a long pass filter is used (rather than a bandpass filter). On the Zeiss 510, we have used the 458-nm line of the argon laser to image flavoprotein autofluorescence. Indeed, using the multitracking facility of the microscope, we have devised a protocol that allows sequential (near simultaneous) measurements of flavoprotein and NADH signals in tandem (Fig. 11).

Measurement of autofluorescence may give a guide to the oxygen consumption at the level of the single cell, a variable that otherwise is hard to measure. Thus, if resting oxygen consumption is high, one would expect that the autofluorescence signals should not be greatly altered by an uncoupler, whereas inhibition of respiration should produce a relatively large change in signal. We have tried to quantify this index by scaling the autofluorescence between 0—for NADH the fully oxidized response to uncoupler—and 1, for NADH, the fully reduced signal in the presence of CN^-. In many cell types that we have examined in this way, the resting signal typically lies roughly midway between the two extremes[36]—we can define the resting level as ∼0.5. In cells known to have a high oxygen consumption, such as the Type I cell of the carotid body, we found a resting level for NADH at ∼0.2, suggesting that the redox state at rest was predominantly oxidized.[38] In contrast, in many cells in culture, oxidative metabolism seems to be suppressed, and the ratio moves toward the reduced, with levels closer to ∼0.8. Indeed, in published records of flavoprotein autofluorescence from cardiomyocytes kept in culture for

[38] M. R. Duchen and T. J. Biscoe, *J. Physiol.* **450,** 33 (1992).

just 1 day, the resting level appears to be almost completely reduced, so that CN^- causes almost no change in signal compared to a huge response to FCCP.[39]

Some caveats/validation. There is very little published material dealing with imaging of autofluorescence signals. We have recently carried out a number of experiments similar to that illustrated in Fig. 11 using a range of different cells. What is gradually emerging is that under conditions that favor either NADH or FAD fluorescence, the proportion of signal clearly attributable to these redox carriers and to mitochondria varies between cell types. Thus, in cardiomyocytes, it seems that almost all the signal in response to UV illumination is clearly mitochondrial— changing as predicted with CN and FCCP, whereas illumination at 458 nm often shows some spotty green fluorescence that is nonmitochondrial and that is in- dependent of mitochondrial respiration. In adipocytes, we have also seen some UV-induced blue fluorescence in fat droplets. We have no idea at present what this derives from, but it does show that one must be careful in making assumptions about the origin of the fluorescence signals. In contrast, changes in signal with changes in mitochondrial respiration can be safely attributed to the mitochondrial redox carriers.

Measuring Mitochondrial Calcium Handling

It has long been clear that isolated mitochondria show a massive capacity to accumulate calcium. The functional importance of that pathway has really become clear only recently. Ca^{2+} is accumulated into mitochondria through the ruthenium red-sensitive uniporter in response to an electrochemical gradient established by a combination of the negative mitochondrial membrane potential and a low in- tramitochondrial calcium concentration ($[Ca^{2+}]_m$) that is maintained by the mi- tochondrial Na^+/Ca^{2+} exchange (see Fig. 1). Calcium plays a central role in the regulation of mitochondrial metabolism, as the three major rate-limiting dehy- drogenases of the TCA cycle are all upregulated by calcium. Mitochondrial calcium uptake also influences the spatiotemporal patterning of $[Ca^{2+}]_c$ signals, acting as a spatial buffering system and regulating microdomains of $[Ca^{2+}]_c$ close to channels and influencing calcium signaling in subtle ways.[40–42] Excessive mi- tochondrial calcium accumulation may be toxic and initiate pathways that lead to cell death.

Measurement of changes in $[Ca^{2+}]_m$ within cells is thus of considerable interest both in physiological and pathophysiological processes. Although such measure- ments can be made with relative ease in preparations of isolated mitochondria, the

[39] B. O. Rourke, B. M. Ramza, and E. Marban, *Science* **265**, 962 (1994).
[40] L. S. Jouaville, F. Ichas, E. L. Holmuhamedov, P. Camacho, and J. D. Lechleiter, *Nature* **377**, 438 (1995).
[41] E. Boitier, R. Rea, and M. R. Duchen, *J. Cell Biol.* **145**, 795 (1999).
[42] M. R. Duchen, *J. Physiol.* **516**, 1 (1999).

functional consequences of a change in $[Ca^{2+}]_m$ can be properly understood only when the mitochondrion is studied in its native environment—the cell.

Imaging $[Ca^{2+}]_m$ with Fluorescent Indicators

When cells are loaded with standard acetoxymethyl ester (AM) derivatives of fluorescent indicators, the dyes inevitably tend to distribute throughout the cell, localizing into both organelle and cytosolic compartments. The degree of localization differs between cell types and between dyes and is dependent on dye concentration and loading temperature. In some cell types, predominant mitochondrial loading may be achieved with little effort. In many cells, standard loading techniques for dyes such as fura-2 may result in as much as 50% of dye within organelles, which can then be exploited to measure changes in $[Ca^{2+}]_m$.[43] This can be developed further by exaggerating those aspects of the loading procedures that increase compartmentalization and encouraging dye loss from the cytosol. For example, Csordás et al.[44] found that in a mast cell line (RBL-2H3 cells), loading with 5 μM fura-2FF/AM at room temperature for 60 min yielded a predominantly mitochondrial signal. In those experiments, the residual cytosolic dye was removed as the cells were permeabilized. The same result can be achieved using whole cell patch clamp recording to dialyze out the residual cytosolic dye.

In other instances, parameters of dye concentration and temperature need to be manipulated to a much greater degree in an attempt to get more selective loading into mitochondria. Griffiths et al.[30] found that isolated rat ventricular myocytes can be loaded with Indo-1 AM under conditions in which about half the dye is concentrated within mitochondria (5 μM, 15 min at 30°). If cells were then incubated for 1.5 hr at 37°, only the mitochondrial fluorescence remained. These protocols exploit the temperature sensitivity of plasmalemmal anion transporters to remove the cytosolic fraction of the indicator, and clearly depend on the expression of the transporters by the cell type being studied. The extent to which protocols involving exposure to higher temperatures for the selective removal of cytosolic dye are applicable to a wider array of fluorescent Ca^{2+} indicators and cell types remains to be determined.

Rhod-2-AM. The AM ester of rhod-2 and the closely related dyes X-rhod-1 and rhod-2FF are currently the only cell-permeant $[Ca^{2+}]$ indicators that carry a delocalized positive charge, and so are preferentially concentrated into polarized mitochondria. On hydrolysis of ester moieties the free acid remains trapped inside the mitochondria in which it reports increased $[Ca^{2+}]_m$ as an increase in fluorescence intensity. In our hands, in almost all cell types, the dye effectively partitions between cytosol and mitochondria so that a significant proportion of dye is left in

[43] S. Ricken, J. Leipziger, R. Greger, and R. Nitschke, *J. Biol. Chem.* **273**, 34961 (1998).
[44] G. Csordás, A. P. Thomas, and G. Hajnoczky, *EMBO J.* **18**, 96 (1999).

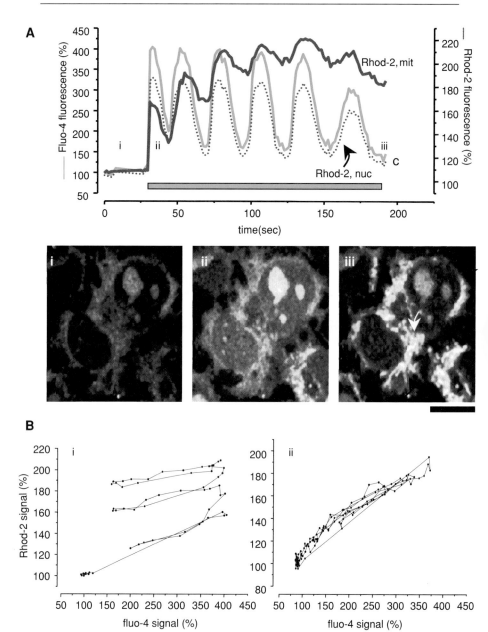

the cytosolic compartment. At first glance, and merely looking at images of quiescent, rhod-2-loaded cells, this may not seem to be much of a problem. Images may show a beautiful mitochondrial localization that can be well resolved, and this suggests that changes in the mitochondrial compartment can be followed and monitored accurately. However, stimulation of the cells to raise $[Ca^{2+}]_c$ to raise $[Ca^{2+}]_m$ usually also causes a significant increase in the fluorescence of rhod-2, which remains in the cytosol[41] (e.g., see Fig. 12). We have found that if the change in $[Ca^{2+}]_c$ is very large, this signal may completely obscure the mitochondrial rhod-2 signal. Thus, in hippocampal neurons stimulated with glutamate, we have found resolution of a clearly identifiable mitochondrial signal very difficult (J. Keelan and M. Duchen, unpublished observations). A major caveat then is that it is essential in using this dye to be able to resolve clearly that the measured signal is coming from identifiable mitochondria or that appropriate measures have been taken to minimize residual cytosolic dye. One solution is to use dyes that have a low affinity for Ca^{2+} on the grounds that the mitochondrial signal is likely to change more than the cytosolic signal,[45] but low-affinity variants of rhod-2 have not been available until very recently and have not been much characterized.

In an attempt to improve the mitochondrial localization of the dye, several groups have developed protocols that involve incubation at 37° for several hours or overnight after initial standard loading of the rhod-2 into the cells at room temperature for 30 min. During the incubation, plasma membrane transporters may eliminate the residual cytosolic dye without affecting the mitochondrial loading. Similar prolonged incubations apparently enhance the mitochondrial localization

---

[45] M. Montero, M. T. Alonso, E. Carnicero, I. Cuchillo-Ibanez, A. Albillos, A. G. Garcia, J. Garcia-Sancho, and J. Alvarez, *Nat. Cell Biol.* **2,** 57 (2000).

FIG. 12. The relationship of changes in mitochondrial and cytosolic $[Ca^{2+}]$ may be explored using separate indicators for each compartment. These images were obtained using HeLa cells coloaded with fluo-4 as a cytosolic indicator and rhod-2 as the mitochondrial indicator, and images were acquired using a confocal imaging system switching sequentially between excitation at 488 nm (measuring at 505–530 nm—fluo-4) and excitation at 543 nm (measuring at >585 nm—rhod-2). The cells were stimulated with 10 μM ATP applied from a nearby puffer pipette. (A) The traces show the intensity of signal with time over a small volume of cytosol including clearly defined mitochondrial structures and excluding as much as possible any contamination of extramitochondrial rhod-2 signal. Note that the cytosolic signal (green) oscillates whereas the mitochondrial signal (red) shows a damped oscillation and a progressive increase as the mitochondria accumulate calcium. The rhod-2 signal recorded over the nucleus is shown as a dotted red line. Images of the rhod-2 signal sampled from the time sequence at times indicated (i, ii, and iii) are shown below. In (B) we plotted the change in rhod-2 signal as a function of the fluo-4 signal over a mitochondrion (i) and over the nucleus (ii). The nuclear signal remains fairly linear (ii), suggesting that nuclear rhod-2 reflects the cytosolic signal, whereas the growth of the mitochondrial signal with each oscillation of the cytosolic signal is clear in (i).

of rhod-2 in myocytes,[46] hepatocytes,[47] and oligodendrocytes,[48] and this approach has clearly been used effectively by several groups. Indeed, we have found that neurons can be loaded with rhod-2 and then incubated at 37° for 12 hr to reveal a signal that is clearly highly localized to mitochondria. Remarkably, however, in our hands, the rhod-2 signal seems to become significantly less responsive to changes in Ca^{2+} after a long incubation, and we have failed to see changes in signal even after massive changes in calcium induced by ionophore.

We have experienced a similar problem when using dihydrorhod-2AM, a nonfluorescent derivative, readily made in the laboratory from rhod-2AM (instructions provided by Molecular Probes, Eugene, OR). This is taken up into mitochondria, in the same way as rhod-2-AM, and the cells are then washed and left in the incubator for several hours or overnight. Oxidation of the dye within the mitochondria produces the fluorescent rhod-2. We have tried this protocol and see clearly selective loading of dye into mitochondria. However, we found these signals completely unresponsive to manipulations that raise $[Ca^{2+}]_c$ massively, and have ceased using this approach. The lack of literature using the dye this way suggests that others have also found it difficult to use, and it does raise questions about the performance of the dye.

Conventional imaging techniques (such as a CCD camera) can be readily applied to image rhod-2 signals in flat thin cells, e.g., fibroblasts or astrocytes, in which mitochondria can be easily resolved.[41] Other cell types, including "fatter" neurons, require the use of confocal microscopy to resolve the rhod-2 fluorescence in different compartments unequivocally. Exactly the same constraints apply as described above in measuring mitochondrial potential, except that the contaminating dye in the cytosol is now Ca^{2+} sensitive, and may be exposed to larger dynamic changes than the mitochondrial signal. Confocal microscopy may permit improved resolution of the mitochondrial rhod-2, and reduce contamination by cytosolic fluorescence, but even this approach is not without problems. Peng et al.[49] have commented that measuring the fluorescence from a region of interest that includes large areas of cytoplasm outside the mitochondria will "dilute" the changes in $[Ca^{2+}]_m$, for example, in response to NMDA, but if the cytosol also contains a significant amount of dye then the measured signal will not be diluted but rather contaminated or distorted.

One of the most powerful features of the confocal system is the ability to acquire simultaneous images at different wavelengths without needing filter changes or time delays. This allows the combination of two indicators to measure the cytosolic and mitochondrial signals as shown in Fig. 12. HeLa cells in culture

---

[46] D. R. Trollinger, W. E. Cascio, and J. J. Lemasters, *Biochem. Biophys. Res. Commun.* **236,** 738 (1997).

[47] A. M. Byrne, J. J. Lemasters, and A. L. Nieminen, *Hepatology* **29,** 1523 (1999).

[48] P. B. Simpson and J. T. Russell, *J. Physiol.* **508,** 413 (1998).

[49] T. I. Peng, M. J. Jou, S. S. Sheu, and J. T. Greenamyre, *Exp. Neurol.* **149,** 1 (1998).

were loaded with both fluo-4 and rhod-2. As shown, the signal from a given mitochondrion can be related to the change in signal from the restricted volume of cytosol surrounding it, and so the local relationship between cytosolic calcium and mitochondrial uptake becomes accessible to study. These cells responded to ATP application with prolonged oscillatory changes in $[Ca^{2+}]_c$, reflected in the mitochondria where the signal was damped, reflecting the slower kinetics of mitochondrial calcium handling. Interestingly, the residual cytosolic rhod-2 signal measured over the nucleus of the cell was superimposable on the fluo-4 signal, reassuring us that differences in the mitochondrial and cytosolic signals do not simply reflect some difference in the behavior of the dye.

A novel and interesting approach to try to separate mitochondrial and cytosolic signals has been described recently by Gerencser and Adam-Vizi,[50] who have employed spatial Fourier filtering to try to resolve separately the component with a high spatial frequency (i.e., the mitochondrial signal) and that with a low spatial frequency component (i.e., the cytosolic fraction). These data look reasonably convincing and can be separated pharmacologically by inhibition of mitochondrial Ca^{2+} uptake. This kind of image processing is a useful approach if it is not possible to separate the signals in other ways.

Validation. It should be evident now that in discussing the validations and limitations involved in measuring $[Ca^{2+}]_m$, one cannot assume that dyes marketed as "cytosolic" or "mitochondrial" actually localize to these compartments when loaded as AM esters, and that adequate controls need to be conducted to be sure that one knows the true origin of the fluorescence signal in a particular preparation. This is particularly important when using fluorometric or conventional imaging techniques where mitochondria may not be adequately resolved.

Fluorescence can be shown to originate primarily from mitochondria if the fluorescence signals can be altered by inhibition of mitochondrial Ca^{2+} uptake. This is best achieved as follows.

1. *Inhibition of mitochondrial Ca^{2+} uptake by ruthenium red and of efflux by clonazepam or CGP37157.* Ruthenium red is a large cationic molecule that should by rights be impermeant. In fact, it has been reported as entering some cells (e.g., cardiomyocytes), but it generally requires specific methods of introduction into the cell, such as microinjection or lipofusion techniques. Ruthenium red is a very nonspecific agent, blocking a range of different classes of calcium channel, including the ryanodine receptor-mediated calcium release channel of sarcoplasmic reticulum. When microinjected, its effects on the plasma membrane are apparently avoided. A derivative, Ru360, which is more selective and is said to be "cell permeant,"[51] is now commercially available, although in our hands this compound

[50] A. A. Gerencser and V. Adam-Vizi, *Cell Calcium* **30**, 311 (2001).
[51] M. A. Matlib, Z. Zhou, S. Knight, S. Ahmed, K. M. Choi, J. Krause-Bauer, R. Phillips, R. Altschuld, Y. Katsube, N. Sperelakis, and D. M. Bers, *J. Biol. Chem.* **273**, 10223 (1998).

altered physiological intracellular Ca^{2+} signaling in astrocytes (J. Jacobson and M. R. Duchen, unpublished observations).

2. *Collapse of the mitochondrial potential, removing the driving force for calcium accumulation.* This is readily achieved using a mitochondrial uncoupler, such as FCCP, to abolish $\Delta\Psi_m$. The problem is to ensure that any changes in $[Ca^{2+}]_c$ signaling are not due to a secondary effect on cellular [ATP] and the consequential inhibition of Ca^{2+} efflux or sequestration via the relevant Ca^{2+}-ATPases. Inclusion of low concentrations of oligomycin (e.g., 2.5 $\mu g/ml$) to inhibit ATP hydrolysis by the F_1F_0-ATPase is more or less obligatory to help limit this problem. Collapse of $\Delta\Psi_m$ may similarly be achieved using a respiratory inhibitor together with oligomycin, and this will help to ensure that effects seen using an uncoupler are not due to collapsing proton gradients in other compartments or to pH shifts.

Note: Rhodamine-2 always seems to concentrate (or to give a brighter signal) in the nucleolus. This can be very bright so that the signal dominates the whole cell signal. At present we do not know what it is about the nucleolar environment that encourages the dye to stay there, but it is a very distinctive feature of the dye labeling (see Fig. 12).

Mitochondrially Directed Transfection of Recombinant Aequorin. It would seem negligent to discuss the measurement of intramitochondrial calcium without inclusion of the technique pioneered by Rizzuto and colleagues in the early 1990s, whereby the Ca^{2+}-sensitive photoprotein aequorin targeted to the mitochondrial matrix is expressed in cells. Cells are transfected with a chimeric cDNA encoding aequorin and a mitochondrial presequence from the mitochondrial enzyme cytochrome oxidase is encoded in order to target the fusion protein specifically to mitochondria.[52,53] This is an elegant technique ensuring specific measurement of Ca^{2+} changes within a discrete compartment. The signal generated by the photoprotein expressed this way is generally too weak to be useful for $[Ca^{2+}]_m$ imaging at the single cell level using current technology, and data are typically gathered from a population of cells. The approaches discussed above are nicely complementary with the use of recombinant aequorin; spatial information is inferred from the selective expression of the reporter protein in different compartments in the cells, and although single cell imaging using fluorescent dyes has limitations of accurate compartmentalization, the approach provides details of the spatiotemporal features of the signaling pathways.

Looking ahead, it seems likely that the specific targeting of protein reporters to specific compartments will be used to overcome many of the problems of specific probe localization to which we have drawn attention. Use of the brighter chameleon green fluorescent protein reporters is a burgeoning technology that may solve many of these problems.[54]

[52] R. Rizzuto, C. Bastianutto, M. Brini, M. Murgia, and T. Pozzan, *J. Cell Biol.* **126,** 1183 (1994).
[53] R. Rizzuto, M. Brini, M. Murgia, and T. Pozzan, *Science* **262,** 744 (1993).
[54] A. Miyawaki, O. Griesbeck, R. Heim, and R. Y. Tsien, *Proc. Natl. Acad. Sci. U.S.A.* **96,** 2135 (1999).

*Expression of Mitochondrially Targeted Proteins Labeled
with Fluorescent Markers*

Tagging with fluorescent adducts has provided a wealth of information on the distribution and trafficking of a variety of intracellular proteins.[55] Such use of the green fluorescent protein (GFP), derived from the jellyfish *Aequorea victoria,* is now widespread, and in combination with confocal microscopy has provided new insights into mitochondrial protein movements and distribution. Moreover, genetically modified mutants of GFP have provided probes with improved spectral properties greatly enhancing the usefulness of the fluorophores. GFPs with more than 20-fold stronger fluorescence that wild-type have been produced by mutagenesis and selection of the products by FACS analysis.[56] Additionally, thermotolerant mutations of GFP have facilitated use of the protein in animal cells as the wild-type protein fluoresces poorly at temperatures over 25°,[57] and fluorescent proteins with altered spectral properties have been used in experiments using fluorescence energy transfer (FRET).[58] The 238 amino acid, 27-kDa protein may be attached to either the C or N termini of proteins[59] and due to its relatively compact nature retains the ability to diffuse throughout the cytosol and enter the nucleus.

Several groups have utilized GFPs attached to mitochondrial targeting sequences to direct the fluorophore to mitochondria. The great majority of mitochondrial proteins are synthesized in the cytosol and much work has identified a host of protein presequences and other targeting sequences that direct proteins to the mitochondrial matrix, intermembrane space, and inner or outer membranes (for extensive review see Neupert[60]). By using known targeting sequences spliced to the cDNA coding for GFP, the GFP protein can be expressed *in vivo* and then targeted to the mitochondrial region of interest. Rizzuto and colleagues have shown how the targeting presequence of human cytochrome *c* oxidase (COX, complex IV of the respiratory chain) can be attached to photoproteins (initially aequorin, however the same technique may be used with GFPs) to allow mitochondrial targeting and import of the functional protein into the mitochondrial matrix.[61] After digestion with restriction enzymes to excise and isolate the targeting sequence, as well as the cDNA encoding a few of the amino acids of mature COX, the resultant cDNA fragment can be ligated to the cDNA encoding the GFP. By including a portion of the mature COX, the authentic cleavage site of the targeting sequence is retained

[55] A. B. Cubitt, R. Heim, S. R. Adams, A. E. Boyd, L. A. Gross, and R. Y. Tsien, *Trends Biochem. Sci.* **20**, 448 (1995).
[56] B. P. Cormack, R. H. Valdivia, and S. Falkow, *Gene* **173**, 33 (1996).
[57] K. R. Siemering, R. Golbik, R. Sever, and J. Haseloff, *Current Biol.* **6**, 1653 (1996).
[58] N. P. Mahajan, K. Linder, G. Berry, G. W. Gordon, R. Heim, and B. Herman, *Nat. Biotechnol.* **16**, 547 (1998).
[59] S. Wang and T. Hazelrigg, *Nature* **369**, 400 (1994).
[60] W. Neupert, *Annu. Rev. Biochem.* **66**, 863 (1997).
[61] R. Rizzuto, A. W. Simpson, M. Brini, and T. Pozzan, *Nature* **358**, 325 (1992).

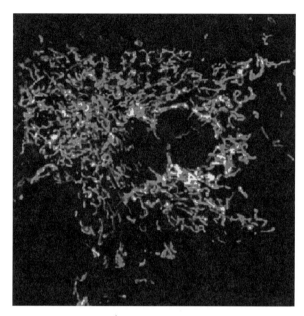

FIG. 13. Enhanced green fluorescent protein (GFP) targeted to mitochondria in primary cortical astrocytes. The N-terminal 20 amino acids of a mitochondrial targeting sequence (mistargeted alanine:glyoxylate aminotransferase) were attached to the N terminus of GFP. Alanine:glyoxylate aminotransferase (AGT) is normally targeted to peroxisomes, however, mistargeting of the enzyme to the mitochondrial matrix results in loss of AGT catalytic activity [M. J. Lumb, A. F. Drake, and C. J. Danpure, *J. Biol. Chem.* **274,** 20587 (1999)]. We have used this targeting sequence to label mitochondria with brightly fluorescent GFP in order to investigate the intracellular distribution of lipophilic cationic dyes, for example, nonyl acridine orange, under conditions in which the $\Delta\Psi_m$ is dissipated. Although mitochondrial protein import is initially dependent upon the $\Delta\Psi_m$, subsequent fluctuations in membrane potential do not induce redistribution of folded mitochondrial proteins. Hence we can use GFPs targeted to the mitochondrial matrix to provide markers for mitochondrial distribution when mitochondria are depolarized.

and then cleaved from the mature photoprotein. Such a technique may be used to examine the distribution of mitochondria in cells in which the use of fluorescent dyes would be impractical. For example, we have used a mitochondrially targeted GFP to establish the location of mitochondria in fixed astrocytes (Fig. 13) in which the fixation process induces redistribution of cationic fluorophores.

Although the targeting sequence for COX is widely used, alternative presequences may be spliced to the GFP cDNA. Thus markers can be directed highly specifically to their subcellular targets allowing, for example, characterization of the signal that directs the Tom 20 protein to the outer membrane,[62] and localization

[62] S. Kanaji, J. Iwahashi, Y. Kida, M. Sakaguchi, and K. Mihara, *J. Cell Biol.* **151,** 277 (2000).

of the key enzyme carnitine palmitoyltransferase 1 to the cytosolic face of the outer mitochondrial membrane.[63]

Validation—a Hobby Horse. To be confident that the transfection really is targeting mitochondria, it is customary to show some colocalization of the GFP signal with some other mitochondrial marker—a mitotracker or a potentiometric indicator in living cells or an antibody to a mitochondrial protein in fixed cells. We are always puzzled that most authors (and journals) consider that showing an image provides sufficient proof of colocalization. In a rigorous approach, one might consider some statistical validation appropriate, especially if the object is to define whether or not a protein really does target mitochondria. In this case, image analysis provides a simple solution in that several image-processing packages allow a simple quantitative analysis of image data on two channels (say red and green) that defines, within an area of interest, how many pixels are red, how many are green, and how many are both red and green. The latter variable can then be expressed as a percentage of the total, i.e., of pixels containing fluorescence signal from the GFP, what proportion of those pixels are also red (e.g., staining with TMRM from mitochondria)? Clearly this allows a statistical validation of the colocalization of signal, which seems far more satisfactory than some vague statement and a single image. Further, the same approach can be used to follow redistribution of a protein—a shift in the proportion of pixels that contains signal from both indicators may provide a very interesting analytical tool, as long as only one of the signals is changing!

CODA

Over the past few years, we have witnessed a remarkable change in our perception of the mitochondrion. It has become generally clear that these organelles are immensely important in health and in disease and so there is a general impetus to explore their behavior in a wide range of physiological and pathophysiological scenarios. Imaging provides a wonderful window to explore the otherwise inaccessible properties of these remarkable structures within living cells, and we will look forward to watching the field develop further in the coming years.

Acknowledgments

We would like to thank the Wellcome Trust, the Medical Research Council, and the Royal Society and INTAS for financial support. We would also like to thank Professor David G. Nicholls for stimulating discussions on the measurement of mitochondrial potential and for making his mathematical model widely available.

[63] F. R. van der Leij, A. M. Kram, B. Bartelds, H. Roelofsen, G. B. Smid, J. Takens, V. A. Zammit, and J. R. Kuipers, *Biochem. J.* **341**, 777 (1999).

[18] Dynamic Imaging of Neuronal Cytoskeleton

By Erik W. Dent and Katherine Kalil

Introduction

Growth cones are the motile tips of growing axons. During development growth cones guide axons along specific pathways to appropriate targets by extending toward or retracting away from attractive or inhibitory guidance cues in their environment. Growth cone motility as well as extension, retraction, and turning behaviors are directed by the neuronal cytoskeleton, principally actin filaments and microtubules. Developing neurons, such as those from the mammalian cerebral cortex, also form connections in the central nervous system (CNS) by extending branches interstitially from the axon shaft. The actin and microtubule cytoskeleton directs growth at the axon tip as well as at axon branch points. Thus, an understanding of how the movement and dynamics of actin filaments and microtubules are regulated and how these two cytoskeletal elements interact is essential for understanding axon pathfinding and formation of connections in the CNS.

Imaging the dynamic cytoskeleton in living neurons from the mammalian CNS presents a number of technical challenges. In comparison with larger neurons from invertebrates or the vertebrate periphery, neurons from the cerebral cortex are relatively small, which makes them particularly difficult to microinject with fluorescent probes for labeling and visualizing the cytoskeleton. Although changes in the location and organization of microtubules and actin filaments can occur rapidly in cortical growth cones, imaging periods of many hours are necessary to visualize how these events are related to protracted events such as development of axon branches. Imaging cortical neurons over many hours at a resolution necessary to detect single fluorescently labeled microtubules or actin filament bundles is difficult to achieve without compromising the viability of the neuron. Even limited photodamage has deleterious effects on growth cone motility and axon branching. Most previous work on dynamic imaging of the neuronal cytoskeleton has used direct injection into large neurons from invertebrates such as *Aplysia, Helisoma,* or grasshopper[1-5] or used injection into the blastula to label neurons from the

[1] J. H. Sabry, T. P. O'Connor, L. Evans, A. Toroian-Raymond, M. Kirschner, and D. Bentley, *J. Cell Biol.* **115**, 381 (1991).
[2] T. P. O'Connor and D. Bentley, *J. Cell Biol.* **123**, 935 (1993).
[3] C. H. Lin and P. Forscher, *Neuron* **14**, 763 (1995).
[4] E. A. Welnhofer, L. Zhao, and C. S. Cohan, *Cell Motil. Cytoskeleton* **37**, 54 (1997).
[5] N. Kabir, A. W. Schaefer, A. Nakhost, W. S. Sossin, and P. Forscher, *J. Cell Biol.* **152**, 1033 (2001).

Xenopus neural tube.[6–9] Typically the approach has been to label and image a single cytoskeletal element such as actin filaments in the living neuron and then fix and stain the same neuron to localize a second cytoskeletal element such as microtubules. This approach, although useful, provides no direct information about dynamic interactions between microtubules and actin filaments that drive growth cone motility and guidance behaviors. Although large flat growth cones from *Aplysia* or *Helisoma* neurons are ideal for imaging cytoskeletal reorganization, the use of neurons from the mammalian CNS has several advantages for studying how the cytoskeleton drives growth cone guidance behaviors. First, a number of guidance molecules has been identified in the mouse, and knockout mice have been developed to test the function of such molecules. A further rationale for studying CNS neurons is that understanding how the cytoskeleton is regulated during axon branching may help to promote regenerative axon sprouting of damaged axons.

In this article we describe techniques for simultaneous imaging of the actin and microtubule cytoskeleton in the processes of living cortical neurons from the rodent brain. All of the imaging hardware and software, the microinjection system, the tissue culture reagents, and the fluorescent cytoskeletal probes are commercially available. Here we emphasize how the use of standard cell culture and imaging techniques can be adapted to permit high-resolution imaging of cytoskeletal dynamics in CNS neurons over the extended time periods necessary to study developmental events such as axon guidance and branching.

Imaging of Cytoskeleton

Preparation of Coverslips and Chambers

Although it may seem like a trivial step we cannot emphasize too strongly the importance of proper preparation of the coverslips, which is necessary for neurons to adhere, survive, and extend processes. We use etched grid glass coverslips to provide landmarks for locating injected neurons for imaging. Without these landmarks it would not be possible to return repeatedly to the same neuron during intermittent imaging. Coverslips are washed extensively following the protocol of Goslin *et al.*[10] This procedure, which takes 4–5 days, requires cleaning in concentrated nitric acid (18–36 hr) followed by extensive washing in tissue culture grade water. After the coverslips are air dried they are attached with a 3 : 1 mixture of

[6] E. M. Tanaka and M. W. Kirschner, *J. Cell Biol.* **115,** 345 (1991).
[7] E. Tanaka, T. Ho, and M. W. Kirschner, *J. Cell Biol.* **128,** 139 (1995).
[8] S. Chang, V. I. Rodionov, G. G. Borisy, and S. V. Popov, *J. Neurosci.* **18,** 821 (1998).
[9] S. Chang, T. M. Svitkina, G. G. Borisy, and S. V. Popov, *Nat. Cell Biol.* **1,** 399 (1999).
[10] K. Goslin, H. Asmussen, and G. A. Banker, *in* "Rat Hippocampal Neurons in Low-Density Culture" (G. A. Banker and K. Goslin, eds.), p. 339. MIT Press, Cambridge, MA, 1998.

paraffin and petroleum jelly[11] to 35-mm plastic culture dishes (Corning, Corning, NY) in which holes 15 mm in diameter have been machine drilled. These chambers are then sterilized under ultraviolet (UV) light. At this point the chambers can be stored indefinitely. Just before use the coverslips are coated with 1.0 mg/ml poly(D-lysine) (30 kDa, Sigma, St. Louis, MO) in borate buffer (76 mM boric acid, 8 mM borax) for at least 1 hr and then rinsed three times with tissue culture grade water.

Cell Culture

Preparing healthy cortical cultures with a minimum of debris is absolutely essential for long-term neuronal survival during live cell fluorescence imaging. To obtain viable cultures from rodent cortex a general rule is the younger the better. For mice and rats, late embryonic stages (E18) work well. At earlier stages of development cortical neurons have fewer processes. Because dissociation shears off axons and dendrites, cultures prepared from cortex before neurons have extended processes have less cellular debris. Another advantage to using younger cortical tissue is that few glial cells are present.

We use 0- to 3-day-old Syrian golden hamsters (*Mesocricetus auratus*) that have only a 16-day gestation period and are born in a less mature state than other rodents. Pups are anesthetized on ice and decapitated. The entire brain is removed and immediately transferred to ice-cold dissection medium [Hanks' balanced salt solution (HBSS) (Gibco, Grand Island, NY) and B27 supplement (Gibco)]. We use the sensorimotor cortex, which has many large pyramidal neurons with long branched efferent axons. The cortex is dissected away from the rest of the brain, stripped of meninges, and cut into small pieces (Fig. 1) with a tungsten needle (Fine Science Tools, Foster City, CA). To avoid cell death it is important to work quickly without attempting to dissect the cortex too finely. Cortical pieces are washed twice with HBSS without Ca^{2+}/Mg^{2+} (Gibco) and digested in HBSS without Ca^{2+}/Mg^{2+} with 0.025% trypsin/EDTA (Gibco) and 0.05% DNase I (Sigma) for 15 min at 37°. Enzymatically digested cortical pieces are washed twice in plating medium [neurobasal medium (Gibco), 5% fetal bovine serum (FBS) (Hyclone, Logan, UT), B27 supplement (Gibco), 0.3% glucose, 1 mM L-glutamine, and 37.5 mM NaCl], dissociated by trituration, and centrifuged at 200 rpm (16g) for 7 min at room temperature. The osmolarity of the culture medium is increased to 300–310 mOsm with NaCl to minimize osmotic shock to the neurons during microinjection. During trituration, fragile cortical neurons are easily damaged and it is therefore important to avoid bubbling the medium. We also use standard 1-ml Eppendorf pipette tips that have a consistent tip diameter. We do not use fire-polished glass pipettes

[11] K. Goslin and G. A. Banker, *in* "Rat Hippocampal Neurons in Low-Density Culture" (G. A. Banker and K. Goslin, eds.), p. 251. MIT Press, Cambridge, MA, 1991.

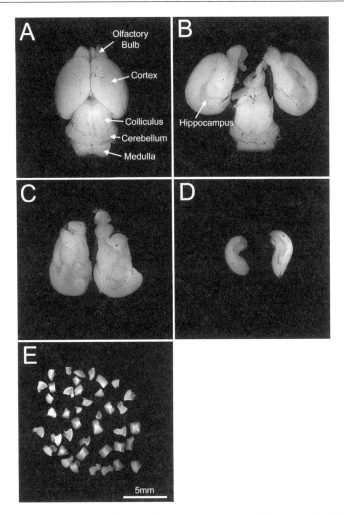

FIG. 1. Dissection of the cortex. (A) Dorsal view of the early postnatal rodent brain. (B) The same brain after removal of the overlying cortex. The location of the hippocampus is shown on the underside of the cortical hemisphere. (C) The cerebral hemispheres shown in the same orientation as (B) with the sensorimotor cortex (D) removed. (E) The sensorimotor cortex after dissection into small pieces prior to dissociation.

because these tip diameters can become constricted and cause cell damage. Pellets are resuspended in serum-containing medium and plated at a density of 1000–8000 cells/cm^2 on etched grid coverslips (Bellco, Vineland, NJ). Neurons fare better at high densities. However, for our purposes, to avoid the complexities of cell–cell interactions among neurons, we culture cells at lower densities. Cells are incubated at

$37°/4\%$ CO_2 and 9% O_2 (Forma model 3140) because we have found that the standard O_2 tension of 18% in standard CO_2 incubators is too high for CNS neurons.[12,13] After 30 min to 1 hr, a 10× volume of serum-free medium (plating medium without FBS) is added to the cultures. Under these culture conditions cortical neurons remain viable for 5–7 days and develop a polarity similar to that observed in cultured hippocampal neurons,[14] which develop a single long axon and several minor processes.[15] These cultures contain very few glial cells (<5%). For studies of events such as synapse formation that require cortical neurons to survive for several weeks, plating neurons at high densities or the use of glial feeder layers is necessary.[11]

Microinjection

The development of green fluorescent protein (GFP) constructs for labeling proteins has made cellular transfection an important strategy for imaging the dynamics and movement of the cytoskeleton.[16–20] This approach has the potential advantage of labeling many cells at a time. However, transfection techniques are not always reliable or efficient. Thus, although microinjection is somewhat laborious and labels only one cell at a time, we have found this method for labeling the actin and microtubule cytoskeleton extremely useful for several reasons. First, one can choose for microinjection neurons of the desired morphology, state of development, and robustness. Second, incorporation of GFP constructs into cytoskeletal elements requires at least 10 hr whereas microinjected cytoskeletal probes are incorporated within 1 hr.[21,22] This is an important advantage for studying precisely timed developmental events such as polarization, axon branching, and growth cone dynamics. Third, if labeling of two or more intracellular elements is required it is uncertain whether transfection strategies will achieve efficient labeling of more than one protein in a single neuron. With microinjection we were able to label both actin filaments and microtubules at the same time in the same neuron with a high degree of reliability. Further, GFP constructs are relatively large molecules in contrast to fluorescent probes such as rhodamine and cyanine dyes, which are therefore less likely to interfere with the normal function of the protein of interest.

[12] D. W. Lubbers, H. Baumgartl, and W. Zimelka, *Adv. Exp. Med. Biol.* **345**, 567 (1994).
[13] G. J. Brewer and C. W. Cotman, *Brain Res.* **494**, 65 (1989).
[14] C. G. Dotti, C. A. Sullivan, and G. A. Banker, *J. Neurosci.* **8**, 1454 (1988).
[15] A. D. de Lima, M. D. Merten, and T. Voigt, *J. Comp. Neurol.* **382**, 230 (1997).
[16] C. Ballestrem, B. Wehrle-Haller, and B. A. Imhof, *J. Cell Sci.* **111**, 1649 (1998).
[17] A. Choidas, A. Jungbluth, A. Sechi, J. Murphy, A. Ullrich, and G. Marriott, *Eur. J. Cell Biol.* **77**, 81 (1998).
[18] M. Fischer, S. Kaech, D. Knutti, and A. Matus, *Neuron* **20**, 847 (1998).
[19] L. Wang, C. L. Ho, D. Sun, R. K. Liem, and A. Brown, *Nat. Cell Biol.* **2**, 137 (2000).
[20] S. Roy, P. Coffee, G. Smith, R. K. Liem, S. T. Brady, and M. M. Black, *J. Neurosci.* **20**, 6849 (2000).
[21] Y. Li and M. M. Black, *J. Neurosci.* **16**, 531 (1996).
[22] E. W. Dent, J. L. Callaway, G. Szebenyi, P. W. Baas, and K. Kalil, *J. Neurosci.* **19**, 8894 (1999).

We prepare pipettes for microinjection from thin-walled micropipettes containing filament (1.0 mm outer diameter, (World Precision Instruments, Sarasota, FL)) that are pulled to tip sizes of approximately 0.5 μm and stored on ice. We use a Sutter P-97 pipette puller, which can be reliably calibrated to reproduce the same shape and diameter. The pipettes are backloaded with the desired probe(s) with microloader pipettes (Eppendorf, Hamburg, Germany). To label microtubules we initially prepared tubulin from bovine brains and fluorescently labeled the tubulin with 5-(and 6)-carboxytetramethylrhodamine succinimidyl ester (Molecular Probes, Eugene, OR). However we now use commercially available rhodamine–tubulin (Cytoskeleton, Inc., Denver, CO). We prepare a stock solution from lyophilized rhodamine by adding ice-cold injection buffer (100 mM PIPES, 0.5 mM MgCl$_2$, pH 6.9) to the desired concentration (4–10 mg/ml). Prior to loading the pipettes this solution is centrifuged at 14,000g for 5 min at 4° to remove particulates. To label actin filaments we have microinjected phalloidin, which binds to actin filaments, coupled to Alexa dyes (Molecular Probes), which are resistant to photobleaching. We use low concentrations of phalloidin because at high concentrations it can potentially stabilize actin filaments. Commercially available rhodamine nonmuscle actin (Cytoskeleton Inc.) can also be used. However, we have found that actin tends to clog the narrow-diameter pipettes that are required for the injection of cortical neurons. The phalloidin is kept as a methanol stock solution at −20°, dried under a nitrogen stream, and resuspended in 0.5 μl dimethyl sulfoxide (DMSO). The desired volume of injection buffer or rhodamine tubulin is then added and the contents are centrifuged as described above.

To microinject cortical neurons we recommend an Eppendorf microinjector 5242/Micromanipulator 5170 mounted on the stage of an inverted microscope (Nikon TE300). This state of the art microinjection system has a micromanipulator capable of very fine movements and the microinjector can be precisely adjusted to specific low pressures (1–2 kPa). The culture dishes are placed on the microscope stage and the cells are located under differential interference contrast (DIC) optics with a 100×/1.4NA Plan Apochromat oil immersion objective and long working distance condenser (Nikon). To locate and position the micropipette close to the cell of interest the objective is switched to a 10× Neofluor objective. During the entire microinjection procedure it is important to maintain positive pressure to the micropipette to avoid backfilling it with media from the culture dish. The 100× objective is swung back into position and the microinjection procedure is visualized directly through the eyepieces. Pyramidal neurons 15–20 μm in diameter are chosen for microinjection not only for their large size but also because pyramidal cells in the cortex have long efferent axons with numerous collateral branches. We usually aim for regions of the cell body that have sufficient cytoplasm to accommodate additional volume from the contents of the micropipette. The micropipette is lowered toward the neuron just until the cell surface is dimpled. Then the side of the microscope is tapped lightly until the pipette tip penetrates the neuron. Almost immediately the cell body begins to swell, indicating that the contents of the pipette are being ejected under constant pressure. It is important that the pipette be withdrawn immediately to prevent bursting of the cell. Removing the pipette at an angle different from the angle of entry can tear the plasma membrane and kill the neuron. To avoid this the Eppendorf microinjection system can be automated. However, we have found that automated microinjection does not work well on cortical neurons because the area of cytoplasm is so small and the elasticity of the neuronal cell membrane tends to deflect the pipette. Damage to the cells by microinjection is immediately apparent by collapse of the membrane and increased vesicular movement inside the cell. We allow the neurons to rest for periods of 30–60 min for cells to incorporate the cytoskeletal probes and recover from the injection procedure. We find that a single pipette can be used to inject about 10 neurons per dish. Because the neurons are injected in neurobasal-based serum-free medium, which contains bicarbonate, we inject only as many neurons as is possible in 15 min (usually about 10 neurons) and then return the dish to the incubator, before the pH rises to levels that are injurious to the cortical neurons.

Choice of Microscope, Camera, and Imaging Software

We chose the Nikon inverted microscope model TE300 Quantum with fluorescence and differential interference contrast (DIC) optics (Fig. 2). The microscope is equipped with $10\times/0.5$NA and $20\times/0.7$NA Neofluor CFI60 series objectives and a $40\times/1.0$NA, $60\times/1.4$NA and $100\times/1.4$NA Plan Apochromat CFI60 series of objectives. Although we also have Zeiss inverted microscopes in the laboratory, only the Nikon Plan Apochromat objectives have the necessary working distance that will allow use of Bellco etched coverslips (0.21 mm nominal thickness). Plan Apochromat objectives are essential for imaging more than one fluorophore in the same focal plane. The microscope has three camera ports. The bottom port, on which the cooled charge-coupled device (CCD) camera is mounted, has the important capability of distributing 100% of the available light to the camera. This feature allows less illumination to be used, which reduces photodamage to living fluorescently labeled cells. The microscope is mounted on an antivibration table (TMC Instruments, Peabody, MA) with a hole cut into the table through which the cooled CCD camera is mounted securely on the bottom port of the microscope. To avoid accidentally bumping the bottom mounted camera we suggest surrounding the camera with a protective aluminum cage (Fig. 2B).

We chose a back-illuminated frame transfer-cooled CCD camera (Micro MAX-512EBFT, Roper Scientific). The frame transfer capability of this camera allows images to be acquired simultaneously with transfer of the previous image to the computer. This important feature decreases the time between acquisition of sequential images and thus allows rapid acquisition of closely spaced sequential

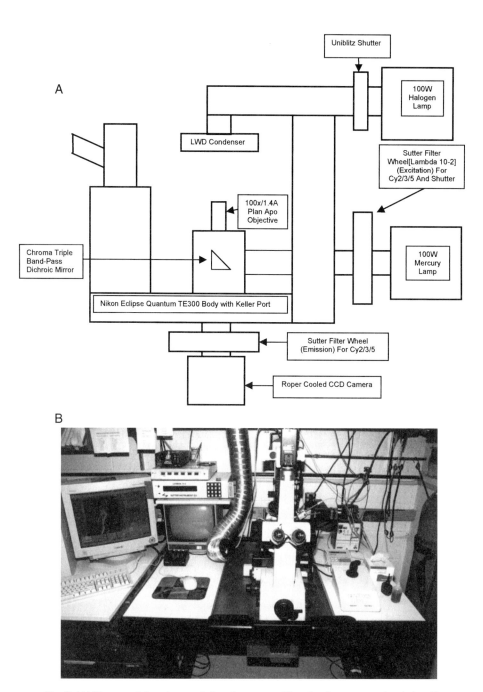

FIG. 2. (A) Diagram of the microscopic imaging system. Note that the camera and emission filter wheel are mounted below the microscope and that the excitation filter wheel is placed between the microscope and the mercury lamp. (B) Photograph of our imaging station illustrating the placement of the major components of the system.

images. The back illumination of the camera makes it one of the most sensitive cameras for collecting visible wavelengths of light (75–85% quantum efficiency). It also has one of the highest resolutions for a back-illuminated camera with a pixel size of 13.0 μm^2. We have found this camera ideal for observing dynamic changes in actin filaments and microtubules at high resolution over extended time periods of many hours. Another useful feature of this camera is that it can be operated in video mode. This allows one to locate cells of interest in DIC and bring them into focus. Note that the camera and eyepieces are not parfocal because of the placement of the emission filter wheel. The microscope is equipped with a Sutter Lambda 10-2 double filter wheel set up for excitation and emission filters plus a standard electronic (Uniblitz) shutter to the halogen light source. The Sutter filter wheels are equipped with Chroma #61005 excitation/emission filter sets capable of exciting Cy2/Cy3/Cy5 wavelength dyes. A DIC polarizer (Chroma) is also placed in the filter wheel. This filter wheel system allows for acquisition of images from at least two different wavelength fluorophores with a DIC (Nomarski) image in quick succession and in almost perfect register. This capability is essential for visualization of actin and microtubules in quick succession within living neurons. Note that because the emission filter wheel is interposed between the bottom port of the microscope and the CCD camera, the camera must be attached to the filter wheel case, which is incapable of supporting the weight of the camera. Therefore, we have mounted the camera on a specially designed aluminum block (Sutter Instruments) attached to the filter wheel case.

There are now a number of commercially available imaging software packages (as well as the NIH Image shareware, which is free) and it is no longer necessary for investigators to write their own software. We use Universal Imaging Metamorph Windows-based image analysis software. This software package performs automated acquisition and analysis of high-resolution digital images, coordinated operation of the shutters and filter wheels, and many postacquisition processing functions (see below). Moreover, this software can be continually upgraded. We strongly advise against built-in imaging systems that cannot be modified or upgraded. At present Metamorph is compatible only with PC-based computers. We recommend using a computer with a fast processor (>1 GHz), as much RAM as possible (at least 512 MB), and several large-capacity hard drives (>30 GB).

Microscopy

Maintaining CNS neurons in a healthy state during long-term imaging presents special challenges. For long-term imaging it is important to avoid a build up of free radicals in the neuron. Therefore, prior to imaging we fill the chambers with an imaging medium containing inhibitors to free radicals that augment those already present in B27. This imaging medium consists of serum-free medium with 1 mM sodium pyruvate (Sigma), 400 μM L-ascorbic acid (Sigma), 10 μM Trolox C (Fluka-Sigma, Milwaukee, WI) and 100 μM butylated hydroxyanisole (BHA, Sigma).[23-26] To have a supply of the medium available at the proper temperature and pH we place a vial containing imaging medium in the incubator overnight. Several hours before imaging a syringe is prepared that contains the imaging medium to which a 1:100 dilution of Oxyrase (Oxyrase, Inc., Mansfield, OH) is added.[27,28] Oxyrase is a solution of oxygen scavenging bacterial membranes that is inert to neurons but will decrease the oxygen tension to almost zero. This low level of oxygen is intolerable to cortical neurons for any length of time. Therefore, immediately before imaging and before sealing the tops of the chambers (see below) the imaging medium from the syringe is filtered through a 0.2-μm filter attached to an 18-gauge needle to remove the Oxyrase membranes.

We use a closed chamber system to maintain pH and to prevent evaporation. This is accomplished by placing an 18-mm-diameter glass ring (Thomas Scientific) over the well in the plastic culture dish containing the etched coverslip on which the cortical neurons are growing (Fig. 3). The top to this chamber consists of a 25-mm round coverslip placed on the glass ring and sealed with silicone grease (Dow Corning).[11] The same grease is used to seal the glass ring onto the plastic culture dish. This sealed system allows one to heat the stage with an airstream incubator (Nicholson Precision Instruments) without evaporating the media in the chamber. To avoid vibration the air steam incubator is placed on a wall-mounted shelf, a 6-foot length of aluminum dryer vent is attached to the blower, and other end of the vent is brought to within about a foot from the microscope stage (Fig. 2B). It is important to maintain the temperature between 35 and 37°. This is measured with a digital traceable thermometer (Fisher Scientific) attached by a flexible wire lead to a small probe (YSI series 400). The temperature is calibrated by attaching the probe to the bottom surface of the chamber itself with a temp heart (YSI) and then measuring the temperature on the microscope stage when the chamber temperature stabilizes at 36°. If the stage temperature goes above this calibrated temperature the control on the airstream incubator can be lowered during the experiment. Maintaining the appropriate temperature is important not only for the viability and motility of the cultured neurons but also for preventing focus drift. Another approach is to keep the entire room at a constant temperature of 36°.

[23] H. S. Chow, J. J. Lynch, III, K. Rose, and D. W. Choi, *Brain Res.* **639,** 102 (1994).
[24] E. Ciani, L. Groneng, M. Voltattorni, V. Rolseth, A. Contestabile, and R. E. Paulsen, *Brain Res.* **728,** 1 (1996).
[25] S. Desagher, J. Glowinski, and J. Premont, *J. Neurosci.* **17,** 9060 (1997).
[26] M. E. Rice, *Methods* **18,** 144 (1999).
[27] C. M. Waterman-Storer, J. W. Sanger, and J. M. Sanger, *Cell Motil. Cytoskeleton* **26,** 19 (1993).
[28] A. V. Mikhailov and G. G. Gundersen, *Cell Motil. Cytoskeleton* **32,** 173 (1995).

A   Cell Culture Chamber

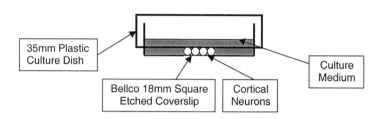

B   Imaging Chamber

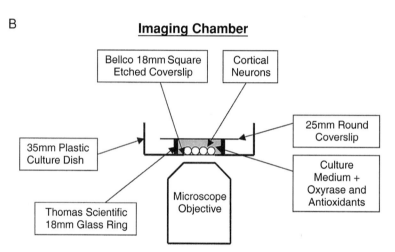

FIG. 3. Diagrams of the cell culture system. (A) Schematic drawing of the plastic culture dish in which cortical neurons are cultured. (B) Schematic drawing of the sealed chamber in which cortical neurons are imaged on the microscope.

Image Acquisition

Several imaging strategies can be used depending on the time course of the dynamic events to be observed. For longer term events, such as the development of branches from the axon shaft or the transformation of a large paused growth cone into a new axon, we have carried out intermittent imaging for periods up to 30 hr. For fluorescently labeled neurons to survive, and to prevent fluorescence from fading, is it clearly impossible to image at frequent intervals over so many hours. Therefore, for these time periods we typically capture images every 30–60 min. To maximize efficiency, approximately five to six dishes are maintained, each containing several viable fluorescently labeled neurons. After an image of each

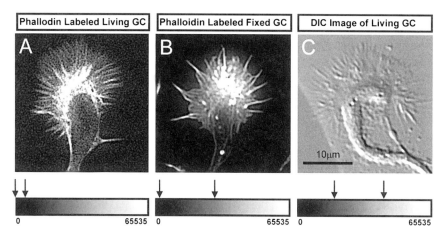

FIG. 4. Examples of different regions of the gray scale used for different types of images. (A) For the living fluorescently labeled growth cone (GC) a small region at the low end of the gray scale is used. (B) For the fixed fluorescently labeled growth cone a broader region of the gray scale can be used because of the greater illumination. (C) For the DIC image of the living growth cone a higher region of the gray scale is used. Numbers at the bottom of figures refer to the complete range of the gray scale and arrows represent regions of the gray scale used in each image.

fluorophore is captured for the neurons in a single dish the dish is returned to the incubator. Here, the etchings on the coverslips are invaluable for following the same neuronal processes over many hours. We have found that neurons tolerate this schedule of imaging such that their development and differentiation proceed normally.[22,29] For more rapid dynamic events, such as microtubule movements or their dynamic growth and shrinkage as well as their interactions with actin filaments, we carry out frequent imaging over a much shorter time period. For monitoring cytoskeletal changes we typically capture images every 5–10 sec for periods up to 30 min. In our experience, dynamic cytoskeletal events rarely survive more than approximately 250 sequential images under low-light level conditions.

Imaging is carried out under low-light level conditions to minimize photo damage to the fluorescently labeled living neurons. This requirement dictates all the other imaging strategies, from choice of the objective to efficiency of the filter sets to resolution and sensitivity of the camera. It is often frustrating when comparing images of fixed and fluorescently labeled neurons, where full illumination provides high resolution of cytoskelctal elements, with images of a living fluorescently labeled neuron where low-level illumination inevitably compromises resolution of individual microtubules and actin filaments (Fig 4). Nevertheless, with patience it is possible to acquire images of the dynamic cytoskeleton that approximate the

[29] E. W. Dent and K. Kalil, J. Neurosci. 21, 9757 (2001).

degree of detail observed in fixed cells. To locate the microinjected neurons of interest (whose positions have been previously recorded in relation to the coverslip etchings) and to determine their viability we scan the coverslip under direct visualization through the eyepieces with DIC optics and the 100× objective. For high-resolution fluorescence imaging we use only the 100×/1.4NA Plan Apochromat Nikon CFI60 objective under oil immersion. Putatively microinjected healthy neurons at the desired stage of development are located under DIC optics and, after removing the DIC slider, a fluorescent image is acquired with the cooled CCD camera under low-light level conditions (5–10% of the mercury light output). Injected neurons are never observed with fluorescence illumination through the eyepieces because the amount of light and the time required to determine if it is injected would cause photodamage that would eventually lead to death of the neuron. Because the images to the eyepieces and the camera are not parfocal the focal plane of the camera must first be located under DIC optics before acquisition of the fluorescence image(s). This scanning procedure is repeated throughout the dish and candidate cells are chosen for further imaging. To minimize light damage to other neurons in the dish and to other regions of the neuron the epifluorescence field diaphragm is closed down to include only the region of interest, which has been centered in the field. Therefore, for most images we are using only about 25% of the area of the camera chip, which has a full dimension of 512×512 pixels. We select this region of the chip with Metamorph software (Universal Imaging).

The imaging procedures are automated through the Metamorph imaging software (Universal Imaging) by choosing specific criteria from the menu such as light level, exposure time, and the interval between successive images of one or more fluorophores. This automation is accomplished by writing journals specifically tailored to each experiment. Typically, we carry out imaging under 5–10% of the available light from the 100-W mercury lamp by means of neutral density filters placed in the light path, mounted either in the Sutter excitation filter wheel (Chroma filters) or in the microscope (Nikon filters). However, the microinjected neurons in our experiments are very bright and dimmer images may require more light. Exposure time also determines the amount of illumination from the specimen to the camera. Thus, for dim fluorescence one can choose a longer exposure. In our experiments with the available light attenuated to 5–10% we use an exposure time of 100–500 ms. The interval between acquired images is determined empirically by the speed of the dynamic events under observation. We have found that analysis of cytoskeletal reorganization in the growth cone requires imaging every 5–15 sec. During the course of the experiment the image is viewed on the computer monitor because the amount of light required for visualization through the eyepieces would kill the cell, whereas the camera is much more sensitive than the eye and can detect the image at much lower light levels. Projection of the image on the monitor also makes it possible for several observers to view the image simultaneously. Because each newly acquired image is projected in succession on the monitor, continual adjustments of the microscope focus can be made. During imaging periods drift in the focus is inevitable because of temperature fluctuations that affect microscope, objective, immersion oil, and specimen. The camera itself determines how many gray levels are available, i.e., a 12-bit camera has 4096 gray levels whereas a 16-bit camera has 65,535 gray levels (our camera). Because it is not possible to view the entire gray scale at once it is necessary to choose the range of gray values best suited for specific images. This principle is illustrated in Fig. 4, which shows that different parts of the gray scale are used for DIC vs fluorescence images of living neurons. The imaging software allows selection of the appropriate range of gray values over which images are viewed. Typically, for a fluorescent image of a living growth cone only about 1000 gray levels are used because using the full range would require either longer exposure times or higher levels of illumination, both of which would damage the cell. In contrast, a DIC image could use more of the gray scale because the transmitted light from the halogen light source tends to be much less injurious to the injected neurons.

Digital images acquired by the camera must now be stored. As mentioned above, we use only about one-quarter of the area of the camera chip to acquire an image. Nevertheless, each image is still relatively large (~150 kb). This presents a problem if hundreds of images are stored on the hard drive of the computer, which has a capacity of only 30–60 Gb. One solution would be to store images on writable CDs, which have a storage capacity of 700 Mb. This translates to about 5000 images. Although this does not present a storage problem per se, manipulation of the images requires much more additional capacity. A further drawback to the use of CDs is that accessing the images for manipulation is very slow compared with the hard drive. For these reasons we store our images directly on Firewire hard drives (Buslink and VST technologies), which have a capacity of up to 100 Gb and the advantages of rapid transfer speeds and unlimited rewritability. The saving in time more than compensates for the added expense of these hard drives. This storage method is also portable and convenient for offline data analysis (see below).

Image Processing

To understand the difference between acquiring and manipulating the image it is useful to consider the analogy from photography where the image is acquired by the camera and manipulated in the darkroom. This analogy points out the importance of first obtaining the very best image with the microscope and the camera before further digital processing of the image. The theoretical limits of resolution are determined primarily by the wavelength of illumination and the numerical aperture of the objective. However, processing of the image by application of functions in the imaging software can significantly enhance the information obtainable from a given image. In our experiments a major problem is background fluorescence within the neuron, resulting from unincorporated or free fluorescent label. This background tends to obscure individual cytoskeletal structures. To enhance the signal-to-noise-ratio and visualize individual microtubules and actin filament bundles we typically use two major image processing functions, unsharp masking and low pass filtering. Note that these functions are applied to the 16-bit images in Metamorph and not to the image after it has been imported as an 8-bit image in Photoshop (Adobe). The effect of these functions on images of the growth cone cytoskeleton is illustrated in Fig. 5. In the original images high levels of unincorporated tubulin and free phalloidin in the neuron, as well as high levels of cytoskeletal labeling, obscure individual fluorescently labeled microtubules and bundles of actin filaments. To overcome these limitations we apply two functions, an unsharp mask followed by a low pass filter. The details of these processing steps are explained in the Metamorph help menu and will not be repeated here. As demonstrated in Fig. 5B the net result of unsharp masking is to sharpen the image by increasing the contrast between neighboring pixels, thereby enhancing the signal-to-noise ratio. Note that this function also elevates the background noise in the image. To attenuate this noise we use a low pass filter. If higher values of unsharp masking are used (Fig. 5C) one can accentuate individual cytoskeletal elements at the expense of more evenly labeled structures such as the lamellipodium and the prominent microtubule loop. This example demonstrates that image processing can be used legitimately to enhance specific structures of interest as long as the processing steps are applied equally to all regions of the image and are clearly specified in the methods.

Data Analysis

Many dynamic biological events, particularly those involving developmental changes over many minutes, require time-lapse imaging to be interpreted. By compressing minutes or even hours of real time into seconds in a time-lapse movie one is able to use the additional dimension of time to visualize events such as retrograde actin flow and microtubule growth and shrinkage, which would be difficult if not impossible to discern in real time. Therefore, we regard the digital movies as a necessary and important component of the experimental data rather than simply an entertaining mode of presentation. Once regions of cytoskeletal dynamics are revealed in the digital movie further quantitative measurements and frame-by-frame analysis can be made. We have found it useful to have an additional computer with Metamorph software so that data analysis can be carried out offline. Prior to making the movie every image or every other image in a file is opened as a stack and processed together. Generally speaking, for rapid dynamic events such as cytoskeletal movements, sequential images should be no further apart than 20 sec or the movie will have a jerky appearance. In our experiments images acquired

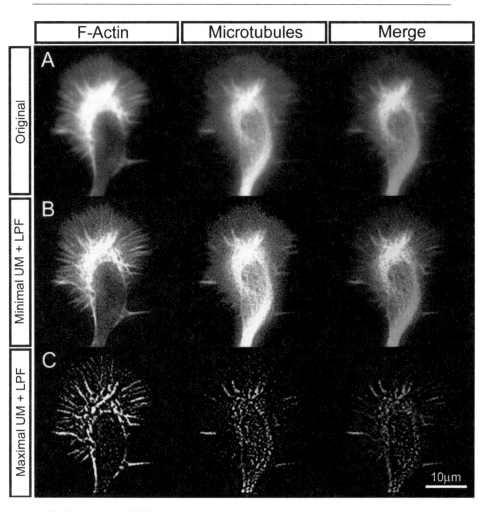

FIG. 5. Examples of different image-processing functions applied to images of a growth cone containing fluorescently labeled F-actin and microtubules. In all rows F-actin is at left, microtubules in the center, and the merged images at right. (A) Original unprocessed images. (B) The same images in (A) processed with minimal unsharp masking (UM) and a low pass filter (LPF). (C) The same images as in (A) processed with maximal unsharp masking and a low pass filter.

5 sec apart produced the smoothest movies. Clearly, in our longer term experiments images acquired at intervals of 30–60 min would not produce a suitable movie. If two cytoskeletal elements such as actin filaments and microtubules are labeled in the same cell a separate movie is made of each one. To analyze their locations with respect to one another the two images can be combined in a color overlay (Fig. 5). By using a triple bandpass dichroic mirror in the light path and changing excitation

and emission filters in the filter wheels we are able to achieve near perfect register of the images in the different wavelengths, thereby ensuring accurate measurements of F-actin/microtubule colocalization. In our experiments the movies were important in suggesting that F-actin and microtubules had the exact same trajectories in the transition region of the growth cone.[29] To distinguish between movement vs polymerization we have used the technique of fluorescent speckle microscopy[30] in which intermittent incorporation of low concentrations of fluorescent tubulin into microtubules results in speckles that serve as fiduciary marks. By using the speckles as landmarks we found that microtubules are capable of anterograde and retrograde movements[22] and that microtubules and actin filaments polymerize and depolymerize in tandem by growing and shrinking at their tips as opposed to moving together.[29] These examples illustrate how the time-lapse movie allows one to visualize various aspects of cytoskeletal dynamics and the frame-by-frame analysis permits one to measure these dynamics accurately.

Illustration and Movies

Presentation of still images and digital movies is important not only for analysis of data (see above) but also for publication and lectures. As mentioned above data analysis should always be performed on movies assembled in the Metamorph program. Movies can be displayed in Metamorph but this requires the Metamorph program to be installed on the computer displaying the movie. Therefore, because Metamorph movies are stored in RAM it is necessary to import the image files into a program that can assemble the files into a QuickTime (.mov format) or Windows Media Player (.avi format) movie accessible on most computers (these movie players are accessible at no cost). For publication and presentation, images are scaled to the appropriate range of the gray scale, copied into an 8-bit format (24 bit for color images) and saved. These images are then imported into Adobe Photoshop for assembling illustrations containing still images or into Adobe Premiere for assembling digital movies. To avoid losing information in prints for publication it is important not to resample the images (make them larger or smaller than the originals) until the final image size is determined and to save the original so that changes in the figure dimensions can be made when necessary. Generally, for publication images should be at least 300 pixels per inch (ppi). Because the movies are displayed only on the monitor they can be 72 ppi. However, we sometimes resample the images that are constructed into a movie to 300 ppi, especially if we digitally zoom areas of interest within the original image. With the advent of web sites for individual laboratories or journals, supplemental movies have become an important way of displaying results in publications.

[30] C. M. Waterman-Storer, A. Desai, J. C. Bulinski, and E. D. Salmon, *Curr. Biol.* **8,** 1227 (1998).

Conclusions

In this methods article we have demonstrated how high-resolution digital imaging in concert with microinjection techniques can be used to study dynamics of the neuronal cytoskeleton. In particular, we have shown how these techniques can be adapted to the study of growth cone guidance of neurons from the developing mammalian CNS. At present, there is intense interest in the intracellular signaling pathways that regulate the actin and microtubule cytoskeleton driving growth cone motility. A number of proteins in these pathways have been identified and with the advent of strategies to fuse proteins to GFP or label purified proteins with fluorescent analogs it is now possible to localize such proteins within growing neurons. However, to determine the role of such cytoskeletal-associated proteins in axon guidance it will be important to follow their intracellular associations concomitant with cytoskeletal reorganization and changes in the behaviors of neuronal growth cones, perhaps in response to extracellular guidance cues. Therefore, we foresee the continued importance of high-resolution live cell digital imaging in understanding the role of specific molecules in axon guidance.

Acknowledgments

Supported by NIH Grant NS14428 to K.K. and a predoctoral training grant award GM07507 to E.W.D. We thank Aileen Barnes for excellent technical assistance with the antioxidant experiments.

[19] Imaging Calcium Dynamics in Developing Neurons

By TIMOTHY M. GÓMEZ and ESTUARDO ROBLES

Introduction

Optical imaging of fluorescent calcium indicators is ideally suited for the study of calcium-dependent mechanisms regulating growth cone motility and guidance. With fluorescently labeled growth cones and proteins, several measures of cellular and molecular behavior can be correlated with Ca^{2+} dynamics. Given the complex morphology of developing and adult neurons, the intracellular calcium concentration ($[Ca^{2+}]_i$) within distinct subdomains of neurons can be estimated. Localized Ca^{2+} signals have important and varied functional consequences on growth cone motility. Ca^{2+} signals localized to cell bodies, axons, branch points, growth cones, growth cone filopodia, dendrites, and dendritic spines can be measured and correlated to a variety of cellular behaviors.[1-6] In addition to providing high spatial resolution, fluorescence imaging also allows for analysis of Ca^{2+} signals over broad temporal resolutions. Depending on the speed of the signals being examined, the capture rate can be varied to monitor changes occurring on a time scale from milliseconds to hours. One limiting factor with high capture rate imaging is the phototoxic effects of the excitation light on growth cone motility. In contrast, low-frequency imaging over long periods of time is limited by dye compartmentalization and growth cone movement away from detectable regions of the tissue. Taking certain precautions can minimize some of these limitations allowing correlation of Ca^{2+} signals with several aspects of growth cone motility. This article will review the techniques we have used to overcome these limitations in order to exploit the many advantages of using optical imaging to study Ca^{2+}-regulated growth cone motility in Xenopus laevis.

Over the past several years using both confocal and conventional fluorescence microscopy we have discovered that growth cones exhibit spontaneous transient $[Ca^{2+}]_i$ elevations over broad temporal dimensions. In early studies using fura-2 or fluo-3 as the reporter, slow spontaneous $[Ca^{2+}]_i$ transients in chick dorsal root ganglion (DRG)[7] and Xenopus spinal neuron growth cones[8] were identified in culture. Ca^{2+} transients in these two systems had similar characteristics including slow rise and decay kinetics, dependence on Ca^{2+} influx through non-voltage-gated channels, and localization to the growth cone. Importantly, the functional effects of growth cone $[Ca^{2+}]_i$ transients was to slow the rate of axon outgrowth in both neuronal types. Using fura-2 as the reporter allowed us to estimate the peak transient amplitude at 150 nM above baseline in 2 mM extracellular Ca^{2+}.[7]

We followed our initial in vitro studies with analysis of growth cone Ca^{2+} transients in identified neurons within the Xenopus spinal cord.[4] These studies required the use of confocal microscopy to eliminate fluorescence from out-of-focus regions of the tissue. We found that growth cones generate transient elevations of $[Ca^{2+}]_i$ as they migrate within the embryonic spinal cord and that the rate of axon outgrowth is inversely proportional to the frequency of transients. Further, the frequency of $[Ca^{2+}]_i$ transients was found to be cell-type specific and dependent on the position of growth cones along their pathway. Using caged-BAPTA or caged-Ca^{2+} allowed us to suppress or mimic $[Ca^{2+}]_i$ transients, respectively, leading us to conclude that growth cone transients are both necessary and sufficient to

[1] H. Komuro and P. Rakic, Neuron 17, 275 (1996).
[2] R. W. Davenport, P. Dou, V. Rehder, and S. B. Kater, Nature 361, 721 (1993).
[3] C. V. Williams, R. W. Davenport, P. Dou, and S. B. Kater, J. Neurobiol. 27, 127 (1995).
[4] T. M. Gómez and N. C. Spitzer, Nature 397, 350 (1999).
[5] T. M. Gómez, E. Robles, M.-m. Poo, and N. C. Spitzer, Science 291, 1983 (2001).
[6] A. Majewska, A. Tashiro, and R. Yuste, J. Neurosci. 20, 8262 (2000).
[7] T. M. Gómez, D. M. Snow, and P. C. Letourneau, Neuron 14, 1233 (1995).
[8] X. Gu and N. C. Spitzer, Nature 375, 784 (1995).

slow axon outgrowth. Our results indicated that environmentally regulated $[Ca^{2+}]_i$ transients control axon growth in the developing spinal cord.

Most recently we have discovered that filopodia generate fast $[Ca^{2+}]_i$ transients independent of global growth cone Ca^{2+} changes.[5] This new form of Ca^{2+} signal was identified by high-speed (8-Hz) confocal imaging of magnified regions of growth cones. The frequency of filopodial $[Ca^{2+}]_i$ transients was substrate dependent and may be due in part to influx of Ca^{2+} through channels activated by integrin receptors. These transients slowed neurite outgrowth by reducing filopodial motility and promoted turning when stimulated differentially within filopodia on one side of the growth cone. These rapid signals appear to serve both as autonomous regulators of filopodial movement and as frequency-coded signals integrated within the growth cone.

This article will be organized into sections covering the *in vitro* and *in vivo* techniques we have developed for work using *Xenopus laevis* as the model system. *Xenopus* is an ideal experimental system due to the ease with which exogenous molecules can be introduced and the robustness of its embryos and cultures that develop normally at room temperature in Ringer's solutions without additives. The growth cones of *Xenopus* spinal neurons are also quite large, facilitating detailed examination of subcellular structures *in vitro* as well as Ca^{2+} imaging *in vivo*. Proteins, fluorescent tracers as well as nucleic acids can be injected into early blastomeres to express them in all neurons, or into late blastomeres to target particular classes of neurons based on established lineage maps. Further, the concentration of injected molecules stays relatively constant due to the modest increase in total volume of embryos by neural tube closure when axon outgrowth begins. The simple structure and rapid rostral to caudal development of the *Xenopus* spinal cord allows analysis of axon outgrowth by identifiable classes of neurons at different points along their pathway. Less than 24 hr postfertilization the first motoneurons (MNs) at the rostral end of the spinal cord initiate axons and begin to form the presumptive ventral fascicle. With a slight delay, the first commissural interneurons (CIs) extend neurites, which are distinguished from MNs by cell body position and axon projection pattern. As development proceeds caudally, new MNs and CIs (as well as all other classes of neurons) initiate axons that are guided in the same stereotypical manner as their predecessors. Therefore, analysis of axon guidance by MNs and CIs can occur at identical decision regions for many different axons over a broad range of development time.

Imaging Growth Cone Calcium in Culture

Spinal Cord Isolation and Culture

Isolation of the *Xenopus* neural tube in tissue culture provides a useful method for eliminating large numbers of nonneuronal cell types derived from the epidermis, somites, and gut. Dissection techniques previously described[9] provide a framework for *Xenopus* spinal neuron tissue culture. However, examination of growth cone motility *in vitro* is aided by the isolation of primary neurons free from potential postsynaptic targets (mainly myocytes) or other cell types that may influence neurite outgrowth.

Toward the end of *Xenopus* gastrulation, induction of the dorsal ectoderm results in the neural plate, a spatially distinct population of neuronal precursors. Around 16 hr postfertilization (hpf) the neural plate begins to invaginate (NF14; Nieuwkoop and Faber), taking on a cylindrical conformation. Closure of the neural tube in the *Xenopus* embryo occurs through stage 20 (22 hpf) and coincides with the earliest pioneer axon outgrowth. The neural tube at this stage is comprised of postmitotic primary neurons, mitotically active neuronal precursors that give rise to secondary neurons, and glial cells (for an in depth description of neuronal proliferation and differentiation in the *Xenopus* spinal cord, see Hartenstein[10]).

Neural tissue can be isolated during the neural plate stage. However it must be considered that at this developmental time point the neural plate consists of a thin bilayer of cells: pigmented superficial cells and a single layer of nonpigmented cells beneath them. For this reason, dissection of young tissue is more difficult and may not fully eliminate epidermal cells originating lateral to the neural plate. In addition, dissection of the neural primordium prior to terminal differentiation raises the possibility that endogenous signals influencing neuronal differentiation may be disrupted, thereby affecting both the number of neurons and phenotypes represented in the culture. For our purposes, isolation of the neural tube between stage 21 and 23 is ideal, providing a self-contained, cylindrical neural tube with sufficient separation from the surrounding epidermis, notochord, and somites.

Dissection Technique

1. Transfer stage 22 embryos (Fig. 1A) raised in 0.1× modified Ringer's (MR) to a petri dish containing 1× MR. All dissections are carried out at 10–50× magnification using a standard dissection microscope equipped with an external light source (Leica Optical).

2. Remove the jelly coat and vitelline membrane surrounding each embryo by using a blunt forceps (Dumont #5-Standard tip) to pin the jelly coat down and a fine forceps (Dumont #55-Biology tip) to pierce and pull away the surrounding membranes.

3. Dissect the dorsal aspect of the embryo by performing a series of precise incisions using fine forceps. First make an incision laterally, perpendicular to the

[9] M.-m. Poo and N. Tabti, *in* "Culturing Nerve Cells" (G. Banker and K. Goslin eds.), p. 137. MIT Press, Cambridge, MA, 1991.
[10] V. Hartenstein, *Neuron* **3**, 399 (1989).

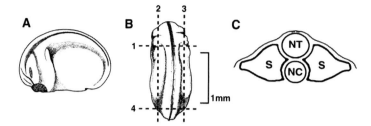

Fig. 1. Schematic diagram of neural tube dissection. (A) Lateral view of a stage 22 *Xenopus* embryo. (B) Dorsal view of a stage 22 embryo with dashed lines indicating the major incisions performed to separate the dorsal aspect of the embryo. (C) Schematic cross section of a dorsal embryo, indicating the location of the neural tube (NT) relative to the somitic mesoderm, notochord (NC), and the surrounding pigmented epidermis (gray line).

longitudinal axis and immediately caudal to the hindbrain (Fig. 1B). This first cut serves to expose the hollow interior of the embryo. Insert one forcep into the interior of the embryo adjacent to the neural tube and pinch the tissue longitudinally to make two subsequent incisions in the posterior direction. A final cut near the posterior end of the embryo results in a rectangular, multilayered piece of tissue containing the neural tube.

4. Transfer all dorsal sections to a 1 mg/ml solution of collagenase B (Boehringer) in 1× MR and incubate for 5–10 min. Subsequent dissection of neural tubes is normally carried out in the collagenase B solution, however if large numbers of neural tubes are being dissected they should be transferred in batches as long-term incubation in collagenase may be detrimental to cell survival.

5. Following collagenase treatment, peel back the dorsal pigmented epithelia by pulling at the edges that have begun to separate from the deeper tissue layers. This can best be achieved using an electrolytically sharpened and hooked tungsten wire.

6. On removal of the epidermis, the neural tube and lateral somitic mesoderm are clearly visible (see Fig. 1C cross section). Somites can be flaked away using forceps or tungsten wire, thus exposing the notochord, a small cylindrical structure ventral to the neural tube.

7. Separate the notochord from the spinal cord by pushing the bent portion of a tungsten wire between these tissues, starting at sites where the collagenase has already loosened the attachment.

8. After complete isolation of intact spinal cords transfer them back into 1× MR using a Pasteur pipette and cut the tissue into ~100 μm pieces. (Alternatively, spinal cords may be disaggregated in a Ca^{2+}-free 1× MR solution containing 10 mM EDTA and streaked onto the culture substrate as dissociated cells using a fire-pulled glass Pasteur pipette with a tip diameter of approximately 0.25 mm.) A typical stage 22 spinal cord is approximately 1 mm in length and 100–200 μm

in diameter. Using a sharpened tungsten wire, cross section this tissue into at least 10 pieces followed by perpendicular cuts to yield ∼20 individual explants.

9. Transfer all the explants onto a culture substrate submersed in 1× MR containing antibiotics (50 μg/ml penicillin/streptomycin and gentamicin, Sigma, St. Louis, MO) with a fire-pulled glass Pasteur pipette (tip diameter of approximately 0.5 mm), being cautious to not allow explants to contact the air–water interface. Explants should be spaced adequately and may be moved within the dish during plating.

Neurite outgrowth is typically observed between 6 and 24 hr following plating, although this varies depending on culture substratum. Perfuse axonal outgrowth will begin within 6 hr on a 10- to 25-μg/ml laminin (Sigma) substratum coated onto acid-washed coverglass for 2 hr at room temperature.

Imaging Calcium in Growth Cone Filopodia

The development of fast and sensitive techniques to image calcium dynamics in individual growth cone filopodia has allowed a detailed exploration of how elemental signaling events regulate complex cellular behaviors such as modulation of adhesion and corresponding cytoskeletal rearrangements. Our current studies of filopodial Ca^{2+} dynamics employ fast scanning confocal microscopy of fluorescent probes and rapid fixation techniques in order to directly explore the signaling cascades associated with these local Ca^{2+} transients.

Fluo-4 (Molecular Probes, Eugene, OR) is the calcium sensor exclusively used in our current studies of filopodial calcium dynamics. Due to its high sensitivity, close excitation match with the 488-nm line of argon laser, and response speed, fluo-4 has proven to be the Ca^{2+} sensor that most reliably detects local $[Ca^{2+}]_i$ elevations. However, one drawback of this dye is its weak fluorescence in the calcium-unbound form, necessitating higher laser power and/or photo-multiplier tube (PMT) gain to visualize dim filopodia.

All cells in a culture dish can be loaded quickly and effectively with the acetoxymethyl ester (AM) form of fluo-4. We use a final loading concentration of 2.5–5 μm by first preparing a 2.5–5 mM fluo-4 AM/10% pluronic acid (Molecular Probes) stock solution in dimethyl sulfoxide (DMSO) that is diluted 1:1000 in culture medium. Uptake and deesterification of fluo-4 AM occur over a 45- to 60-min incubation period at room temperature followed by three changes of fresh culture medium. Neurons may be imaged for up to 2–3 hr after loading, although compartmentalization of the dye into organelles can be detected within 1 hr and compromises the sensitivity of the technique.

For fast Ca^{2+} imaging we use an Olympus Fluoview 500 laser scanning confocal equipped with an argon ion gas laser (488 nm). To maximize the detected signal we completely open the confocal pinhole. Emitted wavelengths are filtered through a 505 long pass filter. The PMT voltage and gain are set high (near 75% of maximum), allowing laser power to be limited to below 5% of maximum. These values will vary for different confocal systems. Images are captured at frequencies of 8–12 Hz at a scan speed of 2 ms/line by reducing the scan region in the vertical (Y) dimension to fewer than 40 horizontal lines. Rapid image capture is necessary to detect brief calcium signals that often persist for less than 200 ms. Acquisition, however, is limited by two factors. First, the duration of the image sequence must be short to prevent phototoxicity (typically less than 1 min) and second, the scan region is necessarily small. We use a high numerical aperture, 100× objective plus additional zoom to acquire images at a resolution of 0.2–0.4 μm/pixel. High-resolution imaging is necessary to resolve localized Ca^{2+} signals in small structures such as individual filopodia. Even higher frequency imaging can be achieved by performing one-dimensional line scans along the length of individual filopodia (see Fig. 4). In this mode, a 2-ms temporal resolution is attained at the sacrifice of spatial resolution and signal to noise.

In some instances ratiometric Ca^{2+} imaging is necessary to rule out the possibility that observed fluorescence changes are due to focus drift or volume changes. Because the ultraviolet (UV)-excited dual-wavelength Ca^{2+} indicators indo-1 and fura-2 cannot be used with a visible light confocal, we use a combination of the highly Ca^{2+} sensitive fluo-4 indicator with Fura-Red AM (Molecular Probes). Fura-Red is a Ca^{2+} indicator that is excited by the argon laser at 488 nm, but has a very long-wavelength emission maximum (~660 nm). Unlike fluo-4, the fluorescence of Fura-Red decreases when it binds Ca^{2+}, however due to its weak fluorescence at baseline, changes in its fluorescence are small relative to the large increase in the fluo-4 signal. Nevertheless, the emitted fluorescence of Fura-Red is used to normalize the fluorescent emission of fluo-4. For these studies, fluo-4 and Fura-Red are coloaded at 5 μM each by the techniques described previously. Imaging is done as described except that fluo-4 and Fura-Red are excited simultaneously with the 488-nm laser line and images are collected at both 515 $\pm$ 10 and at $\geq$610 nm wavelengths.

Rapid fixation of cells after Ca^{2+} imaging allows for colocalization of sites of Ca^{2+} activity with receptors and other key signaling molecules associated with cytoskeletal remodeling. This is done by directly perfusing fixative through the imaging chamber on the microscope stage immediately following sequence acquisition. Commercially available perfusion chambers (Warner Instruments, Hamden, CT) can be employed, although we use a simple and effective hand-made system. For this, cells are cultured onto loose coverslips that are sealed onto glass microscope slides with high-vacuum grease (Dow Corning) using plastic spacers at the corners. Chamber volumes as low as 100 μl can be achieved when using 22-mm coverslips permitting rapid and thorough media exchange. Growth cones are fixed by rapid perfusion of cold 4% (w/w) paraformaldehyde/4% (w/w) sucrose, followed by standard immunocytochemical procedures. Rapid fixation results in only moderate changes in cell morphology during this process but should be performed immediately following live imaging. Because fluo-4 fluorescence is lost on fixation, this wavelength can be used along with others to localize multiple cellular elements by immunofluorescence.

UV Photolysis

Manipulation of $[Ca^{2+}]_i$ by UV photolysis of caged Ca^{2+} is a powerful method to study the effects of various user-defined Ca^{2+} transients on growth cone motility. This approach allows one to examine the downstream effectors of spatially and temporally distinct Ca^{2+} signals. Imposed Ca^{2+} transients are detected by coloading fluo-4, whereas the effects on growth cone behaviors can be studied with differential interference contrast (DIC) optics or with various expressed fluorescent proteins. However, caution must be taken when using caged compounds as growth cones are sensitive to UV light. Therefore, UV-only controls are important to determine the Ca^{2+}-dependent component of behavioral changes in response to photorelease.

UV photolysis of caged compounds is simplified on a confocal microscope as these systems typically have separate light paths for laser scanning imaging and wide-field fluorescence excitation. We use an Olympus AX-70 upright microscope equipped with a fiber launched 100-W mercury arc lamp on our Fluoview 500 laser-scanning confocal system. Excitation and neutral density filters to select and attenuate photolysis wavelengths respectively are positioned in the optical path. We use narrow bandpass filters (either 360 ± 25 or 380 ± 6.5) to efficiently release Ca^{2+} while minimizing UV damage. The region exposed to UV light is adjusted using the field diaphragm, an aperture iris located at the conjugate focal plane between the illuminator and specimen. UV exposure can typically be restricted to areas as small as 10–15 μm in diameter when using a 100× objective and can be determined by imaging a solution of caged fluorescein isothiocyanate (FITC) dextran (Molecular Probes, Brattleboro, VT) sandwiched between coverslips. A dichroic mirror (Chroma, Rochester, NY) reflects light shorter than 400 nm toward the sample while transmitting longer wavelengths. A programmable shutter (Uniblitz) controls both the duration of single UV pulses and the delay between a series of pulses. It is worth noting that the total power of light delivered to the UV spot appears slightly higher toward the center, which may be due to the mechanism of iris opening and increased light defraction near the edge of the field diaphragm.

Photoactivation of caged Ca^{2+} provides higher spatial and temporal control of stimuli compared to alternative techniques (i.e., local or bath perfusion of Ca^{2+} channel activators). We use caged Ca^{2+} to locally generate Ca^{2+} transients at the tips of filopodia to determine the effects of these signals on growth cone guidance. Growth cones typically have many filopodia distributed in a fan shape at the leading edge of the advancing neurite. Thus, a growth cone can be positioned such that filopodia on only one side are exposed to the uncaging stimulus[5] (Fig. 4C). To best assess the spread and kinetics of Ca^{2+} photorelease, we imaged single filopodia with confocal line scans during a train of UV uncaging pulses (Fig. 2). This method provides 2-ms time resolution, but is less sensitive compared to a two-dimensional (2D) scan box. By combining line scanning with UV photorelease of Ca^{2+} we are able to detect local imposed Ca^{2+} transients at the tips of selected filopodia (Fig. 2). However, imposed transients appear to differ from endogenous spontaneous filopodial transients[5] in the extent of retrograde diffusion, possibly due to an increased buffering in the presence of the high-affinity calcium chelator, NP-EGTA.

We use local release of caged Ca^{2+} to determine how imposed Ca^{2+} transients at the tips of filopodia on one side of the growth cone affect the direction of neurite outgrowth. We position NP-EGTA-loaded growth cones so their direction of growth is toward a $\sim$12 μm spot of pulsed UV light. To generate brief Ca^{2+} transients we program the shutter of to open for 100–200 ms at a frequency of 3/min to 6/min. Fluorescent images of green fluorescent protein (GFP) expressing growth cones or DIC images of untransfected growth cones are collected every 15–60 sec for 30 min of growth. Using this paradigm, we find that growth cones loaded with NP-EGTA consistently turn to avoid the region of pulsed UV light and typically do so after only filopodia enter the spot. In contrast, unloaded growth cones do not exhibit significant turning, although some UV-dependent responses are observed.

Imaging Growth Cone Calcium *in Vivo*

Imaging with Single-Wavelength Indicators

The most direct way to image Ca^{2+} in growth cones *in vivo* is to expose the spinal cord after neurite outgrowth has begun and load neurons with a cell-permeant indicator dye. However, loading a fully exposed spinal cord causes the growth cones to become obscured by the brightly labeled cell bodies due to their relatively small size. To overcome this problem we load selected regions of the spinal cord after partial exposure. This approach has the added benefit of allowing semiselective loading of different neuronal types; however, it is important to know the approximate position of growth cones at the time of dissection. We have constructed a map of neural-specific βI–II tubulin-labeled spinal cords at sequential developmental stages to help define the region of the cord to expose and load. A detailed description of how to image descending axons such as MNs and Rohon Beard (RB) neurons follows. The procedures can be applied to load the posterior spinal cord to image ascending neurons as well.

Dissection Technique

1. Embryos at 24 hpf are at a good stage [Nieuwkoop and Faber (NF) 22] to image descending MNs and RB neurons. Prior to dissection embryos must first have their jelly coats and vitelline membranes mechanically removed using fine forceps.

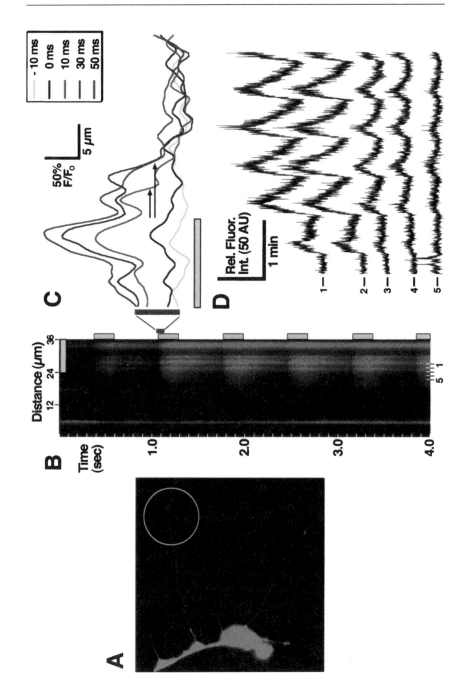

2. Pin an embryo laterally through its abdomen onto a sylgard dish containing 1× MR with two minutian pins bent at right angles (Fig. 3A; fine science tools, 0.1 mm diameter). At this early stage embryos are not capable of movement, but if older embryos (stage 25 or above) are preferred they can be readily immobilized with 1 mg/ml MS222 (Sigma).

3. Using a hooked and electrolytically sharpened tungsten wire, open a window of skin on the dorsal aspect on the anterior half on one side of the embryo exposing the underlying axial myotome (Fig. 3B). This window will define the region of the spinal cord that will be loaded with Ca^{2+} indicator. The region of anterior skin selected should ideally extend up to the rostral extent of pioneering MN axon growth.

4. To help detach the somites from the spinal cord, dilute 1 mg/ml collagenase B 10-fold in the Sylgard dish (0.1 mg/ml final collagenase).

5. Remove the dermatome (outer layer of cells covering somites) by hooking ventrally and pulling dorsally, being cautious to avoid contact with the spinal cord. Shortly after removal of the dermatome the somites will begin to separate from the spinal cord dorsally.

6. Using the bent region of the tungsten wire, separate the somites from the spinal cord down to the notocord still being cautious not to contact the spinal cord (Fig. 3C).

7. Finally, wash the dissected embryo extensively with MR solution to remove collagenase.

Fluo-4 AM Loading

1. Prepare a 10 μM loading solution of calcium indicator prior to the dissection. Due to the reduced access of spinal neurons *in vivo*, it is necessary to load with a

---

FIG. 2. Spatial and kinetic characteristics of Ca^{2+} photorelease at filopodial tips. (A) A fluo-4/NP-EGTA double-loaded growth cone was placed adjacent to a spot within which a single filopodium was exposed to pulsed UV light (yellow circle, 12.5 μm diameter). (B) Pseudocolored linescan through this filopodium collected at 2 ms/line. Two thousand lines were collected over 4 sec during a train of UV pulses. Upper yellow bar indicates the region exposed to UV light pulsed for 200 ms every 700 ms (vertical yellow bars at right). (C) Normalized fluorescence line profiles (to resting levels) along the length of the filopodium at times before and during an imposed Ca^{2+} transient. Each trace represents the average fluorescence of five lines (10 ms) smoothed with a five-point moving average. Time points quantified were selected during the second UV pulse where indicated by the vertical red bar. Imposed Ca^{2+} transients appear to diffuse rapidly over a short distance (arrows). (D) Fluorescence line profiles over time along five lines (average of five lines each) along the filopodium at 1 μm intervals where indicated by dashed lines at bottom of (B). These traces show that photorelease of Ca^{2+} generates invariant transients that rise rapidly to a peak during UV exposure and fall gradually over the subsequent 500 ms. The peak of the imposed Ca^{2+} elevation also diminishes rapidly away from the exposed region.

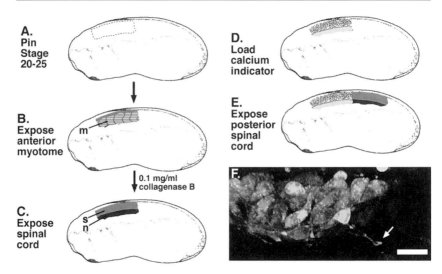

FIG. 3. Partial exposure and fluo-4 AM loading of the *Xenopus* spinal cord. (A) A staged embryo pinned laterally onto a Sylgard dish. Dashed outline indicates area of skin to be removed to label anterior spinal cord. (B) With skin removed the myotome is exposed. (C) With careful dissection and the aid of collagenase B, the myotome is removed to expose the underlying spinal cord and notochord. (D) The embryo is unpinned and placed in an Eppendorf tube containing fluo-4 or other cell permeant dye. (E) After sufficient loading, the embryo is repinned and the remaining skin and somites are dissected with the aid of collagenase B. (F) A maximum projection of a confocal z series showing the discrete loading of the anterior region of the spinal cord. Several axons with growth cones are visible (arrow). Bar: 50 μm.

higher concentration of fluo-4 AM than is normally used *in vitro*. First, prepare a stock solution of 10 mM fluo-4 AM in DMSO.

2. Dilute 10 mM stock fluo-4 1:1 in 20% pluronic acid, then a further 1:500 in 500 μl MR and vortex extensively. The pluronic acid improves cell loading by keeping the dye soluble in an aqueous solution.

3. Transfer partially dissected embryos into the loading solution using a transfer pipette. After 1 hr in fluo-4, transfer the embryo back into the Sylgard dish containing fresh MR solution (Fig. 3D). Pin the embryo out laterally and wash with several changes of MR.

4. Finally, dissect the remaining skin and somites away from the spinal cord as described above using 0.1 mg/ml collagenase B (Fig. 3E). After a final wash the embryo is ready to be transferred to the confocal microscope for imaging. Note that it is important to pin the embryo down as flat as possible longitudinally to maximize the focus field. On the other hand, the tilt of the embryo can be varied to promote the visualization of particular neuronal types. For example, it is best to pin the embryo tilted with the dorsal aspect slightly beyond horizontal to visualize growth cones along the ventral fascicle.

Confocal Imaging

The following descriptions should be used as a guideline for confocal imaging of fluorescence changes in growth cones *in vivo*. Modifications to this general scheme may be necessary based on the types of signals being examined or indicators being imaged. We use a 40× or 60× magnification water immersion lens with high numerical aperture (NA, 0.8 to 0.9) mounted on an inverted microscope (Olympus AX70) to image fluo-4-loaded growth cones *in vivo*. As with *in vitro* imaging, precautions should be taken to limit the phototoxic effects of laser light by maximizing the signal while minimizing excitation light. First, given that fluorescence changes along the z axis are difficult to resolve, we neglect this dimension and open the confocal pinhole as wide as possible to maximize signal above background. Because fluorescence is normally low over unloaded regions of the spinal cord, opening the pinhole does not increase out-of-focus fluorescence significantly in this region. In addition, the PMT voltage and gain are set high at the expense of image resolution. A scan speed of around 2 ms/line (normal mode; Bio-Rad, Hercules, CA) is sufficient to collect high signal-to-noise images in 1 sec for a 512×512 scan box. The size of the scan box should, however, be adjusted down to the necessary field of view to avoid unnecessary laser exposure and collection time. With these conditions set, the intensity of the laser light should be attenuated with ND filters to 1–5%, depending on the brightness of the growth cones. Note that some imaging systems may require more (or less) light depending on the power delivered to the sample. To image long duration ($\sim$1 min) global growth cone calcium transients, we collect images every 15 sec and can do this under the conditions described above for more than 1 hr.

Ratiometric Imaging

The ability to create an image ratio from two wavelengths offers many advantages over single-wavelength imaging especially *in vivo*. Quantification of fluorescence intensity from a single wavelength is susceptible to artifacts due to movement and volume changes. Having a second wavelength to normalize against minimizes these effects and in some circumstances allows calibration of Ca^{2+} concentration. Several options for dual-wavelength indicators or combinations of fluorochromes to use with a laser scanning confocal are available, although each has certain limitations. If a UV laser is available, then dual emission imaging can be performed using indo-1 [excitation(max) = 330−346 nm; Molecular Probes]. Indo-1 is available as both a cell-permeant AM form and as a dextran conjugate. Using a dextran conjugate form of indo-1 is advantageous as it can be introduced by blastomere injection (see below), labeling all cells derived from the originally injected cell. However, indo-1 is not favored due to (1) high autofluorescence of UV-excited tissues, (2) the sensitivity of *Xenopus* cells excited by UV light, and (3) the rapid bleaching of this fluorophore.

An alternative approach we prefer for *in vivo* ratio imaging of Ca^{2+} signals is to combine the highly sensitive fluo-4 Ca^{2+} indicator with a stable fluorochrome that emits and/or is excited by a distinct wavelength. This type of ratio imaging was described above for the combination of fluo-4 with Fura-Red, although Fura-Red does respond moderately to Ca^{2+} changes. Here we describe an alternative combination that involves preinjecting embryos with a nonoverlapping fluorescent dextran conjugate, such as tetramethylrhodamine dextran (Molecular Probes). In addition to allowing for normalization of fluorescent signals, the intense fluorescence of rhodamine-dextran allows visualization of the entire growth cone including filopodia, which improves analysis of behavior. Because the fluorescence intensity of fluo-4 is low at baseline Ca^{2+} levels, the palm region of a growth cone is typically the only visible portion *in vivo*. Therefore, ratiometric Ca^{2+} measurements must be made within regions of the growth cone with significant "double-labeling" such as the palm, whereas behavioral analysis can be conducted using the more intensely fluorescent red signal. Ideally the red fluorescent tracer is targeted to the proper blastomere in order to label specific populations of spinal neurons to aid in both the identification and visualization of growth cones of interest. A drawback to using two separate fluorochromes on different molecules is that calibration is difficult due to variable cell labeling. However, the benefits of dual-excitation ratiometric confocal imaging far outweigh this limitation and the additional steps necessary (see below) to perform these experiments.

Microinjection

1. Embryos injected at the 8–32 cell stage (2.5–3 hpf) typically provide labeling patterns that allow the identification and imaging of isolated growth cones. Embryos need to be dejellied prior to injection. This can be performed chemically on many embryos by swirling in 2% cysteine (Sigma) in $0.1 \times$ MR pH 8.0–8.1 for 5 min. After jelly coats are shed the embryos should be rinsed extensively in $0.1 \times$ MR.

2. Transfer dejellied embryos into 6% Ficoll in $0.1 \times$ MR to shrink their vitelline membranes. To provide stability during injection, the embryos are arranged onto a 1-mm plastic grid (Small Parts, Inc., Miami Lakes, FL).

3. Pull and break omega-dot glass capillary tubing (FHC) to ~1 μm tip diameter and back-fill with 10 mg/ml rhodamine-dextran, 10,000 molecular weight in distilled H_2O (or Alexa Fluor 546 dextran). Rhodamine dextran (1–3 nl) is pressure injected into identified blastomeres according to well-defined fate maps.[11]

4. After an additional 24 hr of development in 6% Ficoll, surviving embryos are separated back into $0.1 \times$ MR for further loading with fluo-4 using the techniques described above.

[11] M. Jacobson and G. Hirose, *J. Neurosci.* **1**, 271 (1981).

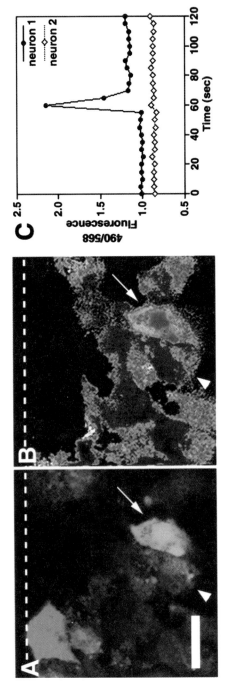

FIG. 4. *In vivo* ratiometric Ca²⁺ imaging using a visible light scanning confocal microscope. (A) Merged image of rhodamine–dextran and fluo-4 signals. Cells positive for rhodamine–dextran or fluo-4 appear only red or green, respectively, whereas double-loaded cells appear yellowish (arrow and arrowhead). (B) Image ratio of fluo-4/rhodamine signals taken at the peak of a 40 m*M* KCl-induced Ca²⁺ transient in neuron 1 (arrow). (C) The dual excitation ratio graph shows a large response in neuron 1 with little response in neuron 2. Dashed line represents the dorsal aspect of the spinal cord. Bar: 20 μm.

Fluo-4 AM is loaded into the exposed spinal cord of embryos that exhibit the desired rhodamine-labeling pattern. The procedures described above should be used to image multiple fluorochromes adding the appropriate laser lines as needed. Images of fluorochromes with significant spectral overlap may need to be captured sequentially to limit fluorescence cross talk. Figure 4 shows an example of ratiometric Ca^{2+} imaging from double-loaded neurons in the spinal cord.

Summary

Here we describe the techniques developed to image Ca^{2+} signals in motile nerve growth cones both in culture and in the developing *Xenopus* spinal cord. We have used these methods to identify two spatially and temporally distinct classes of Ca^{2+} transients in growth cones. Imaging Ca^{2+} in morphologically complex migratory cells allows for analysis and correlation of discrete signals with a wide variety of cellular behaviors. For example, we find that localized Ca^{2+} changes at the tips of individual filopodia correlate with reduced filopodial motility. Further, rapid fixation after Ca^{2+} imaging made it possible to determine that transients occur at integrin receptor clusters that may generate and in turn be regulated by these local signals. We describe the use of caged-Ca^{2+} to locally impose Ca^{2+} transients in individual filopodia and find this treatment sufficient to repel neurite outgrowth.

Calcium signals across broad spatial and temporal dimensions are universal regulators of numerous complex and varied cellular functions.[12] The imaging methods we describe here begin to view growth cones over a range of spatial resolutions and temporal frequencies necessary to detect different types of Ca^{2+} transients, however it is clear that not all dimensions have been examined. In particular, imaging cells more rapidly and at higher magnification may one day allow us to detect more elemental events such as single-channel openings, as has been achieved in nonneuronal cells.[13,14] We also describe techniques used to examine Ca^{2+} signals in growth cones migrating within the spinal cord. These types of studies are ultimately necessary to confirm the relevance of *in vitro* findings. Although designed for the *Xenopus* spinal cord, the methods we outline should be applicable to other tissues and organisms. Finally, we use caged Ca^{2+} as a tool to reproduce very precise changes in cytosolic Ca^{2+} levels. This is a powerful means to test the function of different types of Ca^{2+} transients and assess the downstream regulators of those signals. These types of manipulations can also be used with other types of caged compounds, many of which are commercially available (Molecular Probes) or readily synthesized.

[12] M. Berridge, P. Lipp, and M. Bootman, *Curr. Biol.* **9**, R157 (1999).
[13] M. Bootman, E. Niggli, M. Berridge, and P. Lipp, *J. Physiol.* **499**, 307 (1997).
[14] P. Lipp and E. Niggli, *J. Physiol.* **508**, 801 (1998).

[20] Optical Monitoring of Neural Activity Using Voltage-Sensitive Dyes

By MAJA DJURISIC, MICHAL ZOCHOWSKI, MATT WACHOWIAK, CHUN X. FALK, LAWRENCE B. COHEN, and DEJAN ZECEVIC

Introduction

An optical measurement of membrane potential using a molecular probe can be beneficial in a variety of circumstances. One advantage is the possibility of simultaneous measurements from many locations. This is especially important in the study of nervous systems in which many parts of an individual cell, or many cells, or many regions of the nervous system, are simultaneously active. Furthermore, optical recording offers the possibility of recording from processes of neurons that are too small or fragile for electrode recording.

Several different optical properties of membrane-bound dyes are sensitive to membrane potential including fluorescence, absorption, dichroism, birefringence, fluorescence resonance energy transfer (FRET), nonlinear second harmonic generation, and resonance Raman absorption. However, because the vast majority of applications have involved fluorescence or absorption, these will be emphasized in this article. All of the optical signals described in this article are "fast" signals[1] that are presumed to arise from membrane-bound dye; they follow changes in membrane potential with time courses that are rapid compared to the rise time of an action potential. We begin with a brief characterization of voltage-sensitive dyes and then discuss the methods that are needed to achieve optimal signal-to-noise (*S/N*) ratios. We finish with two examples of results obtained from measurements addressing two quite different scientific problems. The optical signals in the two example measurements are not large; they represent fractional changes in light intensity ($\Delta I/I$) of from 10^{-4} to 3×10^{-2}. Nonetheless, they can be measured with an acceptable *S/N* ratio after paying attention to details of the measurement.

Figure 1 illustrates two qualitatively different areas of neurobiology where imaging membrane potential has been useful. First (Fig. 1, left), to know how a neuron integrates its synaptic input into its action potential output, one needs to be able to measure membrane potential everywhere where synaptic input occurs and at the places where spikes are initiated. This application is the first example discussed below. Second (Fig. 1, right), responses to sensory stimuli and generation of motor output in the vertebrate brain are often accompanied by synchronous activation of many neurons in widespread brain areas; voltage-sensitive dye recordings allow simultaneous measurement of population signals from many areas. This application

[1] L. B. Cohen and B. M. Salzberg, *Rev. Physiol. Biochem. Pharmacol.* **83,** 35 (1978).

PARTS OF A NEURON

Many Detectors - One neuron

Potential changes in dendrites.
Microinject dye: stain one neuron.

POPULATION SIGNALS

One Detector - Many Neurons

Signals are population average.
Vertebrate brain.
Bath applied dye.

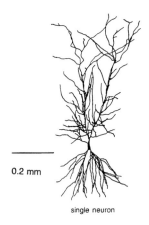

0.2 mm

single neuron

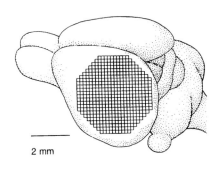

2 mm

FIG. 1. Schematic drawings of the two kinds of measurements used as examples. On the left is shown an individual cortical hippocampal CA1 pyramidal cell. Each pixel of the camera would receive light from a small part of the dendrite, axon, or cell body of the neuron. An optical measurement of membrane potential would provide important information about how the neuron converts its synaptic input into its spike output. The right-hand panel shows a vertebrate brain with a superimposed 464 element photodiode array. In this circumstance each pixel of the array would receive light from thousands of cells and processes. The signal would be the population average of the change in membrane potential in those cells and processes. The image of the hippocampal neuron was taken from Z. F. Mainen, N. T. Carnevale, A. M. Zador, B. J. Claiborne, and T. H. Brown, *J. Neurophysiol.* **76,** 1904 (1996).

is the second example discussed below. In both instances optical recordings have provided information about the function of the nervous system that was previously unobtainable.

Different kinds of staining are used in the two examples described in this article. (1) For studying the membrane potential in individual dendrites of a neuron the dye is released from an intracellular electrode in the soma and then allowed to spread into the dendritic tree. The amount of dye in an individual dendrite will be small and thus we expect dark noise to be a problem. This intracellular staining procedure has the advantage that all of the fluorescence comes from the injected neuron and thus any intensity change can be attributed to that neuron. (2) For the population signals, the olfactory bulb is superfused for 60 min in a solution of the dye. From previous experience we expected that this procedure would stain all of the neurons in the bulb and that the amount of dye, and thus the fluorescence intensity, would be large.

Two kinds of cameras are used; both have frame rates faster than 1000 frames per second (fps). One camera is an 80 × 80 pixel charge-coupled device (CCD) camera and the second is a photodiode array with a total of 464 pixels. Even though the spatial resolution of the two cameras differs rather dramatically, the most important difference is in the range of light intensities over which they provide an optimal S/N ratio. The CCD camera is optimal at low light levels (for single-cell imaging), and the photodiode array is optimal at high light levels (for imaging populations of neurons stained by bulk application of dye).

Voltage-Sensitive Dyes

In both preparations described in this article, fluorescence changes are measured. However, with these voltage-sensitive dyes there are also changes in absorption that accompany activity.[2] In *in vitro* slice preparations where bulk stains are used and populations signals are measured, Jin and Wu[3] found that absorption signals have a larger S/N ratio. Although measurements of transmitted light are not easy in an *in vivo* preparation, larger signals might be obtained if they are used.

The voltage-sensitive dyes used here are membrane-bound chromophores whose absorption or fluorescence properties change in response to changes in membrane potential. Figure 2 shows the results of optical measurements during an action potential (Fig. 2, top) and during voltage clamp steps (Fig. 2, bottom) in a model preparation, the giant axon from a squid. Clearly, these signals are fast, following membrane potential with a time constant of <10 μs.[4] The results in Fig. 2 also indicate that their size is linearly related to the size of the change in potential (see also Gupta *et al.*[2]). Thus, these dyes provide a direct, fast, and linear measure of the change in membrane potential of the stained membranes.

Several voltage-sensitive dyes (e.g., Fig. 3) have been used to monitor changes in membrane potential in a variety of preparations. Figure 3 illustrates four different chromophores (the merocyanine dye, XVII, was used for the measurement illustrated in Fig. 2). For each chromophore, approximately 100 analogs have been synthesized in an attempt to optimize the S/N ratio that can be obtained in a variety of preparations. (This screening was made possible by synthetic efforts of three laboratories: Jeff Wang, Ravender Gupta, and Alan Waggoner then at Amherst College; Rina Hildesheim and Amiram Grinvald at the Weizmann Institute; and Joe Wuskell and Leslie Loew at the University of Connecticut Health Center.) For each of the four chromophores illustrated in Fig. 3, there are 10 or 20 dyes that give approximately the same signal size on squid axons.[2] However,

[2] R. K. Gupta, B. M. Salzberg, A. Grinvald, L. B. Cohen, K. Kamino, S. Lesher, M. B. Boyle, A. S. Waggoner, and C. H. Wang, *J. Memb. Biol.* **58**, 123 (1981).
[3] W. J. Jin and J. Y. Wu, (2001). Submitted.
[4] L. M. Loew, L. B. Cohen, B. M. Salzberg, A. L. Obaid, and F. Bezanilla, *Biophys. J.* **47**, 71 (1985).

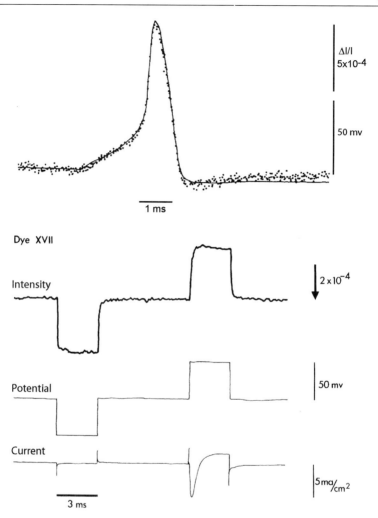

FIG. 2. Changes in absorption (dots) of a squid giant axon stained with a merocyanine dye, XVII (Fig. 3) during a membrane action potential (*top*) and during voltage clamp steps (*bottom*). *Top:* Results of the optical recording are shown in the dots and the electrode recording of the action potential is shown in the smooth trace. The change in absorption and the action potential had the same time course. *Bottom:* Simultaneously recorded intensity change (top trace), membrane potential (middle trace), and ionic currents (bottom trace) are illustrated. The intensity change has a time course similar to the membrane potential change and quite different from the ionic currents. In this and subsequent figures the size of the vertical line (or arrow) represents the stated value of the fractional change in intensity, $\Delta I/I$, or fluorescence ($\Delta F/F$). For the top section, the response time constant of the light measuring system was 35 μs; 32 sweeps were averaged. The dye is available from Nippon Kankoh-Shikiso Kenkyusho, Okayama, Japan. [Reprinted with permission from W. N. Ross, B. M. Salzberg, L. B. Cohen, A. Grinvald, H. V. Davila, A. S. Waggoner, and C. H. Wang, *J. Membr. Biol.* **33**, 141 (1977).]

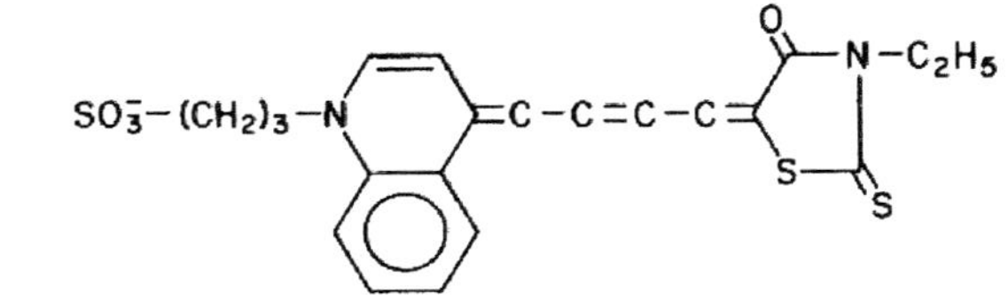

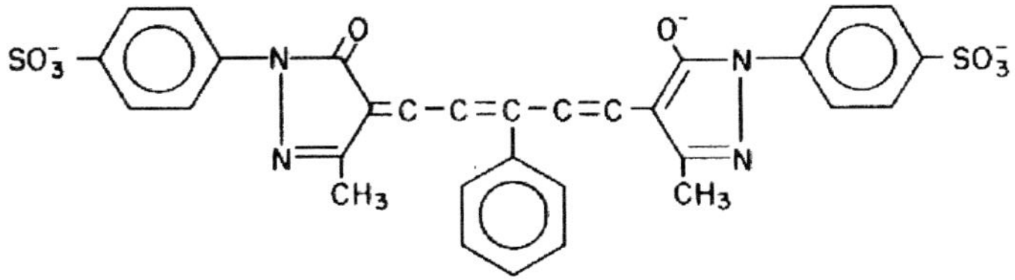

RH155, Oxonol, Absorption

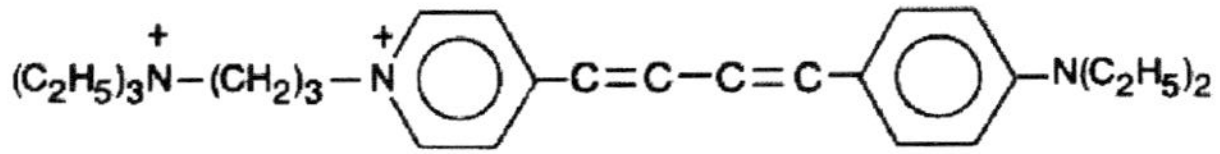

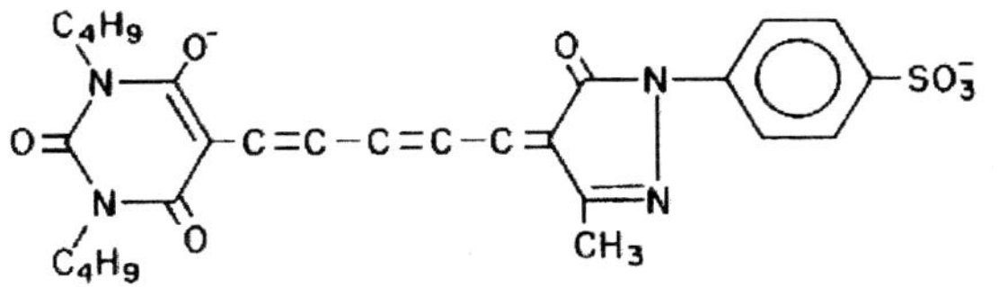

XXV, Oxonol, Fluorescence, Absorption

XVII, Merocyanine, Absorption, Birefringence

RH414, Styryl, Fluorescence

FIG. 3. Examples of four different chromophores that have been used to monitor membrane potential. The merocyanine dye, XVII (WW375), and the oxonol dye, RH155, are commercially available as NK2495 and NK3041 from Nippon Kankoh-Shikiso Kenkyusho Co. Ltd. (Okayama, Japan). The oxonol, XXV (WW781), and styryl, RH-414, are available commercially as dye R-1114 and T-1111 from Molecular Probes (Eugene, OR).

dyes that had nearly identical signal size on squid axons could have very different responses on other preparations, and thus tens of dyes usually have to be tested to obtain the largest possible signal. A common failing is that the dye did not penetrate through connective tissue or along intercellular spaces to the membrane of interest.

The following rules-of-thumb seem to be useful. First, each of the chromophores is available with a fixed charge that is either a quatenary nitrogen (positive) or a sulfonate (negative). Generally the positive dyes have given larger signals in both kinds of experiments discussed in this article. Second, each chromophore is available with carbon chains of several lengths. Relatively hydrophilic dyes (methyl or ethyl) work best in both preparations.

Amplitude of the Voltage Change

Both of the signals discussed in this article are presented as a fractional intensity change ($\Delta F/F$). These signals give information about the time course of the potential change but no direct information about the absolute magnitude. However, in some instances, approximate estimations can be obtained. For example, the size of the optical signal in response to a sensory stimulus can be compared to the size of the signal in response to an epileptic event.[5] Another approach is the use of ratiometric measurements at two independent wavelengths[6] (see below).

Noise in Optical Measurements

Shot Noise

The limit of accuracy with which light can be measured is set by the shot noise arising from the statistical nature of photon emission and detection. The root-mean-square deviation in the number emitted is the square root of the average number emitted (I) over a long measuring period. Thus, the signal in a light measurement will be proportional to I and the noise in that measurement will be proportional to the square root of I. Therefore, the S/N ratio is proportional to the square root of the number of measured photons; more photons measured means a better S/N ratio. The S/N is also inversely proportional to the square root of the bandwidth of the photodetection system.[7] The basis for this square root dependence on intensity is illustrated in Fig. 4. In Fig. 4A the result of using a random number table to distribute 20 photons into 20 time windows is shown. In Fig. 4B the same procedure was used to distribute 200 photons into the same 20 bins. Relative to the average light level there is more noise in the top trace (20 photons) than in the bottom trace

[5] H. S. Orbach, L. B. Cohen, and A. Grinvald, *J. Neurosci.* **5,** 1886 (1985).
[6] E. Gross, R. S. Bedlack, Jr., and L. M. Loew, *Biophys. J.* **67,** 208 (1994).
[7] H. J. J. Braddick, *Rep. Prog. Phys.* **23,** 154 (1960).

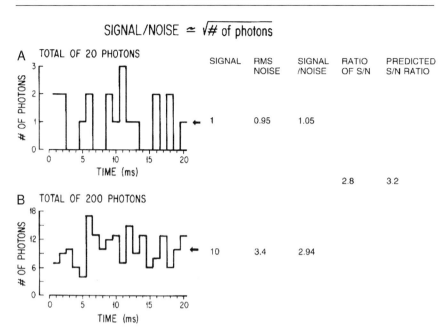

FIG. 4. Plots of the results of using a table of random numbers to distribute 20 photons (A) or 200 photons (B) into 20 time bins. The result illustrates the fact that when more photons are measured the signal-to-noise ratio is improved. On the right, the signal-to-noise ratio is measured for the two results. The ratio of the two signal-to-noise ratios was 2.8. This is close to the ratio predicted by the relationship that the signal-to-noise ratio is proportional to the square root of the measured intensity.

(200 photons). This square-root relationship is indicated by the dotted line in Fig. 5 which plots the light intensity divided by the noise in the measurement (S/N) versus the light intensity. At high light intensities this ratio is large and thus small changes in intensity can be detected. For example, at 10^{10} photons/ms a fractional intensity change of 0.1% can be measured with an S/N ratio of 100 in a single trial. On the other hand, at low intensities the ratio of intensity divided by noise is small and only large signals can be detected. For example, at 10^{4} photons/ms the same fractional change of 0.1% can be measured with an S/N ratio of only 1 only after averaging 100 trials. In a shot–noise limited measurement, improvement in the S/N ratio can be obtained only by (1) increasing the illumination intensity, (2) improving the light-gathering efficiency of the measuring system, or (3) reducing the bandwidth.

Figure 5 compares the performance of two particular camera systems, a photodiode array (blue lines) and a cooled, back illuminated, CCD camera (red lines) with the shot–noise ideal (green line). The photodiode array approaches the shot–noise limitation over the range of intensities from about 3×10^{6} to 10^{10} photons/ms.

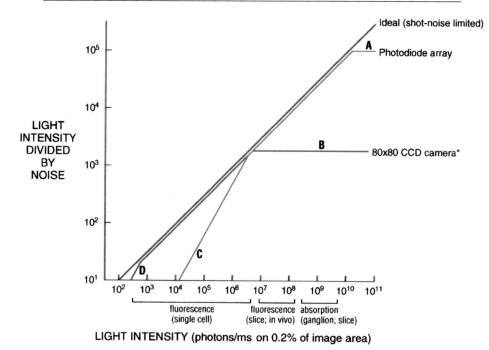

FIG. 5. The ratio of light intensity divided by the noise in the measurement as a function of light intensity in photons/ms/0.2% of the object plane. The theoretical optimum signal-to-noise ratio (green line) is the shot-noise limit. Two camera systems are shown, a photodiode array with 464 pixels (blue lines) and a cooled, back-illuminated, 2-kHz frame rate, 80 × 80 pixel CCD camera (red lines). The photodiode array provides an optimal signal-to-noise ratio at higher intensities and the CCD camera is better at lower intensities. The approximate light intensity per detector in fluorescence measurements from a single neuron or from transported dye is indicated on the left. The fluorescence intensity from a slice or *in vivo* preparation after soaking with the dye is indicated in the middle. The signal-to-noise ratio for the photodiode array falls away from the ideal at high intensities (A) because of extraneous noise and at low intensities (C) because of dark noise. The lower dark noise of the cooled CCD allows it to function at the shot-noise limit at lower intensities until read noise dominates (D). The CCD camera saturates at intensities above 5×10^6 photons/ms/0.2%.

This is the range of intensities obtained in absorption measurements and fluorescence measurements from *in vitro* slices and *in vivo* brains where all of the neurons are stained via direct application of the dye. On the other hand, the cooled CCD camera approaches the shot–noise limit over the range of intensities from about 3×10^2 to 5×10^6 photons/ms. This is the range of intensities obtained from fluorescence measurements from individual dendrites. The discussion that follows will indicate the aspects of the measurements and the characteristics of the two camera systems that cause them to deviate from the shot–noise ideal. The two camera systems illustrated in Fig. 5 are examples with outstanding dark noise, quantum efficiency, and saturation characteristics; cameras that do not have as good a performance would be dark noise limited at higher light intensities, would be farther from ideal at all light levels, and/or would saturate at lower intensities.

Extraneous Noise

A second type of noise, termed extraneous or technical noise, is more apparent at high light intensities where the sensitivity to this kind of noise is high because the fractional shot noise and dark noise are low. One type of extraneous noise is caused by fluctuations in the output of the light source (see below). Two other sources of extraneous noise are vibrations and movement of the preparation. A number of precautions for reducing vibrational noise have been described.[8,9] These include avoiding buildings near roads with heavy traffic and preferring ground-floor rooms.

The pneumatic isolation mounts on many vibration isolation tables are more efficient in reducing vertical vibrations than in reducing horizontal movements. One solution is air-filled soft rubber tubes (Newport Corp., Irvine, CA). Minus K Technology sells Biscuit bench top vibration isolation tables with very low resonant frequencies in both vertical and horizontal directions. Nevertheless, it is difficult to reduce vibrational noise to less than 10^{-5} of the total light. For this reason the performance of the photodiode array system is shown reaching a ceiling in Fig. 5 (segment A, blue line).

Dark Noise

Dark noise will degrade the S/N ratio at low light levels. Because the CCD camera is cooled and the photosensitive area (and capacitance) is small, its dark noise is substantially lower than that of the photodiode array system. The excess dark noise in the photodiode array accounts for the fact that segment C in Fig. 5 is substantially to the right of segment D.

Light Sources

Three kinds of light sources have been used. Tungsten filament lamps are a stable source, but their intensity is relatively low, particularly at wavelengths less than 480 nm. Arc lamps are somewhat less stable but can provide more intense illumination. Measurements made with laser illumination have been substantially noisier.[10]

It is not difficult to provide a power supply stable enough so that the output of a tungsten filament lamp fluctuates by less than 1 part in 10^5. In absorption measurements, where the fractional changes in intensity are relatively small, only tungsten

---

[8] B. M. Salzberg, A. Grinvald, L. B. Cohen, H. V. Davila, and W. N. Ross, *J. Neurophysiol.* **40**, 1281 (1977).

[9] J. A. London, D. Zecevic, and L. B. Cohen, *J. Neurosci.* **7**, 649 (1987).

[10] J. C. Dainty (ed.), "Laser Speckle and Related Phenomena." Springer-Verlag, New York, 1984.

filament sources have been used. On the other hand, fluorescence measurements often have larger fractional changes that will better tolerate noisy light sources, and the measured intensities are lower, making improvements in the S/N ratio from brighter sources attractive. Opti-Quip Inc., (Highland Mills, NY) provides 150- and 250-W xenon power supplies, lamp housings, and arc lamps with noise that are in the range of 1–3 parts in 10^4. The 150-W bulb yielded three times more light at 520 ± 45 nm than a tungsten filament bulb. The extra intensity is especially useful for fluorescence measurements from single neurons or from weakly stained nerve terminals. If the dark noise is dominant, then the S/N ratio improves linearly with the intensity of the light source.

Optics

Numerical Aperture and Depth of Focus

The need to maximize the number of measured photons is a powerful factor affecting the choice of optical components. In epifluorescence, both the excitation light and the emitted light pass through the objective, and thus the intensity reaching the photodetector is proportional to the fourth power of numerical aperture.[11] Thus, the numerical aperture (NA) of the objective is a crucial factor in increasing the number of measured photons. A Macroscope (RedShirtImaging, LLC, Fairfield, CT) provides 0.4 NA at 4×; 0.5 NA objectives can be obtained at 10× and 0.9 NA or higher at magnifications of 20× and above. In addition to determining the intensity reaching the photodetector, NA also determines the depth of focus. Salzberg et al.[8] found that the effective depth of focus for a 0.4 NA objective lens was about 600 μm and Kleinfeld and Delaney[12] found an effective depth of focus of 200 μm using 0.5 NA optics. In some instances (e.g., Zecevic et al.[13]) a lower numerical aperture (and lower S/N ratio) was accepted in order to have a larger depth of focus.

Light Scattering and Out-of-Focus Light

Light scattering can limit the spatial resolution of an optical measurement. Figure 6A shows that when no tissue is present, essentially all of the light (750 nm) from a small spot falls on one detector. Figure 6C illustrates the result when a 500-μm-thick slice of salamander olfactory bulb is present. The light from the small spot is spread to about 200 μm. Mammalian cortex appears to scatter more than the salamander bulb. Thus, light scattering will cause considerable blurring of signals in adult vertebrate preparations.

[11] S. Inoue, "Video Microscopy," p. 128. Plenum Press, New York, 1986.
[12] D. Kleinfeld and K. Delaney, J. Comp. Neurol. **375**, 89 (1996).
[13] D. Zecevic, J. Y. Wu, L. B. Cohen, J. A. London, H. P. Hopp, and C. X. Falk, J. Neurosci. **9**, 3681 (1989).

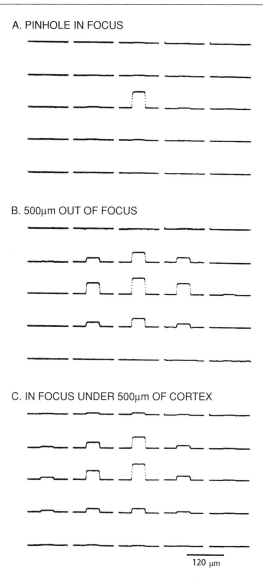

A. PINHOLE IN FOCUS

B. 500μm OUT OF FOCUS

C. IN FOCUS UNDER 500μm OF CORTEX

120 μm

FIG. 6. Effects of focus and scattering on the distribution of light from a point source onto the array. (A) A 40-μm pinhole in aluminum foil covered with saline was illuminated with light at 750 nm. The pinhole was in focus. More than 90% of the light fell on one detector. (B) The stage was moved downward by 500 μm. Light from the out-of-focus pinhole was now seen on several detectors. (C) The pinhole was in focus but covered by a 500-μm slice of salamander cortex. Again the light from the pinhole was spread over several detectors. A 10 × 0.4 NA objective was used. Köhler illumination was used before the pinhole was placed in the object plane. The recording gains were adjusted so the largest signal in each of the three trials would be approximately the same size.

A second source of blurring is signal from regions that are out of focus. For example, if the active region is a cylinder perpendicular to the plane of focus, and the objective is focused at the middle of the cylinder, then the light from the in-focus plane will have the correct diameter at the image plane. However, the light from the regions above and below are out of focus and will have a diameter that is too large. Figure 6B illustrates the effect of moving a small spot of light 500 μm out of focus. The light from the small spot is spread to about 200 μm. Thus, in preparations with considerable scattering or with out-of-focus signals, the actual spatial resolution may be limited by the preparation and not by the number of pixels in the imaging device.

The confocal microscope substantially reduces both the scattered and out-of-focus light that contributes to the image. A recent modification using two-photon excitation of the fluorophore further reduces out-of-focus fluorescence and photobleaching.[14] With both types of microscope one can obtain images from intact vertebrate preparations with much better spatial resolution than can be achieved with an ordinary microscopy. However, at present the sensitivity of these micro-scopes is relatively poor and many milliseconds are required to record the image from a single very thin $x-y$ plane. The kinds of problems that can be approached using a confocal microscope are limited by these factors.

Cameras

Because the S/N ratio in a shot–noise-limited measurement is proportional to the square root of the number of photons converted into photoelectrons, the quantum efficiency of this process is important. Silicon photodiodes have quantum efficiencies approaching the ideal (1.0) at wavelengths at which most dyes absorb or emit light (500–900 nm). In contrast, only specially chosen vacuum photocath-ode devices (phototubes, photomultipliers, or image intensifiers) have a quantum efficiency as high as 0.2. Thus, in shot–noise-limited situations, a silicon diode will have an S/N ratio that is substantially larger. The discussion below considers only silicon diode systems and only those that have frame rates near 1 kfps.

Parallel Readout Arrays

Photodiode arrays with 256–1020 elements are now in use in several laboratories.[13,15–17] These arrays are designed for parallel readout; each detector

[14] W. Denk, D. W. Piston, and W. Webb, in "Handbook of Biological Confocal Microscopy" (J. W. Pawley, ed.), p. 445. Plenum Press, New York, 1995.

[15] T. Iijima, M. Ichikawa, and G. Matsumoto, Abstracts Soc. Neurosci. 15, 398 (1989).

[16] M. Nakashima, S. Yamada, S. Shiono, M. Maeda, and F. Sato, IEEE Trans. Biomed. Eng. 39, 26 (1992).

[17] A. Hirota, K. Sato, Y. Momose-Sato, T. Sakai, and K. Kamino, J. Neurosci. Methods 56, 187 (1995).

is followed by its own amplifier whose output can be digitized at frame rates of >1 kfps. Although the need to provide a separate amplifier for each diode limits the number of pixels in parallel readout systems, this amplification scheme contributes to the very large (10^5) dynamic range that can be achieved. A parallel readout array system is commercially available; NeuroPlex-II (464 pixels), RedShirtImaging, LLC.

Serial Readout Arrays

By using a serial readout, the number of amplifiers is greatly reduced. Furthermore, it is simpler to cool CCD chips to reduce the dark current noise and noise from the measuring amplifier (read noise). In addition, the much smaller pixel size reduces the diode capacitance, which also contributes to the much lower total dark noise of the CCD camera. However, because of saturation, presently available CCD cameras cannot be used with the higher intensities available in some neurobiological experiments. This accounts for the bending over of the CCD camera performance at segment B in Fig. 5. A dynamic range of even 10^3 is not easily achieved with presently available CCD cameras. Thus, these cameras will not be optimal for fluorescence measurements where the staining intensity is high. The light intensity would have to be reduced with a consequent decrease in *S/N* ratio. On the other hand, CCD cameras are close to ideal for measurements where there are fewer than 10^7 photons/ms/0.2% of the image (Fig. 5). The intensity that is obtained from stained dendrites is within the range in which the CCD camera is close to ideal. Table I compares three CCD cameras with frame rates near 1 kfps.

TABLE I

CHARACTERISTICS OF FAST CCD CAMERA SYSTEMS[a]

Camera	Frame rate (fps) full frame	Well size[b] ($\times 1{,}000\ e^-$)	Read noise[c] (electrons)	Back illumination[d]	Bits a-to-d	Pixels
MiCAM 01[e]	1333	—	—	No	12	92×64
Dalsa CA-D1-0128[f]	756	300	360	No	12	128×128
NeuroCCD-SM[g]	2000	300	20	Yes	14	80×80
			5 at 125 Hz			

[a] As reported by the manufacturer.
[b] Number of electrons that can be stored at each pixel before saturation occurs.
[c] The rms read noise; at fast frame rates (1 kfps) the read noise will be the dominant dark noise.
[d] A back-illuminated camera will have a quantum efficiency (number of photoelectrons/photon) of about 0.9. A front-illuminated camera will have a quantum efficiency of <0.4.
[e] www.scimedia.co.jp
[f] www.dalsa.com
[g] www.redshirtimaging.com

Recording from Processes of Individual Neurons

It is widely recognized that dendritic membranes of many (or all) neurons contain active conductances that underlie dynamic patterns of electrical activity. This complicated electrical behavior of branching neuronal processes is impossible to predict by a model in the absence of detailed measurements. To obtain such a measurement one would, ideally, like to be able to monitor, at multiple sites, subthreshold events as they travel from the sites of origin on neuronal processes and summate at particular locations to influence action potential initiation. A major leap forward in studying individual neurons was achieved by the development of the recording method that allowed simultaneous monitoring of voltage transients from two locations on a single neuron (double-patch recording from individual neurons in brain slices).[18] However, simultaneous recording from two locations is a limited tool for assessing rapidly changing spatiotemporal patterns of signal initiation and propagation in complex dendritic structures. Thus, most of the uncertainties in the interpretation of the results from double-patch experiments were due to the insufficient spatial resolution of the measurements.[19-21]

For a more complete description of the functional organization of individual neurons, it is necessary to complement the patch-electrode approach with technologies that permit a massively parallel recording from neuronal processes. This can be achieved by using voltage imaging, a fast multisite optical measurement with intracellular voltage-sensitive dyes.[22-24]

Voltage Imaging as an Indirect Measurement of Membrane Voltage

Voltage-sensitive dye recording of membrane potential transients belongs to a class of indirect measurements. The quantity that is being measured directly, using silicon photodetectors, is light intensity, and the quantity that needs to be monitored (membrane potential) must be derived from a known relationship between the light intensity and the membrane potential.

Calibration. It is convenient that the relationship between light intensity and membrane potential is linear for many voltage-sensitive dyes.[2,25] If the measurements are being made from one place and if an electrical measurement can be done simultaneously with the optical measurements (as in Fig. 2), an absolute calibration is automatically obtained. Then, barring bleaching or changes in staining, the

[18] G. J. Stuart and B. Sakmann, *Nature* **367,** 69 (1994).
[19] W. Chen, J. Midtgaard, and G. Shepherd, *Science* **278,** 463 (1997).
[20] D. A. Hoffman, J. C. Magee, C. M. Colbert, and D. Johnston, *Nature* **387,** 869 (1997).
[21] G. Stuart and M. Hausser, *Neuron* **3,** 703 (1994).
[22] D. Zecevic, *Nature* **381,** 322 (1996).
[23] S. Antic, G. Major, and D. Zecevic, *J. Neurophysiol.* **82,** 1615 (1999).
[24] S. Antic, J. Wuskel, L. Loew, and D. Zecevic, *J. Physiol.* **527,** 55 (2000).
[25] J. Y. Wu and L. B. Cohen, *in* "Fluorescent and Luminescent Probes for Biological Activity" (W. T. Mason, ed.). Academic Press, San Diego, CA, 1993.

electrodes could be removed and all further optical recordings would be precisely calibrated in terms of voltage.

A more complicated situation arises when optical measurements of membrane voltage are made from multiple (e.g., several hundred) sites on a neuron. Usually, most of these sites are not accessible to direct electrical measurements and the simple calibration procedure described above is not possible. In a multisite recording, the fractional change in light intensity is still proportional to voltage, but also, to a different extent at different sites, to two additional factors: (I) The first factor is the amount of dye that contributes to the fluorescent light intensity recorded by each individual detector. The amount of dye will depend on two parameters: (a) the volume of the objects projected by a microscope objective onto each detector (surface area if depth of field is shallow) and (b) the amount of dye bound to a unit volume (or surface area). (II) The second factor is the ratio of the amount of dye that is bound to connective tissue or membranes that do not change potential (inactive dye) to the amount of dye that is bound to the excitable membrane being monitored (active dye). The inactive dye contributes to the resting fluorescence only, and the light from active dye contributes to the resting fluorescence and also carries the signal. Obviously, if all of the dye is in the excitable membrane, the optical signal expressed as the fractional fluorescence change ($\Delta F/F$) will be larger, for the same change in membrane potential, than if there is inactive dye. This is an essential consideration that explains why extracellular dye application has a dramatically lower S/N ratio than intracellular staining in recording from individual nerve cells; extracellular staining generates a large excess of inactive dye that contributes only to the background fluorescence.

If all of the voltage-sensitive dye is bound to excitable membrane, the differences in the sensitivity between detectors are caused solely by the differences in the resting intensity of the light projected to different detectors. In this situation dividing the signal (ΔF) by resting light intensity (F) for each detector will equalize the sensitivity of all elements in the photodetector array. In that case, calibration of the optical signal in terms of membrane potential from any one location by simultaneous optical and electrical recording will be valid for the whole array. One deviation from this rule would arise in situations in which autofluorescence (not related to voltage-sensitive dye) from the preparation is not negligible and contributes to a different degree to the resting fluorescence recorded by each detector. If autofluorescence is not a problem, and if all of the voltage-sensitive dye is bound to excitable membrane, it is possible to use dual-wavelength ratiometric methods, analogous to measurements with cation indicators,[26] as a reliable measure of membrane potential.[27]

However, the situation in which all of the dye is active is rarely found. Generally, there will be both active dye and inactive dye contributing to fluorescence.

[26] G. Grynkiewicz, M. Poenie, and R. Y. Tsien, *J. Biol. Chem.* **260**, 3440 (1985).
[27] V. Montana, D. L. Farkas, and L. M. Loew, *Biochemistry* **28**, 4536 (1989).

For example, if the preparation is stained by the extracellular application of the dye, inactive dye is bound to the connective tissue or other cellular components of the preparation. In experiments utilizing intracellular application of the dye, inactive dye is bound to intracellular membranes and organelles. Furthermore, it is a general rule that the ratio of active dye to inactive dye is unknown. This ratio will be different for different regions of the object. It follows that the sensitivity of a multidetector array in recording from different regions of the object will not be uniform and the calibration of all detectors cannot be done by calibrating the optical signal from any single site. In this situation the amplitude calibration of optical signals in terms of voltage will require separate calibration of each pixel to determine the "sensitivity profile" of the array. Such a calibration is absolute and straightforward if a calibrating electrical signal that has known amplitude at all locations is available. A nondecremental action potential signal is ideal for this purpose. This type of calibration was used to scale the amplitudes of hyperpolarizing subthreshold signals in an interneuron from land snail.[28]

If action potentials are not available, another type of calibrating electrical signal that has known amplitude at all locations is necessary. One idea is to take advantage of the fact that slow electrical signals spread without decrement over relatively long distances in neuronal processes because nerve cells are "electrically compact" for slow voltage changes. This has been documented by direct electrical measurement from two locations on a neuron, for both invertebrate nerve cells[29] and vertebrate neurons.[21] On the other hand, in many measurements the absolute calibration of optical signals in terms of voltage is not critical. Many conclusions depend on the comparison of relative amplitudes and on timing information that can be obtained directly from voltage-sensitive dye recordings.

Methods

Dye Injection

Invertebrate Neurons. Sharp electrodes are used for the injection of the dye in invertebrate neurons. The tip of the microelectrode is backfilled with the solution of a voltage-sensitive dye dissolved in distilled water. The dye solution is filtered before filling to eliminate microscopic particles. We use Millex-GV4 filters with a 0.22-μm pore size (Millipore, Bedford, MA). Dyes are injected by applying repetitive, short-pressure pulses to the microelectrode using a Picospritzer (General Valve Corp., Fairfield, NJ). The microelectrode tip size, pressure settings, and pulse duration are adjusted for different cells (pressure was varied between 5 and 60 psi, pulse duration between 1 and 50 ms, and microelectrodes ranged from 2 to 10 MΩ

---

[28] S. Antic and D. Zecevic, *J. Neurosci.* **15**, 1392 (1995).
[29] L. Tauc, *J. Gen. Physiol.* **45**, 1077 (1962).

when filled with 3 M KCI). After the injection was completed, the ganglia are routinely incubated for 12 hr before making optical recordings to allow for the spread of dye into the distal processes.

Vertebrate Neurons. The dye is applied into vertebrate neurons under visual control using infrared differential interference contrast (DIC) video microscopy and patch electrodes. Simple diffusion is sufficient to load the soma with an adequate amount of the dye and pressure injection is not necessary. The major problem in loading vertebrate neurons from patch pipettes is the leakage of the dye from the electrode. The patching technique requires positive pressure to be applied to the patch pipette during electrode manipulation through the tissue while approaching healthy neurons. This pressure ejects the solution from the electrode. To avoid extracellular deposition of the dye and the resulting large background fluorescence, the tip of the electrode has to be filled with dye-free solution and the electrode back-filled with the voltage-sensitive dye (JPW1114 or its dimethyl analog JPW3028; Molecular Probes, Inc., Eugene, OR). It is possible, with practice, to load neurons without any leakage of the dye to the surrounding tissue. Examples of selective staining are shown in Figs. 7–9.

Optical Recording

Photodiode Array. The recording system for fast, multisite optical monitoring of membrane potential changes used in invertebrate experiments is based on an array of silicon photodiodes.[25,28,30] The preparation is placed on the stage of a microscope and the image of the stained cell is projected onto the photodiode array positioned at the primary image plane. A 250 W xenon, short-gap arc lamp is used as a light source. The best signals with styryl dyes are obtained using an excitation interference filter of 520 ± 45 nm, a dichroic mirror with central wavelength of 570 nm, and a 610 nm barrier filter (a Schott RG610). The optical signals are recorded with a 464 element photodiode array (NeuroPlex, RedShirtImaging, LLC, Fairfield, CT). The output current of each diode is converted to voltage and individually amplified. High-frequency noise in the recording is limited by the 800 Hz cutoff frequency of the low-pass filter. An RC filter with a cutoff frequency of 1.7 Hz was used to limit low-frequency noise. The fastest acquisition rate, limited by a single conversion time of 1.3 μs, is 0.6 ms per full frame with 464 pixels.

CCD Camera. The dominant noise using the photodiode array was the dark noise. High-performance CCD cameras are characterized by a reduced dark noise. A cooled back-illuminated CCD camera has dark noise that is substantially lower than that of the photodiode array system. We showed that under our measurement conditions with intracellular fluorescent dyes, a carefully chosen CCD camera (NeuroCCD, RedShirtImaging, LLC) can improve the *S/N* ratio by a factor of

---

[30] L. B. Cohen and S. Lesher, *Soc. Gen. Physiol. Ser.* **40,** 71 (1986).

approximately 10. This CCD camera is close to ideal for measurements from processes of individual neurons stained with internally injected dyes. The improved sensitivity allows optical recording of both subthreshold and action potential signals from individual processes of vertebrate neurons in brain slices (Fig. 9).

Example Results

Invertebrate Neurons

A typical result of a multisite voltage-sensitive dye recording is shown in Fig. 7. The image of the cell was projected by an objective onto the array of photodiodes as indicated in Fig. 7A. This represents multisite recording of action potential signals, evoked by a transmembrane current step (Fig. 7B), from axonal branches marked Br2, Br3, and Br4. Optical signals associated with action potentials, expressed as fractional changes in fluorescent light intensity ($\Delta F/F$), were between 1% and 2%

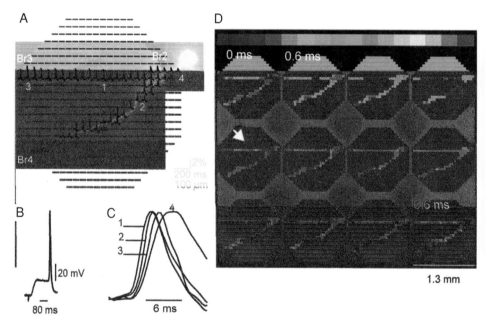

FIG. 7. Voltage imaging from an invertebrate (*Helix pomatia*) neuron. (A) Raw optical recordings of fluorescence signals ($\Delta F/F$) associated with a 85 mV action potential from elements of photodiode array positioned over the CCD image of the fluorescence of the axonal arborizations of a metacerebral cell *in situ*, stained with the voltage-sensitive dye, JPW1114. (B) Synaptically evoked action potential recorded by a microelectrode in the soma. (C) Superimposed recordings from individual detectors from different locations as indicated in (A). (D) Color-coded representation of the spatial and temporal dynamics of the synaptically evoked spike. The peak of the action potential is shown in red. Individual frames are separated by 0.6 ms.

in recordings from neuronal processes. The signal size ($\Delta F/F$) dropped abruptly in the most distal axonal regions (Br4 in Fig. 7A) because the concentration of the dye declined with distance from the soma and the resting light for the detectors that receive light from distal processes becomes dominated by autofluorescence.

It is straightforward to determine the direction and velocity of action potential propagation from this data. In Fig. 7C recordings from different locations, scaled to the same height, are compared on an expanded time scale to determine the site of origin of the action potential. The earliest action potential, in response to synaptic stimulation, was generated near location 1 in Fig. 7A in the axonal branch Br3 situated in the cerebral–buccal connective outside the ganglion. The spike propagated orthodromically from the site of initiation toward the periphery in branch Br3 and antidromically toward the soma and into the branch Br4 in the external lip nerve. Although the difference in the timing of the spikes at locations 1–4 (Fig. 7C) is small, the pattern of propagation is clear from the color-coded representation of the same data (Fig. 7D). Figure 7D shows the potential changes in the branching structure at nine different times separated by 0.6 ms. The red color corresponds to the peak of the action potential. Figure 7D shows the position of the action potential trigger zone (white arrow) and ortho- and antidromic spread of the nerve impulse from the site of initiation. Monitoring spike initiation and the pattern of propagation and interaction of impulses in neuronal processes will improve our understanding of how individual neurons process information.

Vertebrate Neurons

We applied the same technique to dendrites of vertebrate CNS neurons in brain slices. The initial experiments were carried out on pyramidal neurons in neocortical brain slices from P14–18 rats using a 464-element photodiode array. The experiments provided several important methodological results. First, it was established that it is possible to deposit the dye into vertebrate neurons without staining the surrounding tissue. Second, pharmacological effects of the dye were completely reversible if the staining pipette was pulled out, and cell was allowed to recover for 1 or 2 hr. Third, the level of photodynamic damage already allows meaningful measurements and could be reduced further. Finally the sensitivity of the dye was comparable to that achieved in the experiments on invertebrate neurons.[11] In these preliminary experiments, the dye spread approximately 500 μm into dendritic processes within 2 hr.

An example of a recording from a pyramidal neuron using the photodiode array is shown in Fig. 8. The fluorescent image of the cell was projected onto the array as illustrated in Fig. 8A. The neuron was stimulated by depolarizing the cell body to produce a burst of two action potentials. Each trace in Fig. 8B represents the output of one photodiode for 44 ms. As evident from Fig. 8, the optical signals were found in the regions of the array that correspond closely to the

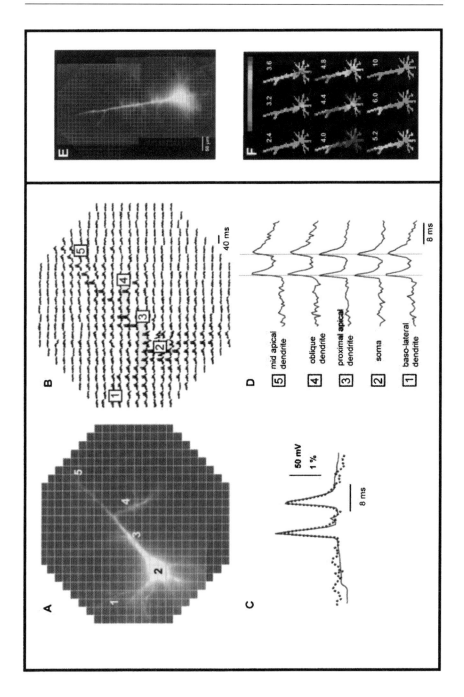

geometry of the cell. Optical signals associated with action potentials, expressed as fractional changes in fluorescent light intensity ($\Delta F/F$), were between 1% and 2% in recordings from neuronal processes. Figure 8C shows a comparison of the electrical recordings from the soma (smooth line) with the optical signals filtered to eliminate high-frequency noise (dotted line). There is a good agreement between time courses of electrical and optical recordings. In Fig. 8D, recordings from different locations, scaled to the same height, are compared on an expanded time scale to establish the pattern of action potential propagation. Each trace is a spatial average from two adjacent detectors. Both spikes in the burst originated near the soma and backpropagated along the apical and basolateral dendrites (see also Fig. 8F).

An improvement in the sensitivity was achieved by using a fast, low noise CCD camera in place of the photodiode array used in experiments shown in Fig. 8. This allowed experiments in which subthreshold synaptic potentials were monitored at the site of origin, in the dendritic tuft of mitral cells. The thin dendrites in the tuft are not accessible to microelectrode recording. Voltage imaging is the only available method for studying signal integration in terminal dendritic branches. An example of voltage imaging with the CCD camera from a mitral cell in a slice of the olfactory bulb of the rat is shown in Fig. 9. In this series of experiments we tested whether the integration of synaptic signals at the site of origin (tuft) could be monitored in the absence of action potentials. Evoked recurrent inhibition was used as a tool to prevent spiking. The image of the distal part of the primary dendrite of a mitral cell was projected onto the CCD camera. The light intensity changes were monitored at a frame rate of 3.7 kHz. The olfactory nerve was stimulated with the intensity that normally resulted in an action potential in the mitral cell. However, if the stimulus to the olfactory nerve was preceded by a spike in the mitral cell [evoked by another stimulating electrode positioned in the external

FIG. 8. Voltage imaging from vertebrate pyramidal neurons. (A) Outline of the 464-element photodiode array superimposed over the fluorescent image of a neuron. (B) Single trial recording of action potential-related optical signals. Each trace represents the output of one diode. Traces are arranged according to the disposition of the detectors in the array. Each trace represents 44 ms of recording. Each diode received light from a 14×14-μm^2 area in the object plane. Spikes were evoked by a somatic current pulse. (C) Comparison of electrical and optical recordings. Spatial averages of optical signals from eight individual diodes from the somatic region (dotted line) are superimposed on the electrical recording from the soma (solid line). (D) Action potentials from individual detectors at different locations along the basal, oblique, and apical dendrites. Traces from different locations are scaled to the same height. The increasing delay between the signal from the somatic region and most proximal dendritic segments reflects the propagation time. (E) Fluorescence image of another pyramidal cell obtained by aligning CCD images taken at the two recording positions. The propagation of the spike evoked by a stimulus delivered to the white matter was monitored. (F) Color-coded representation of the data obtained by synchronizing two measurements to the time of the stimulus. The action potential initiated in the somatic region and back-propagated along apical and basal dendrites.

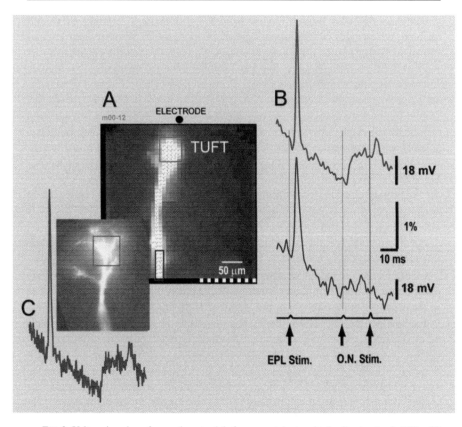

FIG. 9. Voltage imaging of synaptic potentials from a vertebrate mitral cell using the fast 80 × 80 pixel CCD camera. (A) The fluorescence images of a mitral cell stained with voltage-sensitive dye obtained with a conventional high-resolution CCD camera (lower left image) and 80 × 80 CCD camera used for voltage imaging (upper right image). (B) Simultaneous optical recordings of the action potential and synaptic potential signals from the terminal tuft and from a region on the distal primary dendrite, as indicated by colored rectangles on the CCD image. Neuronal impulses were evoked by stimulating the external plexiform layer and olfactory nerve layer by concentric miniature metal electrodes (see text). The data in (B) are filtered digitally using a 1-2-1 smoothing routine 15[9] but only to an extent that did not change the timing information in the slower synaptic signal. Filtering did, however, reduce the amplitude of the action potential. Thus, unfiltered data were used for calibration. (C) Unfiltered data used for the amplitude calibration of optical traces in terms of voltage. The 100 mV action potential was utilized as a calibration signal.

plexiform layer (EPL)] we recorded an excitatory postsynaptic potential (EPSP) alone. Presumably, the previous spike had an inhibitory effect mediated through activity of granule cells in the EPL, and the subsequent activity in the olfactory nerve, evoked by a pair of stimuli in this experiment, resulted in two EPSPs that summated but remained subthreshold (Fig. 9B). The temporal characteristics of action potentials and synaptic potentials recorded from the distal dendritic regions are directly obtained from optical data. The amplitude calibration in terms of membrane potential (mV) can be obtained from optical data in mitral cell experiments using an action potential, measured at soma, as a calibrating signal. It has been established by direct electrical measurements that the action potential in the primary dendrite has constant amplitude.[19]

Summary

Voltage-sensitive dye recording with the present sensitivity and temporal and spatial resolution is a powerful tool in the investigation of the principles of signal integration in single neurons. For example, multisite recording would significantly facilitate experiments on local modulations of excitability in restricted regions of the neuronal dendritic tree[20] or experiments investigating synaptic effects on the precise position of the spike trigger zone.[19]

Recording Population Signals in the Turtle Olfactory Bulb

Odor stimuli have long been known to induce stereotyped local field-potential responses in the olfactory bulb consisting of sinusoidal oscillations of 5–80 Hz riding on top of a slow "DC" signal. This kind of local field-potential signal implies that a population of neurons is somehow synchronously active. Since its first discovery in the hedgehog,[31] odor-induced oscillations have been observed across phylogenetically distant species including locust, frog, turtle, rabbit, monkey, and human. We measured the voltage-sensitive dye signal that accompanies these oscillations in the box turtle. Because the optical measurements have a spatial resolution that is about 25 times better than local field-potential measurements[32] a more detailed visualization of the spatiotemporal characteristics of the oscillations was obtained.

Initial Dye Screening

In initial experiments, several voltage-sensitive dyes were screened using an *in vitro* preparation. The dyes were dissolved in turtle saline (see below); the olfactory bulbs were incubated in the dye solution for 60 min. The bulb was then imaged on the photodiode array (NeuroPlex) and the olfactory nerve was shocked via a suction electrode.[33] Both the signal size and penetration of the dye into the bulb were measured; the results are shown in Table II. The styryl dye, RH414,[34]

[31] E. D. Adrian, *J. Physiol.* **100,** 459 (1942).
[32] M. Zochowski, D. M. Wachowiak, C. X. Falk, L. B. Cohen, Y.-W. Lam, S. Antic, and D. Zecevic, *Biol. Bull.* **198,** 1 (2000).
[33] H. S. Orbach and L. B. Cohen, *J. Neurosci.* **3,** 2251 (1983).
[34] A. Grinvald, E. E. Lieke, R. D. Frostig, and R. Hildesheim, *J. Neurosci.* **14,** 2545 (1994).

TABLE II
RESULTS OF DYE SCREENING ON TURTLE OLFACTORY BULB

Dye	$\Delta F/F$ (%)	Depth of staining (mm)	R	n	X
JPW 1063[a]	0.4–0.9	200	Butyl	2	Et_3N^+
JPW 1113[a]	<0.4	200–300	Ethyl	3	SO_3
JPW 2005[a]	0.6–1.7	100–800	Ethyl	3	Et_3N^+
JPW 3031[a]	0.7–0.9	300–400	Propyl	3[b]	Me_2, $EtOHN^+$
RH 414[c]	1.9–2.4	600–800	Ethyl	3	Et_3N^+
RH 773[c]	1.1–1.6	600–800	Ethyl	3	Me_2, $EtOHN^+$
RH 795[c]	1	Uneven	Ethyl	3[b]	Me_2, $EtOHN^+$

[a] J. P. Wuskell and L. Loew, personal communication.
[b] 2-OH.
[c] A. Grinvald, E. E. Lieke, R. D. Frostig, and R. Hildesheim, *J. Neurosci.* **14,** 2545 (1994).

0.01–0.2 mg/ml in saline, penetrated throughout the thickness of the bulb and had a relatively large signal. The dye staining appeared to be uniform in the different layers of the bulb suggesting that the dye stains all cell types approximately equally.

Methods

Three species of box turtle, *Terepene carolina, T. triunguis,* and *T. ornata,* 300–600 g, were used. The turtles were anesthetized by immersion in wet ice for 2 hr. A craniotomy was performed over the olfactory bulb using a Dremel drill with a small round bit. To facilitate staining, the dura and arachnoid mater were then carefully removed. A segment of polyethylene tubing was inserted into the outlet of the nasal cavity in the roof of the mouth and fixed in place by Krazy Glue and epoxy. Suction applied to this tube was used to draw odorant through the nose.

The turtle was then allowed to warm to room temperature; during this time the bulb was stained by superfusing a solution of the dye in turtle saline on the bulb for 60 min. The solution was changed every 10 min. To reduce movement artifacts during the optical recording, the animals were restrained. The tip of the nose was clamped to the recording apparatus with a piece of flexible plastic and the body was taped to the apparatus. These procedures were approved by the Yale University and the Marine Biological Laboratory Animal Care and Use Committees.

Odor delivery was done with an olfactometer (Fig. 10, bottom left) copied from Kauer and Moulton[35] with minor modifications. The output of the odor

[35] J. S. Kauer and D. G. Moulton, *J. Physiol. (Lond)* **243,** 717 (1974).

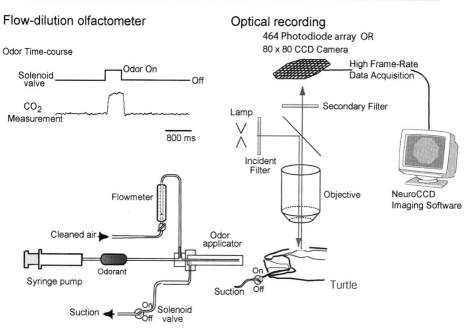

FIG. 10. Schematic diagram of the apparatus used for measuring responses to odorants. *Left Panel*: *Top:* Time course of the odor output from the olfactometer measured by monitoring the CO_2 in the carrier gas. The upper trace shows the time course of the command pulse delivered to the suction solenoid of the outer barrel of the odor applicator. The lower trace is the output of the CO_2 detector probe. There is a delay of about 100 ms between the command pulse and the arrival of the pulse at the CO_2 detector. The odor pulse is approximately square shaped. *Bottom:* Schematic diagram of the olfactometer. Compressed air containing 1% CO_2 was used as the carrier gas. It was cleaned, desiccated, and then mixed with room air saturated with odorant vapor in the odor applicator. The flow rates of the air and the odorant vapor were controlled by a flow meter and a syringe pump respectively. The odor applicator had two barrels; the outer one was normally under suction to remove the odor. Turning off the suction to the outer barrel releases odorant from the end of the applicator. *Right panel:* Schematic diagram of the optical imaging apparatus. The olfactory bulb was illuminated using a 100 W tungsten halogen lamp (voltage-sensitive dye) or a xenon arc lamp (calcium dye). The incident light passed through a heat filter and a ±45 nm bandpass interference filter (520 nm for voltage-sensitive dye and 480 nm for the calcium dye) and was reflected onto the preparation by a long-pass dichroic mirror (580 nm for the voltage-sensitive dye and 510 nm for the calcium dye). For the voltage-sensitive dye experiments the image of the preparation was formed by a 25 mm, 0.95 f camera lens onto a 464-element photodiode array after passing through a 610 nm long-pass secondary filter. For the calcium dye the image of the preparation was formed by a 10.5× or 14× objective lens onto a 80 × 80 CCD camera after passing through a 530 nm long-pass secondary filter. The secondary filter is needed to block reflected incident wavelengths that are transmitted by the dichroic mirror. [Reprinted with permission from Y.-W. Lam, L. B. Cohen, M. Wachowiak, and M. R. Zochowski, *J. Neurosci.* **20**, 749 (2000).]

from the applicator was monitored by measuring the CO_2 in the carrier gas with a CO_2 detector (Beckman Medical Gas Analyzer, LB-2, Schiller Park, IL). The upper trace in Fig. 10 (top left) represents the command pulse sent to the solenoid pump controlling odor delivery; the lower trace is the CO_2 level detected by the Gas Analyzer. The concentration of odorant used in Fig. 11 is 10% cineole; 1 volume of cineole mixed with 9 volumes of clean air. The accuracy and stability of the flow dilution system over the range used were confirmed

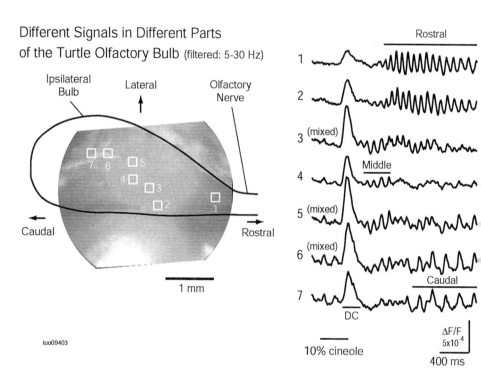

FIG. 11. Simultaneous optical recordings from seven different areas of an olfactory bulb. An image of the olfactory bulb is shown on the left. Signals from seven selected pixels are shown on the right. The positions of these pixels are labeled with squares and numbers on the image of the bulb. All seven signals have a filtered version of the DC signal at the time indicated by the bar-labeled DC. The oscillation in the rostral region has a high frequency and relatively long latency and duration (detectors 1 and 2). The oscillation from the middle region has a high frequency and short latency and duration (detector 4). The oscillation from the caudal region has a lower frequency and the longest latency (detector 7). The signal from detectors between these regions (3, 5, and 6) appears to contain a mixture of two components. The horizontal line labeled "10% cineole" indicates the time of the command pulse to the odor solenoid. The data are filtered by a high-pass digital RC (5 Hz) and low-pass Gaussian (30 Hz) filters. [Reprinted with permission from Y.-W. Lam, L. B. Cohen, M. Wachowiak, and M. R. Zochowski, *J. Neurosci.* **20,** 749 (2000).]

with a photoionization detector. Dedicated lines for each odorant avoided cross-contamination.

Optical imaging was carried out with optics (Fig. 10, right) optimized for light collection efficiency at low magnification. Because the fluorescence intensity in epifluorescence is proportional to the fourth power of the objective numerical aperture and conventional microscope optics have small numerical apertures at low magnifications, we used a 4× Macroscope (RedShirtImaging, LLC) based on a 25 mm focal length, 0.95 f, C-mount, camera lens (with the C-mount end facing the preparation).[36,37] The intensity reaching the photodetector was 100 times larger with the Macroscope than with a conventional 4×, 0.16 NA, microscope lens.

Fluorescence was measured and analyzed using a 464-element photodiode array camera (NeuroPlex) placed at the real inverted image formed by the Macroscope. The preparation was illuminated using a 100 W tungsten–halogen lamp. The excitation filter was 520 ± 45 nm. A 580 nm long-pass dichroic mirror was used to reflect the excitation light onto the preparation. The secondary filter was a RG610 long-pass filter. The bandpass filters in the amplifiers were set to 0.07–125 Hz. The data were recorded at a frame rate of 250 Hz.

Example Result

The recordings of voltage-sensitive dye responses to odorant stimulation from seven selected diodes are shown in Fig. 11 (right). The location of these diodes is indicated by the numbered squares on the image of the olfactory bulb in Fig. 11 (left). In rostral locations (detectors 1 and 2), there was a relatively long-lasting, 15 Hz oscillation with a long latency. On a diode from a middle location (Fig. 11) (detector 4) there was a relatively brief, short-latency oscillation and on a diode from the caudal bulb (detector 7), the oscillation was of a lower frequency and longer latency then rostral. In areas between the two regions, the recorded oscillations were combinations of two signals. In addition, a DC signal, which appears as a single peak after high-pass filtering in Fig. 11, was observed over most of the ipsilateral olfactory bulb. In additional to differences in frequency and latency, the three oscillations also had different shapes—the rostral and caudal oscillations had relatively sharp peaks, while the middle oscillation was more sinusoidal. Thus, as a result of making a simultaneous measurement from many sites using a signal with improved spatial resolution, three independent oscillations were identified in the turtle olfactory bulb.

The noise in the above measurement of fluorescence from a bulk-stained vertebrate brain is consistent with expectations from a shot–noise-limited measurement.

[36] G. Salama, *SPIE Proc.* **94,** 75 (1988).
[37] E. H. Ratzlaff and A. Grinvald, *J. Neurosci. Methods* **36,** 127 (1991).

In the turtle experiments, the photocurrent on each detector was about 2×10^{-8} A (equivalent to 10^8 photons/ms). Because we digitally low-pass filtered the data at 30 Hz, the effective sample period is 33 ms and thus the number of photons/sample period is 3×10^9. The shot noise in this measurement should then be about 2×10^{-5} of the resting intensity (Fig. 5). Consistent with this prediction, the noise in the measurements shown in Fig. 5 is less than 10^{-4} of the resting fluorescence.

Summary and Future Directions

In this article we described methods for making optical measurements of neuron activity with the goal of optimizing the signal-to-noise ratio and thus optimizing the information that can be obtained from the measurement. Although it is clear that attention to a large number of factors is required, many of these factors are fairly well understood.

Because the light-measuring apparatus is already reasonably optimized (see above), any improvement in the sensitivity of these optical measurements will need to come from the development of better dyes and/or investigating signals from additional optical properties of the dyes. The voltage-sensitive dyes in Fig. 3 and the vast majority of those synthesized are of the general class named polyenes.[38] Although it is possible that improvements in signal size can be obtained with new polyene dyes, the signal size on squid axons has not increased in recent years (Gupta et al.[2]; L. B. Cohen, A. Grinvald, K. Kamino, and B. M. Salzberg, unpublished results), and most improvements[28,39–42] have involved synthesizing analogs that work well in new applications or on specific preparations.

The best of the styryl and oxonol polyene dyes have fluorescence changes of 10–20%/100 mV in situations in which the staining is specific to the membrane whose potential is changing.[39,43] Gonzalez and Tsien[44] introduced a new scheme for generating voltage-sensitive signals using two chromophores and energy transfer. Although these fractional changes were also in the range of 10%/100 mV, more recent results are about 30% (J. Gonzalez and R. Tsien, personal communication).

---

[38] F. M. Hamer, "The Cyanine Dyes and Related Compounds." John Wiley & Sons, New York, 1964.

[39] A. Grinvald, R. Hildesheim, I. C. Farber, and L. Anglister, *Biophys. J.* **39**, 301 (1982).

[40] Y. Momose-Sato, K. Sato, T. Sakai, A. Hirota, K. Matsutani, and K. Kamino, *J. Membr. Biol.* **144**, 167 (1995).

[41] Y. Tsau, P. Wenner, M. J. O'Donovan, L. B. Cohen, L. M. Loew, and J. P. Wuskell, *J. Neurosci. Methods* **70**, 121 (1996).

[42] D. Shoham, D. E. Glaser, A. Arieli, T. Kenet, C. Wijnbergen, Y. Toledo, R. Hildesheim, and A. Grinvald, *Neuron* **24**, 791 (1999).

[43] L. M. Loew, L. B. Cohen, J. Dix, E. N. Fluhler, V. Montana, G. Salama, and J. Y. Wu, *J. Membr. Biol.* **130**, 1 (1992).

[44] J. E. Gonzalez and R. Y. Tsien, *Biophys. J.* **69**, 1272 (1995).

However, because one of the chromophores must be hydrophobic and does not penetrate into brain tissue, it has not been possible to measure signals with a fast pair of dyes in intact tissues (T. Gonzalez and R. Tsien, A. Obaid and B. M. Salzberg, personal communications).

Neuron-Type Specific Staining

An important new direction is the development of methods for neuron-type specific staining. Three quite different approaches have been tried. First, the use of retrograde staining procedures has recently been investigated in the embryonic chick and lamprey spinal cords.[43] An identified cell class (motoneurons) was selectively stained. Although spike signals from individual neurons were sometimes measured in lamprey experiments, further efforts at optimizing this staining procedure are needed. The second approach is based on the use of cell-type-specific staining developed for fluorescein by Nirenberg and Cepko.[45] It might be possible to use similar techniques to selectively stain cells with voltage-sensitive or ion-sensitive dyes. Third, Siegel and Isacoff[46] constructed a genetically encoded combination of a potassium channel and green fluorescent protein. When introduced into a frog oocyte, this molecule had a (relatively slow) voltage-dependent signal with a fractional fluorescence change of 5%. More recently, Sakai *et al.*[47] and Ataka and Pieribone[48] developed similar constructs with very rapid kinetics. Neuron-type-specific staining would make it possible to determine the role of specific neuron types in generating the input–output function of a brain region.

Optical recordings already provide unique insights into neuron and organization. Clearly, improvements in sensitivity or selectivity would make these methods more powerful.

Acknowledgments

The authors are indebted to their collaborators Vicencio Davila, Amiram Grinvald, Kohtaro Kamino, David Kleinfeld, Les Loew, Bill Ross, Guy Salama, Brian Salzberg, Alan Waggoner, Jian-young Wu, and Joe Wuskell for numerous discussions about optical methods. We are grateful to the electronics shop of the Department of Physiology, Yale University School of Medicine for technical support. Supported by NIH Grant NS08437-DC05259, NSF Grant IBN-9812301, a Brown-Coxe Fellowship from the Yale University School of Medicine, and an NRSA fellowship, DC 00378.

---

[45] S. Nirenberg and C. Cepko, *J. Neurosci.* **13,** 3238 (1993).
[46] M. S. Siegel and E. Y. Isacoff, *Neuron* **19,** 735 (1997).
[47] R. Sakai, V. Repunte-Canonigo, C. D. Raj, and T. Knopfel, *Eur. J. Neurosci.* **13,** 2314 (2001).
[48] K. Ataka and V. A. Pieribone, *Biophys. J.* **82,** 509 (2002).

[21] Steady-State Fluorescence Imaging of Neoplasia

By Erin M. Gill, Gregory M. Palmer, and Nirmala Ramanujam

Introduction

Fluorescence imaging has emerged as a promising technique for the early detection of precancer and cancer, a capability that is critical for more effective treatment of this disease and improved survival rates. Fluorescence imaging is achieved by exciting fluorophores in tissue with specific wavelength(s) of light and measuring the fluorescence response, thus extracting information about the concentration, location, and environment of the fluorophores. The sensitivity of this technology depends on the ability to establish a source of contrast between the neoplastic and nonneoplastic tissue. This contrast may come from preferential or exclusive presence of the fluorophore within the cancerous tissue, as with exogenous fluorophores, or it may come from more subtle changes of endogenous fluorophores in diseased versus nondiseased tissue. In either case, it is desirable to identify techniques that maximize contrast and improve sensitivity; these methods will be further expounded in the following sections.

Once contrast is established, one needs a system capable of imaging this contrast spatially and, if desired, depthwise. There are a number of approaches to this problem currently being explored, each attempting to ascertain one or both of the two types of information. Spatial information can most directly be acquired using a charge-coupled device (CCD) camera to instantaneously image a surface.[1,2] In this method, the light can be delivered and collected directly, or coupled through fiber optics. An alternative is to use a single channel photomultiplier tube (PMT) to successively image different points and form an image pixel by pixel, sacrificing speed at a substantial savings in cost.[3]

Depthwise imaging is more difficult due to the turbid nature of tissue. One option is to cut thin sections of tissue and image each transversely.[4–6] This method allows one to directly obtain depth information; however, it has the limitation

[1] T. D. Wang, G. S. Janes, Y. Wang, I. Itzkan, J. Van Dam, and M. S. Feld, *Appl. Opt.* **37**, 8103 (1998).
[2] T. D. Wang, J. M. Crawford, M. S. Feld, Y. Wang, I. Itzkan, and J. Van Dam, *Gastrointest. Endosc.* **49**, 447 (1999).
[3] N. Ramanujam, J. Chen, K. Gossage, R. Richards-Kortum, and B. Chance, *IEEE Transact. Biomed. Eng.* **48**, 1034 (2001).
[4] H. W. Wang, J. Willis, M. I. Canto, M. V. Sivak, Jr., and J. A. Izatt, *IEEE Transact. Biomed. Eng.* **46**, 1246 (1999).
[5] C. K. Brookner, M. Follen, I. Boiko, J. Galvan, S. Thomsen, A. Malpica, S. Suzuki, R. Lotan, and R. Richards-Kortum, *Photochem. Photobiol.* **71**, 730 (2000).
[6] R. Drezek, C. Brookner, I. Pavlova, I. Boiko, A. Malpica, R. Lotan, M. Follen, and R. Richards-Kortum, *Photochem. Photobiol.* **73**, 636 (2001).

that it requires excision, and so cannot be used diagnostically *in vivo*. A second approach involves the use of confocal or two-photon microscopy to extract depth information from tissues *in vivo*.[7] However the turbid nature of tissue limits the depths that these methods can resolve. An alternative approach involves the use of a variable aperture fiber-optic probe. It has been demonstrated using Monte Carlo simulations that the sensitivity to depth depends on the aperture size of the fiber optic probe used to deliver the light to and collect the light from the sample.[8] This approach could have the advantage of reducing the cost and complexity of depth-sensitive *in vivo* imaging systems, but further experimentation is needed to demonstrate the effectiveness of this method.

Fluorescence imaging of neoplasia has two primary goals. Both of these goals search for a fundamental understanding of the nature of cancer and mechanisms of its growth. Fluorescence imaging can be implemented in animal models to track the progression of disease.[9-12] This is achieved using molecular reporters such as green fluorescent protein (GFP)[10-12] and luciferase[9,13-17] that are transfected into the genome of the cancerous cells. The other goal of fluorescence imaging methods is to develop a noninvasive diagnostic tool for human precancer and cancer that will enable more accurate classification of cancer at an earlier stage.

In these techniques, the method of exciting the fluorophores can be divided into two categories, single-photon excitation and two-photon or multiphoton excitation. Single-photon excitation excites each fluorophore with a photon of sufficient energy to elevate it to its excited state. Two-photon excitation, on the other hand, uses lower energy photons (longer wavelength), that are individually incapable of exciting the fluorophore. However, when used at high enough luminance, two

[7] Y. S. Sabharwal, A. R. Rouse, L. Donaldson, M. F. Hopkins, and A. F. Gmitro, *Appl. Opt.* **38**, 7133 (1999).

[8] L. Quan and N. Ramanujam, *Opti. Lett.* **27**, 104 (2002).

[9] C. H. Contag, D. Jenkins, P. R. Contag, and R. S. Negrin, *Neoplasia* **2**, 41 (2000).

[10] M. Yang, E. Baranov, P. Jiang, F. X. Sun, X. M. Li, L. Li, S. Hasegawa, M. Bouvet, M. Al-Tuwaijri, T. Chishima, H. Shimada, A. R. Moossa, S. Penman, and R. M. Hoffman, *Proc. Natl. Acad. Sci. U.S.A.* **97**, 1206 (2000).

[11] M. Yang, E. Baranov, A. R. Moossa, S. Penman, and R. M. Hoffman, *Proc. Natl. Acad. Sci. U.S.A.* **97**, 12278 (2000).

[12] M. Yang, E. Baranov, X. M. Li, J. W. Wang, P. Jiang, L. Li, A. R. Moossa, S. Penman, and R. M. Hoffman, *Proc. Natl. Acad. Sci. U.S.A.* **98**, 2616 (2001).

[13] C. H. Contag, S. D. Spilman, P. R. Contag, M. Oshiro, B. Eames, P. Dennery, D. K. Stevenson, and D. A. Benaron, *Photochem. Photobiol.* **66**, 523 (1997).

[14] P. R. Contag, I. N. Olomu, D. K. Stevenson, and C. H. Contag, *Nat. Med.* **4**, 245 (1998).

[15] M. Edinger, T. J. Sweeney, A. A. Tucker, A. B. Olomu, R. S. Negrin, and C. H. Contag, *Neoplasia* **1**, 303 (1999).

[16] A. Rehemtulla, L. D. Stegman, S. J. Cardozo, S. Gupta, D. E. Hall, C. H. Contag, and B. D. Ross, *Neoplasia* **2**, 491 (2000).

[17] P. Tamulevicius and C. Streffer, *Br. J. Cancer* **72**, 1102 (1995).

photons can impinge on same fluorophore nearly simultaneously, and their combined energies can promote the fluorophore to its excited state, enabling the emission of a fluorescent photon. This chapter will focus on single-photon excitation, while Chapter 22 will discuss two-photon excitation.[17a] We will therefore discuss these methods briefly to distinguish their characteristics and uses.

In two-photon fluorescence imaging, the photons must be focused to a point of intense light in order to have an adequate probability of exciting the fluorophore. This requires that the fluorophore be excited by ballistic photons, whereas in single-photon imaging it is primarily excited by scattered photons.[18] Because the relative number of ballistic photons in turbid media is small, the signal intensity limits the penetration depth relative to single-photon excitation, despite the use of longer wavelengths.[19] Also, the turbidity of tissue requires that high-power intensities and signal collection efficiencies be used to deliver and collect sufficient signal, respectively. To achieve this, a high numerical aperture (NA) objective is employed, which at present renders this method endoscopically incompatible. Two-photon imaging, however, has the advantage of exciting fluorophores only in the plane of focus, and is thus inherently suited for optical depth sectioning at high resolution. Unfortunately, the high cost of two-photon imaging components is prohibitive for routine use in clinical applications. Single-photon imaging does not have the image quality or depth sectioning features of two-photon imaging. However, it has a greater penetration depth, is compatible with endoscopic delivery systems, and represents a less expensive approach for bulk tissue imaging.

In focusing on the techniques and applications of single-photon fluorescence imaging, three sections will be presented. The first will detail the sources of fluorescence contrast, using both endogenous and exogenous fluorophores. It will also discuss issues related to the scattering and absorption of light that can make imaging these fluorophores challenging. Second, state of the art instrumentation and imaging strategies will be described in further detail. Finally, the applications of these various techniques will be described. The article will conclude with a section on future prospects of *in vivo* fluorescence imaging.

Tissue Fluorescence Imaging

Contrast

Fluorescence imaging requires that there be one or more fluorescent compounds distributed throughout the medium of interest. These may be present intrinsically or they may be introduced artificially by means of injection, ingestion, or genetic manipulation. The relative concentration and distribution of these

[17a] T. M. Ragan, H. Huang, and P. T. C. So, *Methods Enzymol.* **361**, [22], 2003 (this volume).
[18] X. Gan and M. Gu, *J. Appl. Phys.* **87**, 3214 (2000).
[19] G. Min, G. Xiaosong, A. Kisteman, and M. G. Xu, *Appl. Phys. Lett.* **77**, 1551 (2000).

TABLE I
ENDOGENOUS FLUOROPHORES, EXCITATION AND EMISSION MAXIMA, AND LOCATION IN TISSUE

Type	Endogenous fluorophores	Excitation maxima (nm)	Emission maxima (nm)	Primary tissue location	Ref.
Structural protein	Collagen	325	400, 405	Connective tissue	21
	Elastin	290, 325	340, 400	Connective tissue	22, 23
Electron carrier	FAD	450	535	Cells	24
	NADH	290, 351	440, 460	Cells	24
Heme-related	Porphyrins	400–450	630, 690	Blood	25
Amino acid	Tryptophan	280	350	Proteins	26

fluorophores can then be determined using fluorescence imaging techniques (see Instrumentation and Imaging Strategies). Localized differences in fluorescence intensities at specific excitation and emission wavelengths provide a means of contrast by which the tissue of interest (e.g., cancer) can be identified. The greater the fluorescence contrast, the more rapidly and accurately these differences can be identified. Therefore it would be desirable to identify those fluorophores, whether endogenous or exogenous, that can maximize contrast.

The amount of contrast is determined by the fluorescence efficiency of the fluorophore of interest and the concentration and distribution of that fluorophore within the tissue of interest. Endogenous fluorophores, which have relatively low fluorescence efficiencies, are typically present in both neoplastic and nonneoplastic tissues. However, neoplasia leads to metabolic and morphological changes in the tissue, which can alter the properties, concentration, and distribution of these fluorophores. For instance, the electron carrier, reduced nictoinamide adenine dinucleotide (NADH), is fluorescent, but its oxidized form is not[20] and therefore this fluorophore is sensitive to changes in metabolism. The use of endogenous fluorophores has the additional advantage of obviating concerns over the toxicity and mode of application of the fluorophore, as it is already present in the tissue and ready to be imaged. Endogenous fluorophores that may be useful in the diagnosis of neoplasia are shown in Table I.[21–25,26]

[20] B. R. Masters and B. Chance, in "Fluorescent and Luminescent Probes for Biological Activity: A Practical Guide to Technology for Quantitative Real-Time Analysis" (W. T. Mason, ed.), Ch. 4, p. 44, Academic Press, NY, 1993.

[21] D. Fujimoto, Biochem. Biophys. Res. Commun. 76, 1124 (1977).

[22] Z. Deyl, K. Macek, M. Adam, and O. Vancikova, Biochim. Biophys. Acta 625, 248 (1980).

[23] D. P. Thornhill, Biochem. J. 147, 215 (1975).

[24] B. Chance, B. Schoener, R. Oshino, F. Itshak, and Y. Nakase, J. Biol. Chem. 254, 4764 (1979).

[25] A. M. Kluftinger, N. L. Davis, N. F. Quenville, S. Lam, J. Hung, and B. Palcic, Surg. Oncol. 1, 183 (1992).

[26] J. R. Lakowicz, "Principles of Fluorescence Spectroscopy," 2nd Ed., pp. xxiii, p. 698. Kluwer Academic/Plenum, New York, 1999.

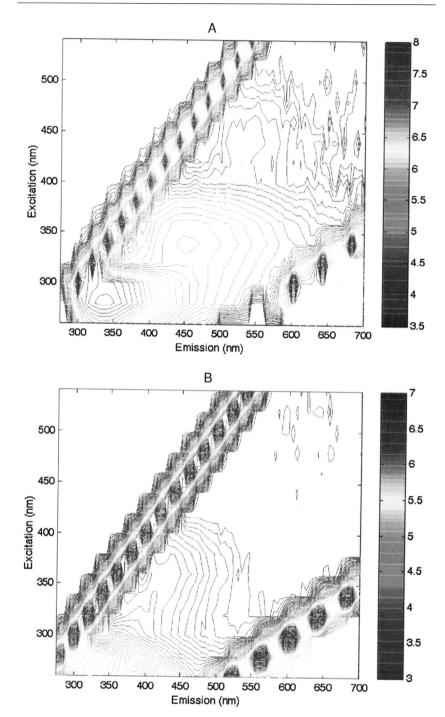

Each fluorophore is generally localized to a specific tissue constituent. Figure 1 shows contour plots of the fluorescence spectra acquired at multiple excitation wavelengths from normal breast (MCF10) cells (Fig. 1A) and collagen I (Fig. 1B). The contour plots shown in Fig. 1 are called fluorescence excitation–emission matrices (EEMs). Each EEM is shown on a log color scale and each contour corresponds to points of equal fluorescence intensity. The meaningful information is straddled by two sets of Rayleigh scattering lines in each EEM. Figure 1A indicates that there are three fluorescence peaks at 280 nm excitation and 340 nm emission (280, 340 nm), 340, 460 nm, and 450, 520 nm. These are attributed to tryptophan, NADH, and flavin adenine dinucleotide (FAD).[24] In Fig. 1B, the fluorescent peaks at 340, 410 nm can be attributed to the cross-link hydroxylysyl pyridoline (HP),[21] and that at 270, 310 nm may be attributed to tyrosine.[26]

The second mode of achieving contrast involves artificially introducing fluorophores to the tissue. This can be done by two methods. The first is done by physically introducing fluorophores to the tissue either locally or systemically. The contrast produced by a fluorophore of this type is determined by how well it can preferentially be taken up by the tissue region of interest, as well as the fluorescence properties of the fluorophore itself. The second method employs genetic manipulation to cause the cells of interest to produce fluorescent or luminescent proteins. This method has an advantage that one can choose to label only the cells of interest (e.g., cancer) and more precisely demarcate the boundaries of the neoplastic tissue. An additional benefit of this method is that it is possible to label specific structures within the cells of interest by encoding these proteins to be attached to an actin molecule,[27] for instance, although that is beyond the scope of this article. Unfortunately, fluorophores of this type are not suitable for diagnosis in humans, because they require genetic manipulation of cancer cells. Useful exogenous fluorophores are listed in Table II.

Sources of Noise

The concentration and distribution of fluorophores determine the amount of contrast attainable under optimal conditions. However, tissue is a turbid medium, and, therefore, light traveling through it can be scattered and/or absorbed, leading to decreased fluorescence contrast.

---

[27] G. Pawalk and D. M. Helfman, *Curr. Opin. Genet. Dev.* **11**, 41 (2001).

FIG. 1. Fluorescence excitation–emission matrices (EEMs) of (A) normal breast cells (MCF10) and (B) collagen I in the extracellular matrix of an organotypic tissue culture. Each EEM is shown on a color log scale, and each contour corresponds to points of equal fluorescence intensity. The meaningful information is straddled by two sets of Rayleigh scattering lines.

TABLE II
EXOGENOUS FLUOROPHORES, EXCITATION AND EMISSION MAXIMA, AND REPRESENTATIVE APPLICATIONS

Introduction	Exogenous fluorophores	Excitation maxima (nm)	Emission maxima (nm)	Representative applications	Ref.
Local/systemic	Aminolevulinic acid (protoporphyrin IX)	375–440	635	Squamous cell carcinoma of the oral cavity in humans	95–98
				Gastrointestinal cancer in humans	95
				Bladder cancer in humans	99
				Squamous cell carcinoma of the rat palatal mucosa	100
				Malignant gliomas in humans	101
				Superficial skin cancers in humans	63
	Carotenohematoporphyrin derivatives (Ref. 102)	475–482	627, 695		
	Indocyanine green and derivatives	550–780	785–812	Human serum albumin and transferrin labeling in mice	103
				Monoclonal antibody labeling	89–91
				Glioma in rats	86
				Angiography	104
				Liver biopsy	81
	Lutetium texaphyrin	460, 474	740, 732, 750	Squamous cell carcinoma in hamsters	85
				Ocular fundus angiography and photodynamic therapy	104
	Lanthanide chelate	270	490, 550, 590	*In vivo* mice–large intestine	105
				Disease diagnosis and bone image enhancement	106
	Ethyl nile blue	633	680	Glioma in mice	107
				Chemically induced mucosal lesions in rats	108
	Aluminum phthalocyanine disulfonate (AlPcS$_2$)	660	670	Squamous cell carcinoma of the rat palatal mucosa	109
				Imaging response to photodynamic therapy in tumor-bearing mice	110
	Foscan (mTHPC)	416, 516, 542, 594, 650	652	Photodynamic therapy of oral cancer in humans	111
Genetic	Luciferase	NA	548–623	Cancer progression in animal models	9, 13–16, 112, 113
	Green fluorescent protein (GFP) mutants	360–516	440–583	Lung cancer metastasis in mice	114–119
				Whole body imaging in mice	10, 12, 114–118

Several sources of absorption are present in tissue. In the ultraviolet and visible spectrum, oxy- and deoxyhemoglobin are the dominant absorbers.[28] Absorption decreases in the near-infrared (NIR).[28] However this is outside the range in which intrinsic fluorophores are excitable, so tissue imaging at these wavelengths requires an extrinsic source of fluorescence. These effects are highly wavelength dependent and result in an attenuation of the excitation and emission light traveling through the tissue. This limits the penetration depth of light and, therefore, the depth of imaging.

Another difficulty in deep tissue imaging is scattering. This occurs due to inhomogeneities in the index of refraction of the tissue, arising from membranes and other tissue structures. This causes blurring of the signal and limits the depth from which localized fluorescence can be detected.[29]

Each of the previous sources of difficulty in fluorescence imaging deals with getting light to and from the fluorophore of interest, but it is also possible for the fluorophore itself to exhibit different properties as a result of interactions with other fluorophores or its environment. One important interaction to consider in tissue imaging is photobleaching. This occurs for both intrinsic and extrinsic fluorophores in a manner and rate that are specific to each individual molecule. For example, fluorescein has been shown to photobleach faster at higher concentrations due to fluorescein–fluorescein interactions,[30] whereas the rate of photobleaching of protoporphyrin IX (PpIX), the fluorophore produced from δ-aminolevulinic acid (5-ALA), has been shown to be predominantly oxygen and irradiance dependent.[31]

Instrumentation and Imaging Strategies

Single-photon fluorescence imaging can in principle yield spatial and depth-wise information. Furthermore, at each pixel and/or voxel, fluorescence intensity can be measured at different wavelengths to provide spectral information, and time-resolved fluorescence decay can be measured to provide the excited state lifetime of the molecules. Ultimately these techniques can merge in order to exploit as much information as possible from fluorescence imaging. In the following sections, fluorescence imaging strategies and the information content of these techniques are outlined for *in vitro* and *in vivo* applications. Additionally, advances in techniques for depth-resolved fluorescence imaging *in vivo* are discussed.

[28] N. Ramanujam, *in* "Encyclopedia of Analytical Chemistry" (R. A. Meyers, ed.), p. 20. John Wiley & Sons Ltd., New York, 2000.
[29] L. Xingde, B. Chance, and A. G. Yodh, *Appl. Opt.* **37,** 6833 (1998).
[30] L. Song, R. P. van Gijlswijk, I. T. Young, and H. J. Tanke, *Cytometry* **27,** 213 (1997).
[31] J. C. Finlay, D. L. Conover, E. L. Hull, and T. H. Foster, *Photochem. Photobiol.* **73,** 54 (2001).

In Vitro Imaging Strategies

To explore the relationship between tissue fluorescence and the presence of dysplasia, a method is needed that can establish the source and distribution of fluorescence contrast within tissue. This is difficult to do *in vivo* because bulk tissue is optically thick or turbid. Therefore, *in vitro* techniques employing optically thin tissue slices/sections represent a logical means for obtaining this information. When choosing a method it is important to consider whether the *in vitro* model realistically represents the *in vivo* environment. This is particularly important for characterizing endogenous fluorophores that are involved in biochemical processes. For example, fluorescence from fluorophores such as NADH and FAD yields information about the reduction–oxidation (redox) state of tissue.[20] If the metabolic and/or oxygenation state of the tissue is altered *in vitro,* the measurement will be affected; therefore preserving the metabolic state of *in vitro* samples or keeping the samples viable in some way is important for fluorescence studies.

Fluorescence Microscopy of Unstained, Frozen Tissue Sections. Conventionally, unstained frozen tissue sections are cut and then imaged at room temperature using fluorescence microscopy. The tissue oxidizes at room temperature, which can affect the metabolic state and hence the fluorescence characteristics of the tissue sample.

Although this method is not ideal, results from studies that examine differences between normal and dysplastic tissue have been able to set a qualitative precedent for *in vivo* studies. Fluorescence microscopy of fluorescent tissue microstructures has been performed on unstained, frozen sections of a variety of tissue types, including the cervix,[32] skin,[33] breast,[34] lung,[35] aerodigestive tract,[36] brain,[37] and colon.[4,38–42] Romer *et al.*[38] observed an increase in the fluorescence intensity of dysplastic epithelial cells relative to that of normal cells of the colon when they

---

[32] W. Lohmann, J. Mussmann, C. Lohmann, and W. Kunzel, *Naturwissenschaften* **76**, 125 (1989).

[33] W. Lohmann and E. Paul, *Naturwissenschaften* **76**, 424 (1989).

[34] W. Lohmann and S. Kunzel, *Naturwissenschaften* **77**, 476 (1990).

[35] W. Lohmann, B. Hirzinger, J. Braun, K. Schwemmle, K. H. Muhrer, and A. Schulz, *Z. Naturforsch.* [C] **45**, 1063 (1990).

[36] A. Fryen, H. Glanz, W. Lohmann, T. Dreyer, and R. M. Bohle, *Acta. Otolaryngol.* **117**, 316 (1997).

[37] G. Bottiroli, A. C. Croce, D. Locatelli, R. Nano, E. Giombelli, A. Messina, and E. Benericetti, *Cancer Detect. Prev.* **22**, 330 (1998).

[38] T. J. Romer, M. Fitzmaurice, R. M. Cothren, R. Richards-Kortum, R. Petras, M. V. Sivak, Jr., and J. R. Kramer, Jr., *Am. J. Gastroenterol.* **90**, 81 (1995).

[39] G. S. Fiarman, M. H. Nathanson, A. B. West, L. I. Deckelbaum, L. Kelly, and C. R. Kapadia, *Dig. Dis. Sci.* **40**, 1261 (1995).

[40] K. Izuishi, H. Tajiri, T. Fujii, N. Boku, A. Ohtsu, T. Ohnishi, M. Ryu, T. Kinoshita, and S. Yoshida, *Endoscopy* **31**, 511 (1999).

[41] G. Bottiroli, A. C. Croce, D. Locatelli, R. Marchesini, E. Pignoli, S. Tomatis, C. Cuzzoni, S. Di Palma, M. Dalfante, and P. Spinelli, *Lasers Surg. Med.* **16**, 48 (1995).

[42] R. S. DaCosta, L. Ligle, J. Kost, M. Cirroco, S. Hassaram, N. Marcon, and B. C. Wilson, *J. Anal. Morphol.* **4**, 192 (1997).

were excited at 350–360 nm. Fiarman *et al.*[39] also observed an increase in the dysplastic cell fluorescence intensity at 488 nm excitation in the colon. Furthermore, Bottiroli *et al.*[41] observed a red fluorescence at around 630 nm when some parts of neoplastic tissue sections of the colon were excited at 366 nm, which was attributed to porphyrins. Romer *et al.*[38] and Fiarman *et al.*[39] observed a decreased fluorescence intensity from the collagen in the lamina propria of the adenomatous colon relative to that of the normal colon at 350–360 nm and 488 nm excitation, respectively. The results of these investigations, which were performed on unstained, frozen tissue sections, support the use of fluorescence imaging to distinguish between optically thick, neoplastic and nonneoplastic tissues *in vivo*.

Fluorescence Microscopy of Short-Term Tissue Cultures. The use of short-term tissue cultures instead of unstained, frozen tissue sections provides a way to avoid problems associated with oxidation by maintaining the viability of the tissue *in vitro*.[5,6]

Brookner *et al.*[5] introduced a novel sample preparation technique that seeks to maintain the viability and metabolic status of biopsied cervical tissue. They prepared short-term tissue cultures from ∼200-μm thin slices of cervical biopsies, and imaged autofluorescence using fluorescence microscopy. Spatial patterns of fluorescence imaged from the transverse samples were strongly correlated with patient age. Addition of potassium cyanide (which impedes oxidative phosphorylation and alters the redox state of the tissue) produced increased fluorescence in the basal and superficial layers of the epithelium, which was attributed to NADH. These results validate the sample preparation method, which provides a preferable alternative to imaging unstained, frozen tissue sections at room temperature. Caveats of this technique are that the thin tissue slices are fragile and difficult to prepare, and must be imaged within a time window of 5 hr or less postbiopsy. Also, variations in the tissue-slice thickness ($\pm 10\%$) make quantitative comparisons of fluorescence intensity difficult.

Drezek *et al.*[6] employed the same sample preparation technique to examine differences in the autofluorescence of normal and dysplastic cervical tissue. An increase in epithelial fluorescence intensity at 380 nm excitation was observed in dysplastic relative to normal tissue. This change was tentatively assigned to a change in concentration of NADH. A decrease in the stromal fluorescence intensity at 380 and 460 nm excitation was observed in dysplastic relative to normal tissue, which was assigned to a decrease in collagen crosslinking. Interestingly, these results suggest that stromal fluorescence is influenced by the presence of dysplasia in the epithelium before stromal invasion occurs. One possible reason for this is increased levels of metalloproteinases that break down the extracellular matrix. This important observation gives evidence that the tissue microenvironment surrounding precancer exhibits biochemical changes that can be detected with fluorescence.

Low-Temperature Fluorescence Imaging of Freeze-Trapped Tissue Blocks. Low-temperature fluorescence imaging involves milling a flat surface of frozen

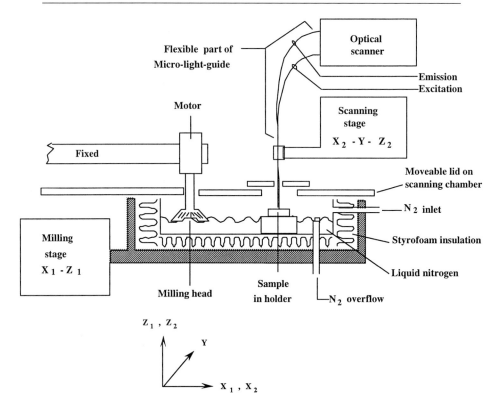

FIG. 2. Low-temperature fluorometer for fluorescence imaging of freeze-trapped tissue blocks. Image adapted with permission from B. Quistorff, J. C. Haselgrove, and B. Chance, *Anal. Biochem.* **148,** 389 (1985).

tissue sections and consequently imaging the fluorescence, while the tissue is in the frozen state. The frozen tissue contains the intact vasculature and the metabolic state is preserved.[24] One caveat is that fluorescence intensities at low temperatures ($-196°$) do not directly represent the fluorescence intensities of *in vivo* tissue (at $37°$). However relative comparisons are indicative of *in vivo* changes.

A low-temperature fluorometer has been designed (constructed at the Johnson Foundation of the University of Pennsylvania)[43] that is uniquely suited for imaging the fluorescence of freeze-trapped tissue blocks at liquid nitrogen temperatures. A schematic of the mechanical part of the apparatus is shown in Fig. 2. The sample is embedded in isopentane or other tissue-freezing medium and placed in the liquid nitrogen chamber. The instrument incorporates a milling head that creates a flat

[43] B. Quistorff, J. C. Haselgrove, and B. Chance, *Anal. Biochem.* **148,** 389 (1985).

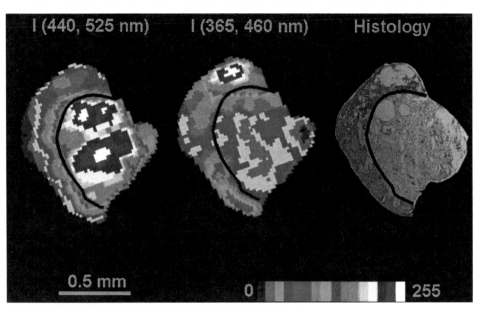

FIG. 3. Hematoxylin and eosin (H&E)-stained section of a freeze-trapped, normal cervical tissue cross section and the corresponding autofluorescence images at two excitation–emission wavelength pairs: 440, 525 and 365, 460 nm (Ref. 44).

imaging surface on the frozen tissue and provides the ability to sequentially image planar sections that can be combined to form a 3D image. The optical scanner consists of a mercury arc lamp, PMT, and excitation and emission filter wheels. Light is transmitted to and detected from the sample via a micro light guide. The distal end of the light guide is precisely positioned at a fixed distance above the imaging surface, typically at a distance of 50 μm. The scanning stage contains a stepper motor that steps the light guide across the imaging surface in discrete steps. Fluorescence intensities are recorded from each discrete pixel on the tissue surface at several excitation–emission wavelength pairs.

Ramanujam et al.[44] have used this instrument to image the autofluorescence from freeze-trapped cervical biopsy cross sections. Figure 3 illustrates fluorescence images from a normal cervical biopsy at excitation–emission wavelength pairs of 440, 525 nm and 365, 460 nm, and the corresponding histological hematoxylin and eosin (H&E)-stained white light image. At 440, 525 nm the fluorescence is primarily attributed to FAD whereas fluorescence at 365, 460 nm is attributed to NADH within the cervical epithelium.[24] The fluorescence from the

[44] N. Ramanujam, R. Richards Kortum, S. Thomsen, A. Mahadevan Jansen, M. Follen, and B. Chance, *Opt. Express* **8,** 335 (2001).

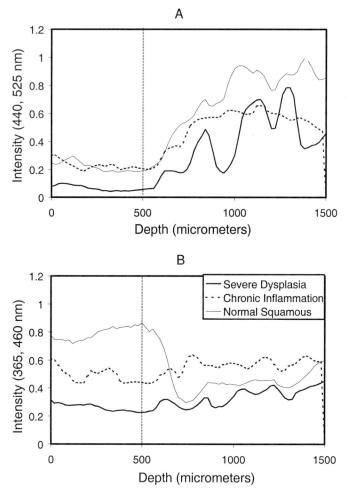

FIG. 4. The average fluorescence intensity as a function of tissue depth at (A) 440, 525 nm and (B) 365, 460 nm for normal, inflammatory, and severely dysplastic freeze-trapped cervical tissue cross sections (Ref. 44).

stroma is attributed to collagen crosslinks. An increase in the color intensity scale corresponds to an increase in the fluorescence intensity. These images indicate that fluorescence intensity at 440, 525 nm is higher in the stromal section whereas fluorescence intensity at 365, 460 nm is higher in the epithelial layer (see H&E image for orientation). Figure 4 displays the average fluorescence intensity as a function of tissue depth at 440, 525 nm (Fig. 4A) and 365, 460 nm (Fig. 4B) for a normal, inflammatory, and severely dysplastic tissue. Evaluation of Fig. 4A indicates that the intensity at 440, 525 nm is significantly greater in the stroma relative to that in the epithelium for all three tissue types. Figure 4B indicates no significant difference in intensity with depth for tissues with inflammation and severe dysplasia; however, it does indicate a difference in intensity with depth for normal tissue.

Tissue Culture Models. Long-lived tissue culture models could provide a nearly ideal, three-dimensional (3D) *in vivo* model for imaging tissue fluorescence and, in particular, changes in fluorescence with dysplasia. One such organotypic culture model has been developed at the University of Wisconsin, Madison and consists of a collagen base plated with the near-diploid immortalized keratinocytes (NIKS) human skin cell line.[45] It has been shown that a fully stratified squamous epithelium is formed, which includes basal, spinous, granular, and cornified layers. Applications for this system are far reaching and include its potential use in transplants for burn victims and in high throughput schemes for applications such as drug discovery. With regard to dysplasia, this system is nearly ideal, because cancer proliferation can be studied in a human tissue culture system without the need for animal models. Cancerous cells such as squamous cell carcinoma (SCC) cells, with and without GFP, can be added to the basal layer to simulate neoplastic disease progression in the epithelium. Additionally, changes in the tissue microenvironment with dysplasia can be studied. Furthermore, 3D fluorescence imaging with depth sectioning can potentially be performed with confocal and/or multiphoton fluorescence imaging techniques. One caveat of this system is the exclusion of a blood supply.

Summary. The strategies described above provide the basis for *in vivo* imaging, by shedding light on the distribution of fluorescence contrast in tissue. Most of these methods have been used to look at endogenous fluorescence contrast. They can also be extended to examine exogenous tissue fluorescence contrast. These studies also establish the importance of measuring depthwise fluorescence to gain an additional dimension of information.[6,44]

In Vivo Imaging Strategies

As mentioned above, one of the challenges imposed on *in vivo* fluorescence imaging is tissue turbidity, or the influence of absorption and scattering on the fluorescence signal. Another challenge is to resolve fluorescence from heterogeneous and depth-dependent sources. In this section we examine instrumentation currently used for *in vivo* fluorescence imaging and review the different imaging strategies associated with these instruments.

The various types of instruments employed for *in vivo* fluorescence imaging essentially have the same basic components, except for a few differences. For a thorough description of the characteristics of individual components, the reader is referred to Ramanujam.[28] A schematic of the basic components of such an instrument is shown in Fig. 5. It consists of either a broadband or monochromatic excitation

[45] B. L. Allen-Hoffmann, S. J. Schlosser, C. A. Ivarie, C. A. Sattler, L. F. Meisner, and S. L. O'Connor, *J. Invest. Dermatol.* **114,** 444 (2000).

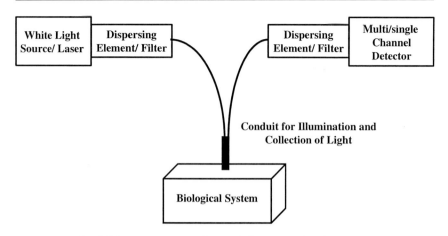

FIG. 5. Primary components of a generic fluorescence imaging system.

light source, a dispersing element or filter to spectrally select the excitation band, a delivery and collection conduit for the delivery of excitation light to, and the collection of the emitted light from the imaging surface of the biological medium, a dispersing element or filter to spectrally select the emitted light, and a multi- or single-channel detector, which measures the spatial and/or spectral distribution of the intensity of the emitted light. Typical systems employ multichannel detectors; however, it is possible in principle to raster-scan a narrow excitation beam over a surface area and collect the emitted light with a single-channel detector.[3]

Multichannel Endoscopic and Nonendoscopic Methods. Several groups have developed endoscopic-compatible[2,46–58] and nonendoscopic-based[59–64]

[46] B. Palcic, S. Lam, J. Hung, and C. MacAulay, *Chest* **99,** 742 (1991).
[47] R. S. DaCosta, B. C. Wilson, and N. E. Marcon, *Gastrointest. Endosc. Clin. N. Am.* **10,** 37 (2000).
[48] M. Zargi, L. Smid, I. Fajdiga, B. Bubnic, J. Lenarcic, and P. Oblak, *Acta. Otolaryngol. Suppl.* **527,** 125 (1997).
[49] G. A. Wagnieres, A. P. Studzinski, and H. E. van den Bergh, *Rev. Sci. Instrum.* **68,** 203 (1997).
[50] R. S. DaCosta, B. C. Wilson, and N. E. Marcon, *J. Gastroenterol. Hepatol.* **17 suppl.,** S85 (2002).
[51] S. Lam, T. Kennedy, M. Unger, Y. E. Miller, D. Gelmont, V. Rusch, B. Gipe, D. Howard, J. C. LeRiche, A. Coldman, and A. F. Gazdar, *Chest* **113,** 696 (1998).
[52] B. Kulapaditharom and V. Boonkitticharoen, *Ann. Otol. Rhinol. Laryngol.* **107,** 241 (1998).
[53] T. McKechnie, A. Jahan, I. Tait, A. Cuschieri, W. Sibbett, and M. Padgett, *Rev. Sci. Instrum.* **69,** 2521 (1998).
[54] K. Svanberg, I. Wang, S. Colleen, I. Idvall, C. Ingvar, R. Rydell, D. Jocham, H. Diddens, S. Bown, G. Gregory, S. Montan, S. Andersson-Engels, and S. Svanberg, *Acta. Radiol.* **39,** 2 (1998).
[55] H. Zeng, A. Weiss, R. Cline, and C. E. MacAualy, *Bioimaging* **6,** 151 (1998).
[56] K. Izuishi, H. Tajiri, M. Ryu, J. Furuse, Y. Maru, K. Inoue, M. Konishi, and T. Kinoshita, *Hepato-gastroenterology* **46,** 804 (1999).
[57] C. S. Betz, M. Mehlmann, K. Rick, H. Stepp, G. Grevers, R. Baumgartner, and A. Leunig, *Lasers Surg. Med.* **25,** 323 (1999).

fluorescence imaging systems. One system originally developed by Palcic et al.[46] for fluorescence bronchoscopy, has led to a commercial light-induced fluorescence endoscopy (LIFE) device (Xillix Technologies Corporation, Richmond, BC, Canada) that is used for fluorescence imaging of relatively large tissue fields (a few centimeters in diameter).

Andersson-Engels et al.[60] describe a nonendoscopic multispectral imaging system for differentiating neoplastic and nonneoplastic tissue. This system uses a pulsed nitrogen laser with 4-ns pulses at an excitation wavelength of 337 nm. The excitation is directed via a dichroic mirror to illuminate a tissue surface area of 10 mm × 10 mm. The fluorescence emitted from the tissue is transmitted through the same dichroic mirror at wavelengths above 370 nm and impinges on a Cassegrainian telescope. This telescope consists of a set of four different bandpass filters, a primary mirror split in four segments, and an output mirror. Each part of the primary mirror is tilted slightly off the optical axis producing four separate images of the same area of tissue in the plane of the imaging detector. Each image occupies one quadrant of the detector and corresponds to one of the four spectral colors. In this manner, the images corresponding to the four spectral channels are recorded simultaneously. An aperture is used in front of the first telescope mirror to improve image resolution. The detection system consists of a dual-microchannel plate image-intensified and Peltier cooled (5°) CCD camera. The image intensifier is gated at 200 ns to integrate the fluorescent light while suppressing the ambient light. The spatial resolution of the instrument was determined to be 0.5 mm. In this study, the filters were chosen based on the spectral characteristics of tissue autofluorescence as well as the photosensitizer, Photofrin (Axcan Scandipharm, Inc., Birmingham, AL).

In a later study, the multispectral imaging system was modified for use with the photosensitizer 5-ALA.[61] Modifications included the addition of a dye laser that emits at 405 nm with a pulse duration of 3 ns, an increased imaging area of 25 mm × 35 mm, different bandpass filters, a shorter 100-ns gating time, and further cooling of the CCD camera to −30°. In both studies, the four images were arithmetically combined to produce maximum contrast between the tumor and surrounding normal skin.

Wang et al.[1,2] describe a colonoscope, modified for autofluorescence imaging of colonic neoplasms. Specifically, they incorporate a fiber-optic excitation probe

[58] J. Haringsma, G. N. Tytgat, H. Yano, H. Iishi, M. Tatsuta, T. Ogihara, H. Watanabe, N. Sato, N. Marcon, B. C. Wilson, and R. W. Cline, *Gastrointest. Endosc.* **53**, 642 (2001).

[59] P. S. Andersson, S. Montan, and S. Svanberg, *IEEE. J. Quant. Electron.* **QE-23**, 1798 (1987).

[60] S. Andersson-Engels, J. Johansson, and S. Svanberg, *Appl. Opt.* **33**, 8022 (1994).

[61] S. Andersson-Engels, R. Berg, K. Svanberg, and S. Svanberg, *Bioimaging* **3**, 134 (1995).

[62] B. W. Chwirot, S. Chwirot, J. Redzinski, and Z. Michniewicz, *Eur. J. Cancer* **34**, 1730 (1998).

[63] J. Hewett, V. Nadeau, J. Ferguson, H. Moseley, S. Ibbotson, J. W. Allen, W. Sibbett, and M. Padgett, *Photochem. Photobiol.* **73**, 278 (2001).

[64] J. Y. Qu, H. Zhijian, and H. Jianwen, *Appl. Phys. Lett.* **76**, 970 (2000).

TABLE III
SINGLE-CHANNEL FLYING SPOT SCANNER[a] COMPARED TO MULTICHANNEL
FLUORESCENCE-IMAGING SYSTEM[b]

Parameters	Single-channel system	Multichannel system
Power (mW)	15	300
Diameter of illuminated area (mm)	10	40
Power density (mW/mm^2)	2	0.239
Working distance (mm)	60	20
Spatial resolution measured using similar resolution targets (mm)	1.0	0.5
Signal-to-noise (3 × 3 pixel area averaged from 6 frames)	25 ± 7	32 ± 5
Frame rate (s)	1	0.033

[a] From N. Ramanujam, J. Chen, K. Gossage, R. Richards-Kortum, and B. Chance, *IEEE Transact. Biomed. Eng.* **48**, 1034 (2001).
[b] Developed by T. D. Wang, G. S. Janes, Y. Wang, I. Itzkan, J. Van Dam, and M. S. Feld, *Appl. Opt.* **37**, 8103 (1998).

into a videocolonoscope that is able to perform conventional colonoscopy as well as fluorescence imaging, in an *in vivo* setting. The modified endoscopic-based imaging system consists of a high-powered argon laser, producing excitation at 356 nm, and an intensified charge injection device (CID) camera. The image is transmitted through a spatially coherent optical fiber bundle and is spectrally filtered with a 400-nm long-pass filter prior to detection. Table III lists parameters for this multichannel imaging system. Autofluorescence images are collected with an excitation wavelength of 356 nm and emission in the spectral band of 400–700 nm.

Single-Channel Methods: The Flying Spot Scanner. As mentioned earlier, it is in principle possible to perform fluorescence imaging with a single-channel detector provided the imaging area is raster scanned. Such an instrument has been developed, the details of which are described.

The flying spot scanner (FSS),[3] constructed at the University of Pennsylvania, consists of three primary components as shown in Fig. 6. A 442-nm wavelength, helium–cadmium laser is the light source. A mechanical positioning assembly directs the excitation light from the laser to the tissue and directs the emitted light from the tissue though a filter wheel into a PMT, which is contained within the mechanical positioning assembly. The assembly also provides three degrees of freedom for illumination and collection: tilt, rotation, and translation. Between the laser and the assembly is a box that contains the electronic and optical components for deflection of the laser illumination in the horizontal and vertical directions, electronics for signal detection and processing, and a computer interface.

The laser light passes through a chopper, two scanning mirrors, and a series of planoconvex lenses and flat mirrors before it is incident on the tissue surface. The excitation light is scanned across the tissue surface and the emitted light from each discrete pixel is collected and focused by a Fresnel lens onto the PMT. Between

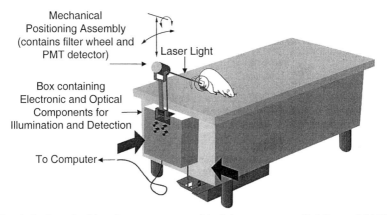

FIG. 6. A schematic of the primary components of the flying spot scanner. (Ref. 3, copyright ©2001 IEEE).

the Fresnel lens and the PMT is a filter wheel that contains four filters. These filters are chosen to selectively transmit fluorescence while rejecting back-scattered excitation light, and also for reflectance measurements. A neutral density filter is used in a second filter wheel slot to ensure that the fluorescence and reflectance signals are within the same dynamic range. The FSS accounts for ambient light conditions and the dark current through a simple real-time subtraction scheme. Table III lists the parameters of the single-channel system.

Figure 7 illustrates an image recorded with the FSS: autofluorescence and reflectance ratio (F/R) images (Fig. 7A) and normalized F/R profiles (Fig. 7B)

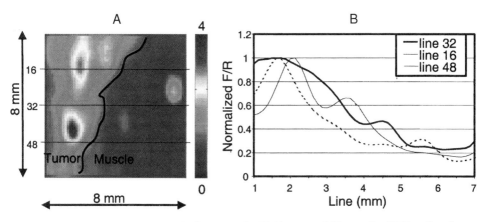

FIG. 7. (a) Autofluorescence and reflectance ratio (F/R) images and (b) normalized F/R profiles of the 9L glioma tumor (9 days) and adjacent muscle tissue in the rat flank. (Ref. 3, copyright ©2001 IEEE).

of a 9L glioma tumor (9 days) and adjacent muscle tissue in the rat flank are shown.

Signal-to-Noise Comparison of Multichannel and Single-Channel Methods. The FSS represents a significant decrease in cost, relative to multichannel, fluorescence-imaging systems. The reduction in cost would result from the use of relatively low-power light sources and single-channel detectors. However, it is important to confirm that the less expensive FSS performs comparably to the multichannel systems. To address this issue, the performance of the FSS was compared to that of the multichannel fluorescence imaging system developed by Wang et al.[1] for the detection of precancers in the colon (described in Multichannel Endoscopic and Nonendoscopic Methods). The characteristics of the two systems are shown in Table III. The multichannel system illuminates a 40-mm-diameter area with a power density of 0.239 mW/mm^2. If the illumination area is reduced to 10 mm in diameter, the power density is increased to 4 mW/mm^2, which is within a factor of two reported for the FSS. A reduction in the working distance by a factor of two doubles the spatial resolution of the FSS, making both the working distance and spatial resolution of the single-channel system comparable to that of the multichannel system. The signal-to-noise ratios (*S/N*) of the two systems are also very similar. However, the frame rate of the multichannel system is 30 times faster compared to that of the FSS. Thus, it can be concluded that the FSS will represent a low-cost alternative to multichannel fluorescence imaging systems in applications in which high frame rates are not required.

Summary. These strategies represent the *in vivo* fluorescence imaging techniques in existence. Endoscopic instruments are convenient for *in vivo* use, whereas nonendoscopic methods are more appropriate for imaging large surface areas. The FSS illustrates the potential for imaging large tissue areas with good image quality and low cost, which is important in the development of clinical diagnostic tools. In all of these methods the excitation light probes an unknown volume of tissue, which contains as yet, unresolved depth-dependent information.

In Vivo Imaging Strategies for Imaging Depthwise Fluorescence

The imaging strategies described above integrate over tissue depth, which loses the depth distribution of fluorescence. The *in vitro* work of Drezek et al.[6] and Ramanujam et al.[44] suggests that depthwise information, especially in the case of endogenous fluorescence, may provide additional contrast in fluorescence imaging.

The next challenge for *in vivo* fluorescence imaging is to develop a technique for measuring depthwise fluorescence. A primary consideration is the illumination and collection geometry. Diffuse reflectance and fluorescence emitted from the tissue surface are nonisotropic due to the filtering effect of absorption and scattering. It can be hypothesized that the probed volume depends on the aperture of the source and detector.

Confocal Microscopic Methods. Confocal microscopy is the only conventional technique extended to obtaining 3D fluorescence images of turbid media. The method employs a high NA objective to create a focal plane of light that can be scanned within a limited distance along the z axis. Sophisticated instruments are able to image fluorescence in real time and have been used for such applications as observing calcium signaling in cardiac muscle.[65]

Several groups have applied confocal microscopy to fluorescence imaging of tissue, *in vivo* and *in vitro*.[7,66–68] Sabharwal *et al.*[7] and Rouse and Gmitro[68] have developed a confocal microendoscope for *in vivo* imaging and further adapted it for multispectral imaging. They measured the performance by imaging fluorescent microspheres, and cultured and stained live cells, stained prostate tissue, and stained peritoneum of a live mouse.[7] This system consists of an argon ion laser, illumination and detection optics, a CCD camera, and a fiber-optic imaging bundle that incorporates a miniature objective lens and focusing mechanism. The overall diameter of the fiber bundle is 1 mm with an active image diameter of 720 μm. The focusing mechanism allows imaging to a depth of 200 μm below the tissue surface. The instrument exhibits a lateral resolution of 3 μm and a depth resolution of 25 μm. Using a slit instead of point aperture makes imaging faster. For a more detailed description of the apparatus the reader is referred to Sabharwal *et al.*[7] To collect spectral information, a prism is added to the detection arm that disperses the emitted light onto the CCD camera. In this configuration, the spectral resolution is 11 nm; however, the spectral resolution can change with prism geometry or with the substitution of a diffraction grating. Spectral information can be processed to isolate emission bands and to provide image reconstruction that enhances contrast or provides specific information about the tissue.

In general, confocal microscopic techniques are limited by the highly scattering nature of tissue, yielding a low signal-to-noise ratio in the absence of strong contrast. Moreover, the depth from which the fluorescence is detected using confocal methods is superficial; that is, the detected volume is on the scale of the mean free scattering path,[66] which is typically about 100 μm in the ultraviolet–visible (UV–VIS).[69–71] Furthermore, this approach requires the use of high-powered

[65] L. A. Blatter and E. Niggli, *Cell Calcium* **23**, 269 (1998).
[66] B. W. Pogue and T. Hasan, *IEEE J. Select. Top. Quant. Electron.* **2**, 959 (1996).
[67] B. W. Pogue and G. Burke, *Appl. Opt.* **37**, 7429 (1998).
[68] A. R. Rouse and A. F. Gmitro, *Opt. Lett.* **25**, 1708 (2000).
[69] A. J. Welch, C. Gardner, R. Richards-Kortum, E. Chan, G. Criswell, J. Pfefer, and S. Warren, *Lasers Surg. Med.* **21**, 166 (1997).
[70] H. Zeng, C. MacAulay, D. I. McLean, and B. Palcic, *Photochem. Photobiol.* **61**, 639 (1995).
[71] G. I. Zonios, R. M. Cothren, J. T. Arendt, J. Wu, J. Van Dam, J. M. Crawford, R. Manoharan, and M. S. Feld, *IEEE Transact. Biomed. Eng.* **43**, 113 (1996).

A

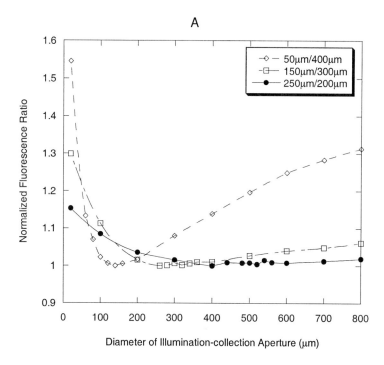

Diameter of Illumination-collection Aperture (μm)

B

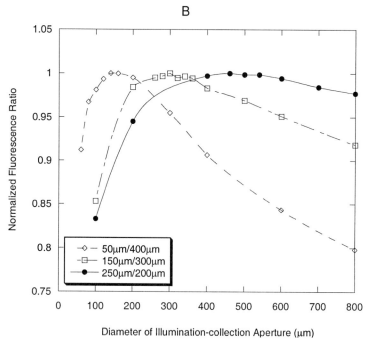

Diameter of Illumination-collection Aperture (μm)

light sources and sensitive detectors, which can make the cost of this technology prohibitive for routine cancer screening and diagnosis. It is therefore important to develop an alternate strategy that can image the depth-dependent distribution of a fluorescent target in a turbid medium with an improved signal-to-noise ratio, increased penetration depth and reduced cost.

Variable Aperture Methods. A possible alternative to 3D imaging with confocal microscopy may be obtained with a particular fiber-optic probe geometry and illumination–collection scheme. Results from Monte Carlo modeling of fluorescent emission from a homogeneous tissue indicate that there is an approximately linear relationship between the optical fiber diameter and the probed depth within the tissue.[72,73] Consequently, it has been suggested that it is possible to measure the depthwise distribution of fluorescence using an optical fiber probe with a variable aperture and a coincident illumination and detection light path.[8]

Quan and Ramanujam[8] have performed Monte Carlo simulations to model the fluorescence from cancerous tissue (SCC) and cancerous tissue tagged with the molecular marker green fluorescent protein (SCC–GFP). The tissue optical properties and fluorescence efficiencies that served as input to the Monte Carlo simulation were estimated from reports in the literature. The fluorescence was calculated as a function of probe aperture diameter and for varying fluorescent target depths and the results are illustrated in Figs. 8 and 9. Figure 8 displays the normalized fluorescence ratio versus aperture diameter for the model containing the SCC layer (without GFP) (Fig. 8A) and the SCC–GFP layer (Fig. 8B). Each curve represents the ratio of two profiles: the fluorescence detected for a model containing the SCC layer and that detected for a model without the SCC layer (equivalent to a normal epithelium). The legend displays the depth and thickness of the SCC layer for each profile. Figure 9 displays the aperture diameter that corresponds to the minimum (for SCC) and maximum (for SCC–GFP) normalized fluorescence ratio (MFR) versus the depth of the SCC layer. A fit to the data indicates that there is a linear relationship between the MFR diameter and depth over the range of 50–300 μm, which implies that the MFR correlates with the depth of the fluorescent target. The next step is to experimentally test the theoretical model

---

[72] T. J. Pfefer, K. T. Schomacker, M. N. Ediger, and N. S. Nishioka, *Appl. Opt.* **41,** 4712 (2002).
[73] Deleted in proof.

FIG. 8. Simulated, normalized fluorescence ratio versus diameter of the illumination–collection aperture for the squamous cell carcinoma (SCC) tissue culture model containing (a) the SCC layer (without GFP) and (b) the SCC–GFP layer. Insets display the depth and thickness of the SCC layer corresponding to each profile (Ref. 8).

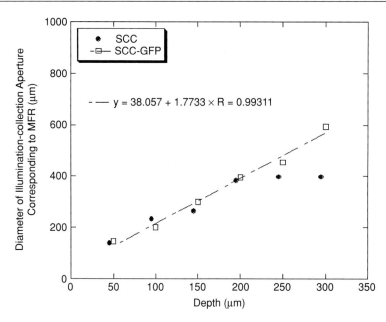

FIG. 9. Diameter of the illumination–collection aperture that corresponds to the minimum (for SCC) and maximum (for SCC–GFP) normalized fluorescence ratio (MFR) versus the depth of the SCC layer. (Ref. 8).

with turbid samples of increasing complexity until it can ultimately be tested *in vivo*.

Other Suggested Methods. It has been suggested[6] that an algorithmic approach for elucidation of depthwise fluorescence is possible. However, there are no analytical algorithms such as the diffusion approximation[74,75] that describe light propagation through tissue that are valid in the ultraviolet–visible regions where tissue absorption and scattering are similar in magnitude. It is feasible that the diffusion approximation will find use in applications that use exogenous fluorophores that emit at wavelengths in the far-red and near-infrared.

Another approach is to use variable wavelengths to resolve depthwise information. However, although it is true that the penetration depth of light in turbid media depends on the wavelength, this approach is inherently confounded by other variables that also have spectral dependencies.

[74] K. Furutsu, *J. Opt. Soc. Am.* **70**, 360 (1980).
[75] A. D. Kim and A. Ishimaru, *Appl. Opt.* **37**, 5313 (1998).

Applications

Autofluorescence Imaging

The goal of imaging autofluorescence in tissue is to be able to accurately and noninvasively diagnose human neoplasia. Table I lists endogenous fluorophores that may be useful in the diagnosis of neoplasia and their properties. These endogenous fluorophores, described earlier in this chapter, generally exhibit weak fluorescence, thus requiring sensitive detection techniques to distinguish differences between dysplastic and normal tissue. The presence of multiple fluorophores with broad and overlapping spectral characteristics makes the choice of the most diagnostically relevant wavelengths nontrivial. For this reason, the sophistication of imaging schemes has evolved over time. Imaging strategies began simply, using a single excitation–emission wavelength pair, and today use several excitation–emission wavelength pairs or spectral bands of fluorescence in addition to diffuse reflectance and white light images.[54,76,77] The multiple fluorescence images provide complementary information to each other and to the other imaging modalities. Typically they are arithmetically combined to maximize image contrast or to provide specific information, such as the tissue metabolic state in redox ratio imaging.[20,44]

Table IV lists recent articles on autofluorescence imaging of neoplasia in a wide variety of organ sites. The reader is referred to the individual references for details on these applications.

Photosensitizer Imaging

The use of photosensitive materials for fluorescence imaging of tissue grew from their use in photodynamic therapy (PDT), an emerging therapeutic technique for treatment of atherosclerosis and dysplasia. Mature lesions take up certain photosensitive agents preferentially with respect to surrounding normal tissue and can be targeted with destructive, high-intensity light. Reasons for the preferential uptake are not fully understood; however, one mechanism that may play a role is the extravasation[78] that occurs in the vasculature of the lesion.

An advantage of using exogenous fluorophores is that the photophysical and pharmocokinetic properties can be selected and are known. Furthermore, exogenous fluorophores are more highly fluorescent than endogenous fluorophores. On the other hand, the disadvantage of using exogenous fluorophores is that issues relating to safety and toxicity of the drug being used have to be addressed. Also, the

[76] S. Andersson-Engels, C. Klinteberg, K. Svanberg, and S. Svanberg, *Phys. Med. Biol.* **42,** 815 (1997).
[77] J. Y. Qu, Z. Huang, and H. Jianwen, *Appl. Opt.* **39,** 3344 (2000).
[78] S. Ito, H. Nakanishi, Y. Ikehara, T. Kato, Y. Kasai, K. Ito, S. Akiyama, A. Nakao, and M. Tatematsu, *Int. J. Cancer* **93,** 212 (2001).

TABLE IV
AUTOFLUORESCENCE IMAGING APPLICATIONS ORGANIZED BY TISSUE SITE AND DISEASE PROCESS

Tissue site and disease process	Model	Application	Imaging technique	Ref.
Oral cancer	Hamster cheek pouch	*In vivo*	Endoscopic imaging	25
	Hamster cheek pouch	*In vivo*	Endoscopic (LIFE[a]) imaging	120
	Hamster, human	*In vivo*	Fluorescence photography	121
Gastrointestinal tract/colon cancer	Human	*In vitro*	Microspectrofluorometry	41
	Human	*In vivo*	Endoscopic imaging	2, 55, 58
	Human	*In vitro*	Confocal microscopy	4, 39, 42
	Human	*In vitro*	Endoscopic imaging	40
	Human	*In vitro*	Microscopy	38
Brain tumor	Human	*In vitro*	Microspectrofluorometry	37
	Rat	*In vivo*	Flying spot scanner	3, 122
Cervical cancer	Human	Culture	Microscopy	5, 6
	Human	*In vivo*	Nonendoscopic imaging	123
	Human	*In vitro*	Low-temperature imaging	44
	Human	*In vitro*	Microscopy	32
Skin melanoma	Human	*In vivo*	Digital imaging	62
	Human	*In vitro*	Microscopy	33
Larynx cancer	Human	*In vivo*	Endoscopic (LIFE[a]) imaging	124
	Human	*In vivo*	Modified endoscopic (LIFE[a]) imaging	125
Lung/bronchus carcinoma	Human	*In vitro*	Microscopy	35
Breast/lymph cancer	Human	*In vitro*	Microscopy	34
	Human	*In vitro*	Multispectral microscopy	126
Head and neck cancer	Human	*In vivo*	Endoscopic (LIFE[a]) imaging	127
	Human	*In vivo*	Modified endoscopic imaging	57

[a] Xillix Technologies Corporation, Richmond, BC, Canada.

selection of the optimal time delay after administration of the drug is nontrivial. Furthermore, there is no specificity to the biochemical aspects of disease process itself.

Table II lists exogenous fluorophores used for imaging neoplasia in tissue. Their spectral characteristics and selected applications are listed as well.

Molecular Imaging

A very exciting and newly emerging field examines cellular- and molecular-specific probes for fluorescence imaging applications. With this approach, the disease of interest can potentially be studied in detail as well as precisely localized within tissue. The high specificity of these techniques provides opportunities to target cellular and molecular changes in the tissue microenvironment with neoplasia, such as in the metastatic process. By these means it may be possible to understand disease mechanisms and accelerate development of therapeutic strategies. In this section we describe current molecular fluorescence imaging techniques, the potential of which have yet to be fully exploited.

Optical reporter genes are transgenes that consist of a luminescent or fluorescent molecule that has been bound to an expressed gene or promoter. Tumor cells may be genetically manipulated to express optical reporter genes. Yang et al. have developed tumor cells that express GFP and have used these cells to monitor tumor growth and metastases in nude mice.[10–12] Fluorescent tumors have also been used to provide contrast for monitoring angiogenesis.[12] The nonluminous capillaries are visible against the bright tumor fluorescence and blood vessel density is quantitatively measured either with intravital or whole body imaging. In these studies, the fluorescent probe is introduced to the live animal model in different ways. The GFP-expressing tumor cells are orthotopically implanted, intravenously injected, or adenoviral GFP is directly injected into the major organs.

Whole-body imaging is achieved with a fluorescent light box and an intensified and/or cooled CCD camera. Alternatively, transilluminated or epifluorescence microscopy is an option. With whole-body imaging of fluorescence from GFP, a 60-μm tumor is detectable at a depth of 0.5 mm, and an 1800-μm tumor is detectable at 2.2 mm below the surface.[10] Figure 10A, C, and D illustrates whole-body images of liver and skull metastases in a mouse model, imaged with a fluorescent light box and cooled CCD camera.[10] The lesions were formed by orthotopically transplanting GFP-labeled tumor cells to live nude mice. The cross-sectional image shown in Fig. 10B was acquired using conventional fluorescence microscopy.

A technique called bioluminescence imaging (BLI) uses the optical reporter firefly luciferase, bound to viral promoters to tag tumor cells.[9,13–16,79] Detection (with an intensified CCD camera) of luciferase-labeled cells is extremely sensitive and can detect as few as 1000 human tumor cells distributed throughout the peritoneal cavity of a mouse.[9,15] This means that the full disease course may be monitored including very early cancer. Edinger et al.[15] developed a stable HeLa cell line that expresses the modified firefly luciferase gene. In a recent study by Rehemtulla et al.,[16] BLI was established as a quantitative tool for assessment of antineoplastic therapies. Quantification of cell kill as a result of chemotherapy was performed with BLI and magnetic resonance imaging (MRI) with comparable results.

The use of optical reporter genes for fluorescence imaging of neoplasia provides an extremely useful tool for understanding the mechanisms of disease and in particular the interaction between neoplastic cells and the surrounding microenvironment. This technique allows the use of live animal models in which regulatory

---

[79] Xenogen Corporation, Alameda, California 94501.

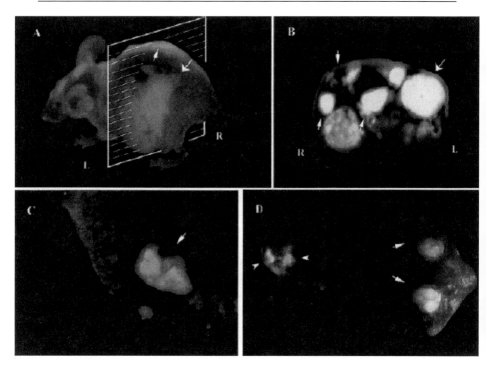

FIG. 10. Whole-body (A, C, D) and fluorescence microscopic (B) images of liver and skull metastases in live nude mice. (A) A whole-body image showing fluorescence from the right and left sides. (B) Fluorescence microscopic image of a cross section, where the level of development of the tumor corresponds to (A). (C) Ventral image of the liver metastases. (D) Dorsal image of the liver and skull metastases. Image reproduced from Ref. 10.

processes and pathways of gene expression are intact. As this research area matures, the predictability of animal models of human disease will improve as study groups can be followed over time. This exciting technique lends itself well to taking advantage of GFP mutants that emit in the near-infrared where the optical transmission through tissue is relatively high.

Another exciting technique is immunofluorescence imaging. In this case, imaging is achieved by binding fluorophores to antibody molecules that interact with specific antigens, thus delivering the fluorophore to the specific site of interest.[80] The fluorophore persists on the time scale of days making it possible to monitor the effect of therapeutic interventions such as radiation therapy and chemotherapy. This technique also has high specificity and does not require the use of potentially

[80] B. Ramjiawan, M. Jackson, and H. Mantsch, in "Encyclopedia of Analytical Chemistry" (R. A. Myers, ed.), p. 5. John Wiley & Sons Ltd., Chichester, 2000.

toxic compounds. *In vivo* immunofluorescence studies are limited by the rate of antibody delivery to the binding site, which is in part determined by the molecular weight of the antibody, typically 150 kDa. For more details on this technique the reader is referred to the excellent article by Ramjiawan *et al.*[80]

Future Prospects

Near-infrared (NIR) absorption and diffuse reflectance spectroscopy are well-established techniques for imaging deep within tissue due to the spectral advantage of being within an optical window in which the absorption is low compared to scattering. Analytical theories such as the diffusion approximation to the transport equation approximate light propagation in this spectral region fairly accurately.[74,75] Major applications of these techniques target perfusion and vascularization, and are not related to the fluorescence characteristics of neoplastic tissue; however, there are some instances of NIR imaging of photosensitizers and molecular reporters in tissue.

The fluorescent dye indocyanine green (ICG) has been used for liver biopsy[81] whereas the photosensitizer lutetium texaphyrin has been used for PDT of cancers and atheromatous plaque.[82–85] New cyanine dyes structurally related to ICG have been synthesized for use as contrast agents for biomedical optical imaging.[86] These dyes exhibit enhanced quantum yield, are faster acting, and last longer relative to ICG. This work demonstrates the potential for engineering dyes to optimize fluorescence imaging in the NIR.

NIR molecular reporters are used in tumor labeling.[87] Becker *et al.* developed a peptide–dye conjugate consisting of cyanine dye and the somatostatin analog octreotate. This approach combines the specificity of ligand/recepor interaction with NIR fluorescence detection. Immunofluorescence imaging has also been performed using cyanine-labeled antibodies.[88–91]

[81] Y. Kimura, T. Higashi, N. Kuwahara, K. Nouso, S. Ohguchi, N. Hino, M. Tanimizu, H. Nakatsukasa, K. Tobe, and T. Tsuji, *Acta. Med. Okayama* **50**, 255 (1996).

[82] K. W. Woodburn, Q. Fan, D. Kessel, M. Wright, T. D. Mody, G. Hemmi, D. Magda, J. L. Sessler, W. C. Dow, R. A. Miller, and S. W. Young, *J. Clin. Laser Med. Surg.* **14**, 343 (1996).

[83] K. W. Woodburn, Q. Fan, D. R. Miles, D. Kessel, Y. Luo, and S. W. Young, *Photochem. Photobiol.* **65**, 410 (1997).

[84] S. W. Young, K. W. Woodburn, M. Wright, T. D. Mody, Q. Fan, J. L. Sessler, W. C. Dow, and R. A. Miller, *Photochem. Photobiol.* **63**, 892 (1996).

[85] M. Zellweger, A. Radu, P. Monnier, H. van den Bergh, and G. Wagnieres, *J. Photochem. Photobiol. B* **55**, 56 (2000).

[86] K. Licha, B. Riefke, V. Ntziachristos, A. Becker, B. Chance, and W. Semmler, *Photochem. Photobiol.* **72**, 392 (2000).

[87] A. Becker, C. Hessenius, K. Licha, B. Ebert, U. Sukowski, W. Semmler, B. Wiedenmann, and C. Grotzinger, *Nat. Biotechnol.* **19**, 327 (2001).

[88] B. Ballou, G. W. Fisher, A. S. Waggoner, D. L. Farkas, J. M. Reiland, R. Jaffe, R. B. Mujumdar, S. R. Mujumdar, and T. R. Hakala, *Cancer Immunol. Immunother.* **41**, 257 (1995).

A new cell-permeant probe, DRAQ5, has been developed for cytometric analysis of the cell cycle and provides 3D nuclear structure and location in live and fixed cells with fluorescence microscopic techniques.[92] New probes developed to image tumor-associated protease activity may detect early stage tumors *in vivo*.[93,94]

[89] B. Ballou, G. W. Fisher, T. R. Hakala, and D. L. Farkas, *Biotechnol. Prog.* **13**, 649 (1997).
[90] S. Folli, P. Westermann, D. Braichotte, A. Pelegrin, G. Wagnieres, H. van den Bergh, and J. P. Mach, *Cancer Res.* **54**, 2643 (1994).
[91] B. Ramjiawan, P. Maiti, A. Aftanas, H. Kaplan, D. Fast, H. H. Mantsch, and M. Jackson, *Cancer* **89**, 1134 (2000).
[92] P. J. Smith, N. Blunt, M. Wiltshire, T. Hoy, P. Teesdale-Spittle, M. R. Craven, J. V. Watson, W. B. Amos, R. J. Errington, and L. H. Patterson, *Cytometry* **40**, 280 (2000).
[93] C. H. Tung, U. Mahmood, S. Bredow, and R. Weissleder, *Cancer Res.* **60**, 4953 (2000).
[94] R. Weissleder, C. H. Tung, U. Mahmood, and A. Bogdanov, Jr., *Nat. Biotechnol.* **17**, 375 (1999).
[95] B. Mayinger, H. Reh, J. Hochberger, and E. G. Hahn, *Gastrointest. Endosc.* **50**, 242 (1999).
[96] A. Leunig, K. Rick, H. Stepp, R. Gutmann, G. Alwin, R. Baumgartner, and J. Feyh, *Am. J. Surg.* **172**, 674 (1996).
[97] A. Leunig, C. S. Betz, M. Mehlmann, H. Stepp, S. Arbogast, G. Grevers, and R. Baumgartner, *Laryngoscope* **110**, 78 (2000).
[98] A. Leunig, M. Mehlmann, C. Betz, H. Stepp, S. Arbogast, G. Grevers, and R. Baumgartner, *J. Photochem. Photobiol. B* **60**, 44 (2001).
[99] N. Lange, P. Jichlinski, M. Zellweger, M. Forrer, A. Marti, L. Guillou, P. Kucera, G. Wagnieres, and H. van den Bergh, *Br. J. Cancer* **80**, 185 (1999).
[100] J. M. Nauta, O. C. Speelman, H. L. van Leengoed, P. G. Nikkels, J. L. Roodenburg, W. M. Star, M. J. Witjes, and A. Vermey, *J. Photochem. Photobiol. B* **39**, 156 (1997).
[101] W. Stummer, S. Stocker, S. Wagner, H. Stepp, C. Fritsch, C. Goetz, A. E. Goetz, R. Kiefmann, and H. J. Reulen, *Neurosurgery* **42**, 518; discussion 525 (1998).
[102] D. Tatman, P. A. Liddell, T. A. Moore, D. Gust, and A. L. Moore, *Photochem. Photobiol.* **68**, 459 (1998).
[103] A. Becker, B. Ricfke, B. Ebert, U. Sukowski, H. Rinneberg, W. Semmler, and K. Licha, *Photochem. Photobiol.* **72**, 234 (2000).
[104] M. S. Blumenkranz, K. W. Woodburn, F. Qing, S. Verdooner, D. Kessel, and R. Miller, *Am. J. Ophthalmol.* **129**, 353 (2000).
[105] D. J. Bornhop, D. S. Hubbard, M. P. Houlne, C. Adair, G. E. Kiefer, B. C. Pence, and D. L. Morgan, *Anal. Chem.* **71**, 2607 (1999).
[106] D. S. Hubbard, M. P. Houlne, G. E. Kiefer, K. McMillan, and D. J. Bornhop, *Bioimaging* **6**, 63 (1998).
[107] D. C. Nikas, J. W. Foley, and P. M. Black, *Lasers Surg. Med.* **29**, 11 (2001).
[108] H. J. van Staveren, O. C. Speelman, M. J. Witjes, L. Cincotta, and W. M. Star, *Photochem. Photobiol.* **73**, 32 (2001).
[109] M. J. Witjes, A. J. Mank, O. C. Speelman, R. Posthumus, C. A. Nooren, J. M. Nauta, J. L. Roodenburg, and W. M. Star, *Photochem. Photobiol.* **65**, 685. (1997).
[110] R. Cubeddu, A. Pifferi, P. Taroni, A. Torricelli, G. Valentini, D. Comelli, C. D'Andrea, V. Angelini, and G. Canti, *Photochem. Photobiol.* **72**, 690 (2000).
[111] M. Zellweger, P. Grosjean, P. Monnier, H. van den Bergh, and G. Wagnieres, *Photochem. Photobiol.* **69**, 605 (1999).
[112] N. Kajiyama and E. Nakano, *Protein Eng.* **4**, 691 (1991).
[113] V. R. Viviani and Y. Ohmiya, *Photochem. Photobiol.* **72**, 267 (2002).

[114] A. B. Cubitt, R. Heim, S. R. Adams, A. E. Boyd, L. A. Gross, and R. Y. Tsien, *Trends Biochem. Sci.* **20**, 448 (1995).

[115] J. Wiehler, J. von Hummel, and B. Steipe, *FEBS Lett.* **487**, 384 (2001).

[116] T. T. Yang, P. Sinai, G. Green, P. A. Kitts, Y. T. Chen, L. Lybarger, R. Chervenak, G. H. Patterson, D. W. Piston, and S. R. Kain, *J. Biol. Chem.* **273**, 8212 (1998).

[117] R. Y. Tsien, *Annu. Rev. Biochem.* **67**, 509 (1998).

[118] Y. Ito, M. Suzuki, and Y. Husimi, *Biochem. Biophys. Res. Commun.* **264**, 556 (1999).

[119] T. Chishima, Y. Miyagi, X. Wang, Y. Tan, H. Shimada, A. Moossa, and R. M. Hoffman, *Anticancer Res.* **17**, 2377 (1997).

[120] I. Pathak, N. L. Davis, Y. N. Hsiang, N. F. Quenville, and B. Palcic, *Am. J. Surg.* **170**, 423 (1995).

[121] K. Onizawa, H. Saginoya, Y. Furuya, and H. Yoshida, *Cancer Lett.* **108**, 61 (1996).

[122] B. R. Silberstein, A. Mayevsky, and B. Chance, *Neurol. Res.* **2**, 19 (1980).

[123] R. J. Nordstrom, L. Burke, J. M. Niloff, and J. F. Myrtle, *Lasers Surg. Med.* **29**, 118 (2001).

[124] M. L. Harries, S. Lam, C. MacAulay, J. Qu, and B. Palcic, *J. Laryngol. Otol.* **109**, 108 (1995).

[125] M. Zargi I. Fajdiga and L. Smid, *Eur. Arch. Otorhinolaryngol.* **257**, 17 (2000).

[126] L. Rigacci, R. Alterini, P. A. Bernabei, P. R. Ferrini, G. Agati, F. Fusi, and M. Monici, *Photochem. Photobiol.* **71**, (2000).

[127] B. Kulapaditharom, V. Boonkitticharoen, and S. Kunachak, *Ann. Otol. Rhinol. Laryngol.* **108**, 700 (1999).

[22] *In Vivo* and *ex Vivo* Tissue Applications of Two-Photon Microscopy

By TIMOTHY M. RAGAN, HAYDEN HUANG, and PETER T. C. SO

Introduction

Two-photon microscopy (TPM) is based on a nonlinear fluorescence excitation process. The simultaneous absorption of two infrared photons promotes the transition of a fluorophore to the excited state. Denk and co-workers first successfully utilized this process for high-resolution, three-dimensional (3D) microscopic imaging.[1] Due to the nonlinear excitation process, there is only sufficient photon density very near the focal region to produce appreciable excitation, which results in an inherent 3D sectioning effect. More detailed explanations of two-photon excitation microscopy can be found in other articles in this volume and in a number of reviews.[2–4]

[1] W. Denk, J. H. Strickler, and W. W. Webb, *Science* **248**, 73 (1990).

[2] W. J. Denk, D. W. Piston, and W. W. Webb, *in* "Handbook of Biological Confocal Microscopy" (J. B. Pawley, ed.), p. 445. Plenum Press, New York, 1995.

[3] C. Xu and W. W. Webb, *in* "Nonlinear and Two-Photon-Induced Fluorescence" (J. R. Lakowicz, ed.), Vol. 5, p. 471. Plenum Press, New York, 1997.

[4] P. T. So, C. Y. Dong, B. R. Masters, and K. M. Berland, *Annu. Rev. Biomed. Eng.* **2**, 399 (2000).

The unique advantages of TPM for deep tissue imaging have been recognized nearly since its inception. TPM tissue imaging applications are proliferating rapidly. In this article, we will discuss the advantages of two-photon imaging in optically thick specimens and the key factors in optimizing the performance of two-photon microscopy in the presence of multiple scattering tissues, and review key areas, such as neurobiology, in which TPM has already made a major impact. Finally, selected emerging areas in two-photon tissue imaging will be considered.

Advantages of Two-Photon Microscopy in Tissue Imaging

TPM is a powerful technique for tissue imaging because of its inherent 3D resolution and long penetration depth. Equally important, this technique provides biochemical information about tissues and causes minimal photodamage. All of these advantages of two-photon microscopy are derived from the underlying photophysics of the two-photon excitation process. The basic photophysics of TPM has been discussed elsewhere in a previous volume.[4a] The optimization of experimental parameters to realize these advantages will be discussed in the following section; this section will provide a perspective on the strengths and weaknesses of TPM in comparison with other biomedical imaging modalities.

Some of the most successful modalities in biomedical imaging include X-ray imaging, magnetic resonance imaging, positron emission imaging, and ultrasound imaging. Most of these methods allow the researcher or clinician to image thick tissues with resolution on the order of millimeters. Using novel contrast agents, limited spectroscopic measurements can be performed to assay tissue biochemical and metabolic states. Although many of these techniques are powerful, some of them have shortcomings, such as limited spatial or temporal resolutions, high costs, the use of ionizing radiations, or issues with limited instrument portability. It is important to develop alternative complementary imaging techniques that can address some of the limitations of these established techniques.

Optical imaging methods are promising alternatives because of their high spatial and temporal resolution, low cost, and use of nonionizing radiation. The major limitation of most optical techniques is a relatively limited penetration depth that ranges from submillimeters to centimeters. TPM is one of these emerging optical tissue imaging techniques, and it is constructive to compare two-photon imaging with other *in vivo* optical tissue imaging techniques. Diffusive optical tomography is an optical technique that features the greatest penetration depth (on the order of 5–10 cm), but with a resolution on the order of 1 mm. It has a high temporal resolution that offers novel opportunities for brain and cognitive studies, and modest spectroscopic diagnostic capabilities, such as the quantification of tissue blood oxygenation and flow velocity. Other major optical imaging techniques include

---

[4a] L. Bagatolli, S. Sanchez, T. Hazlett, and E. Gratton, *Methods Enzymol.* **360**, 481 (2003).

TABLE I
OPTICAL TISSUE IMAGING TECHNIQUES[a]

Parameter	Fluorescence confocal microscopy	Reflected light confocal microscopy	Two-photon fluorescence microscopy	Optical coherence microscopy
Lateral resolution (μm)	0.2	0.3	0.3	0.3
Axial resolution (μm)	0.6	0.8	0.8	5–20
Penetration depth (μm)	10–50	500–1000	500–1000	1000–5000
Achievable frame rate (fps)	30	30	30	30
Potential for spectroscopic biochemical analysis	Modest	Low	High	Low
Photodamage potential	High	Low	Medium	Low
System cost	Medium	Medium	High	Medium

[a] With micron-level resolution.

fluorescence confocal microscopy,[5,6] reflected light confocal microscopy,[7] optical coherence microscopy,[8,9] and two-photon fluorescence microscopy. All these techniques can achieve submicron to a few microns resolution. The relative strengths and weaknesses of these techniques are summarized in Table I. Among all these approaches, TPM is the method of choice for tissue studies that require submicron resolution along with spectroscopic biochemical information, but do not require imaging to a depth beyond a few hundred microns.

Key Factors in Designing Two-Photon Tissue Imaging Experiments

Good optics, high peak power lasers, and high sensitivity detectors are the crucial components in a well-designed two-photon tissue imaging instrument. When these three factors are optimized, good images can be expected for sections near the surface of the sample. However, as we have already mentioned, one of the key factors in a tissue imaging experiment is the effective imaging depth of the system. In almost all practical situations, as one focuses deeper into the sample, the image quality will become worse. To keep the degradation of the image quality to a minimum, it is important to keep four considerations in mind. First, it is essential

[5] P. Corcuff, C. Bertrand, and J. L. Leveque, *Arch. Dermatol. Res.* **285,** 475 (1993).
[6] B. R. Masters, "Selected Papers on Confocal Microscopy." SPIE, Bellingham, 1996.
[7] M. Rajadhyaksha, M. Grossman, D. Esterowitz, R. H. Webb, and R. R. Anderson, *J. Invest. Dermatol.* **104,** 946 (1995).
[8] J. G. Fujimoto, B. Bouma, G. J. Tearney, S. A. Boppart, C. Pitris, J. F. Southern, and M. E. Brezinski, *Ann. N.Y. Acad. Sci.* **838,** 95 (1998).
[9] G. J. Tearney, M. E. Brezinski, J. F. Southern, B. E. Bouma, M. R. Hee, and J. G. Fujimoto, *Opt. Lett.* **20,** 2258 (1995).

BIOPHOTONICS

[22]

to ensure that there is sufficient laser power at the focal volume for efficient two-photon excitation. Second, the detectors in the system must be able to collect the fluorescence signal generated in the interior of the specimen. Third, scattering and optical heterogeneity in the tissue may adversely affect the image resolution and contrast and blur tissue features. Finally, although two-photon excitation does minimize photodamage in thick tissues, potential photodamage is still a danger, particularly if live subjects are involved.

Laser Power Management and Excitation Power in Optically Thick Specimens

The penetration depth of two-photon microscopy is limited by the rapid reduction of laser power delivered to the focal point as the distance from the tissue surface increases. At present, we will assume the image point spread function is invariant with depth; the validity of this assumption will be discussed further in a subsequent section. Three factors govern the effective excitation of fluorophores at the focal point. The first factor is the average laser power at the focal point, which is attenuated exponentially as a function of depth:

$$I(z) = I_0 e^{-[(1-g)\mu_s + \mu_a]z}$$

where $I(z)$ is the intensity at depth z, I_0 is the intensity at tissue surface, μ_s is the scattering coefficient, μ_a is the absorption coefficient, and g is the average cosine of the scattering angle. The scattering coefficient, absorption coefficient, and average cosine for typical tissues are 20–200 mm^{-1}, 0.1–1 mm^{-1}, and 0.7–0.9. This corresponds to a composite photon mean free path of 50–500 μm. Given the quadratic dependence of the fluorescence on the excitation power, the fluorescence signal is also an exponential function of the imaging depth but has a shorter decay length of 25–250 μm. Excitation efficiency can be made independent of depth by increasing average excitation power exponentially with the same decay length.

The second factor affecting the two-photon excitation efficiency is the repetition rate of the laser:

$$F \propto \frac{I^2}{f\tau}$$

where F is the fluorescence intensity (W), f is the repetition frequency, I is the average excitation intensity (W m^{-2}), and τ is the pulse width of the laser pulse train. One approach to enhancing the imaging depth may involve reducing the laser repetition rate while maintaining the average power. Laser pulse repetition can be reduced by a variety of methods without significantly sacrificing average power or pulse width, such as using a cavity dumper, a regenerative amplifier, or an optical parametric amplifier.

The third factor affecting the two-photon excitation efficiency is the laser pulse width. The typical pulse width of titanium–sapphire lasers used in these experiments is 150 fs. Given the time-bandwidth product of these lasers, the potential for further pulse compression is fairly limited. On the other hand, one should be aware that laser pulse dispersion in microscope objectives is significant and pulse width can easily broaden to beyond 500 fs. Some improvement in power can be achieved by using dispersion compensation optics to maintain a short pulse width after the objective.

Given typical fluorophore cross sections and maximum laser power, we expect acceptable imaging depths to remain within five scattering length corresponding to an imaging depth of 1 mm in the best case.

Fluorescence Signal Detection from Multiple Scattering Specimens

It is critical to optimize the detection of fluorescence photons emitted from thick tissues. As with excitation photons, fluorescence photons emitted from the focal volume will be scattered and absorbed by the surrounding tissue. Both absorption and scattering are minimized with increasing wavelength of light. Therefore, in general, fluorophores with longer emission wavelength should result in images with an improved signal-to-noise (S/N) ratio. Furthermore, because the mean free path of visible photons is less than 100 μm in typical tissues, the emitted photon will suffer one or more scattering events for imaging depths beyond the mean free path. However, many of these scattered photons can still be detected as photons are primarily forward scattered in typical tissues. As long as the trajectories of these scattered photons still remain within the collection solid angle of the objective lens, they can be directed into the detection light path of the microscope. Because some of these scattered photons follow more divergent optical paths, maximizing the collection solid angles and physical sizes of the detection optical components can contribute to detecting high signal levels.[10]

Limitations of Image Resolution in Highly Scattering Medium
  with Heterogeneous Index of Refractions

As mentioned in the previous section, biological tissues are generally more optically transparent at the longer wavelengths used in TPM, which makes it possible to image significantly deeper into the sample than is the case with confocal microscopy. However, one still needs to be concerned with both contrast and resolution losses in the optical system as the imaging depth is increased. The two major causes of these effects in TPM are the scattering of the excitation and emission light, and the index of refraction mismatch between the sample and the immersion media of the objective.

Scattering in a tissue medium leads to a degradation of the image due to both the scattering of the excitation light and the emission light. Scattering of the excitation light reduces the intensity levels at the point spread function (PSF); and

---

[10] E. Beaurepaire, M. Oheim, and J. Mertz, *Opt. Commun.* **188**, 25 (2001).

scattering of emission photons can increase the number of photons that fail to reach the detector. TPM is less susceptible to the latter effect as discussed previously. However, scattering of the excitation light is still an appreciable problem that ultimately limits the achievable image depth.

An excellent study investigating the source of the resolution and contrast losses in TPM has been carried out by Dunn et al.,[11] who characterized how the TPM signal intensity decayed exponentially with depth. With Monte Carlo simulations they were able to separate and quantify the reduction of intensity into effects due to either the scattering of the excitation light or scattering of the emission light. The simulations established that the loss of signal intensity was due more to the scattering of the excitation photons rather than the emission photons. Additionally, they were able to conclude that the signal loss is the limiting factor in TPM image quality rather than resolution loss by showing that resolution did not degrade as the image depth was increased, which is in agreement with Centonze and White.[12]

Dong and co-workers (unpublished observations) measured the PSF in TPM directly by imaging 100 nm spheres in agarose gels with different concentration of scatterers. The concentration of scatterers had no effect on the lateral and axial full-width at half-maximum (FWHM) of the PSF for both oil and water immersion objectives. Furthermore, the water immersion objective had no significant degradation on the lateral and axial FWHM of the PSF as the objective was focused more deeply into the sample (Fig. 1). On the other hand, the oil immersion objective suffered significant degradation due the spherical aberration introduced by index of refraction mismatch of the sample, and the lateral and axial FWHM was broadened. One may conclude that for typical tissue thickness on the order of a few hundred microns, resolution loss due to scattering is insignificant but index refraction mismatch can be a major factor.

The use of water-immersion objectives is preferred in biological tissues with a relatively homogeneous index of refraction distribution. However, Dong et al. also investigated the performance of oil versus water immersion objectives for imaging more optically heterogeneous biological tissues such as excised human skin tissue. Human skin is an optically heterogeneous tissue with an index of refraction of 1.47 for the stratum corneum, which decreases to 1.34 for the basal layer, and rises again to 1.41 for the dermal layer. It is found that oil and water immersion objectives perform equally well in imaging dermal structures. The equivalent performance of these objectives results from the fact that neither of these objectives can match all of the refraction indexes of the specimen simultaneously. In this case, the effects of spherical aberration cannot be completely eliminated. In the end, for optically heterogeneous samples, it is optimal to choose an immersion media that best matches the average value of the sample.

[11] A. K. Dunn, V. P. Wallace, M. Coleno, M. W. Berns, and B. J. Tromberg, Appl. Opt 39, 1194 (2000).
[12] V. E. Centonze and J. G. White, Biophys. J. 75, 2015 (1998).

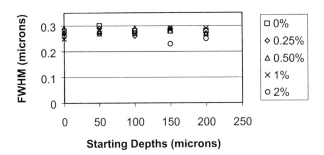

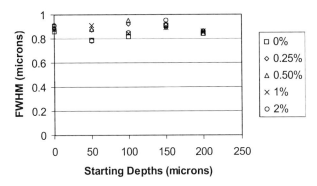

FIG. 1. The full-width at half-maximum of the radial and axial point spread functions for 0.1-μm latex spheres imaged at successive depths. Latex spheres were embedded in agarose gel containing scatterers, Liposyn III, at five different concentrations (0, 0.25, 0.5, 1, and 2%), corresponding to reduced scattering coefficients of 0, 0.38, 0.75, 1.5, and 3 mm^{-1} measured at 680 nm.

Photodamage Mechanisms in Tissues

Two-photon excitation reduces specimen photodamage by localizing the excitation region to within a 1 fl volume. Photodamage is further decreased by the use of near infrared (IR) excitation rather than ultraviolet (UV) and visible radiation. Decreasing the photochemical interaction volume results in a dramatic increase in the viability of biological specimens. The noninvasive nature of two-photon imaging can best be appreciated in a number of embryology studies. Previous work on long-term monitoring of *Caenorhabditis elegans* and hamster embryos using confocal microscopy failed because of photodamage-induced developmental arrest. However, TPM studies indicate that the embryos of these organisms can be imaged repeatedly over the course of hours without observable damage.[13–15] For instance, a hamster embryo was imaged with TPM and then reimplanted into the uterus, and it eventually developed into a normal, healthy, adult animal.

At the focal volume where photochemical interactions occur, TPM can still cause considerable photodamage. Three major mechanisms of two-photon photodamage have been recognized.

1. Oxidative photodamage can be caused by two or higher order photon excitation of endogenous and exogenous fluorophores. The photodamage pathway is similar to that of ultraviolet irradiation. These fluorophores act as photosensitizers in photooxidative processes,[16,17] and the photoactivation of these fluorophores results in the formation of reactive oxygen species that trigger the subsequent biochemical damage cascade in cells. Current studies found that the degree of photodamage follows a quadratic dependence on excitation power indicating that the two-photon process is the primary damage mechanism.[18–22] Experiments have also been performed to measure the effect of laser pulse width on cell viability. Results indicate that the degree of photodamage is proportional to the two-photon-excited fluorescence generated, independent of pulse width. Hence, using a shorter pulse width for more efficient two-photon excitation also produces greater photodamage. An important consequence is that both femtosecond and picosecond light sources are equally effective for two-photon imaging in the absence of infrared one-photon absorbers.[19,21] Additionally we note that flavin-containing oxidases have been identified as one of the primary endogenous targets for photodamage.[22] Photobleaching of fluorophores can result from higher order (three- or four-photon) processes suggesting that higher order photodamage mechanisms may also be present during the imaging of biological specimens.[23]

2. One- and two-photon absorption of the high power infrared radiation can also produce thermal damage. The thermal effect resulting from two-photon

[13] W. A. Mohler, J. S. Simske, E. M. Williams-Masson, J. D. Hardin, and J. G. White, *Curr. Biol.* **8**, 1087 (1998).

[14] W. A. Mohler and J. G. White, *Biotechniques* **24**, 1006 (1998).

[15] J. M. Squirrell, D. L. Wokosin, J. G. White, and B. D. Bavister, *Nat. Biotechnol.* **17**, 763 (1999).

[16] S. M. Keyse and R. M. Tyrrell, *Carcinogenesis* **11**, 787 (1990).

[17] R. M. Tyrrell and S. M. Keyse, *J. Photochem. Photobiol. B.* **4**, 349 (1990).

[18] K. Konig, P. T. C. So, W. W. Mantulin, B. J. Tromberg, and E. Gratton, *J. Microsc.* **183**, 197 (1996).

[19] K. Konig, T. W. Becker, P. Fischer, I. Riemann, and K.-J. Halbhuber, *Opt. Lett.* **24**, 113 (1999).

[20] Y. Sako, A. Sekihata, Y. Yanagisawa, M. Yamamoto, Y. Shimada, K. Ozaki, and A. Kusumi, *J. Microsc.* **185**, 9 (1997).

[21] H. J. Koester, D. Baur, R. Uhl, and S. W. Hell, *Biophys. J.* **77**, 2226 (1999).

[22] P. E. Hockberger, T. A. Skimina, V. E. Centonze, C. Lavin, S. Chu, S. Dadras, J. K. Reddy, and J. G. White, *Proc. Natl. Acad. Sci. U.S.A.* **96**, 6255 (1999).

[23] G. H. Patterson and D. W. Piston, *Biophys. J.* **78**, 2159 (2000).

absorption has been estimated to be on the order of 1 mK for typical excitation power and has been shown to be insignificant.[2,24] However, in the presence of a strong infrared absorber such as melanin,[25,26] there can be appreciable heating due to one-photon absorption. Thermal damage has been observed in the basal layer of human skin in the presence of high average excitation power.[27]

3. Photodamage may also be caused by mechanisms resulting from the high peak power of the femtosecond laser pulses. There are indications that dielectric breakdown occasionally occurs.[18] However, further studies are required to confirm and better understand these effects. An understanding of photodamage mechanisms will be critical for the safe operation of a two-photon optical biopsy instrument.

Major Application Areas of Two-Photon Tissue Imaging

The application of TPM in biology and medicine is proliferating rapidly. Some of the most promising areas utilize the superior penetration power of TPM coupled with its ability to assay tissue biochemical and metabolic states. Today, we find that TPM has been applied to *in vivo* and *ex vivo* imaging in neurophysiology, dermal physiology, pancreatic physiology, and embryology. The primary motivation behind these studies is that investigating the behavior of cells in intact tissues offers several advantages. First, because it is not necessary to extract the cells from the tissue by mechanical or enzymatic means, there is much less opportunity to introduce artifacts into the analysis. For example, cellular morphology is drastically different between cells inside intact tissues and the same cells grown in a 2D tissue culture. Second, the removal of cells from a tissue also necessarily eliminates many of the mechanical and biochemical signal inputs that are known to dramatically affect cellular behavior. Third, the interaction of different cell types inside of an intact tissue can also be an important factor in tissue function that is often impossible to study once the cells have been removed from the tissue.

Neurophysiology

The brain and the peripheral nervous system are highly complex. An understanding of their functions requires the ability to monitor these 3D networks of interconnecting neurons. TPM is well suited to neurophysiology studies because it can capture the 3D neuronal organization *in vivo* and probe neuron communications by assaying action potentials, calcium waves, and neural transmitters using

[24] A. Schonle and H. S. W., *Opt. Lett.* **23,** 325 (1998).
[25] S. L. Jacques, D. J. McAuliffe, I. H. Blank, and J. A. Parrish, *J. Invest. Dermatol.* **88,** 88 (1987).
[26] V. K. Pustovalov, *Kvantovaya Elektron.* **22,** 1091 (1995).
[27] C. Buehler, K. H. Kim, C. Y. Dong, B. R. Masters, and P. T. C. So, *IEEE Eng. Med. Biol.* **18,** 23 (1999).

fluorescence spectroscopy. Studies have been performed to optimize TPM imaging parameters to maximize penetration depths into neuronal tissues.[28]

Most of the TPM studies in neurophysiology have focused either on the remodeling of neuronal dendrictic spines and the subsequent effects these changes have on memory and learning, or on the dynamics of calcium signal propagation. Denk and co-workers pioneered the application of TPM in neurobiology. In 1995, Yuste and Denk first imaged dendritic spines to detect and characterize the dynamics of calcium activity.[29] Svoboda et al. expanded these studies in 1996 using two-photon photobleaching and photoactivation to study the interaction between dendritic shafts and spines, and they were able to measure the electrical resistance at the spine neck.[30] This methodology was further used to relate the dynamics of calcium to sodium.[31]

Many groups have focused on studying the nervous system at the tissue level and have attempted to elucidate the interactions among cells rather than on the remodeling that occurs within a single cell. Svoboda and co-workers studied the in vivo structure of hippocampal tissue[32] (Fig. 2). Using green fluorescent protein (GFP) labeling and knockout mouse technology, the effects of FMR1 deletion on the growth and development of dendritic spines were assayed. They found that in the knockout mice the spine length was substantially longer for the first 2–4 weeks of development. To reach this conclusion, the authors had to obtain statistics on the response of over 16,000 spines in vivo, a feat that is made possible by the advent of TPM, and is not feasible with traditional techniques such as electron microscopy.[33] Another study by Svoboda and co-workers imaged the invasion process of calcium-mediated action potentials in excised rat neocortical slices that utilized the excellent spatial and temporal resolution of TPM.[34] Engert and Bonhoeffer imaged dendritic spine growth and change using superperfusion techniques to create a long-lasting functional enhancement of the CA1 region.[35] Yuste and co-workers acquired time-lapse images in the mouse central nervous system and demonstrated a wide variety of dynamic remodeling of the dendritic protrusions.[36] More recent studies have shown that sensory deprivation decreases motility of dendritic spines.[37]

[28] M. Oheim, E. Beaurepaire, E. Chaigneau, J. Mertz, and S. Charpak, *J. Neurosci. Methods* **111**, 29 (2001).
[29] R. Yuste and W. Denk, *Nature* **375**, 682 (1995).
[30] K. Svoboda, D. W. Tank, and W. Denk, *Science* **272**, 716 (1996).
[31] K. Svoboda, W. Denk, D. Kleinfeld, and D. W. Tank, *Nature* **385**, 161 (1997).
[32] M. Maletic-Savatic, R. Malinow, and K. Svoboda, *Science* **283**, 1923 (1999).
[33] E. A. Nimchinsky, A. M. Oberlander, and K. Svoboda, *J. Neurosci.* **21**, 5139 (2001).
[34] C. L. Cox, W. Denk, D. W. Tank, and K. Svoboda, *Proc. Natl. Acad. Sci. U.S.A.* **97**, 9724 (2000).
[35] F. Engert and T. Bonhoeffer, *Nature* **399**, 66 (1999).
[36] A. Dunaevsky, A. Tashiro, A. Majewska, C. Mason, and R. Yuste, *Proc. Natl. Acad. Sci. U.S.A.* **96**, 13438 (1999).
[37] B. L. Sabatini and K. Svoboda, *Nature* **408**, 589 (2000).

A   B   C

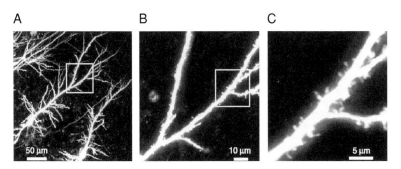

FIG. 2. TPLSM imaging of CA1 pyramidal cell dendrites labeled with EGFP (in a slice 9 days in culture). Images shown are maximum value projections of several sections acquired 0.5 μm apart. (A) CA1 region of a hippocampal slice showing several infected neurons, including dendritic arbors and axons. (B) Apical dendrite and secondary branches from the boxed region in (A). (C) Higher magnification image from the boxed region in (B) showing dendritic protrusions. [Reprinted with permission from M. Maletic-Savatic, R. Malinow, and K. Svoboda, *Science* **283**, 1923 (1999). Copyright © 1999 American Association for the Advancement of Science.]

The retina is an extension of the neuronal system. Although the retina should be an easy organ for optical studies due to the transparency of the anterior of the eye, optical images of the retinal cells have long been hindered by the high optical sensitivity of these cells. Because retinal cells are less sensitive to infrared radiation, TPM typically does not saturate or injure the retinal cells as does imaging using shorter visible and UV wavelengths. Denk and Detwiler were able to successfully investigate calcium signaling pathways in tiger salamander larva retinas using TPM, where standard wide-field fluorescence imaging had failed.[38]

Another important area of neural biology that can be studied using TPM is hemodynamics. Kleinfeld and co-workers used a two-photon microscope to probe red blood cell motion in rat cortical capillaries, and were able to scan 600 μm into the tissue to observe the speed and flux of the red blood cells. In some cases, they were able to detect flow changes in response to externally applied stimuli, such as vibrassa, direct touch, or moderate shock.[39]

Because some cognitive studies often require the use of conscious animals, Helmchen *et al.* constructed a miniature two-photon microscope to image the neuronal activities of freely moving rats via cranial windows. They demonstrated that this miniature microscope was quite capable of tracking calcium transients and capillary blood flow in active rats.[40]

Another major application of TPM is the study of neural pathology. Hyman's group used TPM to examine neural plaques associated with Alzheimer's disease.

---

[38] W. Denk and P. B. Detwiler, *Proc. Natl. Acad. Sci. U.S.A.* **96**, 7035 (1999).
[39] D. Kleinfeld, P. P. Mitra, F. Helmchen, and W. Denk, *Proc. Natl. Acad. Sci. U.S.A.* **95**, 15741 (1998).
[40] F. Helmchen, M. S. Fee, D. W. Tank, and W. Denk, *Neuron* **31**, 903 (2001).

The plaques are composed of amyloid-β peptides and are small enough that conventional imaging techniques cannot resolve them. Hyman's group demonstrated that brain tissue labeled with thioflavin S (a fluorescent marker for plaques) could be imaged in living mice through a mechanically thinned, surgically exposed skull. This type of imaging process is minimally invasive and the animals can be maintained over the long term by surgically closing the wound after TPM imaging. Through long-term studies, it has been demonstrated in a mice model that the size of amyloid plaques remain stable once formed, and that the progression of the disease is associated with the formation of additional plaques.[41,42]

In the future, it is likely that a better understanding of the development and behavior of the nervous system will be gained from studies that use TPM. The suitability of TPM for neuronal studies has been well demonstrated. TPM is clearly the technique of choice for *in vivo* study of cellular communications in the nervous system.

Dermal Physiology

Dermatology has been one of the earliest applications of TPM in tissues. The skin is a natural choice for high-resolution imaging. Not only is it an easily accessible organ, but there are also many potential clinical applications, such as an aid to noninvasive cancer diagnosis, that drive the development of TPM in skin tissues.

The skin is composed of two distinct layers: the epidermis and the dermis. The epidermis is on the order of 50 to 100 μm thick and is composed of several sublayers. The lowermost portion of the epidermis is the basal layer, which is the source of the keratinocytes, which migrate up through the other layers of the epidermis. Directly above the basal layer is the stratum spinosum, followed by the stratum granulosum, and finally the stratum corneum. Figure 3 shows both confocal reflected images and two-photon images of human skin *in vivo* at several depths.[43] As the cells migrate up toward the stratum corneum, they spread and flatten out. By the time they reach the stratum corneum they are no longer alive and eventually slough off the skin. Below the basal layer is the dermal layer, which consists mainly of sparse fibroblasts and structural macromolecules such as collagen, elastin, and proteoglycans, and can be up to a few hundred microns deep.

Laser scanning confocal microscopy has been used to image skin structures using native autofluorescence.[44] However, the imaging depth was limited to

[41] R. H. Christie, B. J. Bacskai, W. R. Zipfel, R. M. Williams, S. T. Kajdasz, W. W. Webb, and B. T. Hyman, *J. Neurosci.* **21**, 858 (2001).
[42] B. J. Bacskai, S. T. Kajdasz, R. H. Christie, C. Carter, D. Games, P. Seubert, D. Schenk, and B. T. Hyman, *Nat. Med.* **7**, 369 (2001).
[43] B. R. Masters and P. T. C. So, *Opt. Express* **8**, 2 (2001).
[44] B. R. Masters, *Bioimages* **4**, 13 (1996).

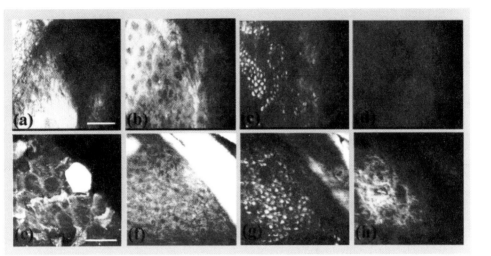

FIG. 3. *In vivo* images of human skin acquired using both confocal reflected microscopy and TPM. The upper images (a–d) are the confocal images and the lower (e–h) are the two-photon images acquired at 780 nm. The images were each taken from regions in the forearm. Images were acquired at 10 μm below the surface in the stratum corneum (a and e), in the stratum spinosum (b and f), in the basal layer (c and g), and within the dermis (d and h). Bar: 50 μm. [Reproduced from B. R. Masters and P. T. C. So, *Opt. Express* **8**, 2 (2001).]

20–50 μm. Masters *et al.* employed TPM to image the autofluorescence of *in vivo* human skin down to a depth of 200 μm.[45] NAD(P)H is the major source of autofluorescence in cells and is located in the cytoplasm of metabolically active cells. As demonstrated in Fig. 3, the NADP(H) fluorescence in corneum, spinosum, and basal layers was clearly seen. Cells were on the order of 15–20 μm in diameter in the upper layers of the epidermis and decreased as the depth was increased. NAD(P)H fluorescence decreases with increasing depth as is expected due to the scattering of the laser and fluorescent light. It was also possible to visualize the elastin in the dermal layer as this macromolecule is excitable by TPM. Additionally second harmonic generation can be used to image collagen structure.

More recent TPM studies of skin tissue have helped characterize the transport properties of the skin.[46,47] A better understanding of this phenomenon will facilitate the development of the transdermal delivery of drugs as an attractive alternative to oral or subcutaneous administration. In general, the skin is a very good barrier to most molecules, and allows only small hydrophobic particles to pass

[45] B. R. Masters, P. T. So, and E. Gratton, *Biophys. J.* **72**, 2405 (1997).
[46] B. Yu, C. Y. Dong, P. T. So, D. Blankschtein, and R. Langer, *J. Invest. Dermatol.* **117**, 16 (2001).
[47] B. S. Grewal, A. Naik, W. J. Irwin, G. Gooris, C. J. de Grauw, H. G. Gerritsen, and J. A. Bouwstra, *Pharm. Res.* **17**, 788 (2000).

through. Researchers are interested in developing ways to permeate the skin for high molecular weight and hydrophobic molecules. Yu *et al.* investigated the transport properties of the hydrophobic probe, rhodamine B hexyl ester, and the hydrophilic probe, sulforhodamine B, in excised human skin tissue. They tested the transport of both the probes in the presence and absence of oleic acid, a well-known chemical enhancer of transport in the skin. By relating the observed intensity gradients of the probes they were able to characterize the vehicle to skin partition coefficient and the skin diffusion coefficient and show that the enhanced permeability for both of the probes arose from different mechanisms.[46]

There are several future directions that TPM microscopy of skin can take. First it will be very helpful to conduct a full spectroscopic analysis of the skin by recording the fluorescence as both a function of emission wavelength and excitation wavelength. This will aid researchers in correlating skin biochemical changes with physiological perturbation such as photoaging. Another direction is to bring TPM into the clinical setting by constructing a hand-held optical biopsy probe. This would be potentially useful for the noninvasive diagnosis of cancer as well as for potentially aiding a surgeon in identifying surgical margins. This will be discussed further in a subsequent section.

Pancreatic Physiology

TPM studies of the pancreas illustrate not only the strengths of TPM, but also the importance of studying cellular function in intact tissues. The first study of the pancreas using TPM was done by Bennet *et al.* who used TPM to investigate whether β cells in pancreas islets displayed heterogeneity in their glucose metabolism pathway.[48] Previous research based on NAD(P)H autofluorescence of dispersed β cells had indicated variability in the levels of metabolic activity. This led some to suggest that this metabolic heterogeneity played a fundamental role in islet insulin secretion; however, other studies indicated that metabolic activity of β cells was relatively homogeneous.

To help resolve this issue, Bennet *et al.* monitored glucose metabolism in β cells of intact pancreatic islet directly by imaging the NAD(P)H autofluorescence using TPM.[48] To do this it was necessary to image through thick tissue sections and still be able to discriminate NAD(P)H autofluorescence from background noise. The high penetration depth of TPM coupled with its thin sectioning effect made it possible to image the relatively faint signal from the NAD(P)H. Figure 4 is an image of an intact pancreatic islet. Furthermore, it was possible to image the tissue for an extended period due to the lack of photodamage and photobleaching in out-of-focus planes. Bennet *et al.* were able to show that NAD(P)H levels of β cells in intact islets were remarkably consistent, which is in contrast to heterogeneity

[48] B. D. Bennett, T. L. Jetton, G. Ying, M. A. Magnuson, and D. W. Piston, *J. Biol. Chem.* **271**, 3647 (1996).

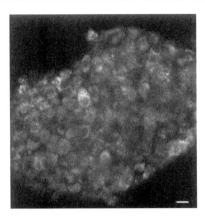

FIG. 4. A two-photon image of an isolated rat islet 40 μm below the surface. The NADP(H) auto-fluorescence is clearly visible, allowing individual β cells and their nuclei to be visualized. [Reproduced from B. D. Bennett, T. L. Jetton, G. Ying, M. A. Magnuson, and D. W. Piston, *J. Biol. Chem.* **271**, 3647 (1996).]

observed in dispersed β cells.[48] This observation has important implications for invalidating models of insulin secretion of islets based on β cell heterogeneity. This underscores how removal of cells from their native tissue environment can significantly change their function or behavior, and how it may be necessary to study the behavior of cells in the native environment.

The same group continued their study of pancreas function by the inactivation of a glucokinase gene allele using a recombinant adenovirus to express Cre recombinase.[49] By combining the results of southern blot analyses and TPM they were able to conclude that intercellular communication played little or no role in intact islets.

Nemoto *et al.* studied exocytosis in pancreatic acini.[50] By employing TPM, they were for the first time able to visualize time sequence images of zymogen-granule exocytosis in exocrine glands. Previous attempts with confocal microscopy failed due to photobleaching and lack of penetration depth. TPM studies were able to shed light on sequential exocytosis, and provide information about a previously unidentified method to replenish secretory granules in pancreatic acini.

Physiology of Other Organs

Other organs and tissues have been investigated using TPM. Piston *et al.* investigated NAD(P)H metabolism in the cornea.[51] The eye is an excellent organ to

---

[49] D. W. Piston, S. M. Knobel, C. Postic, K. D. Shelton, and M. A. Magnuson, *J. Biol. Chem.* **274**, 1000 (1999).

[50] T. Nemoto, R. Kimura, K. Ito, A. Tachikawa, Y. Miyashita, M. Iino, and H. Kasai, *Nat. Cell Biol.* **3**, 253 (2001).

[51] D. W. Piston, B. R. Masters, and W. W. Webb, *J. Microsc.* **178**, 20 (1995).

study with optical techniques as it is optically transparent and it is thus possible to image very deeply into the tissue without significant degradation of the image. Piston *et al.* were able to construct a three-dimensional map of the NAD(P)H levels throughout the entire 400 μm depth of the rabbit cornea, monitor the autofluorescence intensity levels of the basal cells in the epithelium layer as a function of time, and demonstrate that TPM is a suitable tool for investigating how the cornea maintains the ocular surface.

Napadow *et al.* investigated the heterogeneous microscopic structure of mammalian tongue tissue using TPM in conjunction with magnetic resonance imaging (MRI), and thus obtained spatial information on two different length scales.[52] MRI imaging was able to distinguish two populations of tongue fibers, which was confirmed with TPM images of fiber orientation. This provided support for a new model of the myoarchitecture of the tongue that emphasizes the connection between individual fibers and fiber bundle sheets in mammalian tongue tissue.

Embryology

Two-photon imaging has been applied to the study of embryology. Several factors complicate the study of embryos *in vivo* such as the sensitivity of embryos to photodamage. This problem has been largely resolved by using TPM. Furthermore, the relatively small size of the embryos often makes it possible to image the full three-dimensional architecture of the developing embryo and not just the surface layers.

There have been some successful embryological studies using confocal microscopy. Zebrafish (*Danio rerio*) have been popular due to their optical clarity during their development. Isogai *et al.* imaged the development of the vasculature of zebrafish in 3D and were able to identify basic wiring patterns in the progressive interconnections of the vessels.[53] Chang and Lu attempted to identify cellular division in the developing zebrafish using calcium indicators.[54] Another species of embryo that has been examined rather extensively is the fruit fly (*Drosophila melanogaster*). For example, Douglas *et al.* used confocal microscopy to examine cell cycle progression using GFP-labeled kinesin and found that hypoxic (low-oxygen) conditions affect certain portions of the cell cycle.[55]

Whereas embryological studies using confocal microscopy have been relatively successful, recent work using TPM has provided convincing evidence that TPM is a superior method for these studies. Squirrell *et al.* imaged mitochondrias in hamster embryos, and noted that after 24 hr of TPM imaging the blastocysts were

[52] V. J. Napadow, Q. Chen, V. Mai, P. T. So, and R. J. Gilbert, *Biophys. J.* **80**, 2968 (2001).
[53] S. Isogai, M. Horiguchi, and B. M. Weinstein, *Dev. Biol.* **230**, 278 (2001).
[54] D. C. Chang and P. Lu, *Microsc. Res. Tech.* **49**, 111 (2000).
[55] R. M. Douglas, T. Xu, and G. G. Haddad, *Am. J. Physiol. Reg. Integrative Comp. Physiol.* **280**, R1555 (2001).

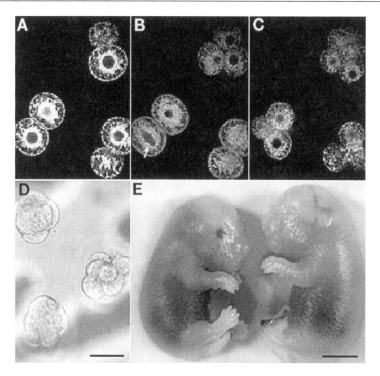

FIG. 5. Two-photon images of mitochondria-labeled embryos at (A) the initiation of imaging, (B) after 8.5 hr of imaging, showing a mitotic spindle (arrow), and (C) the completion of a 24-hr imaging sequence. (D) After imaging, embryos were cultured in the incubator until 82 hr PEA (Nomarski) at which point they were transferred to a recipient female. (E) A black-eyed fetus that developed from one of these imaged embryos is shown next to an albino uterine mate. Bar: 45 m (A–D); 4.75 mm (E). [Reproduced from J. M. Squirrell, D. L. Wokosin, J. G. White, and B. D. Bavister, *Nat. Biotechnol.* **17,** 763 (1999).]

still competent for development (Fig. 5). In contrast, 8 hr of confocal microscopy was sufficient to impair development in the same system.[15]

Phillips *et al.* succeeded in imaging the development of fixed embryonic and mouse kidneys with TPM. By using 3D image analysis and reconstruction, they were able to elucidate the fine structure of the kidneys, such as the nephrons and tubules, at various stages of development. They were also able to image cyst development in *inv/inv* mutant mice kidneys.[56]

A few recent TPM studies focused on the development of nonmammalian species using TPM. Summers *et al.* found that in sea urchin the first cleavage of the developing embryo does not predict the direction of the ultimate bilateral

[56] C. L. Phillips, L. J. Arend, A. J. Filson, D. J. Kojetin, J. L. Clendenon, S. Fang, and K. W. Dunn, *Am. J. Pathol.* **158,** 49 (2001).

axes in the organism.[57] This is a particularly interesting study in that TPM is used to activate caged fluorophores, which allows for the efficient tracking of the developmental lineage of cells in a given tissue structure. A study by Periasamy *et al.* on *Xenopus* (frog) embryos demonstrated that TPM is a viable imaging tool to study this system due to its deep penetration, particularly when deconvolution is used to reduce out-of-focus noise.[58]

Embryology is a field ready for TPM. Many studies using confocal microscopy may be improved by using a two-photon system. This will yield deeper penetration depth and a longer observation time, and should reveal much more about the developmental process of embryos.

Emerging Application Areas of Two-Photon Tissue Imaging

Today, as TPM is approaching maturity, it is being adopted for an increasingly wide range of biomedical applications. In this section, we will review some of the more promising directions that TPM may take, including new instrumentation developments, potential clinical applications, and the investigation of biological systems that have as yet not been extensively studied with TPM.

3D Tissue Spectroscopy Analysis

Multiphoton microscopy allows not only the determination of tissue structures as well as the analysis of the fluorescence signal based on either the excitation spectrum, the emission spectrum, or the fluorescence lifetime, but can also provide complementary biochemical information about the microenvironment. One-photon excitation tissue spectroscopy is commonly used to measure the fluorescence emission signature from bulk tissues. Spectroscopic analysis of endogenous fluorescence emission in tissues has been used to determine the physiological state of tissues.[59–61] In addition to emission spectroscopy, the excitation spectrum of the tissue may also provide valuable clues about the physiology of the tissue, and aid in distinguishing between malignant and benign tissues.[62–65]

---

[57] R. G. Summers, D. W. Piston, K. M. Harris, and J. B. Morrill, *Dev. Biol.* **175**, 177 (1996).

[58] A. Periasamy, P. Skoglund, C. Noakes, and R. Keller, *Microsc. Res. Tech.* **47**, 172 (1999).

[59] R. M. Cothren, R. Richards-Kortum, M. Sivak, M. Fitzmaurice, R. Rava, G. Boyce, M. Doxtader, R. Blackman, T. Ivanc, and G. Hayes, *Gastrointest. Endosc.* **36**, 105 (1990).

[60] R. Richards-Kortum, R. P. Rava, M. Fitzmaurice, L. Tong, N. B. Ratliff, J. R. Kramer, and M. S. Feld, *IEEE Trans. Biomed. Eng.* **36**, 1222 (1991).

[61] R. Richards-Kortum, R. P. Rava, R. E. Petras, M. Fitzmaurice, M. Sivak, and M. S. Feld, *Photochem. Photobiol.* **53**, 777 (1991).

[62] R. Glasgold, M. Glasgold, H. Savage, J. Pinto, R. Alfano, and S. Schantz, *Cancer Lett.* **82**, 33 (1994).

[63] S. P. Schantz, H. E. Savage, P. Sacks, and R. R. Alfano, *Environ. Health Perspect.* **105**, 941 (1997).

[64] K. Licha, B. Riefke, V. Ntziachristos, A. Becker, B. Chance, and W. Semmler, *Photochem. Photobiol.* **72**, 392 (2000).

[65] B. Beauvoit and B. Chance, *Mol. Cell. Biochem.* **184**, 445 (1998).

FIG. 6. A lifetime resolved image of *ex vivo* human dermis. The fluorescence signals contain contributions from elastin fluorescence, collagen second harmonic signal, and collagen fluorescence. Long lifetime regions are elastin rich. Short lifetime regions are collagen rich.

Spectroscopic analysis of bulk tissues yields information about the average spectroscopic properties of the fluorescent constituents in the tissue. However, the decomposition of this mixed spectroscopic signal into individual components is difficult. One solution is to use TPM to restrict the fluorescence signal to femtoliter-sized volumes where there are fewer independent fluorescent species. By reducing the mixing of different fluorescence species, the identity and contributions of each pure biochemical species can be more easily assessed. Although the acquisition of spectral data from femtoliter microscopic volumes allows the extraction of "purer" spectroscopic information at each tissue location, it is not uncommon that the signal from a single image voxel may still contain contributions from two or more endogenous fluorescence species. Therefore, an important recent advance is the application of multivariate curve resolution methods to numerically resolve the spectroscopic contributions from multiple fluorescent species.

An early application of two-photon spectroscopic analysis involved identifying tissue endogenous fluorophores of *in vivo* human skin based on emission spectroscopy and fluorescence lifetime.[45] A similar approach has been applied to quantify endogenous fluorophores in the cornea of excised rabbit eyes.[27] Because these are TPM studies, spectroscopic information is extracted from specific, femtoliter-sized locations in the tissue. Recent work focuses on quantifying spectral signatures from every voxels in a 3D tissue volume allowing better correlation of fluorescence fluorophore distribution with tissue morphology. For instance, excitation spectra, emission spectra, and lifetime maps have been obtained for 3D samples of *ex vivo* human skin[66] (Fig. 6). The invention of high throughput and high sensitivity

[66] L. Hsu, T. M. Hancewicz, P. D. Kaplan, K. Berland, and P. T. C. So, "SPIE," Vol. 4262, p. 294. SPIE Press, San Jose, 1991.

multispectral resolved detectors has further increased the ability to extract spectro-scopic information from 3D tissue specimens.[67] We anticipate that spectroscopic resolved multiphoton microscopy will find important uses in developmental biol-ogy and cancer biology studies in which the physiological consequences of gene expression distribution at the tissue level can be studied using green fluorescence protein and multiphoton microscopic imaging technologies.[68–70]

3D Tissue Image Cytometry

The development of new tools and techniques to investigate cell behavior has played a crucial role in furthering our understanding of basic cell biology. For example, flow cytometry (FCM) has proven to be a very important analytical tool for studying cell populations, and one that is capable of measuring the properties of a very large population, typically on the order of 10^5 to 10^8, of cells. Because of its high throughput, FCM is one of the few techniques capable of generating statistics on a large population of cells and identifying rare events with frequencies as low as 1 cell in 10^7 cells. This is an important technology not only in research areas but also in clinical settings in which the clinician is interested in identifying residual signs of disease, particularly cancer. For a recent review of the technique see Rieseberg *et al.*[71]

Another technique for studying large populations of cells is image cytometry,[72] which was developed in part to address some of the limitation of FCM. A variant of this is laser-scanning cytometry in which a scanned laser beam is used as the excitation source.[73] In image cytometry a population of cells supported on a solid substrate (usually a microscope coverslip) is imaged. Image cytometry is different from conventional microscopy in that a much larger area is imaged (typically on the order of 0.2–1 cm^2) but at a lower resolution. This allows a much larger popu-lation of cells to be assayed than if a smaller area were imaged. It shares many of the strengths of flow cytometry, such as the ability to investigate and classify large cell populations. Despite a generally lower processivity rate, imaging cytometry offers a number of advantages over flow cytometry. First, imaging cytometry in-herently captures morphological features of cell populations. Second, it is possible to reimage cells of interest, which allows time series studies or the relabeling of

[67] Y. L. Pan, P. Cobler, S. Rhodes, A. Potter, T. Chou, S. Holler, R. K. Chang, R. G. Pinnick, and J. P. Wolf, *Rev. Sci. Instrum.* **72,** 1831 (2001).
[68] T. Watabe, M. Lin, H. Ide, A. A. Donjacour, G. R. Cunha, O. N. Witte, and R. E. Reiter, *Proc. Natl. Acad. Sci. U.S.A.* **99,** 401 (2002).
[69] H. S. Stadler, K. M. Higgins, and M. R. Capecchi, *Development* **128,** 4177 (2001).
[70] E. B. Brown, R. B. Campbell, Y. Tsuzuki, L. Xu, P. Carmeliet, D. Fukumura, and R. K. Jain, *Nat. Med.* **7,** 864 (2001).
[71] M. Rieseberg, C. Kasper, K. F. Reardon, and T. Scheper, *Appl. Microbiol. Biotechnol.* **56,** 350 (2001).
[72] C. Cohen, *Hum. Pathol.* **27,** 482 (1996).
[73] Z. Darzynkiewicz, E. Bedner, X. Li, W. Gorczyca, and M. R. Melamed, *Exp. Cell Res.* **249,** 1 (1999).

cells with different probes. Finally, a major advantage of image cytometry is the ability to study cells in intact tissues. However, as was pointed out previously, there are a number of difficulties in obtaining satisfactory images in tissues with conventional confocal or wide-field techniques, which has restricted the suitability of image cytometry for the study of tissues in general. An attractive solution to this problem is to use two-photon excitation. Two-photon image cytometry can combine all the strengths of image cytometry with the advantages of TPM in tissue imaging.

To make two-photon tissue imaging cytometry feasible it is necessary to employ high-speed imaging methodologies. Conventional two-photon microscope systems can typically acquire images at 1 Hz, which is far too slow to image macroscopic volumes in a reasonable period of time. Several groups have developed video-rate two-photon microscopes that allow images to be acquired in as little as 30 ms.[74–78] By incorporating a mechanical stage that can move the sample macroscopic distances into a video-rate microscope system, it is feasible to image large areas in realistic periods of time, and thus allow macroscopic volumes of tissues containing large number of cells to be studied.

This type of instrument can open a number of new research and clinical avenues. For example, the location and frequency of stem cells in many tissue types are not well characterized.[79] By labeling an animal with fluorescent labels, such as bromodeoxyuridine (BrdU), it is possible to identify the slow-cycling stem cells and map their location. Another potential application is to better characterize micrometastasis. Currently, in histological studies it is possible to sample tissues only at very low rates due to the labor-intensive nature of the preparation. With wide-area two-photon imaging it may be possible to dramatically increase the sampling rate and identify cancerous cells that would have otherwise gone undetected. Still another potential application is to characterize neuronal genesis in mammalian brain tissue, which is currently a controversial field. A two-photon image cytometry study of brain tissues may be able to reveal individual neurons that would have otherwise gone undetected using standard microscopy techniques that necessarily employ a lower sampling rate.

As an additional example, two-photon image cytometry can also be used to study rare mitotic recombination events *in situ*.[80] Mitotic recombination plays an

[74] K. H. Kim, C. Buehler, and P. T. C. So, *Appl. Opt.* **38**, 6004 (1999).
[75] G. J. Brakenhoff, J. Squier, T. Norris, A. C. Bliton, W. H. Wade, and B. Athey, *J. Microsc.* **181**, 253 (1996).
[76] J. B. Guild and W. W. Webb, *Biophys. J.* **68**, 290a (1995).
[77] J. Bewersdorf, R. Pick, and S. W. Hell, *Opt. Lett.* **23**, 655 (1998).
[78] A. H. Buist, M. Muller, J. Squire, and G. J. Brakenhoff, *J. Microsc.* **192**, 217 (1998).
[79] F. M. Watt, *Curr. Opin. Genet. Dev.* **11**, 410 (2001).
[80] K. H. Kim, M. S. Stitt, C. A. Hendricks, K. H. Almeida, B. P. Engelward, and P. T. C. So, *SPIE Proc.* (in press).

502
BIOPHOTONICS
[22]

important role in carcinogenesis and a thorough understanding of this process may aid in the prevention and treatment of cancer. A strain of transgenic mice has been engineered that contains two nonfunctional yellow fluorescent protein (YFP) expression cassettes in their genome. Ordinarily, YFP will not be expressed as each YFP cassette is nonfunctional; however, if a mitotic recombination event occurs at the right locus the YPF expression system can become functional. This is a very low probability event with an occurrence on the order of one cell in 10^5 to 10^6 cells, and hence the need to image a large population of cells. Because this experiment is done in intact tissues it is necessary to employ TPM to be able to reliably detect the YFP expression levels above the tissue autofluorescence. 3D image information further allows the correlation between recombination events with cell type and tissue location information.

Applications in Plant Biology

A growing number of botanical applications of TPM have been developed in the past few years. The application of 3D microscopy in plant biology has been limited by the highly scattering cell walls, as well as by the sensitivity of plants to photodamage under UV or visible light. The advent of TPM may partially overcome some of these limitations and provide a new method to study problems in plant physiology, such as nutrient/fluid transport and developmental questions, *in vivo*.

Many studies focus on the cellular level of transport within plant cells. For example, chlorophyll has been imaged in single plant cells using autofluorescence, which is reviewed with other plant work in Konig's article.[81] Tirlapur and Konig also looked at the division of chloroplasts and discovered that the division occurred in under 1 hr and was generally asymmetric (that is, generated daughters of unequal size).[82] Kohler *et al.* reported examining mitochondrial distributions by using a CoxIV GFP fusion protein.[83] With this technique, they were able to show that it is possible to specifically target mitochondria in plant cells, and with the development of more advanced scanning techniques, to explore the distribution and movements of the mitochondria in whole living plant tissues.

There are a number of other pioneering TPM studies of botanical systems at the tissue level. Meiotic events were imaged to a depth of 200 μm in *Agapanthus umbelatus* anthers.[84] Although Feijo and Cox were able to show evidence of cell division at a resolution up to the level of the nuclear dimension, they were unable to detect subnuclear meiotic processes.[84] Meyer and Fricker imaged glutathione and vacuole transport in *Arabidopsis* roots. They found that lowering the temperature of the roots prevented the vacuoles from sequestering.[85]

[81] K. Konig, *J. Microsc.* **200**, 83 (2000).
[82] U. K. Tirlapur and K. Konig, *Planta* **214**, 1 (2001).
[83] R. H. Kohler, W. R. Zipfel, W. W. Webb, and M. R. Hanson, *Plant J.* **11**, 613 (1997).
[84] J. A. Feijo and G. Cox, *Micron* **32**, 679 (2001).
[85] A. J. Meyer and M. D. Fricker, *J. Microsc.* **198**, 174 (2000).

Given the advantage of TPM in imaging the highly opaque tissues that are found in many botanical tissues, and the increasing availability of TPM instrumentation, we foresee a steady increase in the use of TPM in plant biology.

Two-Photon Optical Biopsy and Photodynamic Therapy

Although the clinical potential of TPM has been recognized by many researchers, this potential has yet to be realized. This will likely change in the future as the instrumentation on which TPM is based becomes increasingly more miniaturized and robust. The most promising clinical applications of TPM reside in two areas: diagnosis and treatment.

In terms of diagnosis, optical biopsies can be a valuable supplement to histopathology by providing images of intracellular structures with submicron resolution. The successful design of a minimally invasive optical biopsy instrument must take several issues into consideration, including the lack of contrast agents in the absence of histological stains, and the ability to acquire 3D datasets without resorting to the physical sectioning of the tissue. Additionally, if it is to be successful in a clinical setting, the system must be ergonomic and robust. To be minimally invasive, an optical biopsy technique should make use of any available endogenous contrast, such as index of refraction mismatches or spectroscopic information, that can reveal the biochemical composition of the sample. Generally, histological stains such as H&E will not be available due to the toxicity of most of the types of exogenous dyes. A number of emerging diagnostic methods such as optical coherence tomography (OCT) and diffusive wave imaging are delivering 3D images of tissue structures,[8,86–89] and are potentially very useful in detecting a number of pathologies. Unfortunately, their spatial resolution is too coarse to provide the subcellular details required for definitive pathological diagnosis. TPM with its ability to image subcellular structure and biochemistry may find clinical use by complementing traditional excisional biopsy in diagnosis applications.

We expect that optical biopsy based on TPM will guide, but not supplant, traditional excision biopsy followed by histological analysis. The histologist will always have superior optical resolution as microscopic thin tissue sections provide ideal optical conditions, whereas optical biopsies must contend with thick, highly scattering tissues. However, given that TPM images can be acquired with sufficiently high quality and resolution, noninvasive optical biopsy has several medical benefits: (1) First-line screening may decrease the number of excision biopsies required. (2) Noninvasive tools can be used to precisely select the correct site for the excision biopsy, which can help avoid incorrect diagnoses due to the

[86] D. Huang, E. A. Swanson, C. P. Lin, J. S. Schuman, W. G. Stinson, W. Chang, M. R. Hee, T. Flotte, K. Gregory, C. A. Puliafito, and J. G. Fujimoto, *Science* **254,** 1178 (1991).
[87] J. M. Schmitt, M. J. Yadlowsky, and R. F. Bonner, *Dermatology* **191,** 93 (1995).
[88] K. Suzuki, Y. Yamashita, K. Ohta, and B. Chance, *Invest. Radiol.* **29,** 410 (1994).
[89] E. M. Sevick, C. L. Burch, and B. Chance, *Adv. Exp. Med. Biol.* **345,** 815 (1994).

risk of missing small malignant tissue masses when random sampling is used. (3) Complete surgical removal of malignant tissues is critical and requires accurate demarcation of surgical margins. 3D optical biopsies can allow decisions on pathology to be made on a real-time basis resulting in more accurate determination of surgical margin. Accurate margin determination, the key to Mohs microsurgery used for treatment of skin cancers, will reduce surgical trauma and the necessity for follow-up surgery, minimize the removal of healthy tissue, and improve patient recovery rates.

The use of TPM for noninvasive tissue imaging has been realized in many *ex vivo* tissue systems. A TPM optical biopsy has already been demonstrated for use in imaging cellular morphology and metabolism in the skin of human volunteers. However, the operational complexity, size, imaging speed, and cost have significantly curtailed its clinical application. To circumvent this difficulty, the miniaturization of TPM for clinical imaging is an important instrument developmental challenge. The technical difficulties involved in the development of a two-photon endoscope include the need to deliver high power, femtosecond laser pulses through single mode fiber optics, the miniaturization of the laser scanning optics, and the optimization of the detection system. Although no fundamental limits exist preventing the solution of these problems, significant engineering challenges remain. It should be noted that the development of confocal endoscopy— a similar and complementary technique—is at a more advanced stage,[90] and some of the lessons learned there will aid in the development of two-photon endoscopy.

A complementary technique to two-photon endoscopic biopsy is two-photon photodynamic therapy. Although two-photon endoscopy allows disease detection and diagnosis, two-photon photodynamic therapy is being developed for minimally invasive treatment. As in traditional photodynamic therapy, the elimination of pathological tissues is aided by the selective delivery of photosensitizers. Two-photon illumination of tissues will result in localized and selective damage of cells loaded with photosensitizers. Unlike traditional photodynamic therapy, the two-photon approach may be more spatially selective and less damaging to the surrounding healthy tissues. In two-photon photodynamic therapy, pathological tissues are selected not only by the biochemical specificity of the photosensitizers; but also by the localized two-photon excitation volume, which further restricts the treatment to a specific tissue location. Prasad and co-workers developed one of the earlier two-photon photodynamic therapies. They synthesized dyes that were capable of two-photon absorption and of generating singlet oxygen molecules in the presence of photosensitizers, and thus showed the feasibility of

[90] M. R. Descour, A. H. O. Karkkainen, J. D. Rogers, C. Liang, R. S. Weinstein, J. T. Rantala, B. Kilic, E. Madenci, R. R. Richards-Kortum, E. V. Anslyn, R. D. Dupuis, R. J. Schul, C. G. Willison, and C. P. Tigges, *IEEE J. Quantum Elect.* **38**, 122 (2002).

using two-photon excitation to induce cellular damage.[91] Fisher *et al.* went further and incorporated the psoralen derivative dyes with *Salmonella typhimurium* to demonstrate the feasibility of two-photon photodynamic therapy.[92]

Conclusion

Over the past decade, there has been an explosion in the use of two-photon microscopy for studying biological systems. Much of the work up to now has focused on single-cell imaging; however, tissue-level imaging is likely to become more prevalent, especially as tissue engineering and organ-level physiology develop further. Certain tissues, such as skin, plant roots, and even whole embryos, are excellent candidates for two-photon imaging due to their thinness. Diagnoses of pathological conditions that have surface presentations, such as those in intestines, the tongue, and blood vessel lumens, may also be facilitated by two-photon use via endoscopes. The development of faster computers and larger storage capabilities will further enhance the use of three-dimensional imaging and processing in tissues.

Acknowledgments

The authors acknowledge support from NIH Grants R21/R33, CA84740-01, and P01 HL64858-01A1. We further acknowledge support from the MIT Center for Biomedical Engineering Research Seed Grant program.

[91] J. D. Bhawalkar, N. D. Kumar, C. F. Zhao, and P. N. Prasad, *J. Clin. Laser Med. Surg.* **15,** 201 (1997).

[92] W. G. Fisher, W. P. Partridge, Jr., C. Dees, and E. A. Wachter, *Photochem. Photobiol.* **66,** 141 (1997).

[23] Fluorescence Anisotropy in Pharmacologic Screening

By J. RICHARD SPORTSMAN

Introduction

The enterprise of drug discovery screening has undergone an industrial revolution in the past decade. All major pharmaceutical companies now devote significant resources to high throughput screening (HTS), the process whereby hundreds of thousands or millions of chemical compounds and natural products are screened against pharmacologic targets such as enzymes and receptors. To contain costs, it is necessary to streamline the detection systems and protocols for fast throughput. Fluorescence methods satisfy several requirements of today's high throughput screening enterprise. Fluorescence allows homogeneous formats, important because separation and washing steps are usually too cumbersome, expensive, and slow to be considered competitive. High sensitivity is also required in order to be able to detect biological binding at subnanomolar concentrations, and fluorescence delivers this capability.

It is useful to compare the potential of fluorescence methods against that of traditional screening technology based on radioactivity. If we consider simply the time required to detect a statistically significant number of events, fluorescence has almost one million-fold theoretical advantage over radioactivity. For example, 10 fmol of tritium gives approximately 5 counts per second (cps) in a standard scintillation detector. On the other hand the same mass of fluorescein can deliver approximately 3 million counts per second in an optimized instrument. Of course, background due to scattering and spectral interferences limits fluorescence sensitivity to the point where it is roughly comparable to that of radioactivity, but the capacity for speed of fluorescence is quite clear.

Fluorescence anisotropy (FA, or fluorescence polarization, FP)[1] is now accepted as one of the major techniques suitable for HTS because of its high sensitivity and convenient homogeneous format. Over the past two decades, this technique has already been widely adopted in clinical diagnostics (FP immunoassays). More recently, with the advent of microplate instruments, FA has been successfully used in drug discovery.[2]

Theory of Fluorescence Anisotropy

If a solution of fluorescent molecules is excited by plane polarized light, only those molecules with their absorption dipoles oriented approximately parallel to the electric vector of the exciting light will be excited. Hence, there is a photoselection process that occurs on illumination with polarized light. Then, if no change in the orientation of the molecule occurs during its excited state lifetime, the emitted light or fluorescence emission will also be polarized. Otherwise, if the fluorescent molecule undergoes reorientation during its excited state lifetime, the emission will be depolarized, if it is completely randomized or isotropic. This process is diagrammed schematically in Fig. 1.

Fluorescence anisotropy is technically a rather simple technology, needing little more than a filter fluorometer equipped with polarizing filters. However, additional

[1] The two terms are used interchangeably to describe the technique. FP is most familiar to the screening community, whereas FA is technically advantageous in some applications. See discussion following Eq. (3).
[2] J. C. Owicki, *J. Biomol. Screen.* **5,** 297 (2000).

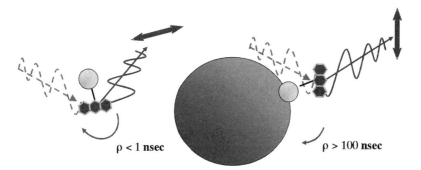

$\rho < 1$ nsec $\rho > 100$ nsec

FIG. 1. Fluorescence anisotropy and molecular binding. A free ligand labeled with a fluorescein-like fluorophore (left) rotates on the subnanosecond time scale (with rotational correlation time ρ) and emits depolarized fluorescence. The same ligand when bound to a large protein or macromolecule (right) rotates much more slowly, hence maintains anisotropy.

challenges are presented when reading anisotropy in microplates, as one cannot read at the preferred 90° angle. Most microplate FP readers therefore use a 180° geometry, as exemplified in Fig. 2. Light of the correct wavelength is linearly polarized, and focused on the sample. The intensity of fluorescent emission is then quantified after passing through a polarizing filter, dichroic mirror, and emission interference filter. Two intensity measurements are made with the emission polarizer set first parallel, then perpendicular, to the direction of the excitation polarizer. These two measurements are referred to as the parallel and perpendicular intensity measurements $I_{\parallel}$ and $I_{\perp}$, respectively. Anisotropy or polarization, represented as r or P, respectively, is then calculated as the difference between these values divided

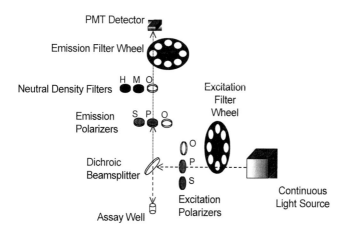

FIG. 2. Optical path of an FP microplate reader: the Molecular Devices' Analyst.

by a sum [Eqs. (1) and (2)]

$$r = \frac{I_\parallel - g I_\perp}{I_\parallel + 2g I_\perp} \tag{1}$$

$$P = \frac{I_\parallel - g I_\perp}{I_\parallel + g I_\perp} \tag{2}$$

Here, g is the so-called grating factor that accounts for the different responsiveness of an instrument to light at the perpendicular orientation of the polarizer. Also, the values of $I_\parallel$ and $I_\perp$ are intended to represent only the fluorescence of the fluorescent species under study, and so should have any residual fluorescence of background subtracted before calculating the anisotropy/polarization. In practice, the value of P is often multiplied by 1000 to yield the convenient unit of mP. The relationship between anisotropy and polarization follows from Eqs. (1) and (2).

$$r = \frac{2P}{3 - P} \tag{3}$$

It should be noted that anisotropy has the advantage of being additive in a simple fashion, so that equations relating anisotropy to binding equilibria and physical phenomena are simpler than those using polarization.[2] Both parameters are equally usable in practice, where the data are used more or less empirically to construct dose–response curves. In this article I shall use the terms interchangeably when referring to the technique.

The relationship between anisotropy, fluorescence lifetime, and rotational diffusion was explored by Perrin and Weber. In terms of anisotropy,

$$r = r_0 \left/ \left(1 + \frac{3\tau}{\rho}\right) \right. \tag{4}$$

where r is the observed anisotropy, r_0 is the limiting anisotropy of a fluorescent molecule (i.e., its anisotropy in the absence of rotation), τ is the excited state lifetime, and ρ is the harmonic mean of the Debye rotational relaxation times about the principal axes of rotation of the molecule. From Eq. (4), it can be seen that anisotropy is dependent on the relation between the excited state lifetime, τ, and the rotational relaxation time, ρ. Because ρ is proportional to molecular weight (from the Stokes–Einstein relation), a high-molecular-weight fluorescent complex will give high anisotropy, as depicted in Fig. 1, whereas the low-molecular-weight form of the same fluorophore will give a lower value. For example, a 40-kDa protein at $20°$ is predicted to have a rotational relaxation time of $\sim$46 ns according to the Stokes–Einstein relation. Therefore, for a fluorescein probe ($\tau = 4$ ns) bound to a 40-kDa protein, and assuming a limiting anisotropy of 0.38 (487 mP) for fluorescein, the expected anisotropy using Eq. (4) is 0.304 (394 mP). By contrast, the anisotropy is 0.018 (27 mP) for free fluorescein.

Two important points should be mentioned here. First, from Eq. (1) it is clear that anisotropy is a ratiometric measurement. Such measurements have an inherent advantage of self-correction for variations due to fluctuations of lamp intensity or interferences due to quenching of fluorescence. A second point is the theoretical boundaries of anisotropy measurements. Fluorophores randomly oriented in solution will be constrained to the range -0.2 to $+0.4$ (-333 to $+500$ mP). A biological assay will rarely show anisotropy less than 0 (0 mP). Thus if a value outside this range occurs, it is a clear sign that there is an error, either in the instrument or, more likely, in the elements comprising the assay sample itself, perhaps a compound that produces spectral interference, for example. Thus FA has a built-in check of assay validity. An additional check of assay validity is provided by the intensity parameter I, which is calculated from the same data used for anisotropy, as shown in Eq. (5).

$$I = I_{\parallel} + 2gI_{\perp} \tag{5}$$

This value of the intensity is proportional to the intensity that would be measured if there were no polarization bias in the measurement. This is not quite the same as the intensity measured without polarizers in place. Even without polarizers the measured intensity has a polarization bias due to observation position[3] (leading to the factor of 2) and instrument bias (leading to the g factor). Because most fluorescence anisotropy assays employ a constant amount of fluorescent tracer added to every sample, this intensity value should not vary outside a certain statistical range. If it does, a fluorescent or quenching compound may be indicated, and a spurious result flagged.

Microplate Readers for FP

Instrumentation for FP

Along with other detection technologies, FP has had to undergo adaptation to the microplate for use in HTS. Although FP has been a standard accessory for spectrofluorometers for years, microplate readers in 96/384- and 1536-well densities have appeared only for the past 6 years. Table I lists current instruments with their respective performance parameters. All the instruments in Table I are capable of 96- and 384-well plate formats. Some are capable of the 1536-well format as well. The prices and performances of the instruments vary, but all are capable of measuring the polarization of fluorescein accurately down to about 5 nM. Some instruments can achieve good precision of polarization (less than 3 mP standard deviation) at concentrations well below 1 nM fluorescein.

[3] J. R. Lakowicz, "Principles of Fluorescence Spectroscopy," 2nd Ed., pp. 291, 298. Kluwer Academic/Plenum Publishers, New York, 1999.

TABLE I
COMMERCIALLY AVAILABLE MICROPLATE READERS FOR FP

Instrument	Manufacturer	Location
Analyst, GT	Molecular Devices Corporation	Sunnyvale, CA
Polarion; Ultra	Tecan	Männedorf, Switzerland
Polarstar; Galaxy	BMG	Offenburg, Germany
Fusion; Envision	PerkinElmer Life Sciences	Boston, MA

Fluorescence Anisotropy in Screening

FP has proven to be a robust assay methodology for high throughput screening. Its chief advantage is that it is a homogeneous assay (i.e., no separation or washing steps are required). This is important in modern assays for drug discovery screening, as costs and delays associated with separation and washing are significant. FP provides the high sensitivity of fluorescence assays, with less interference by quenching or colored compounds than direct fluorescence assays because anisotropy is an intrinsic quantity and fluorescence intensity is not. Of course, there are also the added safety and cost benefits of not using radioactivity.

A complete review of the burgeoning literature on the numerous FP assays in HTS is beyond the intent of this article; for a good review, see that of Owicki.[2] Rather we wish to illustrate the major classes of assays that are now addressed, and technical aspects of their execution.

Assays for Protein Kinases

Protein kinases, with their major role in signal transduction, comprise a significant fraction of the targets screened in high throughput. Several laboratories have used FP competitive immunoassays with success.[4] For tyrosine kinases (TKs), kits from Molecular Devices Corporation and PanVera Corporation are available. These can be used to assay most TKs, with little regard to substrate composition.

For the serine and threonine kinases, the story is more complicated. In contrast to phosphotyrosine, phosphoserine and phosphothreonine do not seem to elicit antibodies that are useful for detection of phosphorylation irrespective of flanking amino acids. Hence antibody-based FP assays are of restricted generality for Ser/Thr kinases. Both Molecular Devices and PanVera Corporation (Madison, WI) offer products that can assay certain Ser/Thr kinases by competitive immunoassay. Other technologies for FP kinases that do not require antibodies have been described,[5] and one example is given here later in this article.

---

[4] R. Seethala and R. Menzel, *Anal. Biochem.* **255**, 257 (1998).
[5] J. Coffin, M. Latev, X. Bi, and T. T. Nikiforov, *Anal. Biochem.* **278**, 206 (2000).

Receptor Binding Assays

There are several reports of FP assays for G-protein-coupled receptors (GPCRs) and other receptors.[6,7] These require a fluorescein-labeled ligand with suitable properties for FP, a challenge that so far has limited the general utility of FP for screening of receptors. Nevertheless there are numerous examples that show the advantages of FP when adequate tracer is available. PanVera has reagents for several nuclear hormone receptors and PerkinElmer Life Sciences has made available fluorescent peptide ligands for a number of hormone receptors.

Other Systems

A variety of other assays of protein–ligand binding,[8] catalysis,[9] and protease activity.[10] are examples of the multiplicity of systems that have found adaptation to the FP format.

Protocols for Microplate FP

In the remaining part of the article I present representative protocols for microplate assays in FP, illustrated by a protein ligand-binding assay (phloxine B binding to human serum albumin), a G-protein-coupled receptor assay (dopamine D2a receptor binding of the fluorescent spiperone derivative BODIPY-NAPS), and two types of protein kinase assay, an FP immunoassay for protein kinase A activity, and a recently described technology for non-antibody-based FP kinase assays called IMAP, illustrated with the serine/threonine kinase Akt. The objective of the former two is to determine affinity constants for ligand or competitor, whereas the kinase assays are designed to detect and characterize kinase activity. All these assays are representative of the kinds of procedures used in HTS. There are some factors to keep in mind in dealing with the special circumstances encountered in screening compound collections, and these are discussed at the end.

Determination of Affinity by FP

The determination of affinity constants is important to understanding the limits of functioning of most fluorescence polarization assays in HTS. The determination of K_i values also requires knowledge of the affinity of the tracer. It is not sufficient to assume that the tracer has an affinity equal to the unlabeled version of itself,

---

[6] M. Allen, J. Reeves, and G. Mellor, *J. Biomol. Screen.* **5,** 63 (2000).

[7] P. Banks, M. Gosselin, and L. Prystay, *J. Biomol. Screen.* **5,** 159 (2000).

[8] S. S. Pin, I. Kariv, N. R. Graciani, and K. R. Oldenburg, *Anal. Biochem.* **275,** 156 (1999).

[9] Z. Li, S. Mehdi, I. Patel, J. Kawooya, M. Judkins, W. Zhang, K. Diener, A. Lozada, and D. Dunnington, *J. Biomol. Screen.* **5,** 31 (2000).

[10] S. Z. Schade, M. E. Jolley, B. J. Sarauer, and L. G. Simonson, *Anal. Biochem.* **243,** 1 (1996).

in contrast to many radiometric assays using tritium labels, where this is indeed usually the case. However, the determination of affinity constants for FP is not at all a lengthy process, and it is much facilitated by use of microplates.

Binding constants require the determination of equilibrium concentrations of bound and unbound fluor. In an FP experiment, one obtains the anisotropy of the tracer, which is approximately proportional to fractional binding. To use commercially available fitting packages to determine parameters such as IC_{50} values or affinity constants, one should calculate fractional tracer binding (B/T) from anisotropy data. An exact relationship can be derived, assuming that there are only two fluorescent states of the species that contribute to the intensity measurements that comprise anisotropy, a bound state and a free state.[11] This relationship is represented in Eqs. (6) and (7).

$$B/T = [\lambda(r_{max} - r)/(r - r_{min}) + 1]^{-1} \qquad (6)$$

$$B/T = \{[\lambda(3000 - P_{min})(P_{max} - P)/((3000 - P_{max})(P - P_{min}))] + 1\}^{-1} \quad (7)$$

Here, r or P is the measured anisotropy (or polarization in mP units) for the sample. The values r_{min}, r_{max} (or corresponding P parameters), and λ are constants that are determined in advance of the experiment. The minimum anisotropy r_{min} is simply the anisotropy of the free tracer, whereas r_{max} is the maximum anisotropy obtained at 100% binding (i.e., at infinite binding protein concentration), a value that can be estimated by extrapolation as described below. The parameter λ

$$\lambda = Q_b / Q_f \qquad (8)$$

is the ratio of the molar fluorescences, Q_b and Q_f, for bound and free tracer, respectively. Q_f is determined in the same measurement made for r_{min}, simply dividing the fluorescence intensity value by the molar concentration of the tracer. Q_b is determined in the same experiment described above for determination of r_{max}, extrapolating to determine the intensity of the tracer at infinite binding protein concentration (I_∞). This limiting intensity is divided by the molar concentration of the tracer to give Q_b.

Procedure. The following procedure illustrates the determination of binding constant for a fluorescent ligand or "fluor." First the parameters r_{min}, r_{max}, and λ are determined. Then the affinity is determined by fixing protein concentration and adding increments of tracer. The procedure is written for using an Analyst FP microplate reader, but other instruments may be used within the limits of their ability to perform similarly. The example chosen is phloxine B, binding to the carrier protein bovine serum albumin (BSA). This is not necessarily a significant pharmacologic target, but rather a convenient system, with readily available reagents, for demonstrating the running of FP assays for binding in microplates, and deriving fundamental information from the data.

---

[11] K. M. Rajkowski and N. Cittanova, *J. Theor. Biol.* **93**, 691 (1981).

DETERMINATION OF Q_f AND r_{min}. These parameters represent the molar fluorescence and anisotropy of the fluor in its free state.

1. Prepare fluor at a relatively high concentration, i.e., one that is predetermined to give at least 50-fold higher fluorescent intensity than buffer alone when diluted 1 : 2 to final concentration. A concentration of 20–50 nM is usually a good choice for most plate readers. For assay buffer, I recommend 20 mM HEPES, pH 7.2, 150 mM NaCl, with 0.01% bovine γ-globulin to suppress adsorptive losses.

2. Add 20 μl of assay buffer to eight wells on the plate. To four of these, add 20 μl of the fluor diluted as above. To the other four, add 20 μl of assay buffer to serve as a blank. Avoid pipetting bubbles into the wells, as these disturb the meniscus and add error.

3. Measure the polarized fluorescence of all wells, collecting the two intensity values that comprise the anisotropy.

4. Compute corrected parallel ($I_{\|c}$) and perpendicular ($I_{\perp c}$) values. Subtract the average of the appropriate background values from the individual parallel ($I_{\|}$) and perpendicular ($I_{\perp}$) values for those wells with tracer in them. Then recalculate the anisotropy and total intensity as show in Table II.

The above procedure is illustrated with data in Table II from an experiment to determine the values of Q_f and r_{min} for the fluorescent xanthene dye phloxine B (Sigma, St. Louis, MO). Here, I_f and r_{min} are computed from background subtracted data by Eqs. (5) and (1), with g set to 1.0 by the instrument; Q_f is computed as $I_f/[\text{fluor}]$ ($= I_f/50$ nM). Averaging these results, we determine $Q_f = 125,000$ cps/nM and $r_{min} = 0.098$.

DETERMINATION OF Q_B, r_{max}. These parameters represent the molar fluorescence and limiting anisotropy of the fluor when 100% bound. Because complete binding is possible only at infinite receptor concentration, we must extrapolate

TABLE II
DATA AND CALCULATIONS FOR DETERMINATION OF Q_f AND r_{min}[a]

Raw data (cps)		Background subtracted		Computed		
$I_{\|}$	$I_{\perp}$	$I_{\|c}$	$I_{\perp c}$	I_f ($= I_{\|c} + 2I_{\perp c}$)	Q_f (cps/nM) ($= I_f/50$)	r_{min}
2,452,382	1,828,027	2,399,937	1,818,671	6,037,279	121,000	0.096
2,621,784	1,943,338	2,569,339	1,933,982	6,437,303	129,000	0.099

[a] Phloxine B, 50 nM in assay buffer, was measured for FA on Analyst plate reader, in a 384-well plate, using rhodamine settings [excitation filter $= 530$ nm (25 fwhm), dichroic mirror 561 nm, emission filter 590 nm (20 nm fwhm) with integration time of 100 ms per well]. Duplicate wells of 40 μl volume were read. Background values on buffer (average $\pm$SD, $N=4$) were $I_{\|} = 52,445 \pm 5476$; $I_{\perp} = 9356 \pm 1271$.

from a titration to determine r_{max} as well as the limiting intensity, I_∞, from which Q_b is calculated.

1. Prepare serial dilutions of the binding protein in assay buffer. This can be done directly in the microwell plate if care is taken not to introduce bubbles that interfere with the accuracy of intensity measurements. The concentration range depends on availability and cost of the protein. Prepare quadruplicates at 20 μl per well, starting with the highest concentration of protein possible. Carry out 2- or 3-fold serial dilutions until the protein concentration is down to 1 nM or ~50 ng/ml.

2. In one set of duplicate dilutions, add 20 μl of fluor at the concentration that was predetermined to give at least 10-fold higher fluorescent intensity than buffer alone (5–20 nM is usually a good choice for most plate readers, depending on the fluorescent properties of the fluor).

3. In the other set of duplicates, add assay buffer in place of fluor. These will serve as background correction for any fluorescence contributed by the protein or buffer itself.

4. After a suitable period of time (20–60 min) read the polarized fluorescence data. Calculate the total intensity and anisotropy as above in Table II, except that average values of the parallel and perpendicular intensities of wells that have protein, but no fluor (i.e., as done in step 3), are subtracted from the individual values from those wells with equivalent concentration of protein with fluor present.

5. From these data, one constructs two plots to extrapolate out to the values of I_∞ and r_{max} (Fig. 3). This can be done by inspection or by use of any competent curve-fitting program.

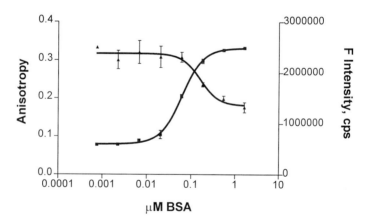

FIG. 3. Data for titration of binding protein (bovine serum albumin, BSA) against the fluorescent molecule phloxine B. Phloxine B (Sigma-Aldrich) was diluted to 18 nM in assay buffer. Serial dilutions of BSA (Sigma) were performed to determine the I_∞ and r_{max}. Instrument settings as in Table III. (■) Anisotropy; (▲) fluoroscense intensity.

Figure 3 is a plot of the data for phloxine B binding by BSA, showing the extrapolation of Q_b and r_{max} using the above method. Note the negative effect of binding on the fluorescent intensity of the tracer. A value for r_{max} of 0.332 and for I_∞ of 1.36×10^6 cps was extrapolated. Thus $Q_b = I_\infty/18$ n$M = 76,000$ cps/nM. From this we have, by Eq. (8), $\lambda = Q_b/Q_f = 68,000/125,000 = 0.61$

Determination of Tracer Affinity. With the parameters λ, r_{min}, and r_{max} in hand, it is now possible to perform the actual affinity determination.

1. Pick a concentration of binding protein that gives submaximal binding of the tracer; from data such as Fig. 3, one could chose 0.2 μM for instance. Prepare 10 wells in triplicates with 20 μl of the chosen protein concentration.

2. Prepare a 2-fold dilution series of the fluorescent tracer, ranging from 0.5 nM to over 128 nM.

3. Add 20 μl of tracer dilution to each of the triplicates prepared in step 1. To one set of triplicates add buffer only for blanks. Incubate 30–60 min (at room temperature) covered.

4. Measure the anisotropy data, subtracting appropriate blanks (wells containing protein without tracer).

5. The corrected anisotropy data can be converted to [B] or B/T by Eq. (6) using a spreadsheet. These data then are suitable for fitting by nonlinear least squares methods (NLLS) to determine affinity values. Graphpad's PRISM V3.0 (Graphpad, San Diego, CA) is one program that does this reasonably well.

Figure 4 shows results for determination of the affinity of phloxine B for BSA. The affinity of this system is 2.1×10^8 liter/mol. The value of B_{max} was 320 μM, or 21 μg/ml, in agreement with the amount of binding protein present and with the notion of a single site on the protein.[12]

Competition Binding Experiments: Determination of K_i for GPCR Ligand

For determination of IC_{50} or K_i values, one measures the displacement of a tracer by one or more nonfluorescent ligands. The method given below illustrates the procedure using the dopamine D_2 receptor and the fluorescent ligand BODIPY-FL-NAPS. Again, this procedure is written for use with the Analyst instrument, or one with similar capabilities (able to read polarization of 1 nM fluorescein with better than 3 mP precision).

Materials

Assay buffer (pH 7.6): 50 mM HEPES containing 150 mM NaCl, 20 mM MgCl$_2$, and 0.1% Tween-20.
Tracer: BODIPY-FL-NAPS (Molecular Probes, Eugene, OR)

---

[12] D. Essassi, R. Zini, and J. P. Tillement, *J. Pharm. Sci.* **79,** 9 (1990).

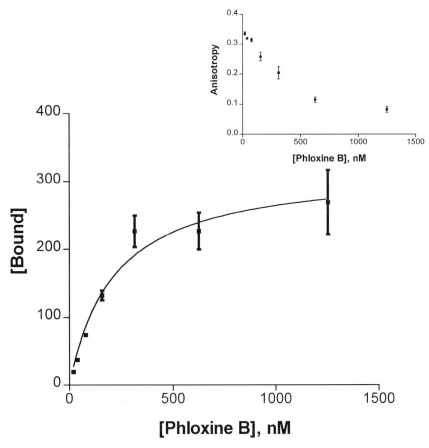

FIG. 4. Determination of protein–ligand binding affinity by FP. Phloxine B was serially diluted according to the scheme in the protocol for the determination of tracer affinity; anisotropy was measured after incubation with 20 μg/ml BSA. *Inset:* Original FA data.

Spiperone (Sigma), 16 μM in assay buffer
Human D_{2a} dopamine receptor (Receptor Biology, Beltsville, MD)
384-Well black microplates (Corning #3710 or equivalent)

Titration of Receptor. The first step involves determining the proper dilution of the receptor for use in the assay. Dilute the tracer to 8 nM in assay buffer. In the example below, the receptor is used at 0, 2.5-, 5-, and 10-fold dilutions using the assay buffer. Pipette the appropriate solutions in the plate using the following template. The total assay volume is 40 μl in a 384-well plate.

1. Add 40 μl assay buffer in column 1, 30 μl in columns 2–6, and 20 μl in columns 7–10.

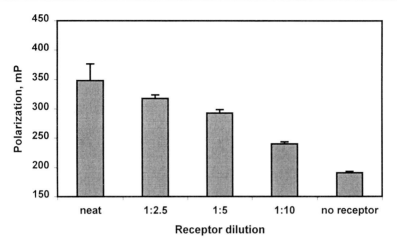

FIG. 5. Optimization of receptor concentration in a membrane receptor FP assay. The receptor was run undiluted or at dilutions of 2.5-, 5-, and 10-fold vs. 2 nM BODIPY-FL-NAPS. Tracer alone ("no receptor") gave 190 mP.

2. Add receptor as follows:
   a. Add 10 μl of the stock solution of the receptor (neat) in columns 2 and 7.
   b. Add 10 μl of a 2.5 -fold dilution to columns 3 and 8.
   c. Add 10 μl of a 5-fold dilution to columns 4 and 9.
   d. Add 10 μl of a 10-fold dilution to columns 5 and 10.
3. Add 10 μl of 8 nM tracer to columns 6 through 10. Incubate the plate for 15–30 min at ambient temperature.
4. Read in Analyst instrument using the fluorescein filter sets in FP mode. Analyze the data after correcting for appropriate blanks, by subtracting intensity values as described in the previous section.

Figure 5 shows the results of this experiment. It indicates that at 5-fold dilution of the receptor, a ΔmP (i.e., bound–free mP) of 102 is obtained. This is adequate to observe displacement of the tracer without needlessly exhausting the expensive receptor supply.

Competition Experiments: Determination of K_i. Using the human dopamine D_{2a} receptor dilution as determined above, one can determine the K_i for agonists of the receptor. The affinity of the tracer BODIPY-FL-NAPS has been reported to be about 1 nM,[13] but recent experiments suggest this value is closer to that for spiperone itself, about 0.05 nM.[14]

---

[13] F. J. Monsma, Jr., A. C. Barton, H. C. Kang, D. L. Brassard, R. P. Haugland, and D. R. Sibley, *J. Neurochem.* **52**, 1641 (1989).
[14] P. G. Strange, personal communication (2001).

TABLE III
DILUTION SCHEME FOR K_i EXPERIMENT[a]

Dilution number	Competitor (nM)	Stock volume (μl)	Buffer volume (μl)	Column no. on plate
1	10	2	798	12
2	5	100 of #1	100	11
3	2.5	100 of #2	100	10
4	1.25	100 of #3	100	9
5	0.62	100 of #4	100	8
6	0.31	100 of #5	100	7
7	0.16	100 of #6	100	6
8	0.08	100 of #7	100	5

[a] The values in the second column above are the final concentrations of the inhibitor in the assay.

1. Dilute the tracer to 8 nM using the assay buffer. Four milliliters is sufficient to run one 384-well microplate plate containing 10 μl of working receptor solution per well with a final assay volume of 40 μl per well.

2. The receptor is diluted to the predetermined optimum level using the assay buffer. The reconstituted receptor solution should be mixed thoroughly and allowed to settle for about 10 min on an ice bath. If sediments are seen, gently remove the clear solution into another vial or test tube without disturbing the sediment. Use this clear solution for running the assay.

3. Prepare the following concentrations of the competitor (spiperone) from the stock solution provided using the serial dilution scheme outlined in Table III for a one plate assay. This serial dilution set provides a bottom and top plateau that can be used with an appropriate nonlinear curve fitting program for data analysis.

4. The following protocol is provided to run a full eight-point inhibition curve with four replicates at each concentration in a 384-well microplate. Pipette the appropriate solutions in the plate using the following template.

  a. Assay buffer: Add 40 μl in column 1, 30 μl in columns 2 and 3, 20 μl in column 4, and 10 μl in columns 5–12.

  b. Receptor: Add 10 μl of the working solution prepared as described above in columns 2 and 4–12.

  c. Calibrator: Add 10 μl of the appropriate concentrations prepared as described in Table III to columns 5–12. For example, the 3.13 nM calibrator working stock solution will be added to column 5, and so on.

  d. Tracer: Add 10 μl of the working solution to columns 3–12.

5. Incubate the plate for 15–30 min at ambient temperature. Read FP in Analyst or a similar FP instrument using a fluorescein filter set.

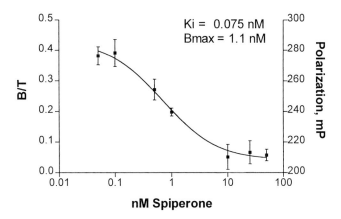

FIG. 6. Dopamine D_2 receptor binding assay. Fluorescent tracer: BODIPY-FL-NAPS (2 nM). Receptor preparation was from Receptor Biology Inc. Right-hand axis is approximate juxtaposition of mP values.

The results of the experiment are given in Fig. 6. In this case the receptor is in the form of a simple membrane preparation that gives rather large fluorescent background. Nevertheless, the specific binding is clearly discernible. The data from this experiment are treated in the same manner as for the phloxine B experiment, subtracting out background fluorescence and computing the corrected anisotropy, then determining fraction of tracer bound. For curve fitting, I use PRISM's one-site model for heterologous competition with depletion, modifying the equation so that it accepts tracer concentration in place of cpm of radioactivity. The affinity of the BODIPY-NAPS tracer is assumed to be 0.05 nM.[13] The value for K_i of the antagonist spiperone is in agreement with that determined by radioreceptor binding assays.[15] In the example above, K_i is about 0.075 nM for spiperone.

FP Immunoassay for Protein Kinase A Activity

The following method uses the Molecular Devices' STX-1 assay for serine/threonine kinases. However, the technique can be adapted to use any antibody raised to a defined phosphopeptide, provided the affinity and anisotropy of the system are relatively strong once the phosphopeptide is labeled with a fluorophore.

The STX-1 HEFP assay measures the activity of serine/threonine kinases via a competitive immunoassay for the product of enzyme activity—a phosphoserine (pSer)-containing peptide. The STX-1 procedure is a two-step process: first a kinase reaction is run, then a fluorescence polarization immunoassay (FPIA) is

[15] http://www.receptorbiology.com/products/1110.htm

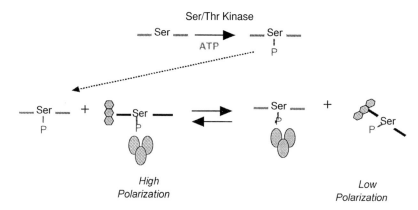

FIG. 7. Principle of the STX-1 FP immunoassay for serine/threonine kinases. A Ser/Thr kinase phosphorylates a peptide substrate in the presence of ATP. When antibody specific for the phosphorylated form of the peptide is added along with a fluoresceinated tracer form of the phosphopeptide, a competitive binding assay is established. The decrease in polarization is a measure of the amount of product formed.

performed on the reaction products to quantify the amount of phosphopeptide formed. The principle of the assay is illustrated in Fig. 7. The phosphoserine (pSer) peptide product of the kinase reaction competes with a fluorescein-labeled pSer-containing peptide (the tracer) for binding sites on the anti-pSer antibody.[16] In the absence of the pSer product, most of the tracer is bound to the antibody. Increasing concentrations of the pSer product competitively decrease the amount of bound tracer.

In the following section we construct a calibration curve for the assay, and then use this to determine the amount of product formed.

Calibration Curve. The assay is calibrated with the phosphorylated form of the substrate. This allows the polarization data to be converted into moles of phosphopeptide.

Materials

Costar flat black 384-well microplates or similar
Analyst Instrument
The following components are taken from the commercial kit:
   STX buffer: 50 mM HEPES with 150 mM NaCl, and 0.1% (w/v) bovine γ-globulin (BGG)
   Tracer: Fluorescein-labeled PKCβ pseudosubstrate-derived phospho-peptide—(5-FAM)-RFARKGS(PO$_3$)LRQKNV; 500 nM stock solution

[16] J. J. Wu, D. R. Yarwood, Q. Pham, and M. Sills, *J. Biomol. Screen.* **5**, 23 (2000).

TABLE IV
DILUTION SCHEME FOR STX CALIBRATION CURVE

Dilution number	Calibrator (nM) (final in well)	Stock volume (μl)	Buffer volume (μl)	Column no. on plate
1	200	20 of 10 μM stock	230	12
2	100	100 of #1	100	11
3	50	100 of #2	100	10
4	25	100 of #3	100	9
5	10	100 of #4	150	8
6	5	100 of #5	100	7
7	1	50 of #6	200	6
8	0.5	100 of #7	100	5

in buffer made from a 1 mM stock in dimethyl sulfoxide (DMSO). Pipette 4 μl of the tracer into 0.996 ml of STX buffer

Antibody: Antibody to PKCβ pseudosubstrate-derived phosphopeptide. A monoclonal antibody is available in the kit.[15] Pipette 10 μl of the antibody into 0.990 ml of STX buffer

Test calibrator: PKCβ pseudosubstrate-derived phosphopeptide–RFARKGS(PO₃)LRQKNV. Use the 10 μM calibrator stock as provided in the kit, or prepare an equivalent solution from a 1 mM stock solution in DMSO, diluting to 10 μM in assay buffer. Prepare the following working dilutions in assay buffer (Table IV). These solutions are prepared in separate tubes and then pipetted into the plate. There is enough of each solution to run an entire plate with eight replicates at each calibrator concentration

Procedure

1. Add 40 μl of buffer per well in column 1, 30 μl in columns 2 and 3, 20 μl in column 4, and 10 μl in columns 5–12.

2. Add 10 μl per well of appropriate concentration of calibrator in columns 5–12, starting with 0.5 nM in column 5.

3. Add 10 μl per well of antibody in columns 2, 4, and 5–12.

4. Add 10 μl per well of tracer in columns 3, 4, and 5–12.

5. Tap plate to even out meniscus in each well.

6. Cover the plate and incubate at room temperature for 20–40 min.

7. Read in FP on Analyst with standard fluorescein settings.

8. Average the replicates and plot mP versus concentration of phosphopeptide calibrator (Fig. 8).

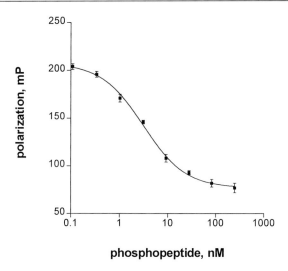

phosphopeptide, nM

FIG. 8. Calibration curve for the STX1 immunoassay for Ser/Thr kinase activity.

FP Assay of Protein Kinase A (PKA) Activity

Materials

The materials listed in the previous section and the following:

Protein kinase A (cAMP-dependent protein kinase catalytic subunit), Promega (Madison, WI); approximately 2.5 mg/ml purified protein with activity about 80,000 U/ml, where 1 U = 1 pmol phosphate transferred per minute to histone H1

400 mM EDTA, pH 7.2; 14.89 g of disodium EDTA in 100 ml distilled water. Adjust the pH to 7.2

Reaction buffer: 40 mM Tris–HCl, PH 7.4, 10 mM MgCl$_2$, 2 mM sodium orthovanadate

10 mM ATP stock: 5.511 mg ATP in 1 ml distilled water; can be kept frozen for 1 year

100 μM ATP (25 μM final concentration of ATP in the assay). Pipette 10 μl of 10 mM ATP into 0.990 ml of enzyme buffer

PKA: Dilute stock enzyme 1 : 200 in assay buffer. Dilute this dilution another 1 : 300 to make a volume of 300 μl of 1 : 60,000 dilution. This is approximately 1 nM or 1.2 U/ml. Prepare 100 μl each of five 3-fold serial dilutions in reaction buffer, i.e., 333, 111, 37, 12, and 4.1 pM

Substrate: PKCβ pseudosubstrate-derived peptide RFARKGSLRQKNV. From a 1 mM substrate stock in DMSO, prepare a 20 μM substrate in assay buffer. Pipette 20 μl of the substrate into 0.980 ml of buffer

Procedure. The kinase reaction is carried out in a 20 μl volume in a 384-well microplate. The remaining volume consists of 10 μl each of the antibody and the tracer. The following procedure can be used. One can also include a calibration curve as above in a separate section of the plate.

1. Pipette the following reagents in the order mentioned below in a 384-well microplate, preparing quadruplicates (e.g., rows A through D). To avoid introducing air bubbles do not blow out the last droplet of reagent while pipetting.
   a. 40, 30, and 20 μl of assay buffer in columns 1, 2, and 3, respectively
   b. 10 μl of assay buffer: column 4–6
   c. 5 μl of assay buffer: all columns 7–12
   d. 5 μl of the 20 μM kinase substrate: columns 4 and 6–12
   e. 5 μl of 1000 pM protein kinase A: columns 5–7
   f. 5 μl of serially diluted PKA: in column 8–12, one column per enzyme dilution
2. To initiate the reaction, add 5 μl of 100 μM ATP (25 μM final concentration in the reaction step) to every well in columns 4, 5, and 7–12. Try to complete this step in as short a time as possible.
3. Cover and incubate the plate at ambient temperature for 1 hr.
4. During the incubation period, prepare the antibody solution as described earlier, except modify the buffer used for dilution by adding disodium EDTA from the 400 mM stock to make 80 mM. Also prepare the tracer solution as described.
5. At the end of the incubation period, stop the reaction by adding 10 μl of the antibody solution to columns 3–12.
6. Add 10 μl of the tracer solution to columns 2–12.
7. Incubate the plate at ambient temperature for 30 min prior to reading in an Analyst instrument using the settings described under the section above
8. Column 1 is the assay blank, column 2 is free tracer, column 3 is maximum antibody binding, column 4 is "no enzyme" control, column 5 is "no substrate" control, column 6 is "no ATP" control, and columns 7–12 are the wells with the serial enzyme dilutions.

The above protocol produces data as shown in Fig. 9. Note that the appearance of the data is different when doses are interpolated from a calibration curve. This illustrates one caution about FP kinase assays based on immunoassay: the actual amount of product formed is not necessarily related to the anisotropy or polarization change in a linear fashion.

In a screen for kinase inhibitors, one would adapt the preceding protocol to use only one enzyme concentration, that which gives about 70–80% of the maximum polarization change, against which library compounds would be tested by adding 5 μl of compound to be tested in place of 5 μl buffer at step 1c.

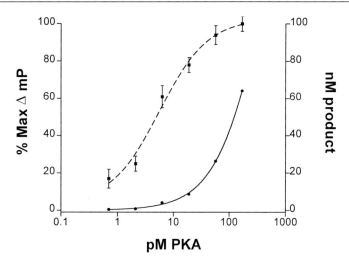

FIG. 9. Titration of activity of protein kinase A by FP immunoassay. 384-Well plate format, 20 μl
kinase reaction, 40 μl final assay volume; 25 μM ATP, 5 μM substrate peptide (RFARKGSLRQKNV).
Data are plotted both as a percentage of maximum polarization change (■) and as nM product (●).
The latter values are interpolated from the standard curve of Fig. 8.

IMAP Assay for Akt Inhibition

The IMAP technology is based on the high-affinity binding of phosphate by
immobilized trivalent metals on nanoparticles. This IMAP "binding reagent" complexes with phosphate groups on phosphopeptides generated in a kinase reaction.
Such binding causes a change in the rate of the molecular motion of the peptide,
and results in an increase in the fluorescence polarization value observed for the
fluorescein label attached at the end of the peptide (Fig. 10). This assay, unlike
antibody-based kinase assays such as the preceding PKA protocol, is applicable
to a wide variety of kinases without regard to the substrate peptide sequences.

The addition of IMAP binding reagent stops the reaction and begins the quantification of phosphorylation of substrate. This procedure can be adapted with
different enzymes and substrates. It is illustrated here with the Ser/Thr kinase Akt.

Akt, or protein kinase B (PKB), is a growth-factor-regulated serine/threonine
kinase. Activation of phosphatidylinositol (PI) 3-kinase leads to translocation of
Akt to the plasma membrane where it is activated by phosphorylation by upstream
kinases including the phosphoinoside-dependent kinase 1 (PDK1). Activated Akt
provides antiapoptotic signals, and also mediates a number of metabolic effects of
insulin.

For assay of Akt, we use the fluorescent substrate FLSN-crosstide, (5-carboxy-
fluorescein)-GRPRTSSFAEG. The kinase reaction is run at concentrations of ATP
and substrate of 5 and 0.1 μM, respectively.

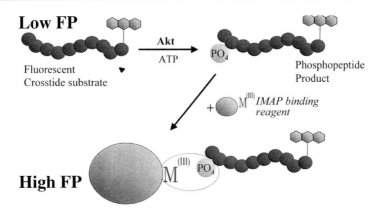

FIG. 10. Principle of the IMAP Akt assay.

Materials (available in the Akt-1 kit, Molecular Devices Corporation)

Binding reagent (400×): Molecular Devices Corporation
Binding buffer: Molecular Devices Corporation
Kinase reaction buffer: 10 mM Tris–HCl, 10 mM MgCl$_2$, 0.1% BSA, 0.05% NaN$_3$, pH 7.2. Other components that can be added without affecting the IMAP system are Mn^{2+} (10 mM), dithiothreitol (DTT), or 2-mercaptoethanol (1 mM), certain detergents up to 0.5%, NaCl (150 mM), and DMSO (up to 10%). EDTA is not compatible with the IMAP system.
Akt1, active (Upstate Biotechnology, Lake Placid, NY)
Fluorescein-labeled   crosstide   substrate,   (5-carboxyfluorescein)-GRPRTSSFAEG (Molecular Devices Corporation)

Procedure

1. Prepare the kinase reaction buffer with DTT (KRB/D) by adding DTT to a final concentration of 1 mM to the kinase reaction buffer. Prepare KRB/D fresh each day of assay.

2. Prepare an Akt enzyme working stock solution of approximately 2.4 units/ml (1 unit = 1 nmol of phosphate transferred to crosstide per minute) in KRB/D. Make 3-fold serial dilutions of this working stock solution to make a dilution range of 2.4, 0.80, 0.27, 0.09, 0.03, and 0.01 units/ml. These concentration values are 4× the final reaction concentration. Also prepare a control solution with no enzyme.

3. Reconstitute the lyophilized substrate in 250 μl KRB/D to make a 20 μM crosstide substrate solution. For best results, incubate the tube for 5 min at room temperature, then vortex gently, making sure that all of the lyophilized substrate

TABLE V
PLATE LAYOUT FOR AKT IMAP ASSAY[a]

	Column							
	1	2	3	4	5	6	7	8
Row A No inhibitor	0.608 U/ml Akt	0.203 U/ml Akt	0.068 U/ml Akt	0.023 U/ml Akt	0.008 U/ml Akt	0.003 U/ml Akt	No Akt	Buffer only
Row B Inhibitor #1 (or replicate of "A")	0.608 U/ml Akt	0.203 U/ml Akt	0.068 U/ml Akt	0.023 U/ml Akt	0.008 U/ml Akt	0.003 U/ml Akt	No Akt	Buffer only
Row C Inhibitor #2 (or replicate of "B")	0.608 U/ml Akt	0.203 U/ml Akt	0.068 U/ml Akt	0.023 U/ml Akt	0.008 U/ml Akt	0.003 U/ml Akt	No Akt	Buffer only

[a] Enzyme concentrations are final.

goes into solution. Prepare a 400 nM substrate working stock solution, i.e., add 20 μl of the 20 μM stock to 980 μl of KRB/D. This working stock is 4× the final reaction concentration of 100 nM substrate.

4. Prepare a 20 μM ATP solution in KRB/D, i.e., add 20 μl of a 10 mM stock of ATP to 980 μl of KRB/D (= 200 μM), then add 500 μl of this 200 μM stock to 4500 μl of KRB/D to make a 20 μM working stock. This solution is 4× the final reaction concentration of 5 μM ATP.

5. Prepare any kinase inhibitors/stimulators (test compounds) in KRB/D at 4× their final reaction concentrations.

6. Leaving the last two empty wells for subsequent use as controls, add 5 μl of each enzyme dilution or "no enzyme" control prepared in step 1 to the appropriate wells of the 384-well plate. A typical plate layout is shown in Table V. Add 5 μl of any inhibitors/stimulators prepared in step 4 to the appropriate wells. Add 5 μl KRB/D to any "no test compound" control wells. Incubate as needed at room temperature to allow for interaction with the enzyme.

7. Add 5 μl of the 400 nM substrate solution prepared in step 3 to the appropriate wells. Add 5 μl of the 20 μM ATP solution prepared in step 4 to the appropriate wells. Each well of the assay should now have 20 μl volume. For the "buffer-only" background control, add 20 μl of KRB/D to the indicated wells.

8. Cover the plate and incubate at room temperature for 60 min.

9. Prepare the IMAP binding solution by diluting the binding reagent 1/400 into binding buffer. This solution should be freshly prepared each day of the assay. Briefly vortex the IMAP binding solution just before adding to the plate. Add 60 μl to each assay well.

10. Cover the plate and incubate at room temperature for 30 min.

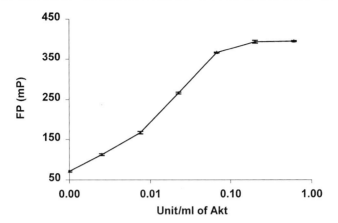

FIG. 11. Titration of Akt with the IMAP system. In black 384-well assay plates, 5 μl of enzyme at various concentrations was reacted in a final volume of 20 μl with 5 μM ATP and 0.10 μM fluorescein-labeled crosstide substrate. After 45 min, 60 μl of IMAP binding reagent was added. Polarization was measured in an Analyst AD multimode reader (Molecular Devices). Enzyme concentrations are those of the 20 μl reaction prior to addition of the binding reagent. Error bars represent 1 standard deviation.

11. Measure the fluorescence polarization on the Analyst AD or HT. The suggested settings include continuous lamp, excitation fluorescein 485 nm–20 fwhm (full width half maximum), emission 530 nm–25 fwhm, fluorescein 505 dichroic, Z height 3 mm, attenuator out, SmartRead or Comparator, sensitivity 0, integration time of 100,000 μs. Other FP readers may be used if they have adequate sensitivity.

12. Calculate the average background (= buffer-only wells) for both parallel and perpendicular data and subtract the appropriate background value from both parallel and perpendicular raw data. Calculate FP and plot FP against Akt enzyme concentration (Fig. 11).

Results. Figure 9 is the titration curve for Akt by the IMAP assay. A large increase in polarization (>300 mP units) occurs over the range of enzyme concentrations, reflecting the conversion of all fluorescent substrate to its phosphorylated form at the highest enzyme concentrations. Considering that the average standard deviation of all points is 2 mP, the signal-to-noise ratio (*S/N*) for the data of Fig. 11 is over 150. Another figure of merit is the so-called z' factor[17] defined as

$$z' = 1 - \left| \frac{(3\sigma_{hi} + 3\sigma_{lo})}{\bar{X}_{hi} - \bar{X}_{lo}} \right| \tag{9}$$

where $\bar{X}_{hi}$ and $\bar{X}_{lo}$ represent the average values of high and low assay controls and σ_{hi} and σ_{lo} are the corresponding standard deviations of these values. A value for

[17] J. H. Zhang, T. D. Chung, and K. R. Oldenburg, *J. Biomol. Screen.* **4,** 67 (1999).

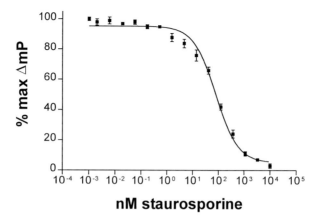

nM staurosporine

FIG. 12. IMAP assay for determination of the IC_{50} for a known kinase inhibitor acting on the Akt kinase. Akt was assayed as in Fig. 9, except that its concentration was fixed at 0.05 U/ml, and staurosporine was added in 3-fold serial dilutions starting at 10,000 nM. Max mP (100%), 364; min mP (0%), 158.

z' of 1 represents a perfect assay, and a value of 0.5 or better is considered to be a "good" assay, one in which active compounds may be expected to be detected with confidence between the limits defined by the assay controls. For the data in Fig. 11, one could realistically report a z' value of 0.92, taking the value at the midpoint of the curve (0.02 U/ml Akt) as the high control and the value at the bottom of the curve (no enzyme) as the low control.

To use the IMAP system to detect or characterize inhibitors, one can adapt the above method to a screening assay. Step 3 is changed to use only one (optimized) dilution of kinase, rather than a dilution series. Step 5 is explicitly used to add either known inhibitors (perhaps at various concentrations or a dilution series to determine an IC50) or compounds from a screening collection.

For the Akt enzyme, we screened a dilution series of the known inhibitor staurosporine. Guided by the enzyme titration data of Fig. 9, we chose an enzyme concentration of 0.05 U/ml. As shown in Fig. 12, an IC_{50} of 78 nM is determined for staurosporine's activity against Akt, close to results determine by other methods.[18]

Additional Considerations to HTS

The enterprise of high throughput screening is a diverse one, with many different formats for compound libraries, plate densities, and target types. Insofar as FP may be used for any of these, it is possible to point to certain features of FP that can be used advantageously in discerning authentic hits from false ones.

[18] G. J. Parker, T. L. Law, F. J. Lenoch, and R. E. Bolger, *J. Biomol. Screen.* **5,** 77 (2000).

Use of Intensity Values

Because anisotropy is computed as a ratio derived from fluorescence intensities, the intensity values are available for inspection separately. This is useful as a check for validity. In HTS, many compounds can interfere with fluorescence by either being themselves fluorescent, or by being colored and therefore potentially absorbing at the wavelengths of the assay.

Therefore, one should always check the fluorescent intensity value individually. This is easily accomplished by screening the data for values that lie outside the expected range.[7]

Check for Anisotropy out of Range

Because anisotropy values always should stay within a range delimited by r_{min} and r_{max}, a value outside that range indicates an artifact. Frequently this arises when there is a large amount of scattering that contributes highly polarized light. In an HTS setting, this can be due to precipitation of compounds being screened.

Spectral Properties of Fluors

Most of the work presented here uses fluors that fluoresce at the standard fluorescein wavelengths. The reason for this has largely to do with habit, as fluorescein is well known and in general is well behaved as a label in anisotropy assays. However, the compound libraries that are commonly used in high throughput screening are populated by many members that contribute spectral interferences in the fluorescein wavelength range. This can produce false positives and negatives that although usually detectable by the above-mentioned checks, nevertheless add significantly to the cost of screening. There is a growing awareness that the use of "red-shifted" fluors such as rhodamines and others can position these assays into spectral regions in which the frequency of compound interferences are much reduced.[19]

---

[19] P. Banks, M. Gosselin, and L. Prystay, *J. Biomol. Screen.* **5,** 329 (2000).

[24] Surface and Printing Effects on Fluorescent Images of Immobilized Biomolecule Arrays

By JONATHAN E. FORMAN, AUDREY D. SUSENO, and PETER WAGNER

Introduction

Arrays of biological molecules immobilized onto functionalized surfaces (bio-chips) are finding widespread use for highly parallel screening applications.[1-4] Great advances in the field of genomics have been realized through the use of nucleic acid arrays that provide the ability to perform genome-wide gene expression monitoring, polymorphism identification over tens of kilobases of sequence, and massively parallel genotyping in single biochip experiments.[1] More recently, arrays of proteins have demonstrated the potential for highly multiplexed proteomic research in the biochip format.

A general format for a biochip is a spatially arrayed surface patterned with 10s, 100s, or even 1000s of features in a two-dimensional coordinate system; each feature contains a population of identical recognition elements (capture probes), such as a specific nucleic acid sequence or antibody (Fig. 1). In a simplified description, one could think of the biochip as a surface and a collection of capture agents.

The surface is a critical component of the biochip, as it is ultimately the surface chemistry that controls the immobilization process for the capture agents. This not only affects the density of capture agents (based on the density of immobilization

[1] (a) M. Schena, ed., "DNA Microarrays, A Practical Approach." Oxford University Press, New York, 1999. (b) M. Schena, ed., "Microarray Biochip Technology." Eaton Publishing, Natick, MA, 2000. (c) B. Phimister, ed., *Nature Genet. Suppl.* **21,** 1 (1999). (d) R. Dhand, ed., *Nature Insight* **405,** 1 (2000).

[2] (a) D. J. Lockhart, H. Dong, M. Byrne, M. T. Follettie, M. V. Gallo, M. S. Chee, M. Mittmann, C. Wang, M. Kobayashi, H. Horton, and E. L. Brown, *Nature Biotechnol.* **14,** 1675 (1996). (b) M. Chee, R. Yang, E. Hubbell, A. Berno, X. C. Huang, D. Stern, J. Winkler, D. J. Lockhart, M. S. Morris, and S. P. A. Fodor, *Science* **274,** 610 (1996).

[3] (a) S. Singh-Gasson, R. D. Green, Y. Yue, C. Nelson, F. Blattner, M. R. Sussman, and F. Cerrina, *Nature Biotechnol.* **17,** 974 (1999). (b) J. H. Butler, M. Cronin, K. M. Anderson, G. M. Biddison, F. Chatelain, M. Cummer, D. J. Davi, L. Fisher, A. W. Frauendorf, F. W. Frueh, C. Gjerstad, T. F. Harper, S. D. Kernahan, and D. Q. Long, *J. Am. Chemi. Soc.* **123,** 8887 (2001). (c) K. Sakai, *Anal. Biochem.* **287,** 32 (2000). (d) M. Caren, and P. Webb, *Nature Genet.* **23,** 21 (1999). (e) D. Englert, in "Microarray Biochip Technology" (M. Schena, ed.), p. 231. Eaton Publishing, Natick, MA, 2000. (f) T. M. Hughes, M. Mao, A. R. Jones, J. Burchard, M. J. Marton, K. W. Shannon, and S. M. Lefkowitz, *Nature Biotechnol.* **19,** 342 (2001). (g) N. Gerry, *J. Mol. Biol.* **292,** 251 (1999).

[4] (a) R. P. Ekins, *J. Pharm. Biomed. Anal.* **7,** 155 (1989). (b) M. F. Templin, D. Stoll, M. Schrenk, P. C. Traub, C. F. Vohringer, and T. O. Joos, *Trends Biotechnol.* **20,** 160 (2001). (c) D. S. Wilson and S. Nock, *Curr. Opin. Chem. Biol.* **6,** 81 (2001).

sites), but also their immobilized conformation. Both factors are important considerations for how efficiently a capture probe can interact with a target molecule that it is designed to recognize in an assay. Surface properties such as wettability (as measured by contact angle) can also have an important impact on the printing of the capture probes, with more wettable (hydrophilic) surfaces leading to larger printed spots (and thus lower densities of immobilized molecules per unit area). The performance of a biochip is thus determined, in part, not only by the immobilization chemistry, but also by the wetting properties of the underlying surface.

In a biochip assay, the chip is exposed to a solution phase sample containing species (targets) that are recognized by the immobilized capture probes.[1] If a target molecule is present in the sample, biological recognition occurs and the target is bound to its complementary capture probe within a specific feature of the array (Fig. 1). Ultimately, the target is detected and identified by a signal generated within the feature that recognized it (using either directly labeled targets or secondary labeled binding agents in a sandwich assay). Data analysis across all the features of the array can provide a significant amount of important biological information on the sample being screened. The quality of this data is dependent on having efficient biological recognition, a result of optimal densities and favorable conformations of the immobilized probes.

Fluorescent imaging is the most widely employed detection method for biochip applications.[5] The ability to generate well-defined and spatially separated features with uniform coverage of immobilized capture agents is an important step toward generating high quality fluorescent images. In this report we will discuss what effect surface properties have on the printing of biological molecules on a biochip surface.

Immobilized Nucleic Acids

Nucleic acid biochips (DNA Chips) represent the most common examples of biochip technologies.[1-3] The two predominant methods used to generate the arrays of features are *in situ* nucleic acid synthesis and the printing of preformed nucleic acids (cDNAs or oligonucleotides) of the desired sequence.[1] *In situ* synthesis employs standard solid-phase nucleotide synthetic methods coupled with either photochemical deprotection or acid-catalyzed dimethoxytrityl deprotection of 5'-hydroxyl groups; in the later case the features must be kept isolated from one another using flow cells, surface chemistries that prevent liquids from spreading, or photoresists.[2,3a,b] The printing of preformed nucleic acids, often referred

[5] (a) T. Basarsky, D. Verdnik, J. Y. Zhai, and D. Wellis, *in* "Microarray Biochip Technology" (M. Schena, ed.), p. 265. Eaton Publishing, Natick, MA, 2000. (b) M. J. Schermer, *in* "DNA Microarrays, A Practical Approach" (M. Schena, ed.), p 17. Oxford University Press, New York, 1999.

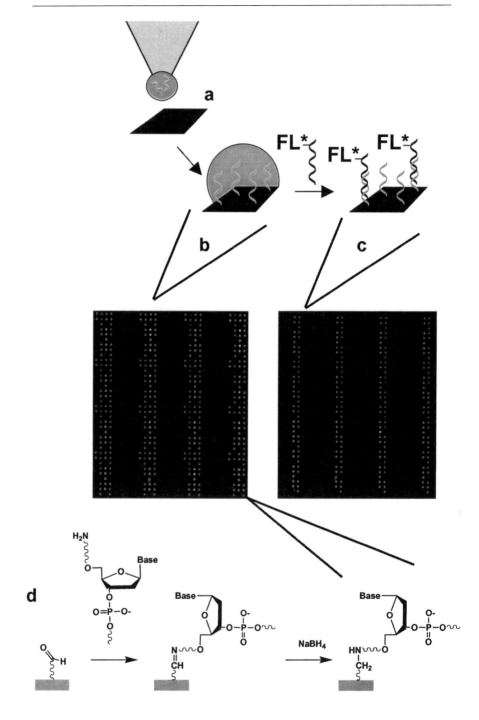

to as "spotting," employs a printing robot (or "arrayer") that deposits picoliter to nanoliter sized volumes of nucleic acid-containing print solutions onto the array through contact of a pin onto the surface.[1] This is followed by a washing and workup process to ensure covalent immobilization of the sample and removal of nonimmobilized probes (Fig. 1). A contact pin can be a solid pin that picks up and deposits a droplet of printing solution,[3c,d] a solid pin that pierces a meniscus of print solution held in a metal ring and deposits a small volume of liquid from that meniscus onto the surface (pin and ring),[6] or a split pin (see Fig. 2) that takes up and deposits an aliquot of printing solution through capillary action.[7] Noncontact (ink jet or pumped volume delivery) printing robots are also available for nucleic acid array printing.[3e–g] Contact pin printing methods are very effective for illustrating surface and printing effects, as the delivered volumes of print solutions and the consistency of printing depend on the interplay between print solutions and the printed surfaces. The quality of the printed features and any damage to the surface from contact with the pin should be immediately evident in fluorescent images.

The examples that follow are from contact pin printing using split pins. Despite this select example of printing methods, the effect of surface properties on printing illustrates a number of general effects that would be expected to manifest themselves with other printing methods (both contact and noncontact).

Surfaces and Immobilization Chemistries for Printed Nucleic Acid Arrays

To maximize the sensitivity of the biochip, surfaces are generally prepared from materials with minimal autofluorescence, such as glass. Reactive surface chemistry for immobilization is introduced by coating processes such as silanation.[8]

---

[6] S. D. Rose, *J. Assoc. Lab. Automation* **3**, 53 (1998).
[7] T. Martinsky and P. Haje, in "Microarray Biochip Technology" (M. Schena, ed.), p. 201. Eaton Publishing, Natick, MA, 2000.
[8] (a) E. P. Plueddemann, "Silane Coupling Agents." Plenum Press, New York, 1982. (b) D. E. Leyden, ed., "Silanes, Surfaces and Interfaces." Gordon and Breach, New York, 1986.

FIG. 1. Printing of a nucleic acid microarray by contact pin methods. A pin brings a drop of nucleic acid-containing print solution to a surface (a) and deposits the droplet to allow immobilization of nucleic acid probes (b). The immobilized probes recognize complementary sequences from a solution in contact with the microarray (c). In the examples above, the printed nucleic acids are 5-amino-modified 20-mer oligonucleotides modified with a fluorescent label (3'-ROX). The amino modifier allows the oligonucleotide to become covalently immobilized by reactive surface functionality [such as aldehyde groups (d)]. The printed and processed (reduction with sodium borohydride in the case of an alehyde surface) array can be imaged for assessing print quality (b) and an assay can be run employing a target sequence with a second fluorophore label (Cy5 in this example); the fluorescent image indicates the presence of sequences that recognize the target. Nonhybridizing printed oligonucleotides do not appear in the hybridization image (c).

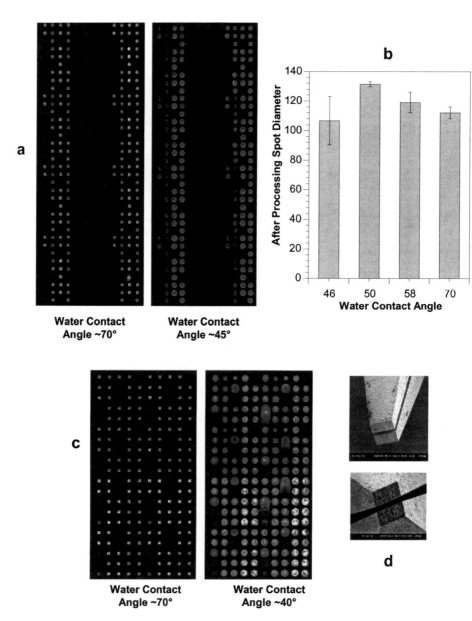

FIG. 2. False color fluorescent images of ROX-labeled oligonucleotides printed on aldehyde surfaces by contact pin printing in 3× SSC buffer (a), illustrating the effect of surface wetting properties on printed spot diameter (b). As print pin sizes increase, hydrophobic surfaces will constrain the spreading droplet to the point where the printed spots take on the square shape of the pin (c); SEM images of the print pin (∼300 μm dimensions) are shown in (d). With hydrophilic surfaces, the droplet spreading is not constrained and the spot sizes actually increase well beyond the dimensions of the print pin (∼560 μm spots printed from a ∼300 μm pin tip), additionally many spots are observed to be "smeared" or "blurred" in the image (c).

There are a multitude of immobilization chemistries that are employed for printed nucleic acid arrays.[9,10] In the simplest immobilization scheme, nucleic acids are physisorbed onto positively charged (amino) surfaces through electrostatic interactions with the negatively charged phosphate backbone of the nucleic acid; this is followed by covalent cross-linking of the thymidine residues in the polynucleotide onto the surface through UV irradiation or heat. The surface is subsequently succinic anhydride blocking of the surface to neutralize the remaining positively charged amines.[10] In more sophisticated immobilization schemes, the nucleic acids are modified with reactive moieties such as amino or thiol groups and immobilized through the modifier to a reactive surface. Figure 1d illustrates the immobilization of amino-modified nucleic acids to an aldehyde surface, a scheme commonly employed with oligonucleotides due to the widespread availability of aldehyde-modified glass surfaces.[9d]

Methods and Materials

Equipment. Contact pin printing is performed with a Genetix Genpak Array21 using microspotting pins from Telechem (Stealth MicroSpotting SMP3, SMP7, or SMP10). Fluorescence scanning is performed using a Packard BioSciences ScanArray 5000XL. Contact angles are measured using a Kruss DSA-10 Drop Shape Analysis System. Scanning electron microscopy is performed with Hitachi S-3500N scanning electron microscope.

Materials. All general reagents were purchased from Sigma (St. Louis, MO). All modified oligonucleotides were purchased from Operon Technologies (Alameda, CA); the amino modification is connected to the 5' end via a 6-carbon spacer; the ROX modifications are 3'; Cy5 modifications are 5'. Capture probes were 20-mer oligonucleotides of sequence 5'-GTCAAGATGCTACCGTTCAG-3'; target sequence was the perfect complement to the capture probe; nonhybridizing control probes were 7-mers of the sequence 5'-TTTTTTT-3'. Aldehyde-modified silanated glass slides were obtained from Zyomyx, Inc. (Hayward, CA), Cel Associates, Inc. (Houston, TX), NoAb BioDiscoveries (Mississauga, Ontario, Canada), Telechem (Sunnyvale, CA), and Xenopore Corporation (Hawthorne, NJ).

[9] (a) C. C. Xiang and Y. Chen, *Biotechnol. Adv.* **18**, 35 (2000). (b) T. Koch, N. Jacobsen, J. Fensholdt, U. Boas, M. Fenger, and M. H. Jakobsen, *Bioconjugate Chem.* **11**, 474 (2000). (c) A. Kumar, O. Larsson, D. Parodi, and Z. Liang, *Nucleic Acids Res.* **28**, 14 (2000). (d) N. Zammatteo, L. Jeanmart, S. Hamels, S. Courtois, P. Louette, L. Hevesi, and J. Remacle, *Anal. Biochem.* **280**, 143 (2000). (e) B. A. Cavic, M. E. McGovern, R. Nisman, and M. Thompson, *Analyst* **126**, 485 (2001). (f) L. A. Chrissey, G. U. Lee, and C. E. O'Ferrall, *Nucleic Acids Res.* **24**, 3031 (1996). (g) D. G. Smyth, O. O. Blumenfeld, and W. Konigsberg, *Biochem. J.* **91**, 589 (1964). (h) M. Boncheva, L. Scheibler, P. Lincoln, H. Vogel, and B. Akerman, *Langmuir* **15**, 4317 (1999).
[10] V. G. Cheung, M. Morley, F. Aguilar, A. Massimi, R. Kucherlapati, and G. Childs, *Nature Genet. Suppl.* **21**, 15 (1999).

Printing and Workup Protocol. Amino-modified probes are suspended in spotting buffers at a concentration of 6 μM with 10% of the total probe concentration modified with a 3′-ROX label. Contact printing was performed under a ∼60% humidity environment.

On completion of printing, the slides are allowed to sit for 2 hr to allow the printed droplets to evaporate. Slide processing begins by immersion in a 0.2% (w/v) sodium dodecyl sulfate (SDS) bath for 2 min at room temperature with agitation, followed by transfer to an agitating water bath for 2 min at room temperature. Following the initial washes, the slides are reduced in a 100 mM solution of sodium borohydride in 20% ethanol in phosphate-buffered saline (PBS) for 10 min at room temperature. Reduction is followed by three successive agitating 0.2% SDS baths (1 min each). Finally, the slides are agitated for 1 min in water and dried by centrifugation or with nitrogen.

Hybridization Assay. Hybridization assays are performed with 5 nM target oligonucleotide at 60° for 16 hr, by immersing the slides, probe side up, in a small tub containing hybridization solution [1 M NaCl, 0.1 M MES (morpholinoethanesulfonic acid), 0.01% Tween 20, pH 6.5]. After hybridization, the slides are placed in an agitating 1 M NaCl bath for 10 min at room temperature, followed by agitation in 1× SSC at 35° for 5 min. Finally, the slides are dried by centrifugation or with a stream of nitrogen.

Scanning. The array was located on the slide using the quick scan method on the Packard scanner (30 μm resolution). After focus, the final scan is taken with 10 μm resolution using detector settings (both PMT and laser) that do not give saturated signals. Correlation plots of data from a given slide scanned at several different detector settings are used to normalize data from different settings onto the same scale for comparison.

Image Analysis. Image analysis is performed with *QuantArray* (Packard Biosciences), employing the adaptive analysis method with an approximate 100 μm spot size. The data are extracted as mean pixel intensities for each spot. All spots are background corrected using the local background measured by *QuantArray*.

Printing Nucleic Acids

Figure 1 illustrates the printing and subsequent hybridization assay of an oligonucleotide array. First a droplet of print solution containing nucleic acids modified to react with the surface is brought in contact with the surface (Fig. 1a), probes are immobilized (Fig. 1b), and a fluorescently labeled sample is hybridized to the array (Fig. 1c).[1] Two fluorescent images are shown, the first (Fig. 1b) is an image of the printed/processed array showing the location of the features in which 10% of the printed probes were modified with a 3′-rhodamine-X (ROX) label. In the second image (Fig. 1c), a Cy5-labeled target sequence has hybridized to the array; it can clearly be seen that only a subset of the total number of features that

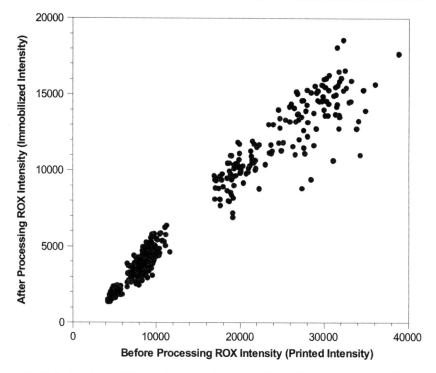

FIG. 3. Scatter plot for 400 printed features of amino-modified oligonucleotides on an aldehyde surface (printed in 3× SSC buffer) showing correlation of postprocessed probe density to printed (preprocessed) probe density. In this case, probe density is taken to be proportional to the signal from the printed ROX-modified oligonucleotides.

are complementary to the target show up in the Cy5 image (there is no signal from the ROX label on the nonhybridizing probes).

Although it is not usual practice to print fluorescently labeled probes, we have found this method to be very useful for evaluating print quality (in a manner that is independent of the assay). The rationale behind using an ROX label for the probe was the ready availability of this modification on the 3′ end of 5′-amino-modified capture probes (for 5′ immobilization to the aldehyde surface). Keeping the labeled probe population at 10% was required to prevent the quenching of fluorescence signals that might result from the close proximity of surface-bound probes.[11] The fluorophore-labeled capture probes also provide a means to evaluate the efficiency of the immobilization process, without having to assume that every feature was printed with the same volume of print solution (this method also provides a means of assessing immobilization performance that is independent of the use of multiple printing pins). An example of this application of labeled probes is given in Fig. 3,

---

[11] P. Bojarski, *Chem. Phys. Lett.* **278**, 225 (1997).

where the slope of a scatter plot of feature intensities taken from fluorescent images of a set of 400 printed features immediately after printing and again after processing provides an estimate of immobilization efficiency of the printing/immobilization chemistry. Notice that at the higher end of the printed (preprocessed) intensity scale, there is considerably greater scatter away from the best straight line that fits the data. This phenomenon is associated with the printing of higher concentrations of capture probe, where smaller fractions of the total amount of printed material actually become covalently immobilized on the surface. The excess capture probe should be washed away in the processing step; however, the washing process becomes less efficient as the density of printed material increases. Observation of the point on the scatter plot at which the data become less consistent provides an estimate of what printing concentration is too high to facilitate efficient processing of the surface after printing.

In the following paragraphs we will describe the effects of surface wetting properties on printing; to do this we will refer to the contact angle of the given surface as a measure of wettability (and surface tension). Because contact angle goniometry is a common tool used in surface analysis, we provide the following definition and references for those readers unfamiliar with the method.[12] Contact angle is defined as the angle between the tangent line drawn from the drop shape to the touch of the solid surface for a liquid drop resting on a solid surface.[12] Higher water contact angles are indicative of surfaces with greater surface tensions and more hydrophobic character.[12] Hydrophilic surfaces have low contact angles (a water droplet will spread itself completely on a surface with a water contact angle near or at $0°$).[12]

Figure 2 illustrates printing effects on surfaces with identical immobilization chemistries (aldehyde) but different wetting properties (contact angles). Figure 2a shows images of printed/processed aldehyde surfaces with water contact angles of $70°$ and $\sim45°$. The surfaces were printed side by side with the same oligonucleotide solutions and the same split pin (designed to print $\sim100\ \mu$m spot diameters), yet the more hydrophilic surface shows printed spots with larger diameters and a larger disparity in spot sizes across the array (see Fig. 2b). Figure 2b further illustrates this effect over four surfaces with different water contact angles; the data are for the average spot diameter of 400 spots printed on the given surface (all spots on all surfaces were printed with the same pin to allow direct comparisons). The lowest contact angle shown, $\sim45°$, has an average spot diameter lower than the others, but much larger disparity in the spot size distribution (higher standard deviation due to much more spreading and inconsistency in the sizes of the printed spots). The higher contact angle surfaces show a tighter distribution of spot sizes, but a decrease in average spot diameter as one moves from a $50°$ to a $70°$ contact angle

---

[12] R. J. Stokes and D. F. Evans, "Fundamentals of Interfacial Engineering." Wiley-VCH, New York, 1997.

surface. It is important to note that all the surfaces illustrated in Figs. 2a and b employ the same immobilization chemistry (aldehyde), but their different surface properties give rise to noticeably different effects on the printing process.

An even more dramatic example of the effect of printing on hydrophobic surfaces is illustrated in Fig. 2c. In Fig. 2c, aldehyde surfaces with contact angles of 70° and 40° were printed with a pin designed to produce ~300 μm spot diameters. The hydrophilic surface shows the expected round spots (however, the spot diameter is actually larger than 300 μm, average diameter ~560 μm), but the hydrophobic surface appears to be square. As the surfaces become more hydrophobic, greater constraint of the spreading of aqueous liquids is observed, in this case the spreading is constrained to the actual area of the pin that contacts the surface. Figure 2d shows scanning electron microscopy (SEM) images of the split pin used to print these arrays; notice that the pin tip is square and it is the spreading of the fluid being printed that produces round spots on the hydrophilic surface. For the hydrophobic surface, the "spot" prints as a square of nearly the same dimensions as the pin tip (the pin acts as a stamp; the surface prevents further spreading). Figure 2c also illustrates a problem often encountered with very hydrophilic surfaces; many of the spots appear smeared or blurred. This is a consequence of the favorable wetting properties of the surface.

In Fig. 4, images of both the processed printed surface (Fig. 4a) and postassay (hybridization) surface (Fig. 4b) are shown. It is clearly evident that defects from the printing process carry over to the hybridization image where they would be expected to affect data quality. These defects include nonuniform signal across a feature, dark or bright centers within a feature, and "smeared" or noncircular spot morphologies.

Just as surfaces with different surface tensions show different printing behavior with a given print buffer, using a set of solutions with a variety of surface tension properties also results in a disparity of spot sizes and qualities (Fig. 4). Printing buffer comparisons carried out in this manner provide a method for selection of print conditions that provide the best consistency of spot morphology and quality on the array. The surface in Fig. 4 has a 70° water contact angle and the majority of the printed spots show consistent spot size and morphology due to the constraint of spreading as described for Fig. 2. However, several large and smeared spots can also be observed in the image; these features are the result of adding surfactants to the printing buffer, a demonstration that the effect of the print buffer is just as important as the surface wetting properties for consistent printing.

The two-color printing/assay system illustrated in Fig. 1 can also be used as a means to assess printed density effects on the biochip. Figure 5 illustrates a plot of target signal intensity (from a Cy5-labeled target) vs. probe signal intensity (from a 10% ROX-labeled probe population). As demonstrated in Fig. 5; the signals generated from the features of the array show a dependence on immobilized probe density; in this case the hybridization signal over 960 printed features on a

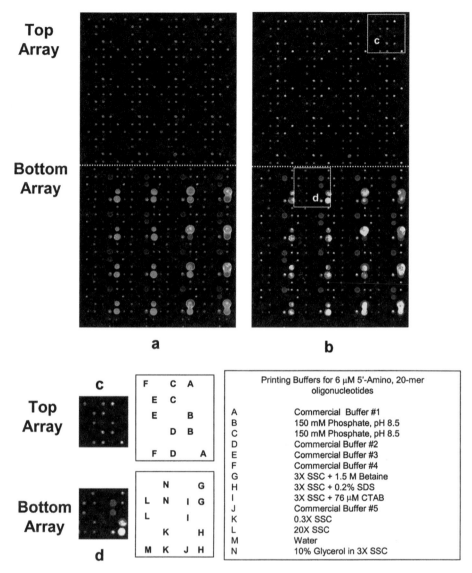

FIG. 4. False color images of an 70° water contact angle aldehyde surface printed with amino-modified ROX-labeled probes (10% labeled probe population) after processing (a) and after a 16-hr hybridization experiment with 5 n*M* Cy5-labeled complementary target sequence in MES buffer with 1 *M* NaCl at 60° (b). The surface has been printed with a variety of print buffers (including several commercially available formulations) to illustrate the effects that the print buffer itself has on array quality. The printing pattern involves two sets of arrays (denoted top and bottom), with each array consisting of 16 subarrays (c and d) in a 4 × 4 pattern as illustrated. A legend for each subarray is given in (c) and (d). In the subarrays there are blank spaces amid the spots; these are places where pure water was spotted between pickups of active print solutions in order to ensure no crossover between prints. In several cases print buffers containing Tween-20 or glycerol inhibited covalent binding of the nucleic acid; these spots are not labeled as they do not show up in the images).

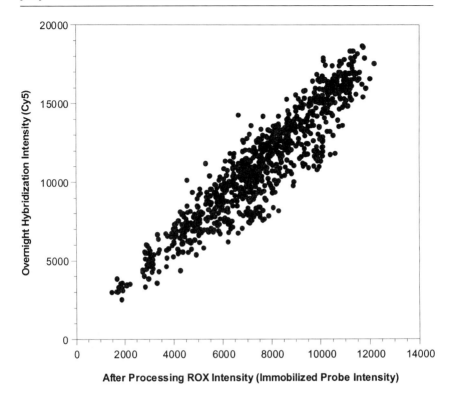

FIG. 5. Scatter plot for 960 printed features of amino-modified oligonucleotides on an aldehyde surface (printed in 3× SSC buffer), showing correlation of hybridization intensity from a Cy5-labeled target sequence (16 hr, 60° assay with 5 nM target sequence in MES buffer with 1 M NaCl) printed (immobilized) probe density. In this case, probe density is taken to be proportional to the signal from the printed ROX-modified oligonucleotides.

single array increases with increasing probe density (as indicated by increasing Cy5 intensity as a function of ROX intensity). Such data are useful for determining the appropriate print concentrations used in array fabrication for more complex biological assays. As the immobilized probe density increases, nonideal conformations of the immobilized probe or crowding effects may begin to reduce the number of available binding sites.[13] In this situation, a plot such as that shown in Fig. 5 would be expected to level off at a ROX intensity corresponding to the higher limit of immobilized probe density.

[13] J. E. Forman, I. D. Walton, D. Stern, R. P. Rava, and M. O. Trulson, in "Molecular Modeling of Nucleic Acids" (N. B. Leontis and J. SantaLucia, eds.), p. 206. American Chemical Society, Washington, D.C., 1998.

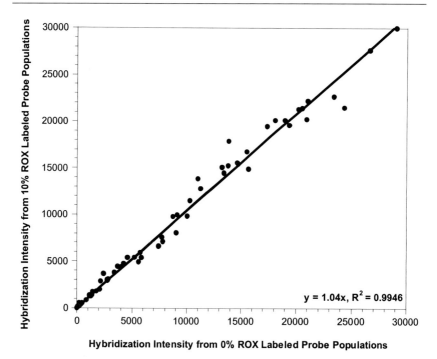

FIG. 6. Correlation of hybridization (5 nM target in MES buffer with 1 M NaCl) signals for populations of printed oligonucleotides modified with 10% or 0% ROX labels. Each point on the scatter plot represents the mean signal intensity of sets of 16 spots printed together on the array. The data are taken from 15 experiments covering a temperature range of 4–75° (see Ref. 14).

In a biologically relevant microarray assay, the probe set would not be labeled. This brings up the question of how appropriate using a two-color method for examining the effects of print quality on the assay really is. That is, does the presence of a label in the probe set induce altered assay performance? Figure 6 demonstrates that this is not the case, as a scatter plot of mean intensities of populations of identically printed labeled and nonlabeled probe sets shows an ~1 : 1 intensity correlation.[14] The data are taken over a range of temperatures (from 4° to 75°) and clearly illustrate that the label has a negligible effect on hybridization performance for the immobilized capture agents.[14]

Immobilized Proteins

With interest in bringing the utility of biochip assays into the realm of proteins,[4] considerations similar to the nucleic acid examples in the preceding section are

[14] A. Suseno, Data presented at the 48th International Symposium of the American Vacuum Society, 2001.

needed for generating high-quality printed arrays. Although the same components that are used to create nucleic acid arrays can also be employed for producing protein arrays, proteins present numerous difficulties that limit the utility of nucleic acid microarray components in many applications.[4a–c]

Critical to sensitivity in an assay is the activity of the immobilized biological molecules within the features of the array. In the case of nucleic acids, it is possible to dry the surface and reconstitute the activity by rehydration. With proteins, such treatment may lead to denaturation of the immobilized proteins, resulting in loss of activity and sensitivity of the assay. Likewise, to keep the immobilized protein in an active conformation, the surface environment should be very hydrophilic.[15]

In the previous section, we saw how printing on surfaces with a greater degree of hydrophilic character produced spots with varying diameters and inconsistent spot morphology. In the examples, the "hydrophilic" surfaces were still relatively hydrophobic (contact angles $\geq 40°$) compared to the ideal surfaces that would be used to produce an array of immobilized proteins. In this section, we will discuss a method of arraying proteins that provides high-quality printing on biochip surfaces optimized toward retention of protein activity.[16]

Methods and Materials

Equipment. Protein microarrays are printed using the Zyomyx, Inc., Parallel Dispensing Chip (PDC) system.[16] In this system, a Zyomyx 1200-pillar chip is inserted into a filled 1200-well PDC; incubation of the top surface of the pillar with the solution in the PDC facilitates transfer of the proteins onto the surface.[16] A detailed description of this technology will be forthcoming in a later publication.[16]

Materials. All general reagents were purchased from Sigma or Aldrich (St. Louis, MO). R-Phycoerythrin conjugated with biotin was obtained from Molecular Probes (Eugene, OR), streptavidin was obtained from Pierce (Rockford, IL), and Cy5-conjugated streptavidin from AmershamPharmacia Biotech (Piscataway, NJ).

Surface Preparation. Plasma-cleaned Zyomyx TiO_2-coated pillar chips are incubated in 100 $\mu g/ml$ of 30% biotinylated poly-L-lysine polyethylene glycol (PLL–PEG)[17] in HEPES buffer at pH 7 for 30 min; after rinsing with PBS, the polymer-coated surface is incubated in a PBS solution of 100 $\mu g/ml$ streptavidin for a minimum of 30 min. Finally, the surface is rinsed with PBS, followed by water, and blown dry with a stream of nitrogen (surfaces were dried just prior to printing).

Printing and Workup Protocol. Using the PDC system described, solutions of biotin-conjugated R-phycoerythrin are incubated for 30 min on streptavidin-functionalized pillar chip surfaces. After incubation, the surface is sprayed with

[15] T. Creighton, "Proteins: Structures and Molecular Properties." Freeman Publishers, New York, 1993.

[16] F. G. Zaugg, P. Oruganti, and P. Lin, unpublished results.

[17] (a) L. Ruiz Taylor and N. Spencer, *J. Phys. Chem. (B)* **104,** 3298 (2000). (b) L. Ruiz Taylor, T. L. Martin, and P. Wagner, *Langmuir* (in press).

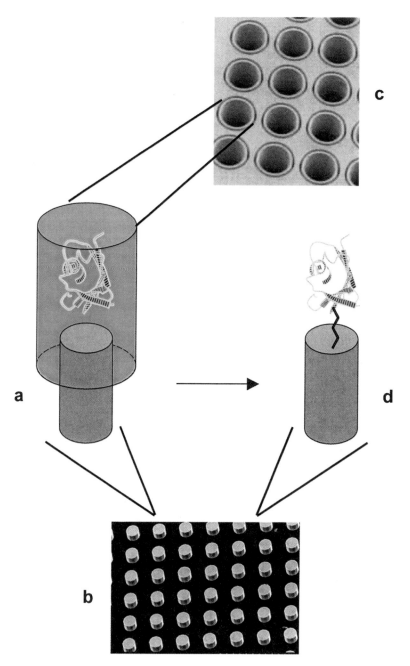

FIG. 7. Illustration of a parallel dispensing scheme for printing proteins on pillar chip surfaces. Pillars rising off the surface are inserted into capillaries containing print buffer with the proteins to be immobilized. (a) Sufficient incubation time allows for immobilization of the proteins to the surface at the top of the pillar (d). SEM images of the pillars (b) and the capillaries (c) illustrate how the two components fit together to facilitate parallel printing of many pillars in a single incubation step (see Ref. 16).

biotin (200 μg/ml PBS) and washed first with PBS followed by water. Prior to scanning, the chip is blown dry with nitrogen.

Scanning and Image Analysis. The protocols employed are essentially identical to those outlined in the nucleic acid section, the exception being that the final scan was done at 5 μm resolution and data extraction employed the fixed circle analysis method with a set spot diameter of 50 μm. Background corrections are made using the average intensity of all nonprinted pillars within the array (global background correction).

Printing Proteins

Contact pin printing methods, as described earlier in this article, print liquid onto the surface in a manner that produces an array of features in a dry environment. For protein applications, a fully hydrated environment is desired to prevent loss of activity due to denaturation.[15] Therefore, an ideal printing method would be to incubate the surface in a bath of print solution. Likewise, constraining the bath from spreading to ensure high-quality spot morphology is necessary for any such printing method.

Such a printing method requires isolation of the features on the array from one another, while covering the surface with fluid in a way that prevents exposure of the feature to an air–water interface. Figure 7 illustrates how this type of deposition can be brought about, employing the use of pillars rising off the surface of the substrate (Fig. 7b), that are inserted into a capillary (Fig. 7c) filled with the printing solution.[16] The printed features are constrained to the area on the top of the pillar. Using microfabrication methods, it is possible to create an array of such pillars and a complementary array of capillaries to allow for completely parallel printing of all the features on the array (a parallel dispensing chip or PDC).[16] Figure 7 illustrates deep reactive ion-etched silicon substrates (Fig. 7b and c) that have been microfabricated into the pillar and dispenser chips just described.[16]

Using polyethylene glycol and biotin-modified polymer coatings (contact angle <35°), a protein immobilization scheme using streptavidin–biotin interactions can be employed.[17] A streptavidin layer is deposited onto the biotin polymer, and subsequent printing of biotinylated proteins produces the array.[17] Figure 8 shows a fluorescent image of such a surface printed with an immobilized fluorescent test protein (phycoerythrin). In the image, the features are clearly defined with no smearing or bleeding of spots as would be expected from a hydrophilic surface printed with aqueous solutions.

Discussion and Conclusions

In this article, we have employed the use of fluorescently labeled biological molecules to probe printing methods for biochip array applications.

FIG. 8. False color fluorescence image of a pillar chip printed with varying surface concentrations of R-phycoerythrin using the parallel dispensing chip (PDC) method described in the text and illustrated in Fig. 7.

Print quality is an important consideration for producing high-quality biochip arrays, where uniform spot diameters and coverages of immobilized molecules are critical for achieving consistent feature-to-feature signals under assay conditions. With a high variation of spot diameter, image analysis can be hampered by having to choose multiple spot sizes for data extraction from a fluorescent image in order to avoid including nonfeature pixels in intensity analysis.[18] This is particularly important for arrays with 100s or 1000s of features, where image analysis and data extraction over the features become cumbersome when small subsets of spots require their own analysis parameters (inconsistent spot sizes, for example).[18] Likewise, poor data quality can arise with poor uniformity of the immobilized capture agent (as observed in Figs. 2 and 4).[18]

The images of Fig. 4 illustrate how poor spot morphology and inconsistent surface coverage of printed capture agents carry over to assay conditions. These

[18] (a) J. C. Mills and J. I. Gordon, *Nucleic Acids Res.* **29**, 15 (2001). (b) C. S. Brown, *Proc. Natl. Acad. Sci. U.S.A.* **98**, 8944 (2001). (c) S. Raychaudhuri, P. S. Sutphin, J. T. Chang, and R. B. Altman, *Trends Biotechnol.* **19**, 189 (2001). (d) G. C. Tseng, M. Oh, L. Rohlin, J. C. Liao, and W. H. Wong, *Nucleic Acids Res.* **29**, 2549 (2001).

images also illustrate how critical the correct choice of printing buffers is for the surface properties of the array substrate. In general, the more hydrophilic the surface, the more the printed liquid is expected to spread, resulting in inconsistent spot diameters and morphologies. For such surfaces, printing solutions that have additives to help constrain spreading are required. For hydrophobic surfaces, spot spreading is constrained by the surface, but too much constraint can also produce poor spot morphology (as observed by "square" spots of Fig. 2). In this case, adding dilute concentrations of surfactants to the printing solution could help to reduce the constraint (see Fig. 4).

In the examples provided for nucleic acid printing, the surfaces had identical immobilization chemistries (aldehyde–amine reaction), but the differences in wetting properties of these surfaces produced significant variations in print quality. Although all of the nucleic acid examples employed contact pin printing with split pins, similar issues related to droplet spreading (and thus consistency of spot diameter and morphology) are expected to manifest themselves with other printing techniques that do not isolate the array features from one another.

Figures 3 and 5 illustrate a general approach toward optimization of microarray surfaces. Using a two-color system for immobilized capture probes and their complementary targets, it is possible to use scatter plots to observe immobilization and assay performance. For immobilization, a scatter plot such as that of Fig. 3 provides information on immobilization efficiency. By observing where the data show a greater degree of scatter off a line, the point at which printing concentrations become high enough to interfere with postprint processing can be determined. With a scatter plot of captured target signal vs. immobilized capture probe signal (Fig. 5), the efficiency of the probe in an assay can be assessed. Probes that interact with their targets with a reduced efficiency (possibly caused by unfavorable immobilized conformations or steric inhibition) will show a leveling off of assay signal at lower immobilized probe density compared to a surface where the capture probes are interacting with high efficiency.

With the hydrophilic surfaces ideal for the printing of protein arrays, alternative methods of printing are desirable. Microfabricated pillar and parallel dispensing chips (Figs. 7 and 8) can be employed to overcome the difficulties associated with spreading liquids and lack of a fully hydrated environment.[16] In the example of Fig. 8, the printing and immobilization of fluorescent proteins illustrate how this method can be used to obtain perfect consistency in spot diameter and morphology, as well as provide a uniform coverage of immobilized protein. Unlike the contact pin printing methods employed for nucleic acid printing, the PDC method isolates all features of the array from one another and allows the deposition of printed proteins to occur in a fully hydrated environment.[16]

We have demonstrated the use of labeled immobilized probe molecules in assessing immobilization efficiency and print quality for biochip surfaces with varying surface properties. Generally, printed capture agents are not fluorescently labeled, so poor spot quality is often associated with poor assays rather than poor printing. The protocols employed in this article allow information useful for optimization of printing processes to be obtained; we feel this will ultimately lead to higher quality assay results.

Acknowledgments

The authors wish to thank Frank Zaugg, Prasad Oruganti, Phil Lin, Nicole Lunceford, Jac Luna, Christina Ho, Hui Zhou, Erik Severin, Mamoru Miyazaki, Beryl Chan, Gee Wan, Joyce DelosReyes, and Ron Fitzgerald and all of Zyomyx, Inc., for assistance in making the experiments described here possible.

Author Index

Mantulin, W. W., 52, 82, 92, 94, 488, 489(18)
Mao, M., 530, 533(3)
Marban, E., 380
Marchesini, R., 460, 461(41), 476(41)
Marcon, N., 466, 466(58), 467, 476(58)
Marder, S. R., 59
Marganski, W. A., 197
Margossian, S. S., 231
Marko, J. F., 158, 161
Marks, T. J., 59
Marquis, H., 327(55), 335
Marriott, G., 323(15), 326, 327(15; 16; 70), 336, 342, 394
Marselle, L. M., 246
Marshall, W. F., 328(89), 336
Marson, N., 460, 476(42)
Marti, A., 458(99), 480
Martin, N. C., 329(96), 337
Martin, P., 197, 327(54), 335
Martin, R. E., 59, 63(61)
Martinez-Arias, A., 197, 327(54), 335
Martinou, J. C., 353
Martinsky, T., 533
Marton, M. J., 530, 533(3)
Maru, Y., 466
Mason, C., 490
Massimi, A., 535
Masters, B. R., 455, 460(20), 475(20), 481, 483, 489, 492, 493, 495, 499(27; 45)
Matheos, D., 328(75), 336
Mathies, R. A., 248
Mathur, A. B., 4, 361, 371
Matlib, M. A., 385
Matsui, T., 212, 215
Matsumoto, G., 434
Matsumura, F., 328(72), 336
Matsutani, K., 450
Matus, A., 326, 327(17), 394
Matusek, G., 43
Matzner, M., 320, 323(35), 324(3), 334
Maxfield, F. R., 340
Maxwell, L., 165
Mayer, M., 315
Mayevsky, A., 476(122), 481
Mayinger, B., 458(95), 480
McAllister, S. S., 340
McAuliffe, D. J., 489
McConnell, H. M., 1
McCormack, K., 305, 307, 314(5)
McCurrach, M., 247

McDonald, K. L., 327(69), 328(69), 336
McElroy, W. D., 215
McGovern, K. N., 328(90), 336
McGovern, M. E., 535
McGuinness, O., 372
McKechnie, T., 466
McKhann, G. M., 91
McKiernan, A. M., 4
McLean, D. I., 471
McMahon, R. J., 59
McMillan, K., 458(106), 480
McNally, E. M., 231
McNiven, M. A., 324(39; 40), 327(65), 334, 336
Medley, A. G., 328(77), 336
Mehdi, S., 511
Mehlmann, M., 458(97; 98), 466, 476(57), 480
Mehta, A. D., 112, 119, 120(17), 124, 124(8), 126, 126(4; 8), 128, 132, 162, 165(1), 172(1), 473
Meisner, L. F., 465
Meister, J. J., 165
Melamed, M. R., 500
Mellor, G., 511
Mellor, J. S., 1, 46
Meng, X., 250
Menzel, R., 510
Merkel, R., 46
Merla, A., 327(61), 335
Merrifield, C. J., 324(36), 334
Merten, M. D., 394
Mertz, J., 7, 50, 52, 52(23), 59, 60(23), 61(23; 24), 65(24), 485, 490
Meseth, U., 82, 95
Messina, A., 460, 476(37)
Mets, Ü., 93, 94, 100(8), 111
Meyer, A. J., 502
Meyer, Ch., 315
Meyer, R. A., 91
Meyer, S. M., 327(59), 335, 341
Meyer-Almes, F. J., 82, 95, 96(17; 18)
Michniewicz, Z., 466(62), 467, 476(62)
Midtgaard, J., 436, 445(19)
Mihara, H., 388
Mikhailov, A. V., 399
Millard, A. C., 47, 48, 49, 52(13), 62(13; 18), 63, 63(13), 65(13), 67(13)
Miller, C. M., 327(52), 335
Miller, R. A., 458(104), 479, 480
Miller, R. K., 328(75), 336
Miller, Y. E., 466

Svoboda, K., 51, 112, 115, 122(13), 126(1; 13), 128(1), 151(3), 156(7), 157(3), 238, 490, 491
Swann, J. W., 43
Swanson, E. A., 503
Swanson, J. A., 4
Sweeney, H. L., 112, 127, 132
Sweeney, T. J., 453, 458(15), 477(15)
Sweier, J. L., 375
Switz, N. A., 140
Sylvestre, V., 330
Szalai, G., 372
Szarowski, D. H., 43
Szebenyi, G., 394, 406(22)
Sziedzic, J., 134

T

Tabak, H. F., 323
Tabti, N., 410
Tachikawa, A., 495
Tait, I., 466
Tajiri, H., 460, 466
Takei, K., 324(39), 334
Takens, J., 389
Tamm, L. K., 4
Tamulevicius, P., 453
Tan, Y., 458(119), 481
Tanaami, T., 185
Tanaka, E. M., 391
Tanaka, H., 228, 233, 239(9)
Taneja, K. L., 246–248, 250, 274(11; 14), 296
Tanimizu, M., 458(81), 479
Tank, D. W., 51, 490, 491
Tanke, H. J., 40, 459
Tanouye, M. A., 305, 307, 314(5), 316
Tans, S., 136
Taroni, P., 458(110), 480
Tashiro, A., 408, 490
Tatematsu, M., 475
Tatman, D., 480
Tatsuta, M., 466(58), 467, 476(58)
Tauc, L., 438
Taylor, D. L., 175, 176, 178, 183, 195(29), 250
Taylor, R. C., 62
Tearney, G. J., 483, 503(8)
Teesdale-Spittle, P., 480

Templin, M. F., 530, 543(4)
Terada, K., 323
Terakawa, S., 3, 305
Terasaki, M., 48, 52(13), 62(13), 63(13), 65(13), 67(13)
Tetin, S. Y., 82
Teukolsky, S. A., 193
Theriot, J. A., 327(55), 335
Thiel, C., 3
Thomas, A. P., 381
Thomas, D. D., 217, 228, 231(3)
Thomas, T. A., 177
Thompson, M., 535
Thompson, N. L., 3, 4, 13(13; 30), 32, 74, 75, 93
Thomsen, S., 452, 461(5), 463, 464(44), 465(44), 470(44), 475(44), 476(5; 44)
Thorn, K. S., 128, 129(30)
Thornhill, D. P., 455
Thoumine, O., 165
Thrall, B., 316
Thyberg, P., 94, 96, 110(21)
Tigges, C. P., 504
Tikhonenko, I., 328(91), 336
Tillement, J. P., 515
Tilton, R. D., 4
Ting, A. Y., 1(12), 3
Tinoco, I., Jr., 136
Tirlapur, U. K., 502
Titus, M. A., 328(82), 336
Tobe, K., 458(81), 479
Todd, I., 1, 36, 46
Toescu, E. C., 365
Tokunaga, M., 231, 233, 234(11; 12), 240, 243, 248, 309, 312(29)
Toledo, Y., 450, 451(42)
Tolksdorf, C., 161
Tomatis, S., 460, 461(41), 476(41)
Tong, L., 498
Toomre, D., 3
Toroian-Raymond, A., 389
Torricelli, A., 458(110), 480
Toyoshima, C., 231
Toyosima, Y. Y., 231
Trang, T. C., 167
Trask, B. J., 246
Traub, P. C., 530, 543(4)
Trautman, J. K., 248
Trautmann, Ch., 315
Tregear, R. T., 125, 239, 240(18)

Subject Index

A

Actin, *see also* F_1-ATPase, single-molecule imaging of rotation; Molecular motors; Neuron
fluorescence *in situ* hybridization, single mesenger RNA molecule detection
criteria for hybridization site identification, 289, 290
dual-wavelength hybridization, 299–300
quantitative analysis, 291, 294
statistical prediction of hybridization behavior, 294–296, 298–299
green fluorescent protein fusion proteins, *see* Green fluorescent protein
Aequorin, mitochondrial transfection and calcium imaging, 386
Akt, IMAP assay for inhibitors
fluorescence polarization measurement, 527–528
incubation conditions, 525–526
materials, 525
principles, 524
Alzheimer's disease, two-photon excitation microscopy studies, 491–492
Anisotropy, *see* Fluorescence polarization
ATP, fluorescent analog turnover by myosin, single fluorophore imaging, 234, 236, 243–244
ATPase, *see* F_1-ATPase
Autocorrelation function, *see* Fluorescence correlation spectroscopy
Autofluorescence
cellular stress measurement, 92
fluorescence correlation spectroscopy interference, 92, 96–97
mitochondrial measurement of NAD(P)H and flavoproteins within cells
caveats and validation, 380
optics and filters
flavoprotein fluorescence, 379–380
pyridine nucleotide fluorescence, 377, 379

signal origins, 375, 377
neoplasia fluorescence imaging, 475
signal-to-noise ratio optimization for biological samples, 268, 270
two-photon excitation microscopy, *see* Two-photon excitation microscopy

B

Biochip, *see* Oligonucleotide array; Protein microarray

C

Caged calcium, photolysis for neuron studies, 414–415
Calcium flux, *see* Mitochondria; Neuron
Cancer, *see* Neoplasia, fluorescence imaging
Caveolin-1, green fluorescent protein fusion effects, 326
Chemotaxis, green fluorescent protein fusion protein studies
actin, 332–333
cyclic AMP receptors, 331–332
cytoplasmic regulator of adenylyl cyclase, 330, 332
Dictyostelium cells, 330–333
Confocal laser scanning microscopy
calcium imaging in neurons, 412–413, 419
embryology studies, 496
neoplasia fluorescence imaging, 470–471, 473
skin studies, 492–494
Cornea, two-photon excitation microscopy studies, 495–496
Correlation-based optical flow
correlation window size, 200–201, 210
correspondence failure detection and correction, 204–205
correspondence problem solution in elastic substrate methods, 198–199
image matrices, 199
image registration artifact correction, 205–207

ISBN 0-12182264-8

9 780121 822644

90051